T0328719

Wavelets in Physics

This book surveys the application of the recently developed technique of the wavelet transform to a wide range of physical fields, including astrophysics, turbulence, meteorology, plasma physics, atomic and solid state physics, multifractals occurring in physics, biophysics (in medicine and physiology) and mathematical physics. The wavelet transorm can analyse scale-dependent characteristics of a signal (or image) locally, unlike the Fourier transform, and more flexibly than the windowed Fourier transform developed by Gabor 50 years ago. The continuous wavelet transform is used mostly for analysis, but the discrete wavelet transform allows very fast compression and transmission of data and speeds up numerical calculation, and is applied, for example, in the solution of partial differential equations in physics. This book will be of interest to graduate students and researchers in many fields of physics, and to applied mathematicians and engineers interested in physical application.

J. C. VAN DEN BERG studied physics and mathematics at the University of Amsterdam. He graduated in high energy physics, doing some work on the automatization of the analysis of bubble chamber films exhibiting the paths of elementary particles in collision experiments. He later took a degree in philosophy of science and logic at the same university, doing his masters thesis on quantum logic. He became a mathematics instructor at Wageningen University in 1973 and is now an Assistant Professor of Applied Mathematics at the Biometris group of Wageningen University and Research Center.

After being interested in the foundations of quantum mechanics for many years, he moved on to non-linear dynamics, especially the concept of multifractals and the difficulties of analysing them. In the writings of Alain Arnéodo on multifractals, he came across the wavelet transform for the first time, taking his first technical course on the subject in 1991 at the CWI in Amsterdam. Soon after, discovering the pioneering works of Marie Farge in turbulence and Gerald Kaiser in electromagnetism, he became convinced that wavelets were important for physics at large. Gradually wavelets overshadowed all his other interests and have remained a main focus ever since. This book is a result of that continuing interest and he hopes it may stimulate others to explore the possibilities of the new tools wavelet analysis continues to deliver.

Wavelets in Physics

Edited by

J.C. VAN DEN BERG

Wageningen University and Research Center,
Wageningen, The Netherlands

CAMBRIDGE UNIVERSITY PRESS
Cambridge, New York, Melbourne, Madrid, Cape Town, Singapore,
São Paulo, Delhi, Dubai, Tokyo, Mexico City

Cambridge University Press
The Edinburgh Building, Cambridge CB2 8RU, UK

Published in the United States of America by Cambridge University Press, New York

www.cambridge.org
Information on this title: www.cambridge.org/9780521533539

© Cambridge University Press 1999, 2004

First published 1999
First paperback edition 2004

A catalogue record for this publication is available from the British Library

ISBN 978-0-521-59311-3 Hardback
ISBN 978-0-521-53353-9 Paperback

Contents

9   The thermodynamics of fractals revisited with wavelets   339
A. Arneodo, E. Bacry and J.F. Muzy

10   Wavelets in medicine and physiology   391
P.Ch. Ivanov, A.L. Goldberger, S. Havlin, C.-K. Peng,
M.G. Rosenblum and H.E. Stanley

Contributors

J.-P. Antoine
Institut de Physique Théorique
Université Catholique de Louvain
Louvain-la-Neuve, Belgium
antoine@fyma.ucl.ac.be

Ph. Antoine
Alcatel
Antwerp, Belgium
philippe.antoine@alcatel.be

A. Arnéodo
Laboratoire de Physique
Ecole Normale Supérieure de Lyon
Lyon, France
alain.arneodo@ens-lyon.fr

E. Bacry
Centre de Mathématiques Appliquées
Ecole Polytechnique
Palaiseau Cedex, France
emmanuel.bacry@polytechnique.fr

J.C. van den Berg (Editor)
Biometris
Wageningen University and Research Center
Wageningen, The Netherlands
hansc.vandenberg@wur.nl

A. Bijaoui
Observatoire de la Côte d'Azur
Dpt. CERGA – UMR CNRS 6527
Nice, France
albert.bijaoui@obs-nice.fr

M. Farge
Laboratoire de Météorologie Dynamique du CNRS
Ecole Normale Supérieure
Paris, France
farge@lmd.ens.fr

A. Fournier
National Center for Atmospheric Research
Climate and Global Dynamics Division
Climate Dynamics and Predictability
Boulder, Colorado, USA
fournier@ucar.edu

A.L. Goldberger
Beth Israel Deaconess Medical Center
Harvard Medical School
Boston, Massachusetts, USA
agoldber@bidmc.harvard.edu

Ch.-A. Guérin
Institut Fresnel
Université Aix-Marseille III
Marseille, France
charles-antoine.guerin@fresnel.fr

S. Havlin
Department of Physics
Bar-Ilan University
Ramat Gan, Israel
havlin@ophir.ph.biu.ac.il

M. Holschneider
Universität Potsdam
Applied and Industrial Mathematics
Potsdam, Germany
hols@rz.uni-potsdam.de

L. Hudgins
Northrop Grumman Electronic Systems
Space Systems Division
Azusa, California, USA
lonnie.hudgins@northropgrumman.com

P.Ch. Ivanov
Center for Polymer Studies, Boston University and
Beth Israel Deaconess Medical Center, Harvard Medical School
Boston, Massachusetts, USA
plamen@argento.bu.edu

J.H. Kaspersen
SINTEF Unimed Ultrasound
7465 Trondheim, Norway
Jon.H.Kaspersen@sintef.no

N.K.-R. Kevlahan
Department of Mathematics and Statistics
McMaster University
Hamilton, Canada
kevlahan@mcmaster.ca

B.Ph. van Milligen
Laboratorio Nacional de Fusión
Asociación EURATOM-CIEMAT
Madrid, Spain
boudewijn, vanmilligen@ciemat.es

J.F. Muzy
Laboratoire SPE, CNRS UMR 6134
Université de Corse
Corte, France
muzy@univ-corse.fr

C.-K. Peng
ReyLab and Harvard Medical School
Boston University
Boston, Massachusetts, USA
peng@chaos.bidmc.harvard.edu

V. Perrier
Laboratoire de Modélisation et Calcul
IMAG
Grenoble, France
Valerie.Perrier@imag.fr

B. Piraux
Laboratoire de Physique Atomique et Moléculaire
Université Catholique de Louvain
Louvain-la-Neuve, Belgium
piraux@fyam.ucl.ac.be

M.G. Rosenblum
Nonlinear Dynamics Group at the Department of Physics
Potsdam University
Potsdam, Germany
MRos@agnld.uni-potsdam.de

K. Schneider
Centre de Mathématiques et d'Informatiques
Université de Provence (Aix-Marseille I)
Marseille, France
kschneid@cmi.univ.mrs.fr

H.E. Stanley
Center for Polymer Studies and Department of Physics
Boston University
Boston, Massachusetts, USA
hes@bu.edu

Preface to the paperback edition

Since the hardback edition of this book was put together wavelets have continued to flourish both in mathematics and in applications in ever more diverse branches of science and engineering. A standard library electronic alert system now easily produces more than fourteen hundred references to papers per year, developing or using wavelet techniques. These are published in a very broad array of journals. Here we can point to only a few of the recently developed methods, in particular as they have been used in physics.

In recent years many variations on the wavelet theme have appeared. One tries to go 'beyond wavelets'. In this context there is a whole family of new animals in the wavelet zoo. Its members carry names like *bandelets, beamlets, chirplets, contourlets, curvelets, fresnelets, ridgelets* . . . These are new bases or frames of functions, customized to handle 2D or 3D data processing better. In [23] for example, it is explained how ridgelets and curvelets can be used in astrophysics. It turns out that noise filtering, contrast enhancement and morphological component analysis of galaxy images are performed much better by a skilful combination of the new transforms than by mere wavelet transforms. More examples can be found on the 'curvelet homepage' [24], maintained by J.L. Starck.

It seems that the applications of the *discrete wavelet transform* (DWT) far outnumber those of the *continuous wavelet transform* (CWT), although the latter started the modern development of wavelet theory in the early eighties. Of the more than two hundred books on wavelet theory that have been published since the early nineties, most are focussed on the DWT and sometimes omit to mention the CWT altogether. This, I think, is unfortunate because both transforms have a lot to offer. A drawback of the CWT is that its computation is much more time consuming than that of the DWT. However, progress has been made in this area too. For example, in [20] a fast algorithm is described for the computation of the CWT at any real scale a and integer time localization b.

The 2D CWT described in detail in Ch. 2 has been further developed by J.-P. Antoine *et al.* and now also covers the case of wavelets living on a sphere instead of on a flat plane [1], [2]. These *spherical wavelets* have been used for instance in astrophysics [4], and also in the recently emerged field of cosmic topology [21], the study of the global shape of the universe. How much richer the world of the 2D transform has become since Ch. 2 was written the reader may see in great detail in the volume especially devoted to this topic [3].

In turbulence studies M. Farge, the earliest promotor of wavelet methods in that field, together with K. Schneider and N. Kevlahan proposed the method of *Coherent Vortex Simulation* (CVS), initially applied to 2D flows, which is already briefly mentioned here in Ch. 4, p. 189. This method was much further developed in the following years, and was recently applied also to 3D flows [7]. More results of Farge and her increasingly productive team, which she set up together with K. Schneider, can be found at [8].

The *Wavelet Transform Modulus Maximim* (WTMM) method and its use for the computation of singularity spectra of multifractals, pioneered by A. Arnéodo's group and described here in Ch. 9, has recently been extended to image analysis [5] and to 3D fields [19]. Another application continuing to produce interesting results is the wavelet-based study of correlations in DNA [6].

The authors of Ch. 10, using wavelet techniques for the study of cardiac dynamics, more recently also adopted the WTMM method [15], [12] to expose the multifractal character of cardiovascular and several other human physiological signals.

It is of interest to note here that M. Haase and B. Lehle [13], using wavelets that are derivatives of the Gaussian function, have been able to derive differential equations for the maxima lines used in the WTMM method. Thus they produce an algorithm for the singularity spectra that is more accurate. More applications can be found at [14].

A. Fournier has advanced the research described in Ch. 7 in at least three ways: by establishing the wavelet-energetics interpretation for idealized fluid models [9], by enlarging the observational dataset to obtain statistically significant results [10] and by inventing customized representations of blocking using 'best shift' wavelets [11].

Let me finish by mentioning some interesting recent examples not directly related to the material of this volume. An application to chaos control was published by G. W. Wei *et al.* [25]. They study a set of chaotic Lorenz oscillators, synchronized by nearest neighbour couplings. Using wavelets to decompose the coupling matrix, they show they can vastly reduce the minimally necessary coupling strength for synchronization to occur.

A. Romeo *et al.* [22] published an appealing N-body simulation of disc galaxies, where N ranges between 10^5 and 9×10^6, in which the initial symmetry is broken after initial fluctuations have been amplified sufficiently by gravitational instability. They show that their use of wavelets to denoise the calculation at each timestep makes their simulations become equivalent to simulations with two orders of magnitude more bodies. Their wavelet method is expected to produce a comparable improvement in performance for cosmological and plasma simulations.

G. Kaiser, well known for his book on wavelets [16], has extended his very interesting programme of finding 'physical wavelets', i.e. wavelets that are also solutions of physical equations such as the Maxwell equations or the wave equation. Initially these were solutions of source-free equations, but now sources have been included in the treatment as well [17].

There are many more interesting recent examples, but reasons of space unfortunately force me to stop here. I hope I have made it clear that wavelets are continuing to inspire physicists in many disciplines to improve existing methods and to explore new territory as well.

At the beginning of this year the wavelet community witnessed the relaunch, after one year of silence, of its popular electronic news bulletin the Wavelet Digest [18], started by Wim Sweldens in 1992, in a modernized format, with an enlarged readership of about 20,000 people, a sure sign, I think, of the vigour of the wavelet enterprise.

HANS VAN DEN BERG

Wageningen University and Research Centre
April 2003

References

[1] J.-P. Antoine and P. Vandergheynst. Wavelets on the 2-sphere: a group-theoretical approach, *Appl. Comput. Harmon. Anal.*, **7**:262–291, 1999.

[2] J.-P. Antoine, L. Demanet, L. Jacques and P. Vandergheynst. Wavelets on the sphere: implementation and approximations, *Applied Comput. Harmon. Anal.*, **13**:177–200, 2002.

[3] J.-P. Antoine, R. Murenzi, P. Vandergheynst and S.T. Ali. *Two-dimensional wavelets and their relatives.* (Cambridge University Press, forthcoming).

[4] J.-P. Antoine, L. Demanet, J.-F. Hochedez, L. Jacques, R. Terrier and E. Verwichte. Application of the 2-D wavelet transform to astrophysical images, *Physicalia Magazine*, **24**:93–116, 2002.

[5] A. Arnéodo, N. Decoster and S.G. Roux. A wavelet-based method for multifractal image analysis. I. Methodology and test applications on isotropic and anisotropic random rough surfaces. *Eur. Phys. J. B*, **15**:567–600, 2000.

[6] B. Audit, C. Thermes, C. Vaillant, Y. d'Aubenton-Carafa, J.F. Muzy and A. Arnéodo. Long-range correlations in genomic DNA: a Signature of the Nucleosomal Structure. *Phys. Rev. Lett.*, **86**(11):2471–2472, 2001.

[7] M. Farge, G. Pellegrino and K. Scheider. Coherent vortex extraction in 3D turbulent flows using orthogonal wavelets. *Phys. Rev. Lett.*, **87**(5):054501, 2001.

[8] wavelets.ens.fr.

[9] A. Fournier. Atmospheric energetics in the wavelet domain. Part I: Governing equations and interpretation for idealized flows. *J. Atmos. Sci.*, **59**:1182–1197, 2002.

[10] A. Fournier. Atmospheric energetics in the wavelet domain. Part II: Time-averaged observed atmospheric blocking. *J. Atmos. Sci.* **60**:319–338, 2003.

[11] A. Fournier. Instantaneous wavelet energetic transfers between atmopsheric blocking and local eddies. *Journal of Climate*, in press 2003.

[12] A.L. Goldberger, L.A. Amaral, J.M. Hausdorff, P.Ch. Ivanov, C.-K. Peng and H.E. Stanley. Fractal dynamics in physiology: alterations with disease and aging. *Proc. Nat. Ac. Sc.* **99** suppl.1:2466–2472, 2002.

[13] M. Haase and B. Lehle. Tracing the Skeleton of Wavelet Maxima lines for the characterization of fractal distributions. In: *Fractals and beyond* (M.M. Novak, Ed.) (World Scientific, Singapore, 1998), pp. 241–250.

[14] www.csv.ica.uni-stuttgart.de/homes/mh/home.html.

[15] P.Ch. Ivanov, L.A. Nunes Amaral, A.L. Goldberger, S. Havlin, M.G. Rosenblum, H.E. Stanley and Z.R. Struzik. From 1/f noise to Multifractal Cascades in Heartbeat Dynamics. *Chaos* **11**:641–652, 2001.

[16] G. Kaiser. *A Friendly Guide to Wavelets*. (Birkhäuser, Boston, 1994).

[17] G. Kaiser. Physical wavelets and their sources: real physics in complex spacetime. arxiv.org/abs/math-ph/0303027, invited 'Topical review' for *Journal of Physics A: Mathematical and General*, www.iop.org/journals/jphysa.

[18] www.wavelet.org.

[19] P. Kestencr and A. Arnéodo. A three-dimensional wavelet based multifractal method: about the need of revisiting the multifractal description of turbulence dissipation data. arxiv.org/abs/cond-mat/0302602.

[20] A. Muñoz, R. Ertlé and M. Unser. Continuous wavelet transform with arbitrary scales and $O(N)$ complexity. *Signal Processing*, **82**:749–757, 2002.

[21] G. Rocha, L. Cayón, R. Bowen, A. Canavezes, J. Silk, A.J. Banday and K.M. Górski. Topology of the universe from *COBE*-DMR; a wavelet approach. arxiv.org/abs/astro-ph/0205155.

[22] A.B. Romeo, C. Horellou and J. Bergh. N-body simulations with two-orders-of-magnitude higher performance using wavelets. arxiv.org/abs/astro-ph/0302343, in press with *Mon. Not. R. Astron. Soc.*

[23] J.L. Starck, D.L. Donoho and E.J. Candès. Astronomical Image Representation by the Curvelet Transform. *Astr. Astroph.*, **398**:785–800, 2003.

[24] www-stat.stanford.eu/~jstarck.

[25] G.W. Wei, M. Zhan and C.-H. Lai. Tailoring wavelets for chaos control. *Phys. Rev. Lett.*, **89**(28):284103, 2002.

Preface to the first edition

Why should physicists bother about wavelets? Why not leave them to the mathematicians and engineers?

Physicists are sometimes reluctant to learn about wavelets because they cannot be interpreted in physical terms as easily as sines and cosines and their frequencies. This is understandable enough: the 'harmonic oscillator' has been with us for more than three centuries, and continues to play its important role. But as we hope to show in the chapters that follow, wavelets can also be of great help in uncovering the presence or absence of certain frequencies in a physical phenomenon. Wavelet analysis is not replacing frequency analysis, but is rather an important refinement and expansion of it: Fourier analysis analyses a signal *globally*, whereas wavelet analysis looks into the signal *locally*.

Let us illustrate this is in musical terms. If you listen to a classical symphony you hear several parts, usually three to four. Each of them has its own main key: e.g. C minor, E♭ major, etc. The Fourier power spectrum of the symphony will of course reveal the dominating keys: groundtones, and their harmonics. Frequencies of other chords which occur more fleetingly during modulations and variations in the piece of music, will also show up. If you would play the parts in a different order, the power spectrum would not change at all, but to the listener it becomes a very different piece, and more so if you interchange parts within the parts, at an ever finer scale: you have changed the musical score drastically. A musical score is a still coarse (the ear catches much more information than the composer writes down in the score) but time-localized frequency analysis of the symphony. This is what a wavelet analysis also supplies you with: it not only gives the main frequencies used, but also, in contrast to the Fourier Transform, indicates *when* they occur, and what

xxii   *Preface to the paperback edition*

their duration is. In the words of Lau and Went (Ch. 1, ref. 18) wavelets 'make a time series sing'.

To be fair, this was already tried with some success in Fourier analysis also: as explained by Antoine in Chapter 1, in 1946 Gabor introduced the Windowed Fourier Transform, by placing a Gaussian time window with constant width over the signal to be analysed, and shifting the window through the signal. Wavelets, springing up in the early 1980s, generalize this in two respects: there is a large and ever growing family of different wavelet functions, and their time resolution is not fixed, but is variable with the frequency, so that high frequencies have a better time resolution. Moreover, one has been able to construct orthonormal bases for many different types of wavelets. Instead of considering signals $f(t)$ to be composed of everlasting oscillations (Fourier Transform) or oscillations within a fixed time window (Windowed Fourier Transform) one considers the signal as being composed of oscillations which arise and die out in time, more rapidly the higher their frequencies. The Wavelet Transform uses a time window which may be shortened or stretched adaptively, thus giving much more flexibility in representing non-stationary signals. This is why the Wavelet Transform is sometimes called a *mathematical microscope*: it allows you to 'zoom' in and out at any desired magnification (inversely proportional to the scale), at any point of time in the signal. It is precisely this kind of flexibility that makes the Wavelet Transform such a useful and efficient analysis tool. Of course the transform can also be performed in two (image analysis) and more dimensions, and even in space-time.

A further reason to learn about wavelets is that wavelets are fast. How fast? For a one-dimensional signal with n data points the Fourier Transform requires $\sim n^2$ operations. This was reduced by the Fast Fourier Transform to $\sim n \log n$, which after its implementation in software packages, made the application of Fourier analysis an industry in many fields of science and technology. Orthornormal wavelets reduce this even further: here one needs only $\sim cn$ computations where the constant c depends only on the type of wavelet used. As already mentioned, wavelets exist in a variety of shapes and one can pick any particular one to work with according to one's need. This is in marked contrast to Fourier analysis, where everything is always analysed in terms of sines and cosines. The computational efficiency is fine for data compression and transmission, and for numerical calculations, and turns out to produce more accurate and/or faster solutions for partial differential equations occurring in physics, as the reader will see for instance in Chapter 4 and Chapter 8.

Every student learning about the Fourier Transform and the power spectrum should now at least be made aware of some of the possibilities wavelets have to offer. From scientists of various disciplines one still sometimes hears the complaint that the mathematics of wavelets is so much more complicated than Fourier analysis that they don't really want to try. This feeling is caused partly because the first generation of good books about the subject is thoroughly mathematical. But the time has arrived that undergraduate books are appearing to serve those people who have only basic mathematical training. To mention only one here: R. Todd Ogden's little book (see Ch. 1), ref. 26). Moreover journals in many fields have published tutorials that deal with the mathematical basics only. Also there are now quite a number of software toolboxes available which can give the beginner a hands-on feeling for the subject without a deep mathematical understanding. The reader will find more on this material in the last paragraphs of Chapter 1 and the references therein.

The first time I myself met wavelets was in 1991, when I read work by Arneodo, Holschneider and others, about the analysis of (multi)fractal measures arising in certain non-linear dynamical systems. My understanding of it was much stimulated by a wavelet course given at the Amsterdam Center of Mathematics and Computer Science at the end of 1991. The closing lecture was given by Michiel Hazewinkel: 'Wavelets Understand Fractals'. He reported on the work by those scientists, and since that lecture I was hooked onto wavelets. Arneodo and Holschneider both contribute to this volume (Ch. 9 and Ch. 11). One of the other speakers in the course was Tom Koornwinder, who later introduced me further into the theory of wavelets. During that period he came up with the suggestion that I produce a book like the present one, for which I am still grateful to him. I soon became aware of the use of wavelets in other areas of physics, in particular by Farge, in turbulence research, and by Kaiser in electromagnetism (applications in radar) and acoustics. Farge and some of her colleagues contribute Chapter 4 of this work, whereas Kaiser's investigations are published in the second part of his fine textbook on wavelets (Ch. 1, ref. 16).

The material you find in this book does not by any means exhaust the applications of wavelets in physics, but I do hope that the reader finds representative examples of good work in this area, and that it stimulates further exploration and application in the fields covered, and elsewhere. Before the book starts, Chapter 0 gives you a brief 'guided tour' through the chapters.

Acknowledgements

First of all I want to thank all the authors for the generous way in which they dealt with my comments and criticisms, and for the patience they sometimes had to have with my lack of understanding of their work. I have learned enormously from going through this process with them. Apart from many discussions with the authors, I have also had the benefit of feedback from other people who read some of the chapters. In particular I want to thank Rob Zuidwijk and Maria Haase for their remarks and suggestions for improvements. Aimé Fournier not only contributed his chapter but offered a number of very welcome suggestions along the way, and some were also supplied by my dear and always critical friend Taco Visser. Many thanks also to Albert Otten at our department, for indefatigably helping me out with my file handling and other computer problems. Without his help I could not have done this job. During the final stages of preparation I was offered the hospitality of the Max-Planck-Institute for the Physics of Complex Systems in Dresden, for which I am very grateful. Continuous encouragement and support I received all the time from Simon Capelin, Publishing Director for the Physical Sciences at Cambridge University Press. Simon, it was a pleasure to work with you and your staff.

August 1998   HANS VAN DEN BERG
Wageningen Agricultural University

0

A guided tour through the book

J.C. VAN DEN BERG

Department of Agricultural, Environment and Systems Technology,
Subdepartment of Mathematics,
Wageningen Agricultural University.

The reader might want to jump right into the book, but I decided to give a guided tour (which one may leave and rejoin at will of course) through the chapters, to whet the reader's taste.

Antoine opens in Chapter 1 with a brief survey of the basic properties of wavelet transforms, both continuous (CWT) and discrete (DWT). In the latter case one learns about the intuitively very appealing concept of *multi-resolution analysis*. Section 1.4 looks ahead to the two- and more-dimensional versions, and summarily brings out connections with well known symmetry groups of physics, and the theory of coherent states.

In the second chapter, also by Antoine, the 2-D wavelet transform is treated. Here the characterization as mathematical microscope must be further qualified, because it misses the new and important property of *orientability* of the 2-D wavelets, which the 1-D case lacks. A real-world microscope is not more sensitive in one direction than in another one, it is 'isotropic'. But the mathematical microscope as embodied in 2-D wavelets has an extra feature: these wavelets can be designed in such a way that they are *directionally selective*. Apart from dilation and translation, one can now also *rotate* the wavelet, which makes possible a sensitive detection of oriented features of a signal (a 2-D image). In many texts the 2-D case is still limited to the DWT, and the wavelets are usually formed by taking tensor products of 1-D wavelets in the x and y-direction, thereby giving preference to horizontal, vertical and diagonal features in the plane. The continuous case is described here in some detail, first because it admits interesting physical applications, such as measuring the velocity field of a 2-D turbulent flow around an obstacle, the disentangling of a superposition of damped plane waves under water produced by a source above the water surface, fault detection in geology, analysis of spectra, contrast enhancement of images. By using the *scale-angle measure* one can exhibit symmetries of objects. Another neat example under development is the

1

detection of Einstein rings by using an annular-shaped wavelet at a fixed scale, leading to, e.g., distance measurements of quasars. The second reason to devote much attention to the 2-D CWT is that the mathematical background, as mapped out in section 2.4, brings out the connections with group representations and coherent states, both used in physics long before wavelets came into the picture. It turns out that wavelets are the coherent states associated to the similitude groups (Euclidean groups with dilations). This section is mathematically somewhat more abstract than the rest of the chapter. The importance of it is that it is shown here, how one can extend the CWT to other 'spaces', such as 3-D space, the sphere, and to space-time ('kinematical' wavelets used in motion tracking, including relativistic effects (using wavelets associated to the Galilei or Poincaré group resp.). Also some applications of the 2-D DWT are indicated.

In Chapter 3 we turn to applications on the largest scale in the Universe: Bijaoui describes a wide variety of applications in astrophysics and observational cosmology. The wavelet transform is a very good tool to study power-law signals, and these occur in many astrophysical sources, such as the light intensity of the solar surface, the brightness of interstellar clouds, or galaxy distributions from galaxy counts. Often the power law behaviour is exhibited by statistical correlation functions, so in many applications there is a combination of statistical techniques with wavelet methods. Cluster analysis of galaxies for instance, was much improved. Image compression is frequently needed in astronomy. Much work was done on Hubble Space Telescope (HST) images and astronomical aperture synthesis. The DWT is not only used in the form resulting from multiresolution analysis, but also by other methods: the 'à trous algorithm', and the 'pyramidal transform' are used for image restoration and analysis. Denoising images also receives a good deal of attention: criteria to establish the notion of 'significant coefficient' were developed. Connected with that is the problem of deconvolving an observed signal (image) to obtain the true object signal: that is the signal before it is convolved with the response function (called the 'point spread function' in optics) of the measuring apparatus. Multiresolution techniques yield a good reduction of resolution here, especially for HST data. To obtain an automated image analysis for astronomical images, one needs a so called 'vision model': a protocol of operations to analyse the image. The classical examples of this were based on edge detection, but this is not adequate to recognize astronomical objects accurately. In a typical image one can see point-like sources (stars), quasi-point-like objects (double stars, faint galaxies...) and complex diffuse structures (galaxies, nebulae, clusters...). The multiscale vision model developed here is able to optimize the detection of objects, because it yields a background mapping adapted to a given object. The earlier methods were only suited to stars of quasi-stellar sources with a slowly varying background. Since in the multiscale approach the notion of subobject is defined, much more complex structures can be analysed.

From astronomical scales down to microscopically small scales one finds turbulence, naturally occurring or man-made. The study of fully developed (high Reynolds number) turbulence by means of the wavelet transform is presented by Farge *et al.* in Chapter 4. The authors argue that part of the reason the subject has not undergone fundamental progress for a long time is that point measurements are used to compute averages in the statistical theory, and also because one keeps thinking in terms of Fourier modes. Thus the presence of coherent structures (here defined as local condensations of the vorticity field which survive much longer than the typical eddy turnover time) is missed, although these are observed in physical space, and their role seems essential in the dynamics. The classical theory of turbulence is not able to see the coherent structures, because they are only felt in the high order statistical moments of the velocity increments in the flow, which have been measured only relatively recently and turn out not to obey Kolmogorov's theory. Wavelets can play a role in separating the coherent components from the incoherent components of turbulent flows, so that one can arrive at new conditional averages, replacing the classical ensemble averages. Fourier space analysis is not capable of this disentanglement, because it averages over space and thus loses local information. The coherent structures correspond to spatio-temporally quasi-singular structures, and thus the use of wavelets to analyse isolated or dense distributions of singularities is brought out, a subject that will be dealt with in extenso in Arneodo *et al.*'s Chapter 9. The separation of coherent structures and random background flow allows new proposals in the modelling of turbulence in which one may expect to be able to explore back and forth transfers of energy between coherent components and the background of the flow. Similar transfers are estimated from real-world global atmospheric data (albeit outside the turbulent regime) by Fournier in Chapter 7. Also in stochastic models of turbulence wavelets are beginning to be used.

Wavelet bases are also increasingly being used to solve partial differential equations numerically. Section 4.6 describes some examples in the literature and presents in some detail algorithms to solve the two-dimensional Navier–Stokes equations.

Coherent structures are also the subject of Chapter 5 by Hudgins and Kaspersen. They focus on the case of cylinder wake flow, and compare the performance of conventional as well as wavelet-based coherent structure detector algorithms. This performance is measured by two statistics: the probability of detection P_D, and the false alarm-rate P_{FA}, that is the probability that a detection will be reported when the relevant event is in fact not present. These quantities are dependent, and this dependence can be parametrized, giving rise to a plot of P_D vs. P_{FA}. The authors test their algorithms on a particular kind of coherent structure called a *burst*: an outrush from the wall, during which the transverse velocity is positive while the streamwise velocity temporarily falls below its mean value. Three conventional detectors are described, and two different wavelet detectors are introduced. Comparison of the results then shows that wavelet methods perform better than the conventional ones, and for high detection rates the second wavelet method outperforms all of the others.

Van Milligen aims at getting a grip on the non-linearity aspect of turbulence in Chapter 6. He defines the notions of *bispectrum* and *bicoherence* based on wavelets. The bicoherence is a measure of the amount of phase coupling that occurs in a signal or between two signals, which means that if two frequencies are simultaneously present in the signal(s) along with their sum (or difference), the sum of the phases of these frequency components is constant in time. Since the wavelet version of these notions is based on integration over a short time interval, temporal variations in phase coupling (intermittent behaviour) can be revealed. Two possible interpretations of the bicoherence are presented: one in terms of coherent structures passing by the observation point, and another one in terms of a coupling constant in a quadratic wave-interaction model. The usefulness of these concepts is first demonstrated in numerical examples: two coupled van der Pol oscillators exhibiting chaos, and then two models for plasma turbulence. It turns out that one can perform detailed spectral analysis on turbulence simulations although only short data series are available (due to CPU-time limitations) rendering Fourier analysis impracticable or impossible. In the last section van Milligen analyses in detail data from torsatron and tokamak plasma experiments.

Turning away from turbulence, in Chapter 7 we find an application, by Fournier, of wavelets to an anomalous state of the earth's atmosphere, namely *blocking*. This is a period of time during which the normal progression (approximately eastward translation) of weather patterns is locally inhibited. It is associated with a quasi-persistent anomalous high pressure system. Fournier reviews the equations derived by Saltzman for the evolution of the mean kinetic energy of eddies. The contributions to this from atmospheric structures of distinct scales are conventionally resolved by (truncated) Fourier series representations. This is replaced here by an analysis in terms of a periodic orthonormal wavelet basis. In terms of these it is possible to construct scale dependent transfer and flux functions of kinetic energy at a certain location. These new concepts are then applied to real-world data from the National Meteorological Center: wind components u (eastward) and v (northward) and the 'geopotential height' Z. Analysis of these data tells us that blocking is largely described by the largest scale part of the multiresolution analysis, and new support is found for the hypothesis that blocking is partially maintained by a particular kind of *inverse* energy cascade (going from smaller to larger scales).

Scaling down to very small distances brings us to applications of wavelets in the domains of atomic and solid state physics. In Chapter 8, Antoine *et al.* start with the case of the generation of light emission resulting from the exposure of atoms to a strong laser pulse. Odd harmonics of the laser frequency are emitted, and in order to understand the mechanism of emission better one would like to know for instance when the harmonics are emitted during the optical cycle, and what the time evolution is during the laser pulse. Standard spectral analysis cannot answer these questions. For atomic hydrogen the emission is investigated by both the Gabor Transform (Windowed Fourier Transform) and the Wavelet Transform, yielding time profiles of each individual harmonic. Analysis of these profiles leads to the conclusion that harmonic emission takes place only when the electron is close to the nucleus. The authors emphasize that in this type of analysis the Gabor transform and the wavelet transform are not each others competitors, but rather they supply complementary information, depending on the exact physical problem one studies. A further development on the basis of these results may be the temporal control of the harmonic emissions by tuning the polarization of the laser, eventually allowing the production of intense attosecond (10^{-18} s) pulses. For the case of multi-electronic wave functions orthogonal wavelet bases on $(0, \infty)$ are being proposed as a basis for the radial part of the wave function, allowing improvements over more conventional Hartree–Fock methods. A combination of wavelet transforms and conventional techniques also allowed a better calculation of energy levels in atoms.

In the second part of the chapter Antoine *et al.* deal with electronic structure calculations in solid state physics. Here both non-orthogonal and orthogonal wavelet bases have been applied successfully, and the recently developed second generation wavelets, used in a biorthogonal basis (see Ch. 1) have been used to solve a 3-D atomic Coulomb problem, namely the Poisson equation for the potential of, for instance, a uranium dimer. The potential is obtained with 6 significant digits throughout the region of interest. In the last section of this chapter the use of 2-D wavelet bases to a 2-D phenomenon, the Fractional Quantum Hall Effect, is explored.

The last three chapters all deal with phenomena in which scaling is of central importance.

Arneodo *et al.* show in Chapter 9 how wavelets can be applied to analyse the scaling properties of multifractal signals which have densely packed singularities of varying strengths. When a signal possesses a single isolated singularity at x_0, with strength, $\alpha(x_0)$ (mathematicians call this the Hölder exponent), this property is reflected in the behaviour of the wavelet transform at that location, and $\alpha(x_0)$ can therefore be extracted from, a log-log plot of the wavelet transform amplitude versus the scale. The dense packing of singularities in a multifractal signal makes straightforward application of this impossible. In order to analyse multifractal signals, a method not involving wavelets, called the *thermodynamical formalism* was developed more than a decade ago. It enables one to calculate the spectrum of singularities, the $f(\alpha)$ spectrum, by statistical means. Before the advent of wavelets this spectrum could be determined for *singular measures* only, but as the authors show, by using wavelets one can extend this to *singular functions* as well, thereby making the method applicable to any experimental signal.

Roughly speaking a (multi)fractal function is non-smooth in all or a large part of its domain, thus making traditional analytical (calculus) methods inadequate to analyse it. Unfolding the function in the wavelet domain restores the applicability of these methods. In particular, the *wavelet transform modulus maxima* (WTMM) are used to obtain a *skeleton* of the function, which provides a partition allowing the merging of the WTMM method with the thermodynamical formalism, so that the singularity spectrum can be determined. This remedies some defects of classical 'box counting' for measures, and of the 'structure function' method used for turbulent signals.

In a further development the WTMM skeleton method is used to address the 'inverse fractal problem': if a fractal object is produced by a dynamical system, can one then extract enough information from the object to recover the dynamical system that produced it? This is a big problem if stated in such generality, but as the authors show, one can solve this for instance in the case when the dynamics is generated by 'cookie-cutter maps'.

Finally the method is applied to the analysis of diffusion-limited-aggregation (DLA) processes, and it is shown how one uncovers the 'Fibonacci multiplicative process' responsible for the branching morphology of the clusters formed by DLA. This is a remarkable result, given the geometrically featureless random walk process that generates the clusters.

The method as described in this chapter is being further developed, as Farge *et al.* mention in Ch. 4, and has also been applied for instance to the analysis of DNA nucleotide sequences (Ch. 10, ref. 81).

The application of wavelets in medicine and biology has proliferated in many different directions, witness the reference list to Chapter 10, by Ivanov *et al.* One area is the study of physiological time series, which generally have a non-stationary character. The specific case analysed here is the comparison between time series of heart beat intervals in healthy human individuals, and in patients suffering from sleep apnea. The authors develop the *cumulative variation amplitude analysis* (CVAA), consisting of a sequential application of the wavelet transform and the Hilbert transform. The first step is to take analysing wavelets that are able to eliminate the influence of linear and low-degree polynomial trends in a signal $s(t)$ (the derivatives of the Gaussian supply wavelets that can do this), keeping only in sight the variations of patterns of a certain duration a of interest. By fixing the scale parameter of the wavelet transform one obtains again a 1-D signal, say $s_a(t)$, expressing how strongly patterns with a certain duration around the value a are present within the signal. It is the variations in this strength which are of interest. The Hilbert Transform, applied to the signal $s_a(t)$ enables one to calculate an 'instantaneous amplitude' of that signal, which is an envelope of it. By counting how often in $s_a(t)$ a given instantaneous amplitude occurs, one obtains a distribution of instantaneous amplitude values which tells one what the relative length (total duration in the entire signal) of an 'a-scale pattern' with a given amplitude is. Every individual has its own amplitude distribution, but it turns out that they are scaling copies of a common distribution, at least in groups of healthy patients. Thus by rescaling individual 'healthy distributions', one can collapse them on their common distribution. Moreover this collapse repeats itself, in healthy individuals, for many different values of a. The collapse fails, however, in groups of subjects suffering from sleep apnea. These two groups can thereby be distinguished from one another. (Applying the Hilbert Transform directly to $s(t)$ itself fails to bring this out.) The authors describe how one may attempt to develop this result further into a tool to separate healthy from abnormal cardiac dynamics for an individual, thus setting up a diagnostic. Finally the relation of the scaling property with the non-linear dynamics of the heartbeat control mechanism is discussed.

The last chapter, by Guérin and Holschneider, concerns the description of intermittency in the time evolution of a system. They define the concept of a *lacunarity dimension* which quantifies the notion of intermittent behaviour. This is the only chapter where detailed mathematical proofs are presented, but we have relegated them to the Appendix so that the flow of the argument is not interrupted. Intermittency is a concept that has been mentioned many times already in previous chapters, but only qualitatively. If you think of a particle recurring intermittently in a region A of phase space, its presence in A can be described by a function $h(t) = \chi_A(x(t))$, where $x(t)$ is the trajectory in phase space, and χ_A is the characteristic function of A, registering whether or not $x(t)$ is in the region A. If one knows the dynamics $x(t)$ over a time interval $[0, T]$, one can calculate the fraction of T the particle spends in A, by taking the time average of $h(t)$ over this interval. If this fraction converges to a finite constant as $T \to \infty$, this limit can be interpreted as a rate of presence in region A. By considering not just the average of $h(t)$, but also its higher moments, the authors find the definition of the lacunarity dimension. So far, no wavelets. This definition is then applied to the case of time evolution of a system obeying the Schrödinger equation with a time independent Hamiltonian. The function $h(t)$ is now the probability to find the system in a certain region of space. The lacunarity dimension can be calculated if $h(t)$ is known over a very large time span, but this may be too long for measurements. It turns out that one can circumvent this by using wavelets to define the *generalized wavelet dimensions* of the Hamiltonian's spectral measure. The latter can be determined from time independent data which are known about the system. The main theorem of the chapter establishes that the lacunarity dimension of the time evolution generated by the Schrödinger equation is obtainable from the generalized wavelet dimensions of the spectral measure of the Hamiltonian. Thus the long time chaotic behaviour of the system and small scale spectral properties of the Hamiltonian are strictly related.

This ends our guided tour. I hope it has aroused your curiosity enough to take a closer look into the chapters that follow.

1

Wavelet analysis: a new tool in physics

J.-P. ANTOINE

Institut de Physique Théorique,
Université Catholique de Louvain, Belgium

Abstract

We review the general properties of the wavelet transform, both in its continuous and its discrete versions, in one or more dimensions. We also indicate some generalizations and applications in physics.

1.1 What is wavelet analysis?

Wavelet analysis is a particular time- or space-scale representation of signals which has found a wide range of applications in physics, signal processing and applied mathematics in the last few years. In order to get a feeling for it and to understand its success, let us consider first the case of one-dimensional signals.

It is a fact that most real life signals are nonstationary and usually cover a wide range of frequencies. They often contain transient components, whose apparition and disparition are physically very significant. In addition, there is frequently a direct correlation between the characteristic frequency of a given segment of the signal and the time duration of that segment. Low frequency pieces tend to last a long interval, whereas high frequencies occur in general for a short moment only. Human speech signals are typical in this respect: vowels have a relatively low mean frequency and last quite long, whereas consonants contain a wide spectrum, up to very high frequencies, especially in the attack, but they are very short.

Clearly standard Fourier analysis is inadequate for treating such signals, since it loses all information about the time localization of a given frequency component. In addition, it is very uneconomical: when the signal is almost flat, i.e. uninteresting, one still has to sum an infinite alternating series for reproducing it. Worse yet, Fourier analysis is highly unstable with respect to perturbation, because of its global character. For instance, if one adds an extra term, with a very small amplitude, to a linear superposition of sine waves, the signal will barely be modified, but the Fourier spectrum will be completely perturbed. This does not happen if the signal is represented in terms of *localized* components.

For all these reasons, signal analysts turn to *time-frequency* (TF) representations. The idea is that one needs *two* parameters: one, called a, characterizes the frequency, the other one, b, indicates the position in the signal. This concept of a TF representation is in fact quite old and familiar. The most obvious example is simply a musical score!

If one requires in addition the transform to be *linear*, a general TF transform will take the form:

$$s(x) \mapsto S(a, b) = \int_{-\infty}^{\infty} \overline{\psi_{ab}(x)}\, s(x)\, dx, \qquad (1.1)$$

where s is the signal and ψ_{ab} the analysing function. Within this class, two TF transforms stand out as particularly simple and efficient: the Windowed or Short Time Fourier Transform (WFT) and the Wavelet Transform (WT). For both of them, the analysing function ψ_{ab} is obtained by acting on a basic (or mother) function ψ, in particular b is simply a time translation. The essential difference between the two is in the way the frequency parameter a is introduced.

(1) *Windowed Fourier Transform:*

$$\psi_{ab}(x) = e^{ix/a}\, \psi(x - b). \qquad (1.2)$$

Here ψ is a window function and the a-dependence is a modulation ($1/a \sim$ frequency); the window has constant width, but the lower a, the larger the number of oscillations in the window (see Figure 1.1 (left))

(2) *Wavelet transform:*

$$\psi_{ab}(x) = \frac{1}{\sqrt{a}}\, \psi\left(\frac{x - b}{a}\right). \qquad (1.3)$$

The action of a on the function ψ (which must be oscillating, see below) is a dilation ($a > 1$) or a contraction ($a < 1$): the shape of the function is unchanged, it is simply spread out or squeezed (see Figure 1.1 (right)).

The WFT transform was originally introduced by Gabor (actually in a discretized version), with the window function ψ taken as a Gaussian; for this reason, it is sometimes called the Gabor transform. With this choice, the function ψ_{ab} is simply a canonical (harmonic oscillator) coherent state [17], as one sees immediately by writing $1/a = p$. Of course this book is concerned

1/a ~ frequency

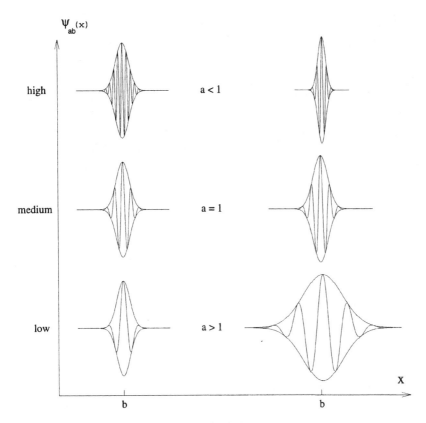

Fig. 1.1. The function $\psi_{ab}(x)$ for increasing values of $1/a \sim$ frequency, in the case of the Windowed Fourier Transform (left) and the wavelet transform (right).

essentially with the wavelet transform, but the Gabor transform will occasionally creep in, as for instance in Chapter 8.

One should note that the assumption of linearity is nontrivial, for there exists a whole class of quadratic, or more properly sesquilinear, time-frequency representations. The prototype is the so-called Wigner–Ville transform, introduced originally by E.P. Wigner in quantum mechanics (in 1932!) and extended by J. Ville to signal analysis:

$$W_s(a, b) = \int e^{-ix/a} \overline{s\left(b - \frac{x}{2}\right)} s\left(b + \frac{x}{2}\right) dx. \tag{1.4}$$

Further information may be found in [6, 11].

1.2 The continuous WT

Actually one should distinguish two different versions of the wavelet transform, the *continuous* WT (CWT) and the *discrete* (or more properly, discrete time) WT (DWT) [10,14]. The CWT plays the same rôle as the Fourier transform and is mostly used for analysis and feature detection in signals, whereas the DWT is the analogue of the Discrete Fourier Transform (see for instance [4] or [29]) and is more appropriate for data compression and signal reconstruction. The situation may be caricatured by saying that the CWT is more natural to the physicist, while the DWT is more congenial to the signal analyst and the numericist. This explains why the CWT will play a major part in this book.

The two versions of the WT are based on the same transformation formula, which reads, from (1.1) and (1.3):

$$S(a, b) = a^{-1/2} \int_{-\infty}^{\infty} \overline{\psi\left(\frac{x - b}{a}\right)} s(x) \, dx, \tag{1.5}$$

where $a > 0$ is a scale parameter and $b \in \mathbb{R}$ a translation parameter. Equivalently, in terms of Fourier transforms:

$$S(a, b) = a^{1/2} \int_{-\infty}^{\infty} \overline{\widehat{\psi}(a\omega)} \widehat{s}(\omega) e^{ib\omega} \, d\omega. \tag{1.6}$$

In these relations, s is a square integrable function (signal analysts would say: a finite energy signal) and the function ψ, the analysing wavelet, is assumed to be well localized *both* in the space (or time) domain and in the frequency domain. In addition ψ must satisfy the following admissibility condition, which guarantees the invertibility of the WT:

$$\int_{-\infty}^{\infty} |\widehat{\psi}(\omega)|^2 \, \frac{d\omega}{|\omega|} < \infty. \tag{1.7}$$

In most cases, this condition may be reduced to the requirement that ψ has zero mean (hence it must be oscillating):

$$\int_{-\infty}^{\infty} \psi(x) \, dx = 0. \tag{1.8}$$

In addition, ψ is often required to have a certain number of *vanishing moments:*

$$\int_{-\infty}^{\infty} x^n \, \psi(x) \, dx = 0, \quad n = 0, 1, \ldots, N. \tag{1.9}$$

This property improves the efficiency of ψ at detecting singularities in the signal, since it is blind to polynomials up to order N.

One should emphasize here that the choice of the normalization factor $a^{-1/2}$ in (1.3) or (1.5) is not essential. Actually, one often uses instead a factor a^{-1} (the so-called L^1 normalization), and this has the advantage of giving more weight to the small scales, i.e. the high frequency part (which contains the singularities of the signal, if any). The choice $a^{-1/2}$ makes the transform unitary: $\|\psi_{ab}\| = \|\psi\|$ and also $\|S\| = \|s\|$, where $\|\cdot\|$ denotes the L^2 norm in the appropriate variables (the squared norm is interpreted as the total energy of the signal).

Notice that, instead of (1.5), which defines the WT as the scalar product of the signal s with the transformed wavelet ψ_{ab}, $S(a, b)$ may also be seen as the convolution of s with the scaled, flipped and conjugated wavelet $\widetilde{\psi}_a(x) = a^{-1/2}\overline{\psi(-x/a)}$:

$$S(a, b) = (\widetilde{\psi}_a * s)(b) = \int_{-\infty}^{\infty} \widetilde{\psi}_a(b - x)\, s(x)\, dx. \qquad (1.10)$$

In other words, the CWT acts as a *filter* with a function of zero mean.

This property is crucial, for the main virtues of the CWT follow from it, combined with the support properties of ψ. Indeed, if we assume ψ and $\widehat{\psi}$ to be as well localized as possible (but respecting the Fourier uncertainty principle), then so are the transformed wavelets ψ_{ab} and $\widehat{\psi}_{ab}$. Therefore, the WT $s \mapsto S$ performs a *local filtering*, both in time (b) and in scale (a). The transform $S(a, b)$ is nonnegligible only when the wavelet ψ_{ab} matches the signal, that is, the WT selects the part of the signal, if any, that lives around the time b and the scale a.

In addition, if $\widehat{\psi}$ has an essential support (bandwidth) of width Ω, then $\widehat{\psi}_{ab}$ has an essential support of width Ω/a. Thus, remembering that $1/a$ behaves like a frequency, we conclude that the WT works at constant *relative* bandwidth, that is, $\Delta\omega/\omega =$ constant. This implies that it is very efficient at high frequency, i.e. small scales, in particular for the detection of singularities in the signal. By comparison, in the case of the Gabor transform, the support of $\widehat{\psi}_{ab}$ keeps the same width Ω for all a, that is, the WFT works at constant bandwidth, $\Delta\omega =$ constant. This difference in behaviour is often the key factor in deciding whether one should choose the WFT or the WT in a given physical problem (see for instance Chapter 8).

Another crucial fact is that the transformation $s(x) \mapsto S(a, b)$ may be inverted exactly, which yields a reconstruction formula (this is only the simplest one, others are possible, for instance using different wavelets for the decomposition and the reconstruction):

$$s(x) = c_\psi^{-1} \int_{-\infty}^{\infty} db \int_0^{\infty} \frac{da}{a^2} \, \psi_{ab}(x) S(a, b), \qquad (1.11)$$

where c_ψ is a normalization constant. This means that the WT provides a decomposition of the signal as a linear superposition of the wavelets ψ_{ab} with coefficients $S(a, b)$. Notice that the natural measure on the parameter space (a, b) is $da\,db/a^2$, and it is invariant not only under time translation, but also under dilation. This fact is important, for it suggests that these geometric transformations play an essential rôle in the CWT. This aspect will be discussed thoroughly in Chapter 2.

All this concerns the continuous WT (CWT). But, in practice, for numerical purposes, the transform must be *discretized*, by restricting the parameters a and b in (1.5) to the points of a lattice, typically a dyadic one:

$$S_{j,k} = 2^{-j/2} \int_{-\infty}^{\infty} \overline{\psi(2^{-j}x - k)} \, s(x) \, dx, \quad j, k \in \mathbb{Z}. \qquad (1.12)$$

Then the reconstruction formula (1.11) becomes simply

$$s(x) = \sum_{j,k \in \mathbb{Z}} S_{j,k} \, \widetilde{\psi_{j,k}}(x), \qquad (1.13)$$

where the function $\widetilde{\psi_{j,k}}$ may be explicitly constructed from $\psi_{j,k}$. In this way, one arrives at the theory of *frames* or nonorthogonal expansions [9, 10], which offer a good substitute to orthonormal bases. Very general functions ψ satisfying the admissibility condition (1.7) described above will yield a good frame, but not an orthonormal basis, since the functions $\{\psi_{j,k}(x) \equiv 2^{j/2} \psi(2^j x - k), j, k \in \mathbb{Z}\}$ are in general not orthogonal to each other!

Yet orthonormal bases of wavelets can be constructed, but by a totally different approach, based on the concept of *multiresolution analysis*. We emphasize that the discretized version of the CWT just described is totally different in spirit and method from the genuine DWT, to which we now turn. The full story may be found in [10], for instance.

1.3 The discrete WT: orthonormal bases of wavelets

One of the successes of the WT was the discovery that it is possible to construct functions ψ for which $\{\psi_{j,k}, j, k \in \mathbb{Z}\}$ is indeed an orthonormal basis of $L^2(\mathbb{R})$.

In addition, such a basis still has the good properties of wavelets, including space *and* frequency localization. Moreover, it yields fast algorithms, and this is the key to the usefulness of wavelets in many applications

The construction is based on two facts: first, almost all examples of orthonormal bases of wavelets can be derived from a multiresolution analysis, and then the whole construction may be transcribed into the language of digital filters, familiar in the signal processing literature.

A *multiresolution analysis* of $L^2(\mathbb{R})$ is an increasing sequence of closed subspaces

$$\ldots \subset V_{-2} \subset V_{-1} \subset V_0 \subset V_1 \subset V_2 \subset \ldots, \tag{1.14}$$

with $\bigcap_{j \in \mathbb{Z}} V_j = \{0\}$ and $\bigcup_{j \in \mathbb{Z}} V_j$ dense in $L^2(\mathbb{R})$ (loosely speaking, this means $\lim_{j \to \infty} V_j = L^2(\mathbb{R})$), and such that

(1) $f(x) \in V_j \Leftrightarrow f(2x) \in V_{j+1}$
(2) there exists a function $\phi \in V_0$, called a *scaling function*, such that the family $\{\phi(x - k), k \in \mathbb{Z}\}$ is an orthonormal basis of V_0.

Combining conditions (1) and (2), one gets an orthonormal basis of V_j, namely $\{\phi_{j,k}(x) \equiv 2^{j/2}\phi(2^j x - k), k \in \mathbb{Z}\}$. Note that we may take for ϕ a *real* function, since we are dealing with signals.

Each V_j can be interpreted as an *approximation space*: the approximation of $f \in L^2(\mathbb{R})$ at the resolution 2^{-j} is defined by its projection onto V_j, and the larger j, the finer the resolution obtained. Then condition (1) means that no scale is privileged. The additional details needed for increasing the resolution from 2^{-j} to $2^{-(j+1)}$ are given by the projection of f onto the orthogonal complement W_j of V_j in V_{j+1}:

$$V_j \oplus W_j = V_{j+1}, \tag{1.15}$$

and we have:

$$L^2(\mathbb{R}) = \bigoplus_{j \in \mathbb{Z}} W_j. \tag{1.16}$$

Equivalently, fixing some lowest resolution level j_o, one may write

$$L^2(\mathbb{R}) = V_{j_o} \oplus \left(\bigoplus_{j \geq j_o} W_j \right). \tag{1.17}$$

Then the theory asserts the existence of a function ψ, called the *mother wavelet*, explicitly computable from ϕ, such that $\{\psi_{j,k}(x) \equiv 2^{j/2}\psi(2^j x - k), j, k \in \mathbb{Z}\}$ constitutes an orthonormal basis of $L^2(\mathbb{R})$: these are the *orthonormal wavelets*.

The construction of ψ proceeds as follows. First, the inclusion $V_0 \subset V_1$ yields the relation (called the scaling or refining equation):

$$\phi(x) = \sqrt{2} \sum_{n=-\infty}^{\infty} h_n \phi(2x - n), \quad h_n = \langle \phi_{1,n} | \phi \rangle. \tag{1.18}$$

Taking Fourier transforms, this gives

$$\widehat{\phi}(\omega) = m_0(\omega/2) \widehat{\phi}(\omega/2), \tag{1.19}$$

where

$$m_0(\omega) = \frac{1}{\sqrt{2}} \sum_{n=-\infty}^{\infty} h_n e^{-in\omega} \tag{1.20}$$

is a 2π-periodic function. Iterating (1.19), one gets the scaling function as the (convergent!) infinite product

$$\widehat{\phi}(\omega) = (2\pi)^{-1/2} \prod_{j=1}^{\infty} m_0(2^{-j}\omega). \tag{1.21}$$

Then one defines the function $\psi \in W_0 \subset V_1$ by the relation

$$\widehat{\psi}(\omega) = e^{i\omega/2} \overline{m_0(\omega/2 + \pi)} \, \widehat{\phi}(\omega/2), \tag{1.22}$$

or, equivalently

$$\psi(x) = \sqrt{2} \sum_{n=-\infty}^{\infty} (-1)^{n-1} h_{-n-1} \phi(2x - n), \tag{1.23}$$

and proves that the function ψ indeed generates an orthonormal basis with all the required properties.

Various additional conditions may be imposed on the function ψ (hence on the basis wavelets): arbitrary regularity, several vanishing moments (in any case, ψ has always mean zero), symmetry, fast decrease at infinity, even compact support. The technique consists in translating the multiresolution structure into the language of digital filters. Actually this means nothing more than expanding (filter) functions in a Fourier series. For instance, (1.19) means that $m_0(\omega)$ is a filter (multiplication operator in frequency space), with filter coefficients h_n. Similarly, (1.22) may be written in terms of the filter $m_1(\omega) = e^{i\omega} \overline{m_0(\omega + \pi)}$. (Notice that this particular relation between m_0, m_1, together with the identity $|m_0(\omega)|^2 + |m_1(\omega)|^2 = 1$, define what electrical engineers call a Quadrature Mirror Filter or QMF.) Then the various restrictions imposed on ψ translate into suitable constraints on the filter coefficients h_n. For instance, ψ has compact support if only finitely many h_n differ from zero.

The simplest example of this construction is the Haar basis, which comes from the scaling function $\phi(x) = 1$ for $0 \leq x < 1$ and 0 otherwise. Similarly, various spline bases may be obtained along the same line. Other explicit examples may be found in [5] or [10].

In practical applications, the (sampled) signal is taken in some V_J, and then the decomposition (1.17) is replaced by the finite representation

$$V_J = V_{j_o} \oplus \left(\bigoplus_{j=j_o}^{J-1} W_j \right). \tag{1.24}$$

Figure 1.2 shows an example (obtained with the MATLAB Wavelet Toolbox [3]) of a decomposition of order 5, namely

$$V_0 = V_{-5} \oplus W_{-5} \oplus W_{-4} \oplus W_{-3} \oplus W_{-2} \oplus W_{-1}. \tag{1.25}$$

As we just saw, appropriate filters generate orthonormal wavelet bases. However, this result turns out to be too rigid and various generalizations have been proposed (see [25] for details).

(i) *Biorthogonal wavelet bases*:
   As we mentioned in Section 1.2, the wavelet used in the CWT for reconstruction need not be the same as that used for decomposition, the two have only to satisfy a cross-compatibility condition. The same idea in the discrete case leads to biorthogonal bases, i.e. one has two hierarchies of approximation spaces, V_j and $\check{V}_j$, with cross-orthogonality relations. This gives a better control, for instance, on the regularity or decrease properties of the wavelets.

(ii) *Wavelet packets and the best basis algorithm*:
   The construction of orthonormal wavelet bases leads to a special subband coding scheme, rather asymmetrical: each approximation space V_j gets further decomposed into V_{j-1} and W_{j-1}, whereas the detail space W_j is left unmodified. Thus more flexible subband schemes have been considered, called *wavelet packets*; they provide rich libraries of orthonormal bases, and also strategies for determining the optimal basis in a given situation [7, 32].

(iii) *The lifting scheme*:
   One can go one step beyond, and abandon the regular dyadic scheme and the Fourier transform altogether. The resulting method leads to the so-called *second-generation wavelets* [31], which are essentially custom-designed for any given problem.

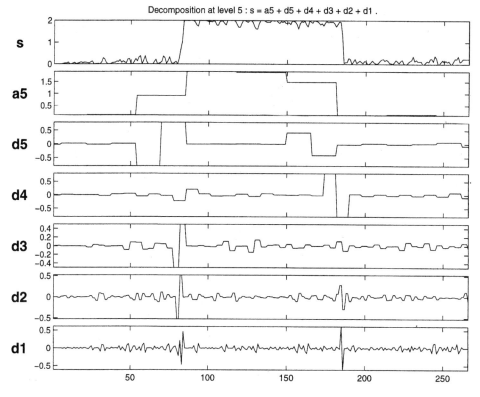

Fig. 1.2. A decomposition of order 5. The signal s lives in V_0 and it is decomposed into its approximation $a_5 \in V_{-5}$ and the increasingly finer details $d_j \in W_{-j}$, $j = 5, 4, 3, 2, 1$.

1.4 The wavelet transform in more than one dimension

Wavelet analysis may be extended to 2-D signals, that is, in *image analysis*. This extension was pioneered by Mallat [19, 20], who developed systematically a 2-D *discrete* (but redundant) WT. This generalization is indeed a very natural one, if one realizes that the whole idea of multiresolution analysis lies at the heart of human vision. In fact, most of the concepts are indeed already present in the pioneering work of Marr [22] on vision modelling. As in 1-D, this discrete WT has a close relationship with numerical filters and related techniques of signal analysis, such as subband coding. It has been applied successfully to several standard problems of image processing. As a matter of fact, all the approaches that we have mentioned above in the 1-D case have been extended to 2-D: orthonormal bases, biorthogonal bases, wavelet packets, lifting scheme. These topics will be discussed in detail in Chapter 2.

However, the continuous transform may also be extended to 2 (or more) dimensions, with exactly the same properties as in the 1-D case [2, 26]. Here again the mechanism of the WT is easily understood from its very definition as a convolution (in the sense of (1.10)):

$$S(a, \theta, \vec{b}) \sim \int d^2\vec{x}\, \overline{\psi(a^{-1}r_{-\theta}(\vec{x} - \vec{b}))} s(\vec{x}), \quad a > 0, 0 \le \theta < 2\pi, b \in \mathbb{R}^2, \quad (1.26)$$

where s is the signal and ψ is the analysing wavelet, which is translated by $\vec{b}$, dilated by a and rotated by an angle θ ($r_{-\theta}$ is the rotation operator). Since the wavelet ψ is required to have zero mean, we have again a filtering effect, i.e. the analysis is *local* in all four parameters $a, \theta, \vec{b}$, and here too it is particularly efficient at detecting discontinuities in images.

Surprisingly, most applications have treated the 2-D WT as a 'mathematical microscope', like in 1-D, thus ignoring directions. This is particularly true for the discrete version. There, indeed, a 2-D multiresolution is simply the tensor product of two 1-D schemes, one for the horizontal direction and one for the vertical direction (in technical terms, one uses only separable filters). However the 2-D *continuous* WT, including the orientation parameter θ, may be used for detecting *oriented features* of the signal, that is, regions where the amplitude is regular along one direction and has a sharp variation along the perpendicular direction, for instance, in the classical problem of *edge detection*. The CWT is a very efficient tool in this respect, provided one uses a wavelet which has itself an intrinsic orientation (for instance, it contains a plane wave). For this reason, a large part of Chapter 2 will be devoted to the continuous WT and its applications.

For further extensions of the CWT, it is crucial to note that the 2-D version comes directly from group representation theory, the group in this case being the so-called similitude group of the plane, consisting of translations, rotations and global dilations [26]. Note that the 1-D CWT may also be derived from group theory [10], in that case from the so-called '$ax + b$' group of dilations and translations of the line.

What we have here is in fact a general pattern. Consider the class of finite energy signals living on a manifold Y, i.e. $s \in L^2(Y, d\mu) \equiv \mathcal{H}$. For instance, Y could be space $\mathbb{R}^n$, the 2-sphere S^2, space-time $\mathbb{R} \times \mathbb{R}$ or $\mathbb{R}^2 \times \mathbb{R}$, etc. Suppose there is a group G of transformations acting on Y, that contains *dilations* of some kind. As usual, this action will be expressed by a unitary representation U of G in the space $\mathcal{H}$ of signals. Then, under a simple technical assumption on U ('square integrability'), a wavelet analysis on Y, adapted to the symmetry group G, may be constructed, following the general construction of coherent states on Y associated to G [1]. This technique has been implemented successfully for extending the CWT to higher dimensions (in 3-D, for instance, one gets a tool for target tracking), the 2-sphere (a tool most wanted by geophysicists) or to space-time (time-dependent signals or images, such as TV or video sequences), including relativistic effects (using wavelets associated to the affine Galilei or Poincaré group). This general approach will be described with all the necessary mathematical details in Chapter 2.

It is interesting to remark that the CWT was in fact designed by physicists. The idea of deriving it from group theory is entirely natural in the framework of coherent states [1, 17], and the connection was made explicitly from the very beginning [12, 13]. In a sense, the CWT consists in the application of ideas from quantum physics to signal and image processing. The resulting effect of cross-fertilization may be one of the reasons of its richness and its success.

1.5 Outcome

As a general conclusion, it is fair to say that the wavelet techniques have become an established tool in signal and image processing, both in their CWT and DWT incarnations and their generalizations. They are being incorporated as a new tool in many reference books and software codes. They have distinct advantages over concurrent methods by their adaptive character, manifested for instance in their good performances in pattern recognition or directional filtering (in the case of the CWT), and by their very economical aspect, achieved in impressive compression rates (in the case of the DWT). This is especially useful in image processing, where huge amount of data, mostly redundant, have to be stored and transmitted.

As a consequence, they have found applications in many branches of physics, such as acoustics, spectroscopy, geophysics, astrophysics, fluid mechanics (turbulence), medical imagery, atomic physics (laser–atom interaction), solid state physics (structure calculations), ... Some of these results will be reviewed in the subsequent chapters. For additional information, see [24].

Thus we may safely bet that wavelets are here to stay, and that they have a bright future. Of course wavelets don't solve every difficulty, and must be continually developed and enriched, as has been the case over the last few years. In particular, one should expect a proliferation of specialized wavelets, each dedicated to a particular type of problem, and an increasingly diverse spectrum of physical applications. This trend is only natural, it follows from the very structure of the wavelet transform – and in that respect the wavelet philosophy is exactly opposite to that of the Fourier transform, which is usually seen as a universal tool.

Finally a word about references. The literature on wavelet analysis is growing exponentially, so that some guidance may be helpful. As a first contact, an introductory article such as [29] may be a good suggestion, followed by the the popular, but highly successful book of Burke Hubbard [4]. Slightly more technical, but still elementary and aimed at a wide audience, are the books of Meyer [25] and Ogden [27]. While the former is a nice introduction to the mathematical ideas underlying wavelets, the latter focuses more on the statistical aspects of data analysis. Note that, since wavelets have found applications in most branches of physics, pedestrian introductions on them have been written in the specialized journals of each community (to give an example, meteorologists will appreciate [18]).

For a survey of the various applications, and a good glimpse of the chronological evolution, there is still no better place to look than the proceedings of the three large wavelet conferences, Marseille 1987 [8], Marseille 1989 [23] and Toulouse 1992 [24]. Finally a systematic study requires a textbook. Among the increasing number of books and special issues of journals appearing on the market, we recommend in particular the volumes of Daubechies [10], Chui [5], Kaiser [16] and Holschneider [14], the collection of review articles in [30] and several special issues of IEEE journals [15,28]. In particular, [3] gives a useful survey of the available software related to wavelets. Another good choice, complete but accessible to a broad readership, is the recent textbook of Mallat [21].

References

[1] S.T. Ali, J.-P. Antoine, J.-P. Gazeau and U.A. Mueller, Coherent states and their generalizations: A mathematical overview, *Reviews Math. Phys.*, **7**: 1013–1104, (1995)

[2] J.-P. Antoine, P. Carrette, R. Murenzi and B. Piette, Image analysis with two-dimensional continuous wavelet transform, *Signal Proc.*, **31**: 241–272, (1993)

[3] A. Bruce, D. Donoho and H.Y. Gao, Wavelet analysis, *IEEE Spectrum*, October 1996, 26–35

[4] B. Burke Hubbard, *Ondes et ondelettes – La saga d'un outil mathématique* (Pour la Science, Paris,1995); 2nd edn. The World According to Wavelets (A.K. Peters, Wellesley, MA, 1998)

[5] C.K. Chui, *An Introduction to Wavelets* (Academic Press, San Diego, 1992)

[6] L. Cohen, General phase-space distribution functions, *J. Math. Phys.*, **7**: 781–786, (1966)

[7] R.R. Coifman, Y. Meyer, S. Quake and M.V. Wickerhauser, Signal processing and compression with wavelet packets, in [23], pp. 77–93

[8] J.-M. Combes, A. Grossmann and Ph. Tchamitchian (eds.), *Wavelets, Time-Frequency Methods and Phase Space (Proc. Marseille 1987)* (Springer, Berlin, 1989; 2d Ed. 1990)

[9] I. Daubechies, A. Grossmann and Y. Meyer, Painless nonorthogonal expansions, *J. Math. Phys.*, **27**: 1271–1283, (1986)

[10] I. Daubechies, *Ten Lectures on Wavelets*, (SIAM, Philadelphia, PA, 1992)

[11] P. Flandrin, *Temps-Fréquence*. (Hermès, Paris, 1993)

[12] A. Grossmann and J. Morlet, Decomposition of Hardy functions into square integrable wavelets of constant shape, *SIAM J. Math. Anal.*, **15**: 723–736, (1984)

[13] A. Grossmann, J. Morlet and T. Paul, Integral transforms associated to square integrable representations. I. General results, *J. Math. Phys.*, **26**: 2473–2479 (1985); id. II. Examples, *Ann. Inst. H. Poincaré*, **45**: 293–309, (1986)

[14] M. Holschneider, *Wavelets, an Analysis Tool* (Oxford University Press, Oxford, 1995)

[15] *IEEE Transaction Theory*, Special issue on wavelet transforms and multiresolution signal analysis, **38**, No.2, March 1992

[16] G. Kaiser, *A Friendly Guide to Wavelets* (Birkhäuser, Boston, Basel, Berlin, 1994)

[17] J.R. Klauder and B.S. Skagerstam, *Coherent States – Applications in Physics and Mathematical Physics* (World Scientific, Singapore, 1985)

[18] K.-M. Lau and H. Weng, Climate signal detection using wavelet transform: How to make a time series sing, *Bull. Amer. Meteo. Soc.*, **76**: 2931–2402, (1995)

[19] S.G. Mallat, Multifrequency channel decompositions of images and wavelet models, *IEEE Trans. Acoust., Speech, Signal Processing*, **37**: 2091–2110, (1989)

[20] S.G. Mallat, A theory for multiresolution signal decomposition: the wavelet representation, *IEEE Trans. Pattern Anal. Machine Intell.*, **11**: 674–693, (1989)

[21] S. Mallat, *A Wavelet Tour of Signal Processing* (Academic Press, London, 1998)

[22] D. Marr, *Vision* (Freeman, San Francisco, 1982)

[23] Y. Meyer (ed.), *Wavelets and Applications (Proc. Marseille 1989)*, (Springer, Berlin, and Masson, Paris, 1991)

[24] Y. Meyer and S. Roques (eds.), *Progress in Wavelet Analysis and Applications (Proc. Toulouse 1992)* (Editions Frontières, Gif-sur-Yvette, 1993).

[25] Y. Meyer, *Les Ondelettes, Algorithmes et Applications*, 2d ed. (Armand Colin, Paris, 1994) ; Engl. transl. of the 1st ed.: *Wavelets, Algorithms and Applications* (SIAM, Philadelphia, PA, 1993)

[26] R. Murenzi, Ondelettes multidimensionnelles et applications à l'analyse d'images, Thèse de Doctorat, Univ. Cath. Louvain, Louvain-la-Neuve, (1990)

[27] R. Todd Ogden, *Essential Wavelets for Statistical Applications and Data Analysis* (Birkhäuser, Boston, Basel, Berlin, 1997)

[28] *Proc. IEEE*, Special issue on Wavelets, **84**, No.4, April 1996

[29] O. Rioul and M. Vetterli, Wavelets and signal processing, *IEEE SP Magazine*, October 1991, 14–38

[30] M.B. Ruskai, G. Beylkin, R. Coifman, I. Daubechies, S. Mallat, Y. Meyer and L. Raphael (eds.), *Wavelets and Their Applications* (Jones and Bartlett, Boston, 1992)

[31] W. Sweldens, The lifting scheme: a custom-design construction of biorthogonal wavelets, *Applied Comput. Harm. Anal.*, **3**: 1186–1200, (1996)

[32] M.V. Wickerhauser, *Adapted Wavelet Analysis: From Theory to Software* (A.K.Peters, Wellesley, Mass., 1994)

2

The 2-D wavelet transform, physical applications and generalizations

J.-P. ANTOINE

Institut de Physique Théorique,
Université Catholique de Louvain, Belgium

Abstract

We begin with a short review of the 2-D continuous wavelet transform (CWT) and describe a number of physical applications. Then we discuss briefly the mathematical background, namely coherent states derived from group representations, and we show how it allows a straightforward extension to more general situations, such as higher dimensions, wavelets on the sphere or time-dependent wavelets. We conclude with a short outline of the 2-D discrete wavelet transform, some generalizations and a few physical applications.

2.1 Introduction

As we have seen in Chapter 1, both the continuous wavelet transform (CWT) and the discrete wavelet transform (DWT) may be extended to two dimensions. Here also, many applications have been developed, in various branches of physics and in image processing. As in the 1-D case, the CWT is better adapted to *analysis*, for instance the detection of specific features in an image. This is true, in particular, for oriented features, if one uses a wavelet which is directionally selective. On the other hand, the strong point of the DWT is *data compression*, notably in transmitting or reconstructing a 2-D signal after processing (e.g. denoising).

We will spend most of the present chapter discussing the 2-D CWT, for two reasons. First, it admits a number of interesting physical applications, that we will describe in Section 2.3. The second motivation is that its mathematical background, namely group representation theory (Section 2.4), suggests a straightforward extension to more general situations, such as wavelets in higher dimensions, or on manifolds (a sphere, for instance), or time-dependent wavelets, a promising tool for motion tracking (Section 2.5). We will also discuss the DWT in 2 dimensions (Section 2.6), but our analysis will be rather brief, because a full treatment requires the language and techniques of signal processing (filter theory), which are in general more familiar to electrical engineers than to physicists. Moreover, the 2-D DWT is mostly used in image processing, which is not the main thrust of the present book.

2.2 The continuous WT in two dimensions

2.2.1 Construction and main properties of the 2-D CWT

We begin by reviewing briefly the basic properties of the CWT in 2 dimensions, which are completely analogous to those discussed in Chapter 1 for the 1-D case (a detailed mathematical discussion is given in Section 2.4).

We consider 2-D signals of finite energy, represented by complex-valued, square integrable functions $s \in L^2(\mathbb{R}^2, d^2\vec{x})$. This condition may be relaxed, to allow, for instance, a plane wave or a δ function. In practice, a black and white image will be represented by a bounded non-negative function: $0 \le s(\vec{x}) \le M$, $\forall \vec{x} \in \mathbb{R}^2$ ($M > 0$), the discrete values of $s(\vec{x})$ corresponding to the level of gray of each pixel.

Given a signal $s \in L^2(\mathbb{R}^2, d^2\vec{x})$, we may transform it by translation, rotation and global dilation [70]. This gives, in position and momentum (or spatial frequency) space, respectively:

$$s_{a,\theta,\vec{b}}(\vec{x}) = a^{-1} s\left(a^{-1} r_{-\theta}\left(\vec{x} - \vec{b}\right)\right), \tag{2.1}$$

$$\widehat{s_{a,\theta,\vec{b}}}(\vec{k}) = a\, e^{-i\vec{b}.\vec{k}}\, \widehat{s}(a r_{-\theta}(\vec{k})). \tag{2.2}$$

In these relations, $\vec{b} \in \mathbb{R}^2$ is the translation parameter, $a > 0$ the dilation, and $r_{-\theta}$ ($0 \le \theta < 2\pi$) denotes the familiar 2×2 rotation matrix. As usual, the hat denotes a 2-D Fourier transform [27]. Clearly, the correspondence $s \mapsto s_{a,\theta,\vec{b}}$ is a unitary map.

By definition, a *wavelet* is a complex-valued function $\psi \in L^2(\mathbb{R}^2, d^2\vec{x})$ satisfying the admissibility condition

$$c_\psi \equiv (2\pi)^2 \int \frac{d^2\vec{k}}{|\vec{k}|^2} |\widehat{\psi}(\vec{k})|^2 < \infty. \tag{2.3}$$

If ψ is regular enough ($\psi \in L^1(\mathbb{R}^2) \cap L^2(\mathbb{R}^2)$ suffices), the admissibility condition simply means that the wavelet has zero mean:

$$\widehat{\psi}(\vec{0}) = 0 \quad \Longleftrightarrow \quad \int \psi(\vec{x}) \, d^2\vec{x} = 0. \tag{2.4}$$

Clearly the map $s \mapsto s_{a,\theta,\vec{b}}$ preserves the admissibility condition (2.3). Hence any function $\psi_{a,\theta,\vec{b}}$ obtained from a wavelet ψ by translation, rotation or dilation is again a wavelet. Thus the given wavelet ψ generates the whole family $\{\psi_{a,\theta,\vec{b}} \; (a > 0, \theta \in [0, 2\pi), \vec{b} \in \mathbb{R}^2)\}$. It is easily seen that the linear span of this family is dense in $L^2(\mathbb{R}^2)$. In the sequel we will denote by G this 4-dimensional parameter space.

As we will see in Section 2.4, the whole construction has a group-theoretical backbone. The parameter space G is in fact the so-called *similitude group* of the plane, composed precisely of translations, rotations and dilations, and the 2-D CWT may be derived from a unitary representation of it.

Given a signal $s \in L^2(\mathbb{R}^2)$, its CWT with respect to the wavelet ψ is:

$$S(a, \theta, \vec{b}) = \langle \psi_{a,\theta,\vec{b}} | s \rangle = a^{-1} \int \overline{\psi(a^{-1}r_{-\theta}(\vec{x} - \vec{b}))} \, s(\vec{x}) \, d^2\vec{x} \tag{2.5}$$

$$= a \int e^{i\vec{b}.\vec{k}} \, \overline{\widehat{\psi}(ar_{-\theta}(\vec{k}))} \, \widehat{s}(\vec{k}) \, d^2\vec{k}. \tag{2.6}$$

The properties of the wavelet transform are best expressed in terms of the map $W_\psi : s \mapsto c_\psi^{-1/2} S$. They may be summarized as follows [6, 69, 70].

- W_ψ is *linear*, contrary, for instance, to the Wigner–Ville transform, which is bilinear.
- W_ψ is *covariant* under translations, dilations and rotations.
- W_ψ *conserves norms*:

$$c_\psi^{-1} \iiint \frac{da}{a^3} d\theta d^2\vec{b} \; |S(a, \theta, \vec{b})|^2 = \int d^2\vec{x} \, |s(\vec{x})|^2, \tag{2.7}$$

i.e., W_ψ is an isometry from the space of signals into the space of transforms, which is a closed subspace of $L^2(G, dg)$, where $dg \equiv a^{-3} da d\theta d^2\vec{b}$ is the natural measure on G.

- As a consequence, the map W_ψ is *invertible* on its range, and the inverse transformation is simply the adjoint of W_ψ. This means that the signal $s(\vec{x})$ may be reconstructed exactly from its transform $S(a, \theta, \vec{b})$:

$$s(\vec{x}) = c_\psi^{-1} \iiint \frac{da}{a^3} d\theta d^2\vec{b} \; \psi_{a,\theta,\vec{b}}(\vec{x}) \, S(a, \theta, \vec{b}). \tag{2.8}$$

In other words, the 2-D wavelet transform provides a decomposition of the signal in terms of the analysing wavelets $\psi_{a,\theta,\vec{b}}$, with coefficients $S(a, \theta, \vec{b})$. As in 1-D [54], one can also reconstruct the signal by resumming only over scales and angles (provided the analysing wavelet satisfies a slightly stronger admissibility condition):

$$s(\vec{x}) \sim \iint \frac{da}{a^2}\, d\theta \, S(a, \theta, \vec{x}). \qquad (2.9)$$

- The projection from $L^2(G, dg)$ onto the range of W_ψ, the space of wavelet transforms, is an integral operator whose kernel $K(a', \theta', \vec{b}'|a, \theta, \vec{b})$ is the autocorrelation function of ψ, also called *reproducing kernel:*

$$K(a', \theta', \vec{b}'|a, \theta, \vec{b}) = c_\psi^{-1} \, \langle \psi_{a',\theta',\vec{b}'}|\psi_{a,\theta,\vec{b}}\rangle. \qquad (2.10)$$

Therefore, a function $f \in L^2(G, dg)$ is the wavelet transform of a certain signal iff it verifies the reproduction property:

$$f(a', \theta', \vec{b}') = \iiint \frac{da}{a^3}\, d\theta d^2\vec{b} \ K(a', \theta', \vec{b}'|a, \theta, \vec{b})\, f(a, \theta, \vec{b}). \qquad (2.11)$$

2.2.2 Interpretation of the CWT as a singularity scanner

In order to get a physical interpretation of the CWT, we notice that in signal analysis, as in classical electromagnetism, the L^2 norm is interpreted as the total energy of the signal. Therefore, the relation (2.7) suggests to interpret $|S(a, \theta, \vec{b})|^2$ as the energy density in the wavelet parameter space.

Assume now, as in 1-D, that the wavelet ψ is fairly well localized both in position space $(\vec{x})$ and in spatial frequency space $(\vec{k})$. Then so does the transformed wavelet $\psi_{a,\theta,\vec{b}}$, with effective support suitably translated by $\vec{b}$, rotated by θ and dilated by a. Because (2.5) is essentially a convolution with a function ψ of zero mean, the transform $S(a, \theta, \vec{b})$ is appreciable only in those regions of parameter space $(a, \theta, \vec{b})$ where the signal is: we get an appreciable value of S only where the wavelet $\psi_{a,\theta,\vec{b}}$ 'matches' the features of the signal s. In other words, the CWT acts on a signal as a *local* filter in all 4 variables $a, \theta, \vec{b}$: $S(a, 0, \vec{b})$ 'sees' only that portion of the signal that 'lives' around $a, \theta, \vec{b}$ and filters out the rest. Therefore, if the wavelet is well localized, the energy density of the transform will be concentrated on the significant parts of the signal. This is the key to all the approximation schemes that make wavelets such an efficient tool.

Let us make more precise the support properties of ψ. Assume ψ and $\widehat{\psi}$ to be as well localized as possible (compatible with the Fourier uncertainty property), that is, ψ has for essential support (i.e. the region outside of which the function is numerically negligible) a 'disk' of diameter T, centred around $\vec{0}$, while $\widehat{\psi}$ has for essential support a 'disk' of diameter Ω, centred around $\vec{k}_o$. Then, for the transformed wavelets $\psi_{a,\theta,\vec{b}}$ and $\widehat{\psi}_{a,\theta,\vec{b}}$ we have, respectively:

. ess supp $\psi_{a,\theta,\vec{b}}$ is a 'disk' of diameter $\simeq aT$ around $\vec{b}$, rotated by r_θ;

. ess supp $\widehat{\psi}_{a,\theta,\vec{b}}$ is a 'disk' of diameter $\simeq \Omega/a$ around $\vec{k}_o/a$, rotated by r_θ.

Notice that the product of the two diameters is constant. Thus the wavelet analysis operates at constant *relative bandwidth*, $\Delta k/k = $ const, where $k \equiv |\vec{k}|$. Therefore, the analysis is most efficient at high frequencies or small scales, and so it is particularly apt at detecting *discontinuities* in images, either point singularities (contours, corners) or directional features (edges, segments).

In addition to its localization properties, the wavelet ψ is often required to have a certain number of vanishing moments. This condition determines the capacity of the WT to detect singularities. Indeed, if ψ has all its moments vanishing up to order $n \geq 1$ (by the admissibility condition (2.4), the moment of order 0 must always vanish),

$$\int x^\alpha y^\beta \, \psi(\vec{x}) \, d^2\vec{x} = 0, \quad 0 \leq \alpha + \beta \leq n, \quad (2.12)$$

then the WT is blind to polynomials of degree up to n, that is, the smoother part of the signal. Equivalently, it detects singularities down to the $(n+1)$th derivative of the signal.

All together, as in the 1-D case, the 2-D wavelet transform may be interpreted as a mathematical, direction selective, microscope, with optics ψ, magnification $1/a$ and orientation tuning parameter θ [19]. Two features must be emphasized here: the magnification $1/a$ is *global*, independently of the direction, and there is the additional property of *directivity*, given by the rotation angle θ.

2.2.3 Practical implementation: the various representations

The first problem one faces in practice is that of visualization. Indeed $S(a, \theta, \vec{b})$ is a function of four variables: two position variables $\vec{b} = (b_x, b_y) \in \mathbb{R}^2$, and the pair $(a, \theta) \in \mathbb{R}_*^+ \times [0, 2\pi) = \mathbb{R}^2 \setminus \{0\}$.

In the 1-D case [37, 61], a^{-1} defines the frequency scale, thus the full parameter space of the 1-D WT, the time-scale half plane, is in fact a phase space, in the sense of Hamiltonian mechanics (for a 1-D mechanical system, the phase space is the time-frequency plane, or the position-momentum plane, with canonical coordinates (q, p)). Exactly the same situation prevails in 2-D: the pair (a^{-1}, θ) plays the role of spatial frequency (or momentum), expressed in polar coordinates, and so the full 4-dimensional parameter space of the 2-D WT may be interpreted as a phase space. This interpretation, which actually extends to higher dimensions (see Section

2.5.1), is borne out by mathematical analysis, using the group-theoretical framework discussed in Section 2.4 (one computes the coadjoint orbits of the similitude group) [4, 9].

Now, to compute and visualize the full CWT in all four variables is hardly possible. Therefore, in order to obtain a manageable tool, some of the variables, a, θ, b_x, b_y must be fixed. In other words, one must restrict oneself to a *section* of the parameter space. There are six possible choices of two-dimensional sections, but the geometrical considerations made above indicate that two of them are more natural: either (a, θ) or (b_x, b_y) are fixed, and the WT is treated as a function of the two remaining variables. The corresponding representations have the following characteristics [4].

(1) The *position or aspect-angle representation:* a and θ are fixed and the CWT is considered as a function of position $\vec{b}$ alone (this amounts to taking a set of snapshots, one for each value of (a, θ)). Alternatively, one may use polar coordinates, in which case the variables are interpreted as *range* $|\vec{b}|$ and *perception angle* α, a familiar representation of images.

(2) The *scale-angle representation:* for fixed $\vec{b}$, the CWT is considered as a function of scale a and anisotropy angle θ, i.e. of spatial frequency. In other words, one looks at the full CWT as through a keyhole located at $\vec{b}$, and observes all scales and all directions at once.

The position representation is the standard one, and it is useful for the general purposes of image processing: detection of position, shape and contours of objects; pattern recognition; image filtering by resynthesis after elimination of unwanted features (for instance, noise). The scale-angle representation will be particularly interesting whenever scaling behaviour (as in fractals) or angular selection is important, in particular when directional wavelets are used. In fact, both representations are needed for a full understanding of the properties of the CWT in all four variables. And both will be seen at work in the various applications described in Section 2.3.

For the numerical evaluation, in particular for exploiting the reconstruction formula (2.8), one has to discretize the WT. In either representation, a systematic use of the FFT algorithm will lead to a numerical complexity of $3N_1N_2\log(N_1N_2)$, where N_1, N_2 denote the number of sampling points in the variables (b_x, b_y) or (a, θ). In the former case, the geometry is Cartesian and a square lattice will give an adequate sampling grid. In the latter, the representation is in polar coordinates, and the discretization must naturally be logarithmic in the scale variable a and linear in the anisotropy angle θ.

In addition to these two familiar representations, there are four other two-dimensional sections, obtained by fixing two of the four variables $(a, \theta, |\vec{b}|, \alpha)$, and analysing the CWT as a function of the remaining two. Among these, the *angle-angle representation* might be useful for applications [8]. Here one fixes the range $|\vec{b}|$ and the scale a and considers the CWT at all perception angles α and all anisotropy angles θ. This case is particularly interesting, because the parameter space is now compact (it is a torus) and the discretization easy (linear) in both variables.

One may also consider three-dimensional sections, for which a single variable is fixed. Suppose, for instance, the anisotropy angle is fixed, or that it is irrelevant, because the wavelet is rotation invariant. Then the transform is a function of position and scale. This representation is optimal for detecting the presence of coherent structures, that is, structures that survive through a whole range of scales. Examples may be found, for instance, in astrophysics (hierarchical structure of galaxy clusters and superclusters) [77] or in the analysis of turbulence in fluid dynamics [47, 48]. Further information on these two topics will be found in Chapters 3 and 4, respectively.

2.2.4 Choice of the analysing wavelet

The next step is to choose an analysing wavelet ψ. At this point, there are two possibilities, depending on the problem at hand, namely isotropic or directional wavelets.

(i) *Isotropic wavelets:*
   If one wants to perform a pointwise analysis, that is, when no oriented features are present or relevant in the signal, one may choose an analysing wavelet ψ which is invariant under rotation. Then the θ dependence drops out, for instance, in the reconstruction formula (2.8). A typical example is the isotropic 2-D Mexican hat wavelet.

(ii) *Anisotropic wavelets:*
   When the aim is to detect directional features in an image, for instance to perform directional filtering, one has to use a wavelet which is *not* rotation invariant. The best angular selectivity will be obtained if ψ is *directional*, which means that its (essential) support in spatial frequency space is contained in a convex cone with apex at the origin. Typical directional wavelets are the 2-D Morlet wavelet or the Cauchy wavelets.

Let us examine in more detail some examples of wavelets of each kind.

2.2.4.1 Isotropic wavelets

The 2-D Mexican hat or Marr wavelet: In its isotropic version, this is simply the Laplacian of a Gaussian:

$$\psi_H(\vec{x}) = (2 - |\vec{x}|^2) \exp(-\tfrac{1}{2}|\vec{x}|^2) = -\Delta \, \exp(-\tfrac{1}{2}|\vec{x}|^2). \tag{2.13}$$

This is a real, rotation invariant wavelet, originally introduced in [64]. There exists also an anisotropic version, obtained by replacing in (2.13) $\vec{x}$ by $A\vec{x}$, where A is an anisotropy matrix. However, this wavelet still acts as a second order operator and detects singularities in all directions and it is of little use in practice. Hence the Mexican hat will be efficient for a fine pointwise analysis, but not for detecting directions. On the other hand, one may also use higher order Laplacians of the Gaussian,

$$\psi_H^{(n)}(\vec{x}) = (-\Delta)^n \, \exp(-\tfrac{1}{2}|\vec{x}|^2). \tag{2.14}$$

For increasing n, these wavelets have more and more vanishing moments, and are thus sensitive to increasingly sharper details. An interesting technique, pioneered in 1-D by A. Arnéodo *et al.* [20], is to analyse the same signal with several wavelets $\psi_H^{(n)}$, for different n. The features common to all the transforms surely belong to the signal, they are not artefacts of the analysis.

Difference wavelets: Many other wavelets (or filters) have been proposed in the literature, often designed for a specific problem. An interesting class consists of wavelets obtained as the difference of two positive functions, for instance a single function h and a contracted version of the latter. If h is a smooth non-negative function, integrable and square integrable, with all moments of order one vanishing at the origin, then the function ψ given by the relation :

$$\psi(\vec{x}) = \alpha^{-2} \, h(\alpha^{-1}\vec{x}) - h(\vec{x}) \quad (0 < \alpha < 1) \tag{2.15}$$

is easily seen to be a wavelet satisfying the admissibility condition (2.4). Such difference wavelets have the additional advantage that they lead to interesting and fast algorithms [46]. We will come back to this point in Section 2.2.5.4.

A typical example is the 'Difference-of-Gaussians' or DOG wavelet, obtained by taking for h a Gaussian

$$\psi_D(\vec{x}) = \alpha^{-2} \, e^{-|\vec{x}|^2/2\alpha^2} - e^{-|\vec{x}|^2/2}, \quad (0 < \alpha < 1). \tag{2.16}$$

The DOG filter is a good substitute for the Mexican hat (for $\alpha^{-1} = 1.6$, their shapes are extremely similar), frequently used in psychophysics works [39, 41]. Notice that h, and thus also ψ, need not be isotropic.

2.2.4.2 *Directional wavelets*

If one wants to detect oriented features (segments, edges, vector field,...), one needs a wavelet which is directionally selective. To be precise, we will say that a given wavelet ψ is *directional* if the effective support of its Fourier transform $\widehat{\psi}$ is contained in a convex cone in spatial frequency space $\{\vec{k}\}$, with apex at the origin, or a finite union of disjoint such cones (in that case, one will usually call ψ *multidirectional*). A review of directional wavelets and their use may be found in [11].

This definition may require a word of justification. According to (2.6), the wavelet acts as a filter in $\vec{k}$-space (multiplication by the function $\widehat{\psi}$). Suppose the signal $s(\vec{x})$ is strongly oriented, for instance a long segment along the x-axis. Then its Fourier transform $\widehat{s}(\vec{k})$ is a long segment along the k_y-axis. In order to detect such a signal, with a good directional selectivity, one needs a wavelet ψ supported in a narrow cone in $\vec{k}$-space. Then the WT is negligible unless $\widehat{\psi}(\vec{k})$ is essentially aligned onto $\widehat{s}(\vec{k})$: directional selectivity demands to restrict the support of $\widehat{\psi}$, not ψ. The corresponding standard practice in signal processing is to design an adequate filter in the *frequency* domain (high pass, band pass, ...). In addition, there are cases (magnetic resonance imaging, for instance) where data are acquired in $\vec{k}$-space (then called the measurement space) and the image space is obtained after a FT: here again directional filtering takes place in $\vec{k}$-space.

According to this definition, the anisotropic Mexican hat is not directional, since the support of $\widehat{\psi}_H$ is centred at the origin, no matter how big its anisotropy is; and, indeed, detailed tests confirm its poor performances in selecting directions [4].

The 2-D Morlet wavelet: This is the prototype of a directional wavelet:

$$\psi_M(\vec{x}) = \exp(i\vec{k}_o \cdot \vec{x}) \, \exp(-\tfrac{1}{2}|A\vec{x}|^2), \tag{2.17}$$

$$\widehat{\psi}_M(\vec{k}) = \sqrt{\epsilon} \, \exp(-\tfrac{1}{2}|A^{-1}(\vec{k} - \vec{k}_o)|^2). \tag{2.18}$$

The parameter $\vec{k}_o$ is the wave vector, and $A = \mathrm{diag}[\epsilon^{-1/2}, 1]$, $\epsilon \geq 1$, is a 2×2 anisotropy matrix. As in 1-D, we should add a correction term to (2.17) and (18) to enforce the admissibility condition $\widehat{\psi}_M(0) = 0$. However, since it is numerically negligible for $|\vec{k}_o| \geq 5.6$, we have dropped it altogether. The modulus of the (truncated) wavelet ψ_M is a Gaussian, elongated in the x direction if $\epsilon > 1$, and its phase is constant along the direction orthogonal to $\vec{k}_o$. Thus the wavelet ψ_M smoothes the signal in all directions, but detects the sharp transitions in the direction perpendicular to $\vec{k}_o$. In Fourier space, the effective support of the function $\widehat{\psi}_M$ is an ellipse centred at $\vec{k}_o$ and elongated in the k_y direction, thus contained in a convex cone, that becomes narrower as ϵ increases. Hence the angular selectivity increases with $|\vec{k}_o|$ and with the anisotropy ϵ and the best selectivity will be obtained by taking $\vec{k}_o$ parallel to the long axis of the ellipse, that is, $\vec{k}_o = (0, k_o)$. The function $\widehat{\psi}_M$ with $\epsilon = 5$, is shown (in perspective and in level curves) in Figure 2.1 (left).

Cauchy wavelets: In order to achieve a genuinely oriented wavelet, it suffices to consider a smooth function $\widehat{\psi}_c(\vec{k})$ with support in a strictly convex cone $\mathcal{C}$ in spatial frequency space, with apex at the origin, and behaving inside $\mathcal{C}$ as $P(k_x, k_y)e^{-\vec{k}\cdot\vec{\eta}}$, with $\vec{\eta} \in \mathcal{C}$, or $P(k_x, k_y)e^{-|k|^2}$, where $P(.)$ denotes a polynomial in two variables. A typical example is the family of Cauchy wavelets, that we now describe.

For simplicity, we consider a cone symmetric with respect to the positive k_x-axis, namely $\mathcal{C} \equiv \mathcal{C}(-\alpha, \alpha) = \{\vec{k} \in \mathbb{R}^2 \mid -\alpha \leq \arg \vec{k} \leq \alpha\}$, the convex cone determined by the unit vectors $\vec{e}_{-\alpha}, \vec{e}_{\alpha}$. The dual cone, also convex, is $\tilde{\mathcal{C}}(-\tilde{\alpha}, \tilde{\alpha}) = \{\vec{k} \in \mathbb{R}^2 \mid \vec{k} \cdot \vec{k}' > 0, \ \forall \vec{k}' \in \mathcal{C}(-\alpha, \alpha)\}$, where $\tilde{\alpha} = -\alpha + \pi/2$, and therefore $\vec{e}_{-\alpha} \cdot \vec{e}_{\tilde{\alpha}} = \vec{e}_{\alpha} \cdot \vec{e}_{-\tilde{\alpha}} = 0$. Given the fixed vector $\vec{\eta} = (\eta, 0), \eta > 0$, we define the Cauchy wavelet in spatial frequency variables [9, 11, 13]:

$$\widehat{\psi}_{lm}^{(C)}(\vec{k}) = \begin{cases} (\vec{k} \cdot \vec{e}_{\tilde{\alpha}})^l \, (\vec{k} \cdot \vec{e}_{-\tilde{\alpha}})^m \, e^{-\vec{k}\cdot\vec{\eta}}, & \vec{k} \in \mathcal{C}(-\alpha, \alpha) \\ 0, & \text{otherwise.} \end{cases} \tag{2.19}$$

The Cauchy wavelet $\widehat{\psi}_{lm}^{(C)}(\vec{k})$ is strictly supported in the cone $\mathcal{C}(-\alpha, \alpha)$ and the parameters $l, m \in \mathbb{N}^*$ give the number of vanishing moments on the edges of the cone. An explicit calculation yields the following result:

$$\psi_{lm}^{(C)}(\vec{x}) = \text{const.} \, (\vec{z} \cdot \vec{e}_\alpha)^{-l-1} \, (\vec{z} \cdot \vec{e}_{-\alpha})^{-m-1}, \tag{2.20}$$

where we have introduced the complex variable $\vec{z} = \vec{x} + i\vec{\eta} \in \mathbb{R}^2 + i\tilde{\mathcal{C}}$. We show in Figure 2.1 (right) the wavelet $\widehat{\psi}_{44}^{(C)}(\vec{k})$ for $\mathcal{C} = \mathcal{C}(-10°, 10°)$; this is manifestly a highly directional filter.

The construction generalizes in a straightforward way to any convex cone $\mathcal{C}(\alpha, \beta)$ [9, 11, 13]. In addition, if one lets $\vec{\eta}$ vary in the dual cone $\tilde{\mathcal{C}}(\tilde{\beta}, \tilde{\alpha})$, then the wavelet $\psi_{lm}^{(C)}(\vec{x})$ is the boundary value of a function $\psi_{lm}^{(C)}(\vec{z})$, holomorphic in the tube $\mathbb{R}^2 + i\tilde{\mathcal{C}}$. This follows from general theorems [78, 79], since the function $\widehat{\psi}_{lm}(\vec{k})$ has support in the convex cone $\mathcal{C} = \mathcal{C}(\alpha, \beta)$ and is of fast decrease at infinity.

Note also that other wavelets, although not directional in the sense of the above definition, may have some capabilities of directional filtering. Such are, for instance, the *gradient* wavelets $\partial_x \exp(-|\vec{x}|^2)$ or $\partial_x \partial_y \exp(-|\vec{x}|^2)$. The latter,

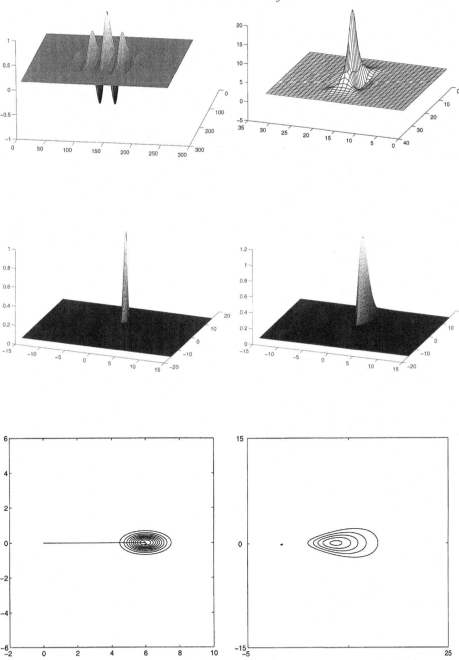

Fig. 2.1. Two 2-D directional wavelets. (Left) The Morlet wavelet ($\epsilon = 5, k_o = 6$). (Right) The Cauchy wavelet $\psi_{44}^{(\mathcal{C})}$ for $\mathcal{C} = \mathcal{C}(-10^\circ, 10^\circ)$. The top row shows the real part of ψ in position space, the other two the wavelet $\widehat{\psi}$ in spatial frequency space, in 3-D perspective (centre) and in level curves (bottom). For the Morlet wavelet (bottom left), the segment represents the vector $\vec{k}_o$.

in particular, looks promising for the detection of *corners* in a contour, as we will discuss in Section 2.3.1.1.

2.2.5 Evaluation of the performances of the CWT

Given a wavelet, what is its angular and scale selectivity (resolving power)? What is the minimal sampling grid for the reconstruction formula (2.8) that guarantees that no information is lost? The answer to both questions resides in a *quantitative* knowledge of the properties of the wavelet, that is, the tool must be *calibrated.*

To that effect, one takes the WT of particular, standard signals. Three such tests are useful, and in each case the outcome may be viewed either at fixed (a, θ) (position representation) or at fixed $\vec{b}$ (scale-angle representation).

- *Point signal:* for a snapshot at the wavelet itself, one takes as the signal a delta function, i.e. one evaluates the impulse response of the filter:

$$\langle \psi_{a,\theta,\vec{b}} | \delta \rangle = a^{-1} \overline{\psi(a^{-1} r_{-\theta}(-\vec{b}))}. \tag{2.21}$$

- *Reproducing kernel:* taking as the signal the wavelet ψ itself, one obtains the reproducing kernel K, which measures the *correlation length* in each variable $a, \theta, \vec{b}$:

$$c_\psi K(a, \theta, \vec{b} | 1, 0, \vec{0}) = \langle \psi_{a,\theta,\vec{b}} | \psi \rangle = a^{-1} \int \overline{\psi(a^{-1} r_{-\theta}(\vec{x} - \vec{b}))} \, \psi(\vec{x}) \, d^2\vec{x}. \tag{2.22}$$

- *Benchmark signals:* for testing particular properties of the wavelet, such as its ability to detect a discontinuity or its angular selectivity in detecting a particular direction, one may use appropriate 'benchmark' signals.

2.2.5.1 The scale and angle resolving power

Suppose the wavelet ψ has its effective support in spatial frequency in a vertical cone of aperture $\Delta\varphi$, corresponding to $\vec{k}_o = (0, k_o)$. The width of $\widehat{\psi}$ in the x and y directions is given by $2w_x$, resp. $2w_y$:

$$w_x = \frac{1}{\|\widehat{\psi}\|} \left[\int d^2\vec{k} \, k_x^2 |\widehat{\psi}(\vec{k})|^2 \right]^{1/2}, \quad w_y = \frac{1}{\|\widehat{\psi}\|} \left[\int d^2\vec{k} \, (k_y - k_o)^2 |\widehat{\psi}(\vec{k})|^2 \right]^{1/2}. \tag{2.23}$$

Then the wavelet $\widehat{\psi}$ is concentrated in an ellipse of semi-axes w_x, w_y, and its radial support is $k_o - w_y \leq |\vec{k}| \leq k_o + w_y$. Thus the scale width or scale resolving power (SRP) of ψ is defined as:

$$SRP(\psi) = \frac{k_o + w_y}{k_o - w_y}. \tag{2.24}$$

In the same way, one defines the angular width or angular resolving power (ARP) by considering the tangents to that ellipse. Then a straightforward calculation yields:

$$ARP(\psi) = 2\cot^{-1} \frac{\sqrt{k_o^2 - w_y^2}}{w_x} \simeq \Delta\varphi. \tag{2.25}$$

For instance, if ψ is the (truncated) Morlet wavelet (2.17), one obtains:

$$SRP(\psi_M) = \frac{k_o\sqrt{2} + 1}{k_o\sqrt{2} - 1}, \quad ARP(\psi_M) = 2\cot^{-1}\sqrt{\epsilon(k_o^2 - 1)}, \tag{2.26}$$

and, for $k_o \gg 1$:

$$ARP(\psi_M) = 2\cot^{-1}(k_o\sqrt{\epsilon}). \tag{2.27}$$

This last expression coincides with the empirical result of [4]: the angular sensitivity of ψ_M depends only on the product $k_o\sqrt{\epsilon}$. Notice also that the SRP is independent of the anisotropy factor ϵ.

If ψ is the Cauchy wavelet (2.19) with support in the cone $\mathcal{C}(-\alpha, \alpha)$, the ARP is simply the opening angle 2α of the supporting cone.

2.2.5.2 *The reproducing kernel and the resolving power of the wavelet*

A natural way of testing the correlation length of the wavelet is to analyse systematically its reproducing kernel. Let the effective support of the wavelet ψ in spatial frequency be, in polar coordinates, $\Delta\rho$ and $\Delta\varphi$. Then an easy calculation [9] shows that the effective support of K is given by $a^{min} = (\Delta\rho)^{-1} \le a \le a^{max} = \Delta\rho$ for the scale variable, and $-\Delta\varphi \le \theta \le \Delta\varphi$ for the angular variable. Thus we may define the wavelet parameters (or resolving power) $\Delta\rho$, $\Delta\varphi$ in terms of the parameters Δa, $\Delta\theta$ of K, as:

. scale resolving power (SRP): $\quad \Delta\rho = \sqrt{\Delta a} = \sqrt{a_{max}/a_{min}}$;
. angular resolving power (ARP): $\quad \Delta\varphi = \frac{1}{2}\Delta\theta$.

In this way, one may design a wavelet filter bank $\{\widehat{\psi}_{a_j, \theta_\ell}(\vec{k})\}$, which yields a complete tiling of the spatial frequency plane, in polar coordinates [6, 9]. Clearly this analysis is only possible within the scale-angle representation. Thus it requires the use of the CWT, and it is outside of the scope of the DWT, which is essentially limited to a Cartesian geometry (see Section 2.6).

2.2.5.3 Calibration of a wavelet with benchmark signals

The capacity of the wavelet at detecting a discontinuity may be measured on the (benchmark) signal consisting of an infinite rod (see [4] for the full discussion). The result is that both the Mexican hat and the Morlet wavelet are efficient in this respect.

For testing the angular selectivity of a wavelet, one computes the WT of a semi-infinite rod, sitting along the positive x-axis, and modelled as usual with a delta function:

$$s(\vec{x}) = \vartheta(x)\,\delta(y), \tag{2.28}$$

where $\vartheta(x)$ is the step function. Let us take first a Morlet wavelet with $\epsilon = 5$, oriented at an angle θ, and compute the CWT of s as a function of x. The result is that ψ_M detects the orientation of the rod with a precision of the order of $5°$. Indeed, for $\theta < 5°$, the WT is a 'wall', increasing smoothly from 0, for $x \leq -5$, to its asymptotic value (normalized to 1) for $x \geq 5$. Then, for increasing misorientation θ, the wall gradually collapses, and essentially disappears for $\theta > 15°$. Only the tip of the rod remains visible, and for large θ ($\theta > 45°$), it gives a sharp peak.

Essentially the same result is obtained with a Cauchy wavelet supported in a cone of opening angle $ARP = 20°$. Conversely, for a fixed misorientation angle $\theta = 20°$, the Cauchy wavelet yields the same selectivity for $ARP \leq 20°$ (Figure 2.2). On the contrary, the same test performed with an anisotropic Mexican hat gives a result almost independent of θ. The conclusion is that the Morlet and the Cauchy wavelets are highly sensitive to orientation, but the Mexican hat is not.

Let now the signal be a segment. If one uses a Morlet or a Cauchy wavelet as above, the WT reproduces the segment if the misorientation $\Delta\phi$ between the signal and the wavelet is smaller than $5°$, but the segment becomes essentially invisible for $\Delta\phi > 15°$, except for the tips (these are point singularities). In the end, the image of the segment reduces to two peaks corresponding to the two endpoints. This is exactly the property used crucially in the measurement of the velocity field of a turbulent fluid (see Section 2.3.2.1 below).

One may remark that the precision mentioned here is obtained with the modulus of the WT. In fact, if the wavelet is complex (like ψ_M), one may also exploit the phase of the WT, and it gives a higher precision yet [4]. But this is practical only on academic signals, real data are in general too noisy and only the modulus is useful.

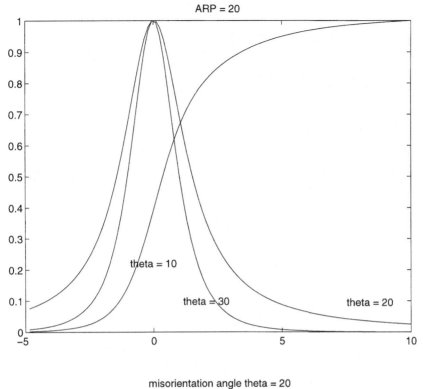

ARP = 20

theta = 10

theta = 30

theta = 20

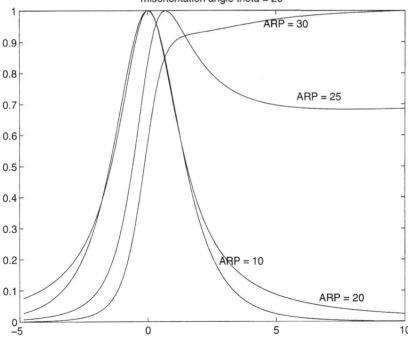

misorientation angle theta = 20

ARP = 30

ARP = 25

ARP = 10

ARP = 20

Fig. 2.2. Testing the angular selectivity of the Cauchy wavelet $\widehat{\psi}_{44}$ with the semi-infinite rod signal. The two figures show the modulus of the CWT as a function of $\vec{x}$. (Top) For fixed $ARP = 20°$ and various values of the misorientation angle θ. (Bottom) For a fixed misorientation angle $\theta = 20°$ and various values of the ARP.

Another way of comparing the angular selectivity of the two wavelets is to analyse a directional signal in the angle-angle representation (α, θ) described above. The result confirms the previous one [8].

2.2.5.4 Discretization of the CWT

The reproduction property (2.11) means that the information contained in the WT $S(a, \theta, \vec{b})$ is highly redundant. This redundancy may be eliminated (this is the basic idea behind the *discrete* WT), or exploited, either under the form of interpolation formulas or for discretizing the reconstruction formula (2.8), as needed for numerical evaluation. The integral is replaced by a sum over a discrete (but infinite) family of wavelets $\psi_{a_j, \theta_k, \vec{b}_l}$, which can be chosen in such a way that no information is lost:

$$s(\vec{x}) = \sum_{jkl} \psi_{a_j, \theta_k, \vec{b}_l} (\vec{x}) \, S(a_j, \theta_k, \vec{b}_l). \tag{2.29}$$

Such an overcomplete family is called a *frame*, according to the terminology introduced by Duffin and Schaefer [43] in the context of nonharmonic Fourier series. Its existence for specific wavelets may be proven along the same lines as in the 1-D case [36, 37, 38] with similar results [69, 70]. In practical applications, the infinite sum will be truncated (a few terms will often suffice) and the approximate reconstruction so obtained is numerically stable [37, 38].

The problem, of course, is how to choose the sampling grid in an optimal fashion. The 2-D wavelet transform too obeys sampling theorems, that give lower bounds on the density of sampling points, like the standard Shannon theorem of signal analysis, only more complicated. Nevertheless, in practice, the sampling points are quite often fixed empirically. For the (a, θ) variables, in particular, they are mostly chosen on the basis of biological considerations or symmetry requirements [39, 60, 61, 63]. Now the CWT described here offers a *quantitative* solution of this sampling problem. As we have seen above, a systematic exploitation of the reproducing kernel K leads to a minimal dicretization grid as needed for the numerical evaluation of the reconstruction integral (2.8).

Besides the full discretization described here, and *a fortiori* the discrete WT that we will describe in Section 2.6, there is an intermediate procedure, introduced in [46], under the name of (continuous) wavelet packets. It consists of discretizing the scale variable alone, on an arbitrary sequence of values (not necessarily powers of a fixed ratio). This leads to difference wavelets, as mentioned in Section 2.2.4.1, but more important, to fast algorithms that could put the CWT on the same footing as the DWT in terms of speed and efficiency, for example in reconstruction problems.

2.3 Physical applications of the 2-D CWT

The 2-D CWT has been used by a number of authors, in a wide variety of physical problems [34, 67, 68]. In all cases, its main use is for the *analysis* of images. It can be used for the detection of specific features, such as a hierarchical structure, edges, filaments, contours, boundaries between areas of different luminosity, etc. Of course, the type of wavelet chosen depends on the precise aim. An isotropic wavelet (Mexican hat) suffices for pointwise analysis, but an oriented wavelet (Morlet, Cauchy) is more efficient for the detection of oriented features in the signal, that is, regions where the amplitude is regular along one direction and has a sharp variation along the perpendicular direction.

2.3.1 Pointwise analysis

2.3.1.1 Contour detection, character recognition

Exactly as in the 1-D case, the WT is especially useful to detect *discontinuities* in images, for instance the *contour* [4, 70] or the *edges* of an object [53, 65]. For that purpose, an isotropic wavelet may be chosen, such as the radial Mexican hat ψ_H given in (2.13). In that case the effect of the WT consists of smoothing the signal with a Gaussian and taking the Laplacian of the result. Thus large values of the amplitude will appear at the location of the discontinuities, in particular the contour of objects (which is a discontinuity in luminosity).

In order to test this property, we compute the WT of a simple object, namely a set with the shape of a thick letter 'A', represented by its characteristic function, for various values of the scale parameter a (Figure 2.3). Then, for large values of a, the WT sees only the object as a whole, thus allowing the determination of its position in the plane. When a decreases, increasingly finer details appear. In this simple case, the WT vanishes both inside and outside the contour, since the signal is constant there, and thus only the contour remains, and it is perfectly seen at $a = 0.075$. Of course, if one takes values of a that are too small, numerical artefacts (aliasing) appear and spoil the result. We notice that the exterior contour is a sharp negative 'wall', whereas the interior contour is a positive one. The same effect would appear in 1-D if one would consider, for instance, the full WT of a square

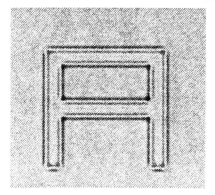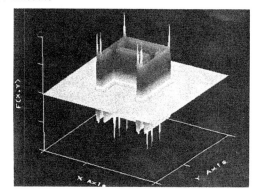

Fig. 2.3. CWT of a thick letter 'A', with a Mexican hat and $a = 0.075$, in level curves and in 3-D perspective.

pulse: the jump from 0 to 1 gives a negative minimum followed by a sharp positive maximum, and the jump from 1 to 0 gives the opposite pattern. Note also that the corners of the figure are highlighted in the WT by sharp peaks. The amplitude is larger at these points, since the signal is singular there in *two* directions, as opposed to the edges. In addition the WT detects the *convexity* of each corner. The six convex corners give rise to positive peaks, whereas the concave ones yield a negative peak. Here we see the advantage of using a real wavelet and plotting the WT itself, *not* its modulus, which is a frequent practice.

This exercise leads to an algorithm for automatic character recognition [8]. The letter 'A', for instance, is entirely characterized by the succession of its 12 corners and a logical flag (concavity or convexity) for each of them. The algorithm consists in locating the local maxima of the CWT and eliminating everything else by thresholding, and it is able to detect an 'A' unambiguously. Actually, since only the corners are needed, we may as well use a wavelet that sees *only* the corners, not the edges. Typically, a directional wavelet (when misaligned), or a real wavelet such as the gradient wavelets $\partial_x \exp(-|\vec{x}|^2)$ or $\partial_x \partial_y \exp(-|\vec{x}|^2)$.

This simple technique may be further improved by adding some denoising and inclusion of a second wavelet capable of dealing with letters of arbitrary shape (for instance, a ring-shaped wavelet sensitive to circular shapes). In addition, the automatic recognition device will need some *training*. An elegant solution would then be to use the simple wavelet treatment as a pre-processing for some sort of 'intelligent' device, such as a neural network.

2.3.1.2 Analysis of 2-D fractals

By definition, a fractal, be it in 1-D or in 2-D, is self-similar under dilation, either globally (genuine fractal) or locally (multifractal). Hence the CWT is a natural tool for analysing it, and there is an abundant literature on the subject. Notice that the *continuous* version of the WT is essential here, since the characteristic scaling ratio is unknown *a priori*.

In fact, a fractal is in general a very irregular object (for instance, its support may be a Cantor-like set), hence it should be represented by a *measure*, rather than a function. Fortunately the CWT may be extended correspondingly [18, 19]. Let μ be a fractal measure on $\mathbb{R}^2$. Then its CWT with respect to the wavelet ψ is defined as

$$T(a, \theta, \vec{b}) = \int \overline{\psi(a^{-1} r_{-\theta}(\vec{x} - \vec{b}))} \, d\mu(\vec{x}). \qquad (2.30)$$

Assume now that the measure has the following scaling behaviour around the point $\vec{x}_o$:

$$\mu(\mathcal{B}(\vec{x}_0, \lambda\epsilon)) \sim \lambda^{\alpha(\vec{x}_o)} \mu(\mathcal{B}(\vec{x}_0, \epsilon)), \quad \lambda > 0, \qquad (2.31)$$

where $\mathcal{B}(\vec{x}_0, \epsilon)$ is a ball of radius ϵ around $\vec{x}_o$ and $\alpha(\vec{x}_o)$ is the local scaling exponent. Then it is easily shown that the WT scales in the same way:

$$T(\lambda a, \theta, \vec{x}_o + \lambda\vec{b}) \sim \lambda^{\alpha(\vec{x}_o)} T(a, \theta, \vec{x}_o + \vec{b}), \quad \lambda \to 0^+. \qquad (2.32)$$

This relation is the key to the wavelet analysis of fractals. For instance, the local exponent $\alpha(\vec{x}_o)$ may be obtained by plotting $\log | T(a, \theta, \vec{b}) |$ vs. $\log a$, for a small enough. This would suffice for an exact (global) fractal, such as a numerical snowflake. But most fractals exhibit the scaling behaviour (2.31) only in the average. Thus a second essential ingredient in fractal analysis is the use of techniques borrowed from statistical mechanics (as a matter of fact, the standard 'box counting' method is already of a statistical nature). This leads to the so-called *thermodynamical formalism* of fractal analysis developed systematically by Arnéodo and his group in Bordeaux, and which is the subject of Chapter 9.

This approach has been applied successfully to a wide variety of examples [18, 19], that cover both artificial fractals (numerical snowflakes, diffusion limited aggregates) and natural ones (electrodeposition clusters, various arborescent phenomena, clouds). In addition to the standard numerical method, these authors have designed an ingenious hardware version, called the Optical WT and based on Fraunhofer diffraction, a familiar tool in optics. This approach amounts to obtaining the WT with a binary approximation to the isotropic Mexican hat.

With both techniques, the method permits the measurement of the fractal dimensions and the unravelling of universal laws (mean angle between branches, azimuthal Cantor structures, etc.). It should be remarked that the analysis uses exclusively an isotropic wavelet (usually a 2-D Mexican hat), and thus there is no θ dependence in (2.32). However, this may not be the end of the story. Indeed we shall exhibit in Section 2.3.2.2 below a fractal ('twisted snowflake') whose structure requires a *directional* wavelet for its complete determination.

2.3.1.3 Shape recognition and classification of patterns

The characterization of a 2-D shape from its outlines is an important problem in several applications of image analysis, such as character recognition, machine parts inspection for industrial applications, characterization of biological shapes such as chromosomes and neural cells, and so on. Furthermore, in the field of human vision and perception, 2-D shape analysis also plays a central role in psychophysics and neurophysiology.

There are two general approaches to shape characterization: *region based*, which deals with the region in the image corresponding to the analysed object; and *boundary based*, where the shape is characterized in terms of its silhouette. In both cases, 2-D wavelets may be used directly, as discussed above. But, for the second approach, there is an alternative, which consists of representing the shape by the complex signal that describes its boundary, and applying the 1-D CWT to this signal [10]. This leads to the so-called *W-representation*, which allows an easy way of performing a number of standard tasks (for instance, in machine vision), such as detection of dominant points, shape partitioning, natural scales analysis. Notice that an essential ingredient of the analysis is the wavelet-based fractal analysis discussed above.

2.3.1.4 Analysis of astronomical images

Astronomical images have two characteristics. They superpose objects living at very different distances (nearby stars, galaxies, quasars, galaxy clusters), and they are very noisy (in particular the bright sky represents noise). A 2-D wavelet analysis is useful on both counts and it has been exploited systematically by A. Bijaoui and his group in Nice. Applications include the unravelling of the hierarchical structure of a galactic nebula, or that of the universe itself (galaxy counts, detection of galaxy clusters or voids), and the removal of the background sky, with a technique similar to that used in 1-D for the subtraction of unwanted lines or noise in spectra [55]. Here too, statistical techniques play an essential rôle. A systematic presentation will be found in Chapter 3.

A new application under development [24] is the detection of Einstein gravitational arcs in cosmological pictures. Whenever the light from a distant bright object (a quasar) is seen through a galaxy, the latter behaves as a gravitational lens, so that the point source appears as a ring, or a portion of a ring ('arclet'), if the alignment is not exact. By measuring the radius of that ring, one may infer the distance of the source. This may be done in two steps. The centre of the arc is obtained with an annular-shaped wavelet,

$$\widehat{\psi}_\kappa(\vec{k}) \sim e^{-(|\vec{k}|^2 - \kappa)}, \tag{2.33}$$

used at a rather large scale (e.g. $a = 2$). This determination is quite robust to noise, in particular spurious bright points, that mimic nearby stars. The arc itself is obtained with a Mexican hat, at a smaller scale (e.g. $a = 0.5$). By superposing the two transforms and applying a severe thresholding (up to 95%) for eliminating the noise, one obtains an image with three bright spots: two points of the arc, around the endpoints, and the centre of the corresponding circle. From this one can reconstruct the arc unambiguously, and thus one obtains a tool for measuring in a simple way the distance of quasars, for instance.

2.3.2 Applications of directional wavelets

As a consequence of their good directional selectivity, the Morlet and Cauchy wavelets are quite efficient for directional filtering. In order to illustrate the point, we analyse in Figure 2.4 a pattern made of rods in many different directions (top). Applying the CWT with a fixed direction, here horizontal, selects all those rods with roughly the same direction (bottom left), whereas the other ones, which are misaligned, yield only a faint signal corresponding to their tips, in agreement with the behaviour discussed above. Since this is in fact noise, one performs a thresholding to remove it, thus getting a clear picture (bottom right). In this way, one can count the number of objects that lie in any particular direction.

2.3.2.1 Application in fluid dynamics

Wavelets have been successfully applied to the analysis of 2-D developed turbulence in fluids, especially localization of small scales in the distribution of energy or enstrophy [48]. This topic is described in Chapter 4. We describe here two other applications of 2-D wavelets in fluid dynamics, which both rely on the possibility of directional filtering with directional wavelets as described above.

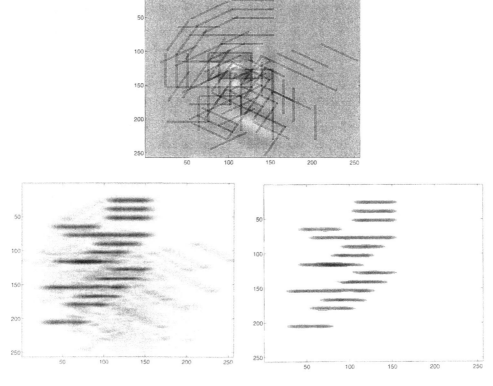

Fig. 2.4. Directional filtering with a Cauchy wavelet ($ARP = 20°$) oriented at $\theta = 0°$. (Top) the pattern; (bottom left) the CWT; (bottom right) the same after thresholding at 25%.

Measuring a velocity field: In the first example [82, 83], the aim is to measure the velocity field of a 2-D turbulent flow around an obstacle. Velocity vectors are materialized by small segments, by the technique of discontinuous tracers. Tiny plastic balls are seeded into the flow and two successive pictures are taken, both with a short exposure time, but the first one shorter. In this way one gets a 'dot-bar' signature for each tracer, which materializes the direction and the length of the local velocity. Then the WT with a Morlet wavelet is computed twice. First the WT selects those vectors that are closely aligned with the wavelet. Then the analysis is repeated with a wavelet oriented in the orthogonal direction, thus completely misoriented with respect to the selected vectors. Now the WT sees only the tips of the vectors and their length may be easily measured. The same two operations are then repeated with various successive orientations of the wavelet. Using appropriate thresholdings, the complete velocity field may thus be obtained, in a totally automated fashion, with an efficiency sensibly better than with more traditional methods. Notice that the analysis gives in principle both the modulus and the phase of the WT. But here, contrary to the simple applications like contour detection [4], the phase cannot be exploited, the data are too noisy. Thus one loses some precision on the orientation. Nevertheless, the method is remarkably efficient.

Disentangling of a wave train: A second example originates from underwater acoustics. When a point source emits a sound wave above the surface of water, the wave hitting the surface splits into several components of very different characteristics (called respectively 'direct', 'lateral' and 'transient'). The resulting wave train is represented by a linear superposition of damped plane waves, and the goal is to measure the parameters of all components. This phenomenon has been analysed successfully with the WT both in 1-D [75] and in 2-D [9], and the extension to a 3-D version is straightforward. The signal representing the underwater wave train is taken as a linear superposition of damped plane waves:

$$f(\vec{x}) = \sum_{n=1}^{N} c_n \, e^{i \vec{k}_n \cdot \vec{x}} \, e^{-\vec{l}_n \cdot \vec{x}}, \tag{2.34}$$

where, for each component, $\vec{k}_n$ is the wave vector, $\vec{l}_n$ is the damping vector, and c_n a complex amplitude. Then, using successively the scale-angle and the position representations described in Section 2.2.3, one is able to measure all the $6N$ parameters of this signal with remarkable ease and precision.

The method proceeds in three steps and uses explicitly the phase space interpretation. First one computes the CWT of the signal (2.34) with a Morlet wavelet. By linearity, the result is the linear superposition of the contributions of the various components. Moreover, each component is the product of two factors, where the first one depends on $\vec{b}$ only and the second one on (a, θ) only:

$$F(a, \theta, \vec{b}) = \sum_{n=1}^{N} c_{\vec{b},n} \, \check{F}_n(a, \theta). \tag{2.35}$$

Now we go to the scale-angle representation and consider the WT (2.35) for fixed $\vec{b}$. Then a straightforward calculation shows that, for each term in this superposition, $a^{-1} \check{F}_n(a, \theta)$ admits a unique local maximum. Suppose that these local maxima are well separated. Then, barring some interference effects (which may often be alleviated by increasing the selectivity of the wavelet), one may write:

$$|F(a, \theta, \vec{b})| \simeq \sum_{n=1}^{N} |c_{\vec{b},n}| \, |\check{F}_n(a, \theta)|. \tag{2.36}$$

One then reverts to the position representation, choosing for (a, θ) successively each of the maxima. Then the filtering effect of the CWT essentially eliminates all components except the nth one, which is then easy to treat. In this way, one is able to measure easily all the $6N$ parameters of the signal.

2.3.2.2 Detection of symmetries

The directional selectivity of a wavelet may also be used for evaluating the symmetry of a given pattern. Let $S(a, \theta, \vec{b})$ be the wavelet transform of an object with respect to the Cauchy wavelet. Define the following positive valued function, called the *scale-angle measure* of the signal:

$$\mu_s(a, \theta) = \int d^2\vec{b} \, |S(a, \theta, \vec{b})|^2. \tag{2.37}$$

This quantity may also be viewed as a partial energy density in the scale and angle variables, that is, in spatial frequency space, according to the phase space interpretation of the CWT given in Section 2.2.3. This is different from the scale-angle representation, where the position parameter $\vec{b}$ is fixed [9]. Here, on the contrary, μ_s averages over all points in the plane, thus eliminating the dependence on the point of observation. For any signal of finite energy, it is clear that μ_s is a continuous bounded function of (a, θ).

We begin with a simplified version and eliminate the scale dependence by integrating over a, thus ending with a function α_s of the rotation angle only, called the *angular measure* of the object. In general, $\alpha_s(\theta)$ is 2π-periodic. But when the analysed object has rotational symmetry n, that is, it is invariant under a rotation of angle $\frac{2\pi}{n}$, then α_s is in fact $\frac{2\pi}{n}$-periodic. Note that, for $n = 2$, there are two different operations of order 2, rotation of π and reflection (mirror symmetry), which may also be seen as a rotation of π around an axis lying in the plane of the figure (Ox or Oy). To give a simple example, consider three geometrical figures: a square, a rectangle and a regular hexagon [13, 15]. The square and the hexagon have symmetry $n = 4$ and $n = 6$, respectively, and thus their angular measures show four, resp. six equal peaks. The width of these peaks is simply the aperture of the support cone (i.e. the ARP) of the wavelet (Figure 2.5). The case of the rectangle is more interesting. It has symmetry $n = 2 \times 2$ (two mirror symmetries, or rotations

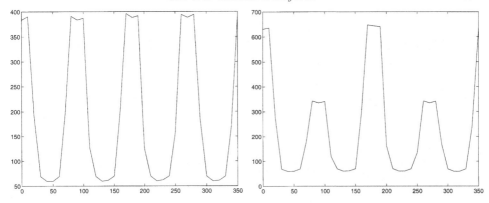

Fig. 2.5. Angular measure of regular figures obtained with a Cauchy wavelet ($ARP = 20°$): a square (left) and a rectangle, with side ratio 2:1 (right).

by π around both Ox or Oy), and that is reflected on the graph of its angular measure: there are two large peaks corresponding to the longest edges and two smaller peaks corresponding to the shortest ones, and the ratio 2:1 between the two equals that of the lengths of the corresponding edges (to be sure, the wavelet catches the direction of the *edges*, not that of the corners, so that indeed the maxima of α_S are again at $\theta = 0°, 90°, 180°, 270°$, just as for the square, but now the amplitudes are different).

This technique also allows one to identify the symmetries of quasi-lattices or tilings. For instance, the angular measure of a Penrose tiling reveals its local 10-fold symmetry, namely α_P is a $\pi/5$-periodic function of θ (Figure 2.6, top left). Actually one may go further and uncover the combined rotation-dilation symmetry of the tiling, using the full scale-angle measure μ_P. This function is again $\pi/5$-periodic in θ, which reflects the 10-fold symmetry (Figure 2.6, top right). But there are in fact *two* sets of ten maxima, for two different scales $\log a_1 = -2.6$ and $\log a_2 = -2.3$, and shifted by $36°$. This means that the tiling has, in addition to its 10-fold symmetry, a combined rotation-dilation symmetry. It is invariant under a rotation by $36°$, followed by a dilation by a factor a_1/a_2. In order to determine the two characteristic scales, one may use the skeleton of μ_P (lines of local maxima [40]), or use a wavelet which is sharply peaked in frequency, such as a Gaussian Cauchy wavelet, as was done in Figure 2.6 (bottom). To illustrate the point, we show in Figure 2.7 the analysis of a 'twisted snowflake'. This means a mathematical snowflake [18, 19] with the following modified construction rule: upon each downscaling by a factor of 3, the figure is turned by $36°$. The scale-angle measure of this object shows precisely the combined

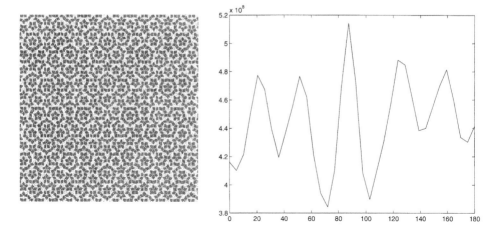

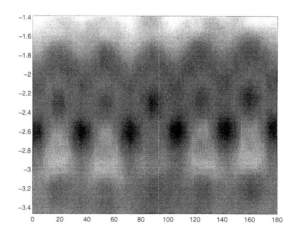

Fig. 2.6. (Top left) A Penrose tiling. (Top right) The angular measure $\alpha_P(\theta)$ reveals the 10-fold local symmetry, through the $\pi/5$-periodicity. (Bottom) The full scale-angle measure $\mu_P(a, \theta)$ shows the combined rotation-dilation symmetry. Both measures are obtained with a Gaussian Cauchy wavelet, and only a half-cycle $[0, \pi]$ is shown.

symmetry. The set of 4 maxima at a given scale a_o is reproduced, at scale $a_o/3$, but translated in θ by $36°$. And reconstructing the WT at the values (a, θ) corresponding to these maxima yields successive approximations of the original signal.

Incidentally, these examples show why it is safer to integrate over all scales in order to isolate the angular behaviour, rather than to fix a certain scale $a = a_o$ and consider $\mu_s(a_o, \theta)$. If a_o coincides with one of the characteristic scales, $a_1, a_2, \ldots$, the result is correct, but if a_o falls in between, no maximum

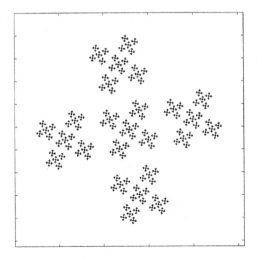 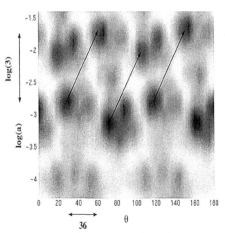

Fig. 2.7. Analysis of the twisted snowflake. (Left) the pattern; (Right) the scale-angle measure; the links indicate pairs of points which are related by the combined rotation (36°) and inflation symmetry by a factor of 3.

will be seen, and the symmetry is not detected. The effect is shown in Figure 2.8 for the Penrose tiling of Figure 2.6.

This technique permits one to determine, in a straightforward way, the (possibly hidden) symmetries of a given pattern. This applies to a genuine lattice, but also to a quasi-lattice, for which the symmetry is only local, for instance the diffraction spectrum of a quasi-crystal. Thus we expect interesting physical applications, either in the field of crystallography, or in texture analysis and classification.

2.3.2.3 Geophysics: fault detection

An interesting application of directional wavelets to geology has been initiated recently [71]. The object to be analysed is a system of geological faults, with shows a self-similar behaviour over scales from a few metres to hundreds of kilometres. This explains the use of the multifractal formalism for analysing such a system. What the authors propose here is a continuous wavelet analysis, with directional wavelets, combined with a multifractal analysis. The motivation for this choice is that the relevant information to be measured is the anisotropy of the fault field, and the variation of this anisotropy with scale. Unfortunately, the authors use only an anisotropic Mexican hat, which has rather poor directional selectivity, and this makes their results less convincing. Clearly such an analysis should be performed

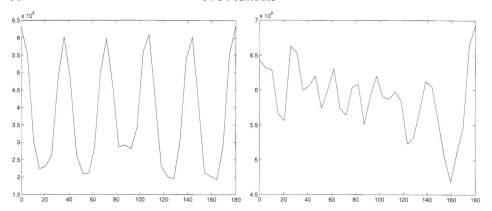

Fig. 2.8. The scale-angle measure of the Penrose tiling from Figure 2.6, for fixed
values a_o of the scale: (Left) for $\log a_o = -2.6$, the periodicity is obvious; (Right) for
$\log a_o = -2$, between two lines of maxima, the symmetry is not seen.

with a genuine directional wavelet, such as a Morlet or a Cauchy wavelet,
and a much better precision is likely to result.

2.3.2.4 Determination of textures

The determination and classification of textures in images is an old and
difficult problem, with many potential applications. Because most textures
are oriented, it is natural to try and use 2-D directional wavelets for attacking
the problem. Actually, some proposals have been made with the discrete WT,
but, since directions are essential, the CWT is certainly better adapted.
Indeed some progress on the texture problem has been achieved recently
along these lines [51, 59]. One of the key steps is the generalization to 2-D
of the algorithm for measuring the instantaneous frequency of the signal
(which becomes here the local wave vector) and the systematic use of the
ridge or skeleton of the CWT, both familiar in the analysis of spectra (asymptotic signals) [40]. Although the results belong more to image processing, the
method *per se* is interesting, which justifies its presence in this review.

2.3.3 Local contrast: a nonlinear extension of the CWT

The intensity of light around us varies considerably, in fact by several orders
of magnitude. Our visual system is well adapted to this situation. Indeed it
analyses the spatial organization of the luminous field by relying on the
contrast of objects and figures contained in the images. Intuitively, contrast
is defined as the ratio between a variation of luminance and a reference level

of luminance, i.e. a quantity of the form $\Delta L/L$, where L is the luminance level. The problem is to find a quantitative definition of contrast.

To that effect, one notices two facts. First the concept of multiscale analysis with functions of constant shape is commonly used in vision research [64]. This suggests to use the wavelet transform for describing the variations of luminance. Now the WT is a space-scale analysis, and the spatial extension of the wavelets is characterized explicitly by their scale factor. Thus it is possible to define at each scale a different normalization, similar to a local average. So one is led to the notion of *local contrast*, defined by combining the wavelet transform with an *adaptive* normalization [3, 44]. The latter will be obtained by projecting the signal, at a given scale, on a local weight function, chosen with the same localization properties as the wavelets. This local mean value will be called *luminous level:* this is the background against which luminance variations are measured, and the WT may be interpreted as a representation of these luminance variations within an image. The resulting contrast analysis is *nonlinear*, but it presents several advantages. It is particularly well adapted to the processing of positive signals. It also yields a multiplicative reconstruction process, which preserves positivity. Let us give some details and an example of application.

Let h be a nonnegative, rotation invariant, function $h \in L^1(\mathbb{R}^2) \cap L^2(\mathbb{R}^2)$, normalized to $\|h\|_{L^1} = 1$. An image is represented by a nonnegative function f. Then the luminous level with respect to the weight function h is defined as

$$M_a[f](\vec{b}) = \langle \tilde{h}_{a,\vec{b}} | f \rangle, \quad \tilde{h}_{a,\vec{b}}(\vec{x}) = a^{-2} h\left(a^{-1}(\vec{x} - \vec{b})\right). \tag{2.38}$$

Note that we use throughout the L^1 normalization, that is, $\tilde{h}_{a,\vec{b}}$ instead of the usual $h_{a,\vec{b}}$. This is more natural in this context, since all the functions $\tilde{h}_{a,\vec{b}}$ have the same L^1 norm.

Then we define the *local contrast* as the ratio of the CWT to the corresponding luminous level (the wavelet ψ is assumed to be also rotation invariant):

$$C_a[f](\vec{b}) = \frac{F_a(\vec{b})}{M_a[f](\vec{b})} = \frac{\langle \tilde{\psi}_{a,\vec{b}} | f \rangle}{\langle \tilde{h}_{a,\vec{b}} | f \rangle} = \frac{\langle \psi_{a,\vec{b}} | f \rangle}{\langle h_{a,\vec{b}} | f \rangle}, \tag{2.39}$$

where again $F_a(\vec{b}) = \langle \tilde{\psi}_{a,\vec{b}} | f \rangle$ is the CWT of f with the L^1 normalization (but the local contrast is independent of the normalization). In order to make sense, this definition requires that the support of ψ be contained in the support of h. The local contrast is nonlinear, but its behaviour is controlled by an integral condition. Large absolute values of contrast imply the existence of a region where the luminance signal is very small. A typical example, very natural in the study of vision, is to take for h a Gaussian and for ψ a Mexican hat.

But one can do better and take for ψ the difference wavelet associated to h, as given in (2.15). Then the local contrast becomes

$$C_a[f](\vec{b}) = \frac{\langle \tilde{h}_{a\alpha,\vec{b}} | f \rangle}{\langle \tilde{h}_{a,\vec{b}} | f \rangle} - 1, \tag{2.40}$$

and the existence condition is simply that the support of h be star-shaped.

This formula in turn leads to a multiplicative reconstruction scheme. Indeed, estimates of the luminous level at smaller and smaller scale factors a may be considered as smoothened versions of the image with progressively contracted weight functions h. Then, as for the WT, the approximation of a function at a given scale may be written in terms of the approximation at a larger scale and the complementary signal :

$$\begin{aligned} M_{a\alpha}[f] &= M_a[f] \cdot (C_a[f] + 1), \\ M_{a\alpha^2}[f] &= M_{a\alpha}[f] \cdot (C_{a\alpha}[f] + 1) \\ &= M_{a\alpha}[f] \cdot (C_a[f] + 1) \cdot (C_{a\alpha}[f] + 1), \end{aligned} \tag{2.41}$$

and by recurrence:

$$M_{a\alpha^n}[f] = M_{a\alpha}[f] \cdot (C_a[f] + 1) \dots (C_{a\alpha^{n-1}}[f] + 1). \tag{2.42}$$

$M_{a\alpha^n}[f]$ is the nth resolution approximation of f, it is the image as seen through the smoothing function h contracted by a factor $a\alpha^n$ ($a < 1$). One notices the obvious analogy with the usual multiresolution analysis (Section 2.6). The formalism may be generalized further to the so-called *infinitesimal contrast analysis* developed in [46].

This technique may be applied for improving the contrast in any kind of images. An example of application to a photograph was given in [3]. Here we show one with a medical image (Figure 2.9). The image f is decomposed over N contrast levels, as in (2.42), using the couple Gaussian-DOG. For each level j, one defines the contrast chart as the modulus of the local contrast,

$$M_j(\vec{b}) = |C_{2^j}[f](\vec{b})|, \ j = 1, \dots, N. \tag{2.43}$$

Then one interprets the product of the N charts, $S(\vec{b}) = \prod_{j=i}^{N} M_j(\vec{b})$ as a measure of the correlation between the successive scales of the image at the point $\vec{b}$. After thresholding, one obtains a binary image or mask. The latter is used in medical imagery, for instance, as a preprocessing to more sophisticated algorithms. It is taken as *a priori* knowledge and helps to reduce the amount of computation.

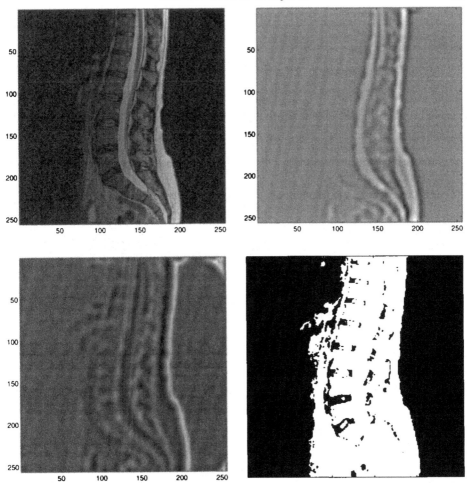

Fig. 2.9. Contrast analysis of a medical image. (Top left) The original image. (Top right) The CWT with a Mexican hat ($j = -1$). (Bottom left) The contrast chart $M_j(\vec{b})$, $j = -1$. (Bottom right) The resulting binary image. Many more details are seen on the two bottom images than on the ordinary CWT.

2.4 Continuous wavelets as affine coherent states

2.4.1 A general set-up

As we have seen in Chapter 1, the natural geometry of the (a, b)-half-plane $\mathbb{R}_+^2$ is not the usual Euclidean one. Indeed the measure $da\,db/a^2$ is invariant not only under time translation, but also under dilation. The reason behind these facts and the nice properties described above is to be found in group representation theory. The natural operations on a 1-D signal are precisely time translations and dilations, and these together constitute the affine group G_{aff} of the line. Then the relation:

$$(U(a, b)f)(x) \equiv f_{ab}(x) = a^{-1/2} f(a^{-1}(x - b)), \quad a \neq 0, \ b \in \mathbb{R}, \tag{2.44}$$

defines a unitary irreducible representation of G_{aff} in the Hilbert space $L^2(\mathbb{R})$ of finite energy signals. This means that, for every $g \equiv (a, b) \in G_{aff}$, $U(g)$ is a unitary operator and one has, for any $g, g' \in G_{aff}$:

- $U(g)U(g') = U(gg')$
- $U(g^{-1}) = U(g)^\dagger$
- $U(e) = I$, where $e = (1, 0)$ denotes the unit element of G_{aff}.

In addition, $L^2(\mathbb{R})$ contains no subspace invariant under U, except the trivial one $\{0\}$. Furthermore, and this is the crucial feature, the representation U is *square integrable*, that is, there exists at least one (and in fact a dense set of) admissible vectors, i.e. vectors ψ such that the matrix element $\langle U(a, b)\psi | \psi \rangle$ is square integrable over the group, with respect to the natural measure, namely $da\,db/a^2$. Now a straightforward calculation shows that

$$\int_{G_{aff}} |\langle U(a, b)\psi | \psi \rangle|^2 \, \frac{da\,db}{a^2} = 2\pi \|\psi\|^2 \int_{-\infty}^{+\infty} |\widehat{\psi}(\omega)|^2 \, \frac{d\omega}{|\omega|}. \tag{2.45}$$

Comparing (2.45) with Eq. (1.7) in Chapter 1, one sees that the two notions of admissibility we have introduced indeed coincide.

Of course, true dilations should be positive, i.e. one should restrict oneself to $a > 0$. This defines a subgroup of G_{aff}, called the 'ax + b' group or connected affine group G_{aff}^+ of the line. When restricted to G_{aff}^+, the representation U becomes reducible and splits into two irreducible components, corresponding to the subspaces $\mathcal{H}_\pm = \{f \in L^2(\mathbb{R}), \widehat{f}(\omega) = 0 \text{ for } \omega \lessgtr 0\}$, called Hardy subspaces in the mathematical literature. Then a function $f \in \mathcal{H}_+$ (resp. $\mathcal{H}_-$) has an analytic extension into the whole upper (resp. lower) half-plane [27, 78]. An element of $\mathcal{H}_+$ is called an *analytic signal*, and a *progressive* one if, in addition, $\widehat{f}(\omega)$ is real.

Choosing $\mathcal{H}_+$, the restriction U_+ of the representation U is unitary, irreducible and square integrable, and from this fact follow all the mathematical properties of the 1-D CWT similar to those described in Section 2.2.1: covariance, norm conservation (2.7), inversion formula (2.8), reproducing kernel (2.11). This is of course no accident! It simply reflects the fact that the 1-D CWT is a particular case of the general theory of coherent states associated to group representations [1, 2]. This observation is of central importance, for it is this approach that allows a natural and easy extension of the 1-D CWT to higher dimensions.

Now the question is, where does the appropriate group come from? As so often in physics, the answer lies in the notion of *symmetry*. Suppose indeed that the signal possesses certain symmetry properties. It is natural to build these into the wavelet transform itself, and this clearly requires the use of the continuous approach. From this there emerges a general pattern, that we now describe.

Consider the class of finite energy signals living on a manifold Y, i.e. $s \in L^2(Y, d\mu) \equiv \mathcal{H}$. For instance, Y could be space $\mathbb{R}^n$, the 2-sphere S^2, space-time $\mathbb{R} \times \mathbb{R}$ or $\mathbb{R}^2 \times \mathbb{R}$, etc. Quite naturally, the measurement of a signal is represented by a continuous linear functional on the space of signals, that is, in the present case, an inner product $s \mapsto \langle \psi \mid s \rangle$. Notice that, if we were to restrict the signals to smooth functions on Y, measurements would be represented by distributions of some kind over Y.

Suppose there is a group G of transformations acting (transitively) on Y: $y \mapsto g[y]$, with $g[g'[y]] = gg'[y]$, $e[y] = y$, and for any pair $y, y' \in Y$, there is at least one $g \in G$ such that $g[y] = y'$. Assume the group G acts linearly on signals. Then the very notion of symmetry requires that U should be a unitary representation of G in the space $\mathcal{H}$ of signals:

$$\langle U(g)s \mid U(g)s' \rangle = \langle s \mid s' \rangle, \ \forall g \in G, \ s, s' \in \mathcal{H}. \tag{2.46}$$

Then, in order to get a wavelet analysis on Y, adapted to the symmetry group G, three conditions must be met:

(1) G contains *dilations* of some kind.
(2) U is irreducible.
(3) U is *square integrable*, i.e. there exists at least one nonzero vector $\psi \in \mathcal{H}$, called admissible, such that the matrix element $\langle U(g)\psi \mid \psi \rangle$ is square integrable as a function on G.

Under these three conditions, a *G-adapted wavelet analysis* on Y may be constructed, following the general construction of coherent states on Y associated to G, that we now sketch (see [1, 2] for details).

2.4.2 Construction of coherent states from a square integrable group representation

2.4.2.1 Definitions and main properties

Let $\mathcal{H} \equiv L^2(Y, d\mu)$ be the space of finite energy signals on a manifold Y, and assume there is a transformation group G acting on Y, with a continuous unitary irreducible representation U in $\mathcal{H}$. Assume furthermore that the representation U is square integrable.

Choose a fixed admissible vector $\psi \in \mathcal{H}$ (the analysing wavelet). Then the wavelets are the vectors $\psi_g = U(g)\psi \in \mathcal{H}$ ($g \in G$), and the corresponding continuous wavelet transform (CWT) is defined as:

$$S_\psi(g) = \langle \psi_g | s \rangle \tag{2.47}$$

Introduce again the linear map $W_\psi : \mathcal{H} \to L^2(G, dg)$ given by $(W_\psi s)(g) \equiv c_\psi^{-1/2} S_\psi(g)$, where

$$c_\psi = \int_G |\langle U(g)\psi | \psi \rangle|^2 \, dg \tag{2.48}$$

and dg denotes the natural measure on G. Then the CWT has the following properties [1, 2], that match exactly those described in Section 2.2.1.

(1) *Norm conservation:*

$$c_\psi^{-1} \int_G |S_\psi(g)|^2 \, dg = \int_Y |s(y)|^2 \, d\mu(y), \tag{2.49}$$

i.e. W_ψ is an isometry; hence its range, the space of wavelet transforms, is a *closed* subspace $\mathcal{H}_\psi$ of $L^2(G, dg)$.

(2) By (1), W_ψ may be inverted on its range by the transposed map, which gives the *reconstruction formula:*

$$s(y) = c_\psi^{-1} \int_G S_\psi(g)\psi_g(y) \, dg. \tag{2.50}$$

(3) The projection from $L^2(G, dg)$ onto $\mathcal{H}_\psi$ is an integral operator with kernel $K(g, g') = c_\psi^{-1} \langle \psi_g | \psi_{g'} \rangle$, that is, the auto-correlation function of ψ, also called a *reproducing kernel*; in other words, a function $f \in L^2(G, dg)$ is a WT iff it satisfies the reproducing relation:

$$f(g) = c_\psi^{-1} \int_G \langle \psi_g | \psi_{g'} \rangle f(g') \, dg'. \tag{2.51}$$

(4) The CWT is *covariant* under the action of the group G:

$$W_\psi[U(g)s](g_o) = (W_\psi s)(g^{-1}g_o), \ \forall g \in G. \tag{2.52}$$

Now it may happen that the analysing wavelet ψ has a nontrivial isotropy subgroup H_ψ, up to a phase, i.e.

$$U(h)\psi = e^{i\alpha(h)}\psi, \ h \in H_\psi. \tag{2.53}$$

In this case, the whole construction may be performed [1, 2] under a slightly less restrictive condition (the representation U need only to be square integrable on the coset space $X = G/H_\psi$). Then one obtains wavelets indexed by the points of X, namely $\psi_x = U(\sigma(x))\psi$ ($x \in X$), where $\sigma : X \to G$ is an *arbi-*

trary section. We will encounter this situation both in the 2-D and in the 3-D case. In fact one can go one step further, and extend the whole construction to the case of an arbitrary coset space $X = G/H$, where H is *not* the stability subspace of any vector ψ, but this will not concern us in this chapter. The interested reader may find the detailed theory in the review [1] and papers quoted there.

As a final remark before discussing examples, we may add that the whole machinery rests upon the postulated existence of an admissible vector ψ, taken as analysing wavelet, but nothing so far tells us how to choose it. In the case of coherent states associated to simple Lie groups, Perelomov [72] gives a criterion, in terms of maximal weight vectors, familiar in the representation theory of Lie algebras (for instance, in the case of the rotation group $SO(3)$, this method yields the extreme spherical harmonics $Y_{ll}(\theta, \phi)$, which then lead to the spin coherent states used in quantum optics [58]). In general, however, there is no systematic result and the best clue is to try and mimic the familiar wavelets, such as the Mexican hat or the Morlet wavelet, as we shall see in several cases below.

2.4.2.2 Examples: the 1-D and 2-D CWT

This formalism is general enough to design a symmetry-adapted CWT in all cases of physical interest, while, of course, reproducing the familiar 1-D CWT discussed above. First, one should notice that the Weyl–Heisenberg group, which consists of phase space translations (translations and modulations), yields the WFT. Indeed the relation (1.2) in Chapter 1 defines a unitary irreducible representation of that group into the space $L^2(\mathbb{R}, dx)$ of finite energy signals, and that representation is square integrable, as can be shown by a direct verification. The corresponding wavelets are called *gaborettes* in the wavelet community, while quantum physicists call them *canonical coherent states* [1, 2, 58].

As for the 2-D case, the relevant group is the so-called *similitude* group of the plane (or Euclidean group with dilations), $SIM(2) = \mathbb{R}^2 \rtimes (\mathbb{R}_*^+ \times SO(2))$ which consists of translations, rotations and global dilations (technically, $\rtimes$ denotes a semidirect product). Then the relation

$$(U(a, \theta, \vec{b})s)(\vec{x}) = s_{a,\theta,\vec{b}}(\vec{x}) = a^{-1} s(a^{-1} r_{-\theta}(\vec{x} - \vec{b})), \qquad (2.54)$$

defines the natural representation of $SIM(2)$ in the Hilbert space $L^2(\mathbb{R}^2, d^2\vec{x})$, and it is unitary and irreducible (it is actually the only one, up to unitary equivalence, as can be shown by the familiar method of induced representations). Furthermore, U is also square integrable (with respect to the natural measure $dg \equiv a^{-3}da\,d\theta\,d^2\vec{b}$), and one has the relation

$$\iiint \frac{da}{a^3} d\theta d^2\vec{b}|\langle U(a, \theta, \vec{b})\psi|\psi\rangle|^2 = c_\psi\|\psi\|^2, \qquad (2.55)$$

where c_ψ is the constant defined in (2.3). From this, we see that all the properties of the 2-D CWT described in Section 2.2.1 are simply the particularization to the group $SIM(2)$ of those listed above. Notice that, if the wavelet ψ is isotropic, its stability subgroup H_ψ is the rotation group $SO(2)$, and the wavelet transform is a function of $(a, \vec{b}) \in SIM(2)/SO(2)$ only. Thus, as announced, all the aspects of the 2-D CWT are indeed rooted in group representation theory.

In the following sections, we will apply the same technique and obtain the extension of the CWT to 3 space dimensions, to the 2-sphere and similar manifolds, and also to space-time (time-dependent signals or images, such as TV or video sequences), including relativistic effects (using wavelets associated to the affine Galilei or Poincaré group).

2.4.2.3 Application: minimal uncertainty wavelets

As is well-known [58, 72], the canonical coherent states have the characteristic property of *minimal uncertainty*, which means that they saturate the inequality in the Heisenberg uncertainty relations, and this is interpreted as a semi-classical behaviour. What about wavelets, which are the coherent states associated to the similitude groups?

According to the standard discussion in quantum mechanics textbooks [32, 52], two observables of a quantum system, represented by self-adjoint operators A and B, obey the uncertainty relation

$$\Delta A.\Delta B \geq \tfrac{1}{2}|\langle[A, B]\rangle|, \qquad (2.56)$$

where $\Delta A \equiv \Delta_\phi A = \sqrt{\langle A^2\rangle - \langle A\rangle^2}$ denotes the variance of A in the state ϕ and $\langle C\rangle = \langle\phi|C\phi\rangle$ is the average of the operator C in the state ϕ. The state ϕ is said to have *minimal uncertainty* if equality holds in (2.56), which happens iff

$$(A - \langle A\rangle)\phi = -i\lambda_o(B - \langle B\rangle)\phi, \qquad (2.57)$$

for some $\lambda_o > 0$.

In order to apply this concept to 2-D wavelets, we consider the infinitesimal generators of the transformation (2.54) or its equivalent (2.2) in $\vec{k}$-space, and denote them by P_1 and P_2 for translations, D for dilations and J for rotations. Among these, there are four non-zero commutators, namely

$$[D, P_1] = iP_1, \quad [J, P_2] = -iP_1, \quad [D, P_2] = iP_2, \quad [J, P_1] = iP_2, \quad (2.58)$$

but the first two transform into the last two under a rotation by $\pi/2$. Thus it is enough to consider the uncertainty relations for the first pair:

$$\Delta D . \Delta P_1 \geq \tfrac{1}{2} |\langle P_1 \rangle|, \quad \Delta J . \Delta P_2 \geq \tfrac{1}{2} |\langle P_1 \rangle|. \quad (2.59)$$

Then, according to (2.57), a vector $\widehat{\psi}$ saturates these inequalities iff it satisfies the following system of equations

$$
\begin{aligned}
(D + i\lambda_1 P_1)\widehat{\psi}(\vec{k}) &= (\langle D \rangle + i\lambda_1 \langle P_1 \rangle)\widehat{\psi}(\vec{k}) \\
(J + i\lambda_2 P_2)\widehat{\psi}(\vec{k}) &= (\langle J \rangle + i\lambda_2 \langle P_1 \rangle)\widehat{\psi}(\vec{k})
\end{aligned}
\quad (\lambda_1, \lambda_2 > 0). \quad (2.60)
$$

Solving this system of partial differential equations in polar coordinates, one finally obtains that a real wavelet $\widehat{\psi}$ is minimal with respect to the commutation relations (2.58) iff it is of the form

$$\widehat{\psi}(\vec{k}) = c\, \chi_c(\vec{k})\, |\vec{k}|^\kappa\, e^{-\lambda\, \vec{k}\cdot\vec{e}_1} \quad (\kappa > 0,\ \lambda > 0), \quad (2.61)$$

where χ_c is the characteristic function (possibly smoothed) of a convex cone $\mathcal{C}$ in the half-plane $k_x > 0$. We may now impose some degree of regularity (vanishing moments) at the boundary of the cone, by taking an appropriate linear superposition of such minimal wavelets $\widehat{\psi}$. Thus we obtain finally:

$$\widehat{\psi}^c(\vec{k}) = c\, \chi_c(\vec{k})\, F(\vec{k})\, e^{-\lambda\, \vec{k}\cdot\vec{e}_1}, \quad (\lambda > 0) \quad (2.62)$$

where $F(\vec{k})$ is a polynomial in k_x, k_y, vanishing at the boundaries of the cone $\mathcal{C}$, including the origin. Clearly a Cauchy wavelet with $\vec{\eta} = \vec{e}_1$ is of this type.

Other minimal wavelets may be obtained if one includes commutators with elements of the enveloping algebra, i.e. polynomials in the generators. For instance, taking the commutator between D and the Laplacian $-\Delta = P_1^2 + P_2^2$, one finds a whole family of minimal isotropic wavelets, among them all powers of the Laplacian, Δ^n, acting on a Gaussian [12]. For $n = 2$, this gives the 2-D isotropic Mexican hat [35].

2.5 Extensions of the CWT to other manifolds

2.5.1 The three-dimensional case

Some physical phenomena are intrinsically multiscale and three-dimensional. Typical examples may be found in fluid dynamics, for instance the appearance of coherent structures in turbulent flows, or the disentangling of a wave train in (mostly underwater) acoustics, as discussed above. In such cases, a 3-D wavelet analysis is clearly more adequate and likely to yield a deeper understanding [21]. Hence we will also describe briefly the 3-D CWT, follow-ing the general pattern of the previous section.

Given a 3-D signal $s \in L^2(\mathbb{R}^3, d^3\vec{x})$, with finite energy, one may act on it by translation, dilation and rotation:

$$s_{a,\gamma,\vec{b}}(\vec{x}) = [U(a, r(\gamma), \vec{b})s](\vec{x}) = a^{-\frac{3}{2}}s(a^{-1}r(\gamma)^{-1}(\vec{x} - \vec{b})), \qquad (2.63)$$

where $a > 0$, $\gamma \in SO(3)$, $\vec{b} \in \mathbb{R}^3$ and $r(\gamma) \in SO(3)$ is a 3×3 rotation matrix. The element $\gamma \in SO(3)$ may be parametrized, for instance, in terms of three Euler angles. These three operations generate the 3-D Euclidean group with dilations, that is, the similitude group of $\mathbb{R}^3$, $SIM(3) = \mathbb{R}^3 \rtimes (\mathbb{R}^+_* \times SO(3))$. Then (2.63) is a unitary representation of $SIM(3)$ in $L^2(\mathbb{R}^3, d^3\vec{x})$, which is irreducible and square integrable, hence it generates a CWT exactly as before.

Wavelets are taken in $L^2(\mathbb{R}^3, d^3\vec{x})$ and the admissibility condition is now

$$\int |\widehat{\psi}(\vec{k})|^2 \frac{d^3\vec{k}}{|\vec{k}|^3} < \infty. \qquad (2.64)$$

Also the two familiar wavelets have a 3-D realization.

- *The 3-D Mexican hat* is given by

$$\psi_H(\vec{x}) = (3 - |A\vec{x}|^2)\exp(-\tfrac{1}{2}|A\vec{x}|^2), \qquad (2.65)$$

  where $A = \text{diag}[\epsilon_1^{-1/2}, \epsilon_2^{-1/2}, 1]$, $\epsilon_1 \geq 1, \epsilon_2 \geq 1$, is a 3×3 anisotropy matrix. We distinguish three cases:
  (1) if $\epsilon_1 \neq \epsilon_2 \neq 1$, one has the fully anisotropic 3-D Mexican hat (the stability subgroup H_ψ is trivial);
  (2) if $\epsilon_1 = \epsilon_2 = 1$, one has the isotropic, $SO(3)$-invariant, 3-D Mexican hat ($H_\psi = SO(3)$);
  (3) if $\epsilon_1 = \epsilon_2 \equiv \epsilon \neq 1$, the wavelet is *axisymmetric*, i.e., $SO(2)$-invariant, but not isotropic ($H_\psi = SO(2)$).
- *The 3-D Morlet wavelet* is given by

$$\psi(\vec{x}) = \exp(i\vec{k}_o \cdot \vec{x})\exp(-\tfrac{1}{2}|A\vec{x}|^2), \qquad (2.66)$$

  where A is the same 3×3 anisotropy matrix as in the first example. Here again, for $\epsilon_1 = \epsilon_2 \equiv \epsilon \neq 1$ and $\vec{k}_o$ along the z-axis, the wavelet ψ is invariant under $SO(2)$.

Then, given a signal $s \in L^2(\mathbb{R}^3)$, its CWT with respect to the admissible wavelet ψ is given as

$$S(a, \gamma, \vec{b}) = a^{-3/2}\int \overline{\psi(a^{-1}r(\gamma)^{-1}(\vec{x} - \vec{b}))}\,s(\vec{x})\,d^3\vec{x}. \qquad (2.67)$$

As compared with (2.5), the only differences are in the normalization factors and the rotation matrices. Since the structure of the formulas is the same as before, so is the interpretation and the consequences (local filtering, reproducing kernel, reconstruction formula, etc.). Thus the CWT (2.67) may be interpreted as a mathematical *camera* with *magnification* $1/a$, *position* $\vec{b}$ and *directional selectivity* given, in the axisymmetric case, by the rotation parameters $\varpi \equiv (\theta, \varphi)$. As for the visualization, the full CWT $S(a, \gamma, \vec{b})$ is a function of 7 variables. However, if the wavelet ψ is chosen *axisymmetric*, i.e. $SO(2)$-invariant, S depends on 6 variables only, $a > 0$, $\varpi \in S^2 \simeq SO(3)/SO(2)$, the unit sphere in $\mathbb{R}^3$, and $\vec{b} \in \mathbb{R}^3$. In this case again, (a^{-1}, ϖ) may be interpreted as polar coordinates in spatial frequency space. This is in fact true in any number of dimensions. It follows that, here too, there are two natural representations for the visualization of the WT: the position representation (a, ϖ fixed) and the scale-orientation (or spatial frequency) representation ($\vec{b}$ fixed). Of course, there are many other posssible representations that may be useful.

In conclusion, let us discuss briefly a simple example, the detection of 3-D objects in a cluttered medium. We consider a scene with 3-D objects (targets) immersed in a cluttered medium, modelled by the signal:

$$s(\vec{x}) = \sum_{m=1}^{N} s_m(\vec{x}) + n(\vec{x}), \tag{2.68}$$

where $s_m(\vec{x})$ denotes the density of the target m, and $n(\vec{x})$ the density of the medium. Since the density of the targets is very different from that of the medium, there will be a high density gradient at the boundary between the objects and the medium. In this situation, the wavelet transform $S(a, \theta, \varphi, \vec{b})$ may be used to extract the 3-D objects and determine their characteristics, position (range and orientation) and spatial frequency. Further details may be found in [7], where a detailed strategy is explained for the 2-D version of the same problem.

2.5.2 Wavelets on the 2-sphere

There are several applications where data to be analysed are defined on a sphere, in geophysics or astronomy, of course, but also in statistics. If one is interested only in very local features, one may ignore the curvature and work on the tangent plane. But when global aspects become important (description of plate tectonics on the Earth, for instance), one needs a genuine generalization of wavelet analysis to the sphere. Several authors have attacked this problem, with various techniques. On the discrete side, an efficient solution has been obtained by Schröder and Sweldens [76] with the so-called lifting scheme (see Section 2.6.2.4 below), but this obviously misses the particular symmetry of the sphere. A continuous approach was developed by Holschneider [56], with several *ad hoc* assumptions. It turns out that the general formalism developed in [1, 2] and sketched in the previous section yields an elegant solution to the problem [16], and in particular allows one to derive all the assumptions of [56].

Although the discussion is too technical to be described here, it is interesting to outline the main ideas, because they lead to significant generalizations. As usual, finite energy signals are taken as square integrable functions on the 2-sphere $S^2 \simeq SO(3)/SO(2)$. The natural operations on such signals are translations (on the sphere) and local dilations. The former are given by rotations from $SO(3)$. Dilations around the North Pole are obtained by considering ordinary dilations in the tangent plane and lifting them to S^2 by stereographic projection from the South Pole. As for dilations around any other point, it suffices to bring it to the North Pole by a rotation, perform the dilation and go back by the inverse rotation. Obviously translations and dilations do not commute. However, the only group that can be obtained by combining only $SO(3)$ and the dilation group $\mathbb{R}_*^+$ is their direct product, which cannot be the 'similitude' group of the sphere.

A way out of this difficulty [16] is to embed the two groups into the Lorentz group $SO_o(3, 1)$, which acts transitively on S^2. Then one is in the general situation described in Section 2.4.2.1 and the machinery developed in [1, 2] may be used. In this way one can indeed set up a theory of wavelets on S^2, which coincides with that of [56]. In addition, this CWT on the sphere has the expected Euclidean limit, that is, as the radius of the sphere increases to ∞, the whole wavelet analysis on the sphere goes into the usual wavelet analysis in the plane (in this case, the tangent plane at the North Pole). Moreover, the limiting process may be performed entirely in group-theoretical language, using the technique known as group contraction.

The whole scheme may be generalized to higher dimensions, essentially verbatim. It can also be extended to other setups, for instance a CWT on a two-sheeted hyperboloid. In $\mathbb{R}^3$, this means $H^2 = SO_o(2, 1)/SO(2)$, and the stereographic projection from either 'pole' is available, mapping one sheet onto the interior of the unit disk in the plane tangent to the other pole, and the other sheet onto the exterior. Now this suggests a further generalization. In both cases, S^2 as well as H^2, the unit disk, image of one sheet or one hemisphere, is a classical domain. Also the stereographic projection has a group-theoretical origin [72]. This paves the way to the generalization of the

CWT to a whole class of homogeneous spaces (Riemannian symmetric spaces).

2.5.3 Wavelet transform in space-time

2.5.3.1 Kinematical wavelets

An important aspect of signal and image processing is the analysis of time-dependent or moving signals, e.g. in television, and the CWT may be extended to this case too [45]. We consider first motion on the line. Finite energy signals are taken as functions $s(x, t) \in L^2(\mathbb{R} \times \mathbb{R}, dx\, dt)$. The natural transformations on such a signal are translations and dilations in space and time independently, $(x, t) \mapsto (a_1 x + b_1, a_0 t + b_0)$. However it is more convenient to replace the two independent dilations a_1, a_0 by a global dilation a and a so-called speed-tuning transformation c, defined as:

$$s(x, t) \mapsto a^{-1} s(a^{-1}x, a^{-1}t), \quad a > 0;$$
$$s(x, t) \mapsto s(c^{1/2}x, c^{-1/2}t), \quad c > 0. \tag{2.69}$$

This transformation comes from the physiological characteristics of motion perception by our visual system: in order to be visible, fast moving objects must be wide, and narrow objects must move slowly (for a typical example, think of the inscriptions on a departing train carriage).

Combining the transformation (2.69) with space and time translations, we obtain the affine group of space-time. This group has a natural unitary irreducible representation in $L^2(\mathbb{R} \times \mathbb{R}, dx\, dt)$:

$$[U(a, c, b_0, b_1, \epsilon)s](x, t) = \frac{1}{a}\, s\left(\frac{\sqrt{c}}{a}(x - b_1), \frac{(-1)^\epsilon}{a\sqrt{c}}(t - b_0)\right), \tag{2.70}$$

where (b_0, b_1) denote space-time translations and $\epsilon \in \{0, 1\}$ corresponds to time-reflection (this additional operation is needed for irreducibility). In addition, the representation U is square integrable. A wavelet ψ is admissible iff it satisfies the condition

$$\iint \frac{|\widehat{\psi}(k, \omega)|^2}{|k||\omega|}\, dk\, d\omega < \infty. \tag{2.71}$$

From here on, everything follows exactly the general pattern. Thanks to the filtering property in a and c, the resulting CWT (called *kinematical*) is efficient in detecting moving objects: the dilation parameter a catches the size of the target, while the new parameter c adjusts the speed of the wavelet to that of the target. Thus the spatio-temporal CWT is a tool for *motion tracking*.

Clearly there are plenty of applications in which such a technique might be used.

The extension of these considerations to higher dimensions is straightforward. First, in n dimensions, the dilation and speed tuning operations (2.69) become:

$$x \mapsto a^{-1}c^{1/(n+1)}x, \quad t \mapsto a^{-1}c^{-n/(n+1)}t. \tag{2.72}$$

Then one has to add rotations, as usual, and follow the general pattern.

2.5.3.2 Relativistic wavelets

The kinematical wavelets just described may not always be sufficient, depending on the type of signal to be analysed. One may wish to consider a specific form of movement, i.e. choose a particular relativity group. Three examples may be of interest (we begin again with one space dimension).

   (i) *Galilean wavelets:* here we add to the transformations discussed above the Galilei boosts, thus getting $(x, t) \mapsto (a_1 x + vt + b_1, a_0 t + b_0)$. The resulting group G_1^{aff}, called the affine Galilei group, is quite complicated. It has a natural unitary representation in the space of finite energy signals, which splits into the direct sum of four irreducible ones. And each of these is square integrable, so that wavelets may be constructed in the usual way. In addition, more restricted wavelets may be obtained by taking as parameter space various quotient spaces G_1^{aff}/H, where H is *not* the stability subgroup of the basic wavelet. Again this construction requires the more general formalism, described in Section 2.4.2.1 [1,2].
  (ii) *Schrödinger wavelets:* one obtains an interesting subclass of the previous one by imposing the relation $a_0 = a_1^2$, so that the transformations leave invariant the Schrödinger (or the heat) equation. Again there are two unitary irreducible representations, both square integrable on the (Schrödinger) subgroup. Thus again a CWT is at hand, which may prove useful for describing, for instance, the motion of quantum particles on the line.
 (iii) *Poincaré wavelets:* in order to get a CWT in the relativistic regime, it suffices to replace Galilei transformations by Poincaré ones, while of course imposing the relation $a_0 = a_1$ to space and time dilations. The resulting affine Poincaré group has a square integrable unitary irreducible representation, defined on the solid future light cone. The Poincaré wavelets might be useful, for instance, in the presence of electromagnetic fields.
       Of course, this analysis extends in a straightforward way to higher dimensions, just by adding rotations.

2.6 The discrete WT in two dimensions

As mentioned in Chapter 1, a key step in the success of the 1-D *discrete* WT was the discovery that almost all examples of orthonormal bases of wavelets may be derived from a multiresolution analysis, and furthermore that the whole construction may be translated into the language of (QMF) filters. In the 2-D case, the situation is exactly the same, as we shall sketch in this section. Further information may be found in [38] or [66].

2.6.1 Multiresolution analysis in 2-D and the 2-D DWT

The simplest approach consists of building a 2-D multiresolution analysis simply by taking the direct (tensor) product of two such structures in 1-D, one for the x direction, one for the y direction. If $\{V_j, j \in \mathbb{Z}\}$ is a multiresolution analysis of $L^2(\mathbb{R})$, then $\{\tilde{V}_j = V_j \otimes V_j, j \in \mathbb{Z}\}$ is a multiresolution analysis of $L^2(\mathbb{R}^2)$. Writing again $\tilde{V}_j \oplus \tilde{W}_j = \tilde{V}_{j+1}$, it is easy to see that this 2-D analysis requires one scaling function : $\Phi(x, y) = \phi(x)\phi(y)$, but three wavelets:

$$\Psi^h(x, y) = \phi(x)\,\psi(y)$$
$$\Psi^v(x, y) = \psi(x)\,\phi(y) \tag{2.73}$$
$$\Psi^d(x, y) = \psi(x)\,\psi(y).$$

As the notation suggests, Ψ^h detects preferentially horizontal edges, that is, discontinuities in the vertical direction, whereas Ψ^v and Ψ^d detect vertical and oblique edges, respectively. Indeed, for $j = 1$, the relation $V_1 = V_0 \oplus W_0$ yields:

$$
\begin{aligned}
\tilde{V}_1 &= V_1^{(x)} \otimes V_1^{(y)} \\
&= (V_0^{(x)} \oplus W_0^{(x)}) \otimes (V_0^{(y)} \oplus W_0^{(y)}) \\
&= (V_0^{(x)} \otimes V_0^{(y)}) \oplus (V_0^{(x)} \otimes W_0^{(y)}) \oplus (W_0^{(x)} \otimes V_0^{(y)}) \oplus (W_0^{(x)} \otimes W_0^{(y)}) \\
&= \tilde{V}_0 \oplus \tilde{W}_0,
\end{aligned}
$$

where $\tilde{V}_0 = V_0^{(x)} \otimes V_0^{(y)} \ni \phi(x)\phi(y)$ and $\tilde{W}_0$ is the direct sum of the three other products, generated by the three wavelets given in (2.73), respectively.

From these three wavelets, one gets an orthonormal basis of $\tilde{V}_j$ by defining $\{\Phi_{kl}^j(x, y) = \phi_{j,k}(x)\phi_{j,l}(y), k, l \in \mathbb{Z}\}$, and one for $\tilde{W}_j$ in the same way, namely $\{\Psi_{kl}^{\alpha,j}(x, y), \alpha = h, v, d$ and $k, l \in \mathbb{Z}\}$. Clearly this construction enforces a Cartesian geometry, with the horizontal and the vertical directions playing a preferential role. This is natural for certain types of images, such as in television, but is poorly adapted for detecting edges in arbitrary directions. Other solutions are possible, however (see below).

As in the 1-D case, the implementation of this construction rests on a pyramidal algorithm introduced by Mallat [60, 61]. The technique consists of translating the multiresolution structure into the language of QMF filters, and putting suitable constraints on the filter coefficients h_n. For instance, ψ has compact support if only finitely many h_n differ from zero.

2.6.2 Generalizations

It turns out that the scheme based on orthonormal wavelet bases is too rigid for most applications and various generalizations have been proposed. We discuss some of them.

2.6.2.1 Biorthogonal wavelet bases

In the CWT, the wavelet used for reconstruction need not be the same as that used for decomposition, they have only to satisfy a cross-compatibility condition [38]. The same idea in the discrete case leads to biorthogonal bases [31], i.e. one has two hierarchies of approximation spaces, $\{V_j\}$ and $\{\check{V}_j\}$, with cross-orthogonality relations. In 1-D, the construction goes as follows, and the extension to 2-D proceeds as above. Start with a scale of closed subspaces $\{V_j\}$, assuming only the existence of a scaling function $\phi \in V_0$ such that its integer translates $\{\phi_k(x) \equiv \phi(x - k), k \in \mathbb{Z}\}$ form a Riesz (or unconditional) basis of V_0. Then, instead of orthogonalizing this basis, which would lead to the construction of an o.n. wavelet basis, one takes the *dual* basis $\{\check{\phi}_k\}$, that is, the vectors defined by the relation $\langle \phi_k | \check{\phi}_l \rangle = \delta_{kl}$. Let $\check{V}_0$ denote the closed subspace generated by $\{\check{\phi}_k, k \in \mathbb{Z}\}$. Then the same construction is repeated for each j, using the dilation invariance of the scale $\{V_j\}$. The outcome is a multiresolution scale $\{\check{V}_j\}$, with exactly the same properties. Next, for each $j \in \mathbb{Z}$, one defines a subspace W_j by the two conditions $W_j \subset V_{j+1}$ and $W_j \perp \check{V}_j$, and similarly $\check{W}_j \subset \check{V}_{j+1}$ and $\check{W}_j \perp V_j$. In this way one obtains two sequences of subspaces $\{W_j\}$ and $\{\check{W}_j\}$, with bases $\{\psi_{j,k}, j, k \in \mathbb{Z}\}$, $\{\check{\psi}_{j,k}, j, k \in \mathbb{Z}\}$, respectively, which are mutually orthogonal:

$$\langle \psi_{j,k} | \check{\psi}_{j',k'} \rangle = \delta_{jj'} \delta_{kk'}. \tag{2.74}$$

In terms of these bases, one gets two types of expansion formulas, for any $f \in L^2(\mathbb{R})$:

$$f = \sum_{j,k \in \mathbb{Z}} \langle \check{\psi}_{j,k} | f \rangle \, \psi_{j,k}$$

$$= \sum_{j,k \in \mathbb{Z}} \langle \psi_{j,k} | f \rangle \, \check{\psi}_{j,k}. \tag{2.75}$$

The resulting scheme is much more flexible and is probably the most efficient one in practical applications. For instance, it gives a better control on the regularity or decrease properties of the wavelets [31].

2.6.2.2 Wavelet packets and the best basis algorithm

As mentioned already in Chapter 1, the construction of orthonormal wavelet bases leads to a special subband coding scheme, rather asymmetrical: each approximation space V_j gets further decomposed into V_{j-1} and W_{j-1}, whereas the detail space W_j is left unmodified. Thus more flexible subband schemes have been considered, called *wavelet packets*, where both subspaces V_{j-1} and W_{j-1} are decomposed at each step [33, 66, 81]. Such a scheme provides rich libraries of orthonormal bases, and also strategies for determining (using entropic criteria) the best basis in a given situation. Another generalization of the strict orthonormal wavelet scheme has been developed by Coifman and Meyer, starting from the so-called Malvar wavelets (see [29] or [66]). This scheme, in a sense, is halfway between the wavelet and the windowed Fourier transforms, and it offers also more flexibility and efficiency, for instance in the analysis of speech signals.

2.6.2.3 More isotropic 2-D wavelets

The tensor product scheme privileges the horizontal and the vertical directions; more isotropic wavelets may be obtained, either by superposition of wavelets with specific orientation tuning [64], as we did above with the CWT, or by choosing a different way of dilating, using a nondiagonal 2-D dilation matrix, which amounts to dilating by a noninteger factor [37]. Consider, for instance, the following dilation matrices:

$$D_0 = \begin{pmatrix} 2 & 0 \\ 0 & 2 \end{pmatrix}, \quad D_1 = \begin{pmatrix} 1 & 1 \\ 1 & -1 \end{pmatrix}, \quad D_2 = \begin{pmatrix} 1 & 1 \\ -1 & 1 \end{pmatrix}. \tag{2.76}$$

The matrix D_0 correspond to the usual dilation scheme by powers of 2, whereas D_1 and D_2 lead to the so-called 'quincunx' scheme [49]. In the standard scheme, a unit square is dilated, in the transition $j \rightarrow j+1$, to another square, twice bigger, with the same orientation. This means that three kinds of additional details have to be supplied, horizontal, vertical and oblique (see Figure 2.10, left). By contrast, the same operation in the

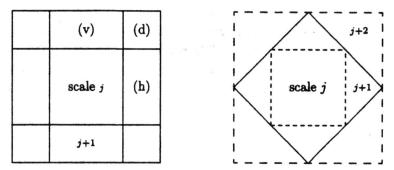

Fig. 2.10. Unit cell at successive resolutions: (Left) for the 'Cartesian' scheme; (Right) for the 'quincunx' scheme.

'quincunx' scheme leads to a square circumscribed to the original one, that is, rotated by 45° and larger by a factor $\sqrt{2}$, so that only one kind of additional details is necessary (Figure 2.10, right). Indeed only one wavelet is needed in this scheme, instead of three. This is consistent with a result of Meyer, according to which the number of independent wavelets needed in a given multiresolution scheme equals $(|\det D| - 1)$, where D is the dilation matrix used.

2.6.2.4 Second-generation wavelets

As indicated in Chapter 1, one can go further and abandon the regular dyadic scheme and the Fourier transform altogether. Using the 'lifting scheme', one obtains the so-called *second-generation wavelets* [80]. The same scheme applies in 2-D as well. For instance, Schröder and Sweldens [76] have applied it to the design of wavelets on the sphere, with a very convincing application to the reproduction of coastlines on a terrestrial globe.

2.6.2.5 Integer wavelet transforms

In their standard numerical implementation, the classical (discrete) WT converts floating point numbers into floating point numbers. However, in many applications (data transmission from satellites, multimedia), the input data consists of integer values only and one cannot afford to lose information: only lossless compression schemes are allowed. Recent developments have produced new methods that allow one to perform all calculations in integer arithmetic [30].

2.6.3 Physical applications of the DWT

As with other methods, wavelet bases may be applied to all the standard problems of image processing. The main problem of course is data compression, and for achieving useful rates one has to determine which information is really essential and which one may be discarded with acceptable loss of image quality. Significant results have been obtained in the following directions:

- Representation of images in terms of *wavelet maxima* [63], as a substitute for the familiar zero-crossing schemes [64];
- In particular, application of this maxima representation to the detection of edges, and more generally detection of local singularities [62];
- Image compression and coding using vector quantization combined with the WT [17];
- Image compression, combining the previous wavelet maxima method for contours and biorthogonal wavelet bases for texture description [50];
- Image and signal denoising, by clever thresholding methods [42].

Some applications are less conventional. For instance, a technique based on the biorthogonal wavelet bases [31] has been adopted by the FBI for the identification of fingerprints. The advantages over more conventional tools are the ease of pattern identification and the superior compression rates, which allows one to store and transmit a much bigger amount of information in real time. The full story may be found in [28]. Another striking application is the deconvolution of noisy images from the Hubble Space Telescope, by a technique combining the DWT with a statistical analysis of the data [25, 26, 73]. The results compare favourably in quality with those obtained by conventional methods, but the new method is much faster. One should also quote a large amount of work under development in the field of High Definition Television, where wavelet techniques are being actively exploited; here again the huge compression rates make them specially interesting.

As for applications of the multidimensional DWT more specifically oriented to physics, we like to mention two. The first one is in quantum field theory (although it was done before the wavelet techniques were born): various perturbation expansions (the so-called 'cluster expansion') used in the analysis of Euclidean field theory models are in fact discrete wavelet expansions [22]. Actually the summation over scales, indexed by j, was originally motivated by renormalization group arguments. In the same domain, we may note that wavelet bases have been used also ([23] and references therein) for estimating the time evolution of solutions of some wave equations (Klein-Gordon, Dirac, Maxwell or the wave equation), or even to expand solutions of the equations in terms of dedicated 'wavelets' (although the functions introduced in the last case seem rather far away from genuine wavelets [57]).

The other application resorts to solid state physics, namely the Quantum Hall Effect (quantization of the electric conductivity) that occurs when a 2-D electron gas is submitted to a strong transverse magnetic field. Here orthonormal wavelet bases may be used for generating localized orthonormal bases for the lowest Landau level, a necessary step towards the analysis of the Hall effect [5]. This is discussed in detail in Chapter 8.

2.7 Outcome: why wavelets?

As in 1-D signal analysis, wavelet techniques have become an established tool in image processing, both in their CWT and DWT incarnations and their generalizations. We want to emphasize here that the CWT and the DWT have almost opposite properties, hence their ranges of application differ widely too. The CWT is very efficient at detecting specific features in signals or images, such as in pattern recognition or directional filtering. On the other hand, the DWT and its generalizations are extremely fast and economical, they yield for instance impressive data compression rates, which is especially useful in image processing, where huge amounts of data, mostly redundant, have to be stored and transmitted.

Both are powerful and flexible tools, and have become a significant element in the standard toolbox of image processing. Indeed they find their way into increasingly many reference books and software codes. In addition, they have found applications in many branches of physics, such as acoustics, geophysics, astrophysics, fluid mechanics (turbulence), medical imagery, solid state physics, quantum field theory, ...

What distinguishes wavelet analysis from more conventional techniques are its simplicity and its adaptive character. The algorithm is as simple, and mathematically justified, as the familiar Fourier transform and its variants (WFT). It is also extremely economical, thanks to the automatic zoom effect. The WT selects the most signicant parts of the signal (in position scale and direction) and is negligible elsewhere. As a consequence, it is extremely stable against approximations. Clearly wavelets are here to stay, and one should expect an increasingly diverse spectrum of physical applications. An important point is that wavelets should not be taken as a replacement of conventional techniques, but as an additional tool, that reveals different aspects of a problem. The most probable trend for the future is towards more merging of wavelet ideas with traditional ones, resulting in specialized tools, optimized for a particular type of problem. This aspect will appear in many of the subsequent chapters.

Acknowledgements

It is a pleasure to thank Romain Murenzi and Pierre Vandergheynst for a long and fruitful collaboration, and in particular for many discussions that are reflected in the present chapter.

References

[1] S.T. Ali, J.-P. Antoine, J.-P. Gazeau and U.A. Mueller, Coherent states and their generalizations: A mathematical overview, *Reviews Math. Phys.*, **7**: 1013–1104, (1995)

[2] S.T. Ali, J.-P. Antoine and J.-P. Gazeau, *Coherent States, Wavelets and their Generalizations* (Springer, New York and Berlin, 1999, to appear)

[3] J.-P. Antoine, M. Duval-Destin, R. Murenzi and B. Piette, Image analysis with 2D wavelet transform : detection of position, orientation and visual contrast of simple objects, in [67], pp.144-159

[4] J.-P. Antoine, P. Carrette, R. Murenzi and B. Piette, Image analysis with two-dimensional continuous wavelet transform, *Signal Proc.*, **31**: 241–272, (1993)

[5] J.-P. Antoine and F. Bagarello, Wavelet-like orthonormal bases for the lowest Landau level, *J. Phys. A: Math. Gen.*, **27**: 2471–2481, (1994)

[6] J.-P. Antoine and R. Murenzi, The continuous wavelet transform, from 1 to 3 dimensions, in *Subband and Wavelet Transforms: Design and Applications*, pp. 149–187, ed. A.N. Akansu and M.J.T. Smith (Kluwer, Dordrecht,1995)

[7] J.-P. Antoine, P. Vandergheynst, K. Bouyoucef and R. Murenzi, Target detection and recognition using two-dimensional continuous isotropic and anisotropic wavelets, *Proc. Conf. "Automatic Object Recognition V", SPIE's 1995 Symposium on Optical Engineering/Aerospace Sensing and Dual Use Photonics*, **2485**: 20–31, (1995)

[8] J.-P. Antoine, P. Vandergheynst, K. Bouyoucef and R. Murenzi, Alternative representations of an image via the 2D wavelet transform. Application to character recognition, *Proc. Conf. "Visual Information Processing IV", SPIE's 1995 Symposium on Optical Engineering/Aerospace Sensing and Dual Use Photonics*, **2488**: 486–497, (1995)

[9] J.-P. Antoine and R. Murenzi, Two-dimensional directional wavelets and the scale-angle representation, *Signal Proc.*, **52**: 259–281, (1995)

[10] J.-P. Antoine, D. Barache, R.M. Cesar Jr. and L. da F. Costa, Multiscale shape analysis using the continuous wavelet transform, *Proc. 1996 IEEE Intern. Conf. on Image Processing (ICIP-96)*, Vol. I, pp. 291–294; ed. P. Delogne (IEEE, Piscataway, NJ, 1996); Shape characterization with the wavelet transform, *Signal Proc.*, **62**: 265–290, (1997)

[11] J.-P. Antoine, R. Murenzi and P. Vandergheynst, Two-dimensional directional wavelets in image processing, *Int. J. of Imaging Systems and Technology*, **7**: 152–165, (1996)

[12] J.-P. Antoine and R. Murenzi, Two-dimensional continuous wavelet transform as linear phase space representation of two-dimensional signals, in *Wavelet Applications IV*, SPIE 1997 Symposium on Aerospace/Defense Sensing, Simulation, and Controls (Orlando, April 1997), **3078**: 206–217, (1997)

[13] J.-P. Antoine, R. Murenzi and P. Vandergheynst, Directional wavelets revisited: Cauchy wavelets and symmetry detection in patterns, *Applied Comput. Harm. Anal.* (to appear)

[14] J.-P. Antoine and P. Vandergheynst, Contrast enhancement in images using the two-dimensional wavelet transform, *Proc. IWISP '96 (3rd Int. Workshop Image & Signal Processing*, pp. 65-68; ed. B.G. Mertzios and P. Liatsis (Elsevier, Amsterdam, 1996)

[15] J.-P. Antoine and P. Vandergheynst, 2-D Cauchy wavelets and symmetries in images, *Proc. 1996 IEEE Intern. Conf. on Image Processing (ICIP-96)*, Vol. I, pp. 597–600; ed. P. Delogne (IEEE, Piscataway, NJ, 1996)

[16] J.-P. Antoine and P. Vandergheynst, Wavelets on the 2-sphere: A group-theoretical approach, *Applied Comput. Harm. Anal.* (to appear); Wavelets, on the n-sphere and related manifolds, *J. Math. Phys.*, **39**: 3987–4008, (1998)

[17] M. Antonini, M. Barlaud, P. Mathieu and I. Daubechies, Image coding using wavelet transform, *IEEE Trans. Image Proc.*, **1**: 205–220, (1992)

[18] F. Argoul, A. Arnéodo, J. Elezgaray, G. Grasseau and R. Murenzi, Wavelet analysis of the self-similarity of diffusion-limited aggregates and electrodeposition clusters, *Phys. Rev. A*, **41**: 5537–5560, (1990)

[19] A. Arnéodo, F. Argoul, E. Bacry, J. Elezgaray, E. Freysz, G. Grasseau, J.F. Muzy and B. Pouligny, Wavelet transform of fractals, in [67], pp. 286–352

[20] A. Arnéodo, E. Bacry, P.V. Graves and J.F. Muzy, Characterizing long-range correlations in DNA sequences from wavelet analysis, *Phys. Rev. Lett.*, **74**: 3293–3296, (1996)

[21] D. Astruc, L. Plantié, R. Murenzi, Y. Lebret and D. Vandromme, On the use of the 3-D wavelet transform for the analysis of computational fluid dynamics results, in [68], pp. 463–470

[22] G. Battle, Wavelets: A renormalization group point of view, in [74], pp. 323–349

[23] G. Battle, Klein-Gordon propagation of Daubechies wavelets, *J. Math. Phys.*, **34**: 1095–1109, (1993)

[24] P. Bosch, private commun. and Thèse annexe, Univ. Cath. Louvain (1998)

[25] K. Bouyoucef, Sur des aspects multirésolution en reconstruction d'image. Application au télescope spatial de Hubble. Thèse de doctorat, Université Paul Sabatier – Toulouse III (1993)

[26] K. Bouyoucef, D. Fraix-Burnaix and S. Roques, Interactive Deconvolution with Error Analysis (IDEA) in astronomical imaging: Application to aberrated HST images on SN1987A, M87 and 3C66B. *Astron. Astroph., Suppl. Series*, **121**: 1–6, (1997)

[27] R.N. Bracewell, *The Fourier Theory and its Applications* (McGraw-Hill, New York, 1986)

[28] C.M. Brislawn, Fingerprints go digital, *Notices Amer. Math. Soc.*, **42**: 1278–1283, (1995)

[29] B. Burke Hubbard, *Ondes et ondelettes – La saga d'un outil mathématique* (Pour la Science, Paris, 1995); 2nd Ed. *The World According to Wavelets* (A.K. Peters, Wellesley, MA, 1998)

[30] A.R. Calderbank, I. Daubechies, W. Sweldens and B.L. Yao, Wavelets that map integers to integers, *Applied Comput. Harm Anal.*, **5**: 332–369, (1998)

[31] A. Cohen, I. Daubechies and J.-C. Feauveau, Biorthogonal bases of compactly supported wavelets, *Comm. Pure Appl. Math.*, **45**: 485–560, (1992)

[32] C. Cohen-Tannoudji, B. Diu and F. Laloë, *Mécanique Quantique, Tome I* (Hermann, Paris, 1977)

[33] R.R. Coifman, Y. Meyer, S. Quake and M.V. Wickerhauser, Signal processing and compression with wavelet packets, in [68], pp. 77–93

[34] J.-M. Combes, A. Grossmann and Ph. Tchamitchian (eds.), *Wavelets, Time-Frequency Methods and Phase Space (Proc. Marseille 1987)* (Springer, Berlin, 1989; 2d Ed. 1990)

[35] S. Dahlke and P. Maass, The affine uncertainty principle in one and two dimensions, *Comp. Math. Appl.*, **30**: 293–305, (1995)

[36] I. Daubechies, A. Grossmann and Y. Meyer, Painless nonorthogonal expansions, *J. Math. Phys.*, **27**: 1271–1283, (1986)

[37] I. Daubechies, The wavelet transform, time-frequency localization and signal analysis, *IEEE Trans. Inform. Theory*, **36**: 961–1005, (1990).

[38] I. Daubechies, *Ten Lectures on Wavelets*, (SIAM, Philadelphia, PA, 1992)

[39] J.G. Daugman, Complete discrete 2-D Gabor transforms by neural networks for image analysis and compression, *IEEE Trans. Acoust., Speech, Signal Proc.*, **36**: 1169–1179, (1988)

[40] N. Delprat, B. Escudié, Ph. Guillemain, R. Kronland-Martinet, Ph. Tchamitchian and B. Torrésani, Asymptotic wavelet and Gabor analysis: Extraction of instantaneous frequencies, *IEEE Trans. Inform. Theory*, **38**: 644–664, (1992)

[41] R. De Valois and K. De Valois, *Spatial Vision* (Oxford Univ. Press, New York, 1988)

[42] D.L. Donoho, Nonlinear wavelet methods for recovery of signals, densities, and spectra from indirect and noisy data, in *Different Perspectives on Wavelets*, pp. 173–205; Proc. Symp. Appl. Math. **38**, ed. I. Daubechies (Amer. Math. Soc., Providence, RI, 1993)

[43] R.J. Duffin and A.C. Schaefer, A class of nonharmonic Fourier series, *Trans. Amer. Math. Soc.*, **72**: 341–366, (1952)

[44] M. Duval-Destin, Analyse spatiale et spatio-temporelle de la stimulation visuelle à l'aide de la transformée en ondelettes, Thèse de Doctorat, Université d'Aix-Marseille II, (1991)

[45] M. Duval-Destin and R. Murenzi, Spatio-temporal wavelets: application to the analysis of moving patterns, in [68], pp. 399–408

[46] M. Duval-Destin, M.-A. Muschietti and B. Torrésani, Continuous wavelet decompositions, multiresolution, and contrast analysis, *SIAM J. Math. Anal.*, **24**: 739–755, (1993)

[47] M. Farge, Wavelet transforms and their applications to turbulence, *Annu. Rev. Fluid Mech.*, **24**: 395–457, (1992)

[48] M. Farge and Th. Philipovitch, Coherent structure analysis and extraction using wavelets, in [68], pp. 477–481

[49] J.-C. Feauveau, Analyse multirésolution par ondelettes non orthogonales et bancs de filtres numériques, Thèse de Doctorat, Université Paris-Sud, (1990)

[50] J. Froment and S. Mallat, Arbitrary low bit rate image compression using wavelets, in [68], pp. 413–418, and references therein

[51] C. Gonnet and B. Torrésani, Local frequency analysis with two-dimensional wavelet transform, *Signal Proc.*, **37**: 389–404, (1994)

[52] K. Gottfried, *Quantum Mechanics. Vol. I: Fundamentals* (Benjamin, New York and Amsterdam, 1966)

[53] A. Grossmann, Wavelet transform and edge detection, in *Stochastic Processes in Physics and Engineering*, pp.149–157; ed. S. Albeverio, Ph. Blanchard, L. Streit and M. Hazewinkel (Reidel, Dordrecht, 1988)

[54] A. Grossmann and J. Morlet, Decomposition of functions into wavelets of constant shape, and related transforms, in *Mathematics + Physics, Lectures on Recent Results. I*, pp.135–166; ed. L. Streit (World Scientific, Singapore, 1985)

[55] P. Guillemain, R. Kronland-Martinet and B. Martens (1991). Estimation of spectral lines with the help of the wavelet transform. Applications in N.M.R. spectroscopy, in [67], pp. 38–60

[56] M. Holschneider, Continuous wavelet transforms on the sphere, *J. Math. Phys.*, **37**: 4156–4165, (1996)

[57] G. Kaiser and R.F. Streater, Windowed Radon transform, analytic signals, and the wave equation, in *Wavelets: A Tutorial in Theory and Applications*, pp. 399–441; ed. C.K. Chui (Academic Press, Boston, 1992)

[58] J.R. Klauder and B.S. Skagerstam, *Coherent States – Applications in Physics and Mathematical Physics* (World Scientific, Singapore, 1985)

[59] C.-S. Lu, W.-L. Hwang, H.-Y.M. Liao and P.-C. Chung, Shape from texture based on the ridge of continuous wavelet transform, *Proc. 1996 IEEE Intern. Conf. on Image Processing (ICIP-96), Lausanne, Sept. 1996* Vol. I, pp. 295–298; ed. P. Delogne (IEEE, Piscataway, NJ, 1996); W.-L. Hwang, C.-S. Lu and P.-C. Chung, Shape from texture: Estimation of planar surface orientation through the ridge surfaces of continuous wavelet transform, *IEEE Trans. Image Proc.* **7**: 773–780, (1998)

[60] S.G. Mallat, Multifrequency channel decompositions of images and wavelet models, *IEEE Trans. Acoust., Speech, Signal Proc.*, **37**: 2091–2110, (1989)

[61] S.G. Mallat, A theory for multiresolution signal decomposition : the wavelet representation, *IEEE Trans. Pattern Anal. Machine Intell.*, **11**: 674–693, (1989)

[62] S. Mallat and W.-L. Hwang, Singularity detection and processing with wavelets, *IEEE Trans. Inform. Theory*, **38**: 617–643, (1992)

[63] S. Mallat and S. Zhong, Wavelet maxima representation, in [67], pp. 207–284

[64] D. Marr, *Vision* (Freeman, San Francisco, 1982)

[65] D. Marr and E. Hildreth, Theory of edge detection, *Proc. R. Soc. Lond. B*, **207**: 187–217, (1980)

[66] Y. Meyer, *Les Ondelettes, Algorithmes et Applications*, 2d ed. (Armand Colin, Paris, (1994) ; Engl. transl. of the 1st Ed.: *Wavelets, Algorithms and Applications* (SIAM, Philadelphia, PA, 1993)

[67] Y. Meyer (ed.), *Wavelets and Applications (Proc. Marseille 1989)*, (Springer, Berlin, and Masson, Paris, 1991)

[68] Y. Meyer and S. Roques (eds.), *Progress in Wavelet Analysis and Applications (Proc. Toulouse 1992)* (Ed. Frontières, Gif-sur-Yvette, 1993).

[69] R. Murenzi, Wavelet transforms associated to the n-dimensional Euclidean group with dilations : signals in more than one dimension, in [33], pp. 239–246

[70] R. Murenzi, Ondelettes multidimensionnelles et applications à l'analyse d'images, Thèse de Doctorat, Univ. Cath. Louvain, (1990)

[71] G. Ouillon, D. Sornette and C. Castaing, Organization of joints and faults from 1-cm to 100-km scales revealed by optimized anisotropic wavelet coefficient method and multifractal analysis, *Nonlin. Proc. in Geophys.*, **2**: 158–177, (1995)

[72] A. Perelomov, *Generalized Coherent States and their Applications*, (Springer, Berlin, 1986)

[73] S. Roques, F. Bourzeix and K. Bouyoucef, Soft-thresholding technique and restoration of 3C273 jet, *Astrophys. Space Sc.*, Nr.**239**: 297–304, (1996)

[74] M.B. Ruskai, G. Beylkin, R. Coifman, I. Daubechies, S. Mallat, Y. Meyer and L. Raphael (eds.), *Wavelets and Their Applications* (Jones and Bartlett, Boston, 1992)

[75] G. Saracco, A. Grossmann and Ph. Tchamitchian, Use of wavelet transforms in the study of propagation of transient acoustic signals across a plane interface between two homogeneous media, in [33], pp.139–146

[76] P. Schröder and W. Sweldens, Spherical wavelets: Efficiently representing functions on the sphere, *Computer Graphics Proc. (SIGGRAPH95)*, ACM Siggraph 1995, pp.161–175

[77] E. Slezak, A. Bijaoui and G. Mars, Identification of structures from galaxy counts. Use of the wavelet transform, *Astron. Astroph.*, **227**: 301–316, (1990) and in [67], pp.175–180

[78] E.M. Stein and G. Weiss, *Introduction to Fourier Analysis on Euclidean Spaces* (Princeton Univ. Press, Princeton, NJ, 1971)

[79] R.F. Streater and A.S. Wightman, *PCT, Spin and Statistics, and all That* (Benjamin, New York, 1964)

[80] W. Sweldens, The lifting scheme: a custom-design construction of biorthogonal wavelets, *Applied Comput. Harm. Anal.*, **3**: 1186–1200, (1996)

[81] M.V. Wickerhauser, *Adapted Wavelet Analysis from Theory to Software* (A.K. Peters, Wellesley, MA, 1994)

[82] W. Wisnoe, Utilisation de la méthode de transformée en ondelettes 2D pour l'analyse de visualisation d'écoulements, Thèse de Doctorat ENSAE, Toulouse, (1993)

[83] W. Wisnoe, P. Gajan, A. Strzelecki, C. Lempereur and J.-M. Mathé, The use of the two-dimensional wavelet transform in flow visualization processing, in [68], pp. 455–458.

3

Wavelets and astrophysical applications

ALBERT BIJAOUI

Observatoire de la Côte d'Azur,
Dpt CERGA – UMR CNRS 6527,
BP 4229 – 06304 Nice Cedex 4, France

Abstract

The wavelet transform is used in astrophysics for many applications. Its use is connected to different properties. The Time-Frequency analysis results from the two-dimensional feature of this transform. Some interesting applications were performed on nonstationary astrophysical signals. Many astrophysical results were obtained by this analysis, either on quasi regular variables, and on chaotic light curves. Solar time series have been also carefully analysed by the wavelet transform. New results have been obtained for series with identified periods (sunspots, diameter, irradiance, chromospheric oscillations) and for chaotic signals (magnetic activity).

Astronomers have exploited the wavelet transform for image compression. Many packages are proposed with significant gains. Some full sky surveys are available now with images compressed by the wavelet transform. Filtering and restorations are derived from this scale-space analysis. Some thresholding rules furnish adapted filtering. The restoration is connected to an approach for which we progressively extract the most energetic features. This may be related to the notion of multiscale support. Many applications were done for *Hubble Space Telescope* (HST) images or for astronomical aperture synthesis. The ability of the wavelet transform to localize an object in scale-space led also to applying this transform to the detection and to the analysis of astronomical sources. A *multiscale vision model* was developed by our group, which allows one to detect and to characterize all the sources of different sizes in an astronomical image. Many applications of image analysis were performed on different astrophysical sources, and specifically the ones having a power-law correlation, i.e. a fractal-like behaviour: molecular clouds, infrared cirrus, clumpy galaxies, comets, X-ray clusters, etc. Our main applications are related to the study of the Large-Scale structure of the Universe, generally from galaxy counts. The nature of the data leads us to develop specific statistical analyses. In the Universe, the distribution of matter has a correlation function with a power law, so the wavelet transform is well suited to analyse it. Many astronomical sources have this behaviour, and fractal analyses were also applied. The same method was successfully applied to the determination of asteroid families. The wavelet transform was also used for the determination of the singularity spectrum. A new statistical indicator for testing cosmological scenarios was obtained from the morphology in the wavelet space at different scales.

The wavelet transform is a tool widely used today by astrophysicists, but they do not apply only the discrete transform resulting from the multiresolution analysis but a large range of discrete transforms: Morlet's transform, for time-frequency analysis, the *à trous* algorithm and the pyramidal transform for image restoration and analysis, pyramidal with Fourier transform for synthesis aperture imaging. Physical constraints generally play an important part in applying a given discrete transform.

3.1 Introduction

The wavelet transform was originally designed to study nonstationary signals. The usual astrophysical signals are quite regular so that the need of this kind of transform was not evident. Astronomers possessed many other tools for image processing, so that it did not seem necessary to implement this new transform for their needs. But, thanks to its covariance under dilations, the wavelet transform appeared rapidly as the best tool to study power-law signals. These can be observed in different situations: the light intensity of the solar surface, the brightness of interstellar clouds, or galaxy distribution from counts. This last item plays an important role in cosmological research so that it was soon found interesting to apply the wavelet transform to this subject.

Galaxy distribution is studied from galaxy counts, which are neither a signal nor an image but a list (a catalogue in astronomical literature) of positions in two or three dimensions. We have to process these catalogues in order to extract the different components, ranging from groups to clusters of galaxies. The existence of a power law for the two-point correlation reflects the observation of a hierarchical structure, small groups are contained in larger ones, and so. The wavelet transform would be able to detect and characterize these groups.

This application of the wavelet transform to cluster analysis was new when I started with my collaborators on this problem [85]. Previously, the wavelet transform was only applied to signals but rarely on images. This has led us to

examine many problems related to the coefficient statistic, the object definition and their reconstruction. We saw that this point of view allowed us to
develop a new way to process images for different cases of noise, with many
applications to image registration, restoration and analysis.

Since our pioneer paper in astrophysics, many groups of astrophysicists
have applied the wavelet transform for many other purposes, times series,
image restoration, image compression, object detection, fractal analysis, etc.
In this chapter I will try to give an overview of these applications, but I will
also describe some of our own specific work on image restoration, image
analysis and the study of the large scale structure of the Universe.

3.2 Time–frequency analysis of astronomical sources

3.2.1 The world of astrophysical variable sources

The night sky seems to be immutable but an important fraction of astrophysical sources are variable. They can be variable in radiation flux, but they can
also exhibit a regular change of their radial velocity or line profiles of their
spectra. Different behaviours were found in the variations, from a large
increase of the flux by a factor 10^6 (supernovae) to about 10^4 (classical
novae), smaller and very regular variations (cepheids, eclipse binaries) or
intermediate situations with variables showing quasi regular variations (RR
Lyrae variables) to eruptive variables without any periodic behaviour.

This wide range of situations is the same for all the observed wavelengths.
Pulsars show very periodic variations in radio wavelengths while quasars
have a chaotic behaviour. The Gamma Ray Bursts (GRB) are one of the
greatest astrophysical enigmas today. They are characterized by a large flux
of gamma rays detected during a few seconds. They seem to be sources
located at cosmological distances.

Temporal variations can be also detected in solar signals. With very accurate measurements, the spatial instrument ACRIM has detected irregular
variations of the flux [108]. The solar radius seems to show also faint variations [107]. The classical solar cycle is associated with the number of solar
spots. A quasi regular cycle of 11 years was found in the nineteenth century,
but irregularities exist too. The radial velocity exhibits also faint variations, a
5 mn. oscillation phenomenon was discovered in the 1960s [59]. With very
accurate techniques many lines around this frequency were separated, giving
fundamental information about the internal structure of the Sun. This technique was also applied to other astrophysical objects, Jupiter, Procyon and
solar-like stars, and similar results were obtained [67].

3.2.2 The application of the Fourier transform

The first purpose for analysing astrophysical signals has to do with the determination of the periods. If a signal is strictly periodic, and if it is observed regularly during many periods, there is no problem to determine the period correctly. One estimates the time between the first and the last maximum and divides it by the number of periods. But this situation is quite idyllic in astronomy since:

- astronomical signals are almost never strictly periodic;
- measurements are noisy;
- the sampling is very irregular. This is due to the night and day alternation, to weather conditions, and to other observational constraints. Generally the sampling is done in a window $W(t)$, which is constituted by a set of square functions.

The Fourier transform is the tool used for analysing the variable astrophysical signals. Its accuracy is limited by the sampling and the noise. Let us consider a function $f(t)$ in a window $W(t)$. We get the observed function

$$F(t) = f(t).W(t), \qquad (3.1)$$

which corresponds in the Fourier space to:

$$\hat{F}(v) = \hat{f}(v) \star \hat{W}(v), \qquad (3.2)$$

where $\star$ is the convolution symbol. Let us consider now the case of a single square window of size T. $\hat{W}(v)$ is the function $\frac{\sin \pi T v}{\pi T v}$, so the original Fourier spectrum is smoothed with this function. We make the following remarks.

- If the signal is really periodic, $\hat{f}(v)$ is a sum of Dirac distributions $\delta(v - v_i)$, but we observed a set of functions $\frac{\sin \pi T (v - v_i)}{\pi T (v - v_i)}$. The observed peaks have a width of $1/T$. The window size limits the capability for observing the deviation from periodicity.
- The sinc function is known to have a very slow convergence, and, as a consequence, many bumps are observed near a bright peak. They can be reduced by applying a nonsquare window $W(t)$ to the data. Many windows were designed to reduce the bumps, but they cause a faint loss in frequency resolution. A Gaussian window is considered to be a good compromise between bumps and resolution.
- If the Fourier transform of a signal $F(t)$ shows peaks of $1/T$ width the signal can be considered as periodic on this time interval T. On the other hand, if the width is greater than $1/T$ that means that the signal is not periodic, the observation time is greater than the coherence time: the phase changes slowly, or the amplitude decreases and a new wave packet appears, etc. The width could be also due to the existence of real secondary peaks around the main one.

In case the Fourier analysis displays peaks compatible with the observational window, the application of another tool does not seem to be useful. For peaks wider than $1/T$, this analysis does not take into account all the information, and we have to apply a transform which takes fully into account the observed phenomena.

3.2.3 From Gabor's to the wavelet transform

The Gabor transform [38] was the first designed operator which enables one to perform a time–frequency analysis. With this transform, a windowed Fourier transform is performed around each time t, which leads to a $2D$ function $\hat{G}(t, v)$. By fixing t, the frequency content at this time is obtained, and by fixing v one gets the variations of the signal at this frequency with the time. In order to reduce the bumps, a sliding Gaussian window is applied. The window width T_w delimits the time resolution, while the frequency resolution is equal to $1/T_w$. When we increase T_w, we reduce the precision in time, but we increase it in frequency, while the product of the two resolutions is constant. The time–frequency plane is divided in identical tiles of $(T_w, \frac{1}{T_w})$ size.

In the Gabor transform the time resolution is independent of the frequency. If the peak widths in the observed signal do not have the same size, this analysis is not optimal. For many signals the width depends on the frequency; the larger the period, the larger must be the window size. With this assumption we are led to the Morlet continuous wavelet transform [66]. With this transform the time–frequency plane is divided in tiles the size of which varies with the frequency. If the frequency resolution δv is kv, where k is a free parameter which depends on the required resolution, then the time resolution is $\frac{1}{\delta v}$.

Taking into account its derivation, the wavelet transform is interesting for a signal for which the natural width of the frequency peaks is proportional to the frequency. If this is not the case, the resulting analysis would not be optimal.

3.2.4 Regular and irregular variables

The first astrophysical paper on the application of the Morlet wavelet transform to astrophysical time series was an analysis of pulsating white dwarfs [41]. The authors found amplitude variations for most detected oscillations with periods of modulation as long as or greater than the time intervals of the observation windows. For this case, where the spectral lines are quite well defined, a Gabor transform should lead to the same conclusions. Nevertheless a better time–frequency analysis was required, and Morlet's wavelet carries out such an analysis with a constant shape pattern. Their analysis allows them to deduce a periodic transfer of energy between two pulsation modes.

This type of analysis was further developed by a group working on so-called *semiregular variables* (SR) [101], a class of stars the luminosity of which exhibits irregular oscillations. After applying a Fourier analysis to determine the periods, they analysed the flux intensity variations with the wavelet transform. Using simulations they have investigated the link between the time-dependent phenomena and physical variations (amplitude or frequency modulations) [102]. They had first shown by a wavelet analysis that the two shorter frequencies of a semiregular variable were unstable, as observed in several other SR-type variable stars [101]. In a recent paper [100] they confirm the interest of this analysis on another semiregular variable V Bootis. Using the ridge technique [43] they showed that the amplitude of the longer period strongly decreased while the amplitude of the shorter one seems to remain stable.

The previous works on semiregular variables could have been also done by the Gabor transform. This tool is being used, with a confusion often occurring between the Gabor and the wavelet transforms [19] [20]. The authors processed the light curve of a peculiar A star. The transform shows unusual evolution over the duration of the observation, which cannot be explained by a beating mechanism of nearby frequencies. The last application on quite regular variable stars concerns cataclysmic variables observed with the high speed UV photometer of the Hubble Space Telescope [84]. A low-frequency flickering was detected by this technique, but the main astrophysical result concerns the detection of rapid UV quasi-periodic oscillations in the star VV Pup, related to shock oscillations in its accretion column.

3.2.5 *The analysis of chaotic light curves*

Previous time series processing has shown the necessity of analysing some light curves with a time-frequency tool. Some light curves of astrophysical sources often exhibit chaotic behaviour, characterized by long-term dependencies and a so-called $\frac{1}{f}$ spectrum over a wide energy range. Due to the slow decay of the correlation function, information is present at all scales, and, consequently, the wavelet transform is always the favourite tool.

The first application was done on the flux data of a double quasar by Hjorth *et al.* [47]. The quasar is seen double due to the gravitational lensing of an angularly close galaxy. The time delay between the two fluxes is due to the travel path difference. With an available model of the lensing, its measure allows one to estimate the Universe expansion rate (i.e. the Hubble constant). From the wavelet analysis the authors have shown that the discrepancies between the previous analyses could be due to the choice of the reduction method. In this signal, no frequency can be determined, so that the wavelet transform is well suited in allowing us a better temporal resolution and localization of the multiple scales of the signal.

Scorpio X-1 is a chaotic variable source in X-ray. Scargle *et al.* [83] have studied the flux variations with an orthogonal wavelet transform, the Haar transform. They have computed the *scalogram*, i.e. the mean energy of the wavelet coefficients for each scale, and they have shown that the chaotic variability of the source Scorpio X-1 agrees well with the accretion model called *dripping handrail*. The detected quasi-periodic oscillations and the very low frequency noise are produced by radiation from blobs with a wide size distribution, resulting from accretion and subsequent diffusion of hot gas.

Another original work was done by Norris *et al.* [70] on the flux data of the Gamma Ray Bursts (GRB), in order to test a cosmological time dilation. In this application the wavelet transform allowed them to rescale all bursts to fiducial levels of peak intensity. They have shown that the dilation operates over a broad range of time scales. If the results are consistent with bursts being at cosmological distances, they conclude that alternative explanations arising from the nature of the physical processes are still possible.

The GRB flux curves exhibit a large variation in time scale, from seconds to minutes; the zero crossings in the wavelet transform have been used to classify GRB [6]. This information, added to a set of other characteristics, leads to separating these objects into 2–3 classes using a self organizing neural network.

3.2.6 Applications to solar time series

Using a wavelet analysis, new variations in the solar cycle were displayed [71]. The periods determined by this approach are in agreement with the ones previously detected. Some long periods in the cycle were also studied.

A comparison between the Fourier transform and wavelets was done by Vigouroux and Delache [107]. They studied the solar radius measurements obtained at the CERGA solar astrolab. These measurements were noisy and not regularly sampled. After resampling and estimating the error bars, they processed them by a Monte-Carlo method to determine the distribution of the Fourier and the wavelet coefficients after applying a Daubechies ortho-gonal transform [27]. They showed that the description which leads to the minimum of parameters corresponds to the wavelet transform rather than the

Fourier transform. They extended their analysis to the historical sunspot numbers [108]. They slightly modified the method and showed the interest of the wavelet transform which has the capability of taking care of unequal error bars.

Bocchialini and Baudin [18] obtained new information on chromospheric oscillations from the application of a Morlet wavelet analysis to observations of the quiet Sun. The temporal behaviour was described in two kinds of regions, a magnetic element network, and a nonmagnetic intra-network cell. Thanks to the wavelet transform they determined the duration of the chromospheric wavetrains. They estimated the correlation between the oscillations in two spectral lines.

In the above papers on the solar time-series, the wavelet transform is applied in different forms, and its advantages compared to the Fourier or the Gabor transforms are not trivial. But solar series may show also a chaotic behaviour. This is the case for the magnetic activity. Komm [52] and Lawrence *et al.* [56] applied the wavelet transform to measure the scaling and the intermittency properties. Komm applied to the series an orthogonal wavelet transform and determined a fractal dimension of 1.7, while Lawrence *et al.* carried out a wavelet spectral analysis with a Morlet wavelet transform. The number of samples was large, 16 000, spread over many decades. They have shown a power-law variation of the magnetic activity on more than two decades (2 years to 2 days or less). They interpret this result as an indication of a generic turbulence structuring the magnetic fields as they rise through the convective zone.

3.3 Applications to image processing

3.3.1 *Image compression*

Richter [37] [78] introduced the old Haar transform [45] to compress astronomical images a decade before the emergence of the wavelet transform and the multiresolution analysis [63]. The corresponding wavelet function is equal to 1 on $[0, \frac{1}{2}[$ and to -1 on $[\frac{1}{2}, 1[$. The Haar transform was applied to image processing in the 1970s [75], but its $2D$ extension was different from the multiresolution one. Richter has developed the two-dimensional transform resulting from the multiresolution analysis and called it the *H-Transform*. Their approach was simple, the values corresponding to closest pixels are correlated, so that the main information is carried by the differences. They developed a tool named HCOMPRESS which is based on the following elements.

- Compute the $2D$ Haar transform.

- Estimate the variance due to the noise at each scale.
- Divide the wavelet coefficients by a value which depends on the variance and on the required compression factor.
- Keep only the integer part with optimal storage using a 4-bit code [112] [50].

This method was applied to compress the *Digital Sky Survey*, which results from a scan done at the *Space Telescope Science Institute* of images of the whole sky [113]. Press [76] has introduced the Daubechies filter of length 4. The compression and uncompression algorithms take more time than HCOMPRESS and the quality of the resulting measurements is generally less than those obtained with the simple Haar transform for astronomical images. This could be due to the characteristics of these images, mainly compound of peaks due to the stars. The correlation length is very short, and it is not relevant to process the data with long filters.

For control during astronomical observations it is essential to have a correct idea of the transmitted images as fast as possible. HCOMPRESS is well adapted to this progressive transmission of the information. The image is restored, and displayed, scale by scale, from the largest scale to the smallest one. Such a modified software was implemented [111] and can be used for remote observing or for access to remote image archives. After less than 1% of the data has been received, the image is visually similar to the original, allowing the user to verify it, and to stop the transmission if necessary.

Unfortunately the restored images display large fields (blocks) of connected pixels having the same value with discontinuities between them. These *blocking effects* are due to null values of the thresholded coefficients of the Haar transform. Richter proposed an improvement using a Kalman filtering [79], while White [112] solved this problem by interpolation. With Bobichon [17] we have shown that another improvement is obtained by restoring taking into account a regularization constraint. The inversion is not the inverse Haar transform, but an iterative algorithm resulting from the constraints. We have also implemented more regular biorthogonal transforms, resulting from the B-spline scaling function. Taking into account the regularization constraint, the quality of the results is also less than the one obtained by the Haar transform.

The pyramidal median transform [96] is similar to the pyramidal wavelet transform [11] (see appendix B), the low-pass linear filtering being replaced by a median filtering. Starck *et al.* have achieved a method based on this transform to perform simultaneously a noise suppression and a compression [96]. This technique is also well adapted for progressive transmission.

3.3.2 *Denoising astronomical images*

3.3.2.1 *First approaches*

Since the first papers on the wavelet transform appeared, it has been shown that this tool was very useful for denoising, thanks to its capabilities to locally separate the signal from the noise. Astronomical images are particularly noisy, so that this advantage seems to be very important, and much work was done to create software for denoising astronomical images using the wavelet transform. Many different strategies were applied in order to take into account the noise properties and the aliasing.

Cappacioli *et al.* [24] developed a method based on the Haar transform for filtering spectrograms and images (see also [79] [61]). Kalman filtering was introduced in order to reduce the block effects resulting of zeroing wavelet coefficients. Donoho's wavelet shrinking [31] was based on a multiresolution analysis [63] and a Daubechies orthonormal basis [27]. We have applied this method and found a lot of artifacts, generated by the aliasing introduced in this approach. The multiresolution analysis is an unredundant transform. At each scale the sampling is divided by two in order to keep the same number of coefficients. The signal is perfectly restored with the full set of data. If we threshold, at each scale we have an aliasing and that will lead necessarily to artifacts near discontinuities and peaks. Lorenz *et al.* [61] reduced these features by applying a Kalman filtering, but it is possible to remove them by a regularization constraint such as we have proposed for the decompression with the Haar transform [17]. These procedures are complicated, some artifacts always remain from this analysis, making a redundant transform preferable.

Denoising depends on the noise statistic, and the coefficient thresholding must take this into account. This led our group to introduce the notion of *significant coefficients*.

3.3.2.2 *Decision theory and significant coefficients*

If the image is locally uniform we can compute the probability density function (PDF) $p(W)$ of the wavelet coefficient W. Then we introduce a statistical meaning of the observed value from classical decision theory [44]. $\mathcal{H}_0$ is the hypothesis that at the scale i the image is constant in the neighbourhood of the pixel (k, l). For a positive coefficient W, the $\mathcal{H}_0$ rejection depends on the probability $p(W)$:

$$P = Prob(w > W(i, k, l)) = \int_{W(i,k,l)}^{+\infty} p(w)dw \qquad (3.3)$$

For a negative coefficient we examine:

$$P = Prob(w < W(i,k,l)) = \int_{-\infty}^{W(i,k,l)} p(w)dw \qquad (3.4)$$

We fix a decision level ϵ. If $P > \epsilon$, $\mathcal{H}_0$ is not excluded at level ϵ, and therefore the coefficient value may be due to the noise. On the other hand, if $P < \epsilon$, we cannot conclude that the value results only from the noise and $\mathcal{H}_0$ must be rejected at this decision level. We say that we have detected a *significant coefficient*.

The number of significant coefficients depends on the decision level ϵ. If we choose $\epsilon = 0.001$, 0.1% of the coefficients are statistically identified as significant even if we have only noise. These false detections will generate artifacts on further processing. If we choose a fainter decision level, i.e. 10^{-6}, very few artifacts will remain, but a large part of the real information will be missed. We must find a compromise between the false alarms and the misses. This compromise depends on the cost of a false alarm and of a miss. Generally we have chosen $\epsilon = 10^{-4}$ in our processing.

3.3.2.3 The PDF of the wavelet coefficients

The PDF of the wavelet coefficients depends on the noise process. We assume generally that we have stationary Gaussian white noise for the image. In the case of Poisson noise, we can transform the pixel intensity n by Anscombe's Transform [2]:

$$x = 2\sqrt{n + \tfrac{3}{8}} \qquad (3.5)$$

Then we process the data x as a Gaussian variable of variance 1. This transform gives correct results for photon counts greater than about 10 per pixel, which is generally the case. For galaxy counts, we have also examined the case of a fainter density for which Anscombe's transform is not available. Another method was given for determining the PDF [86].

For CCD observations, the noise is described by the sum of a Gaussian and a Poisson variable, and we generalize Anscombe's transform, which leads also to a variable with a constant variance 1 which is processed as a Gaussian one [68].

$$x = \frac{2}{\alpha}\sqrt{\alpha n + \frac{3}{8}\alpha^2 + \sigma^2 - \alpha g} \qquad (3.6)$$

α is the coding step, g is the background value and σ the standard deviation of the Gaussian noise.

We have examined the case of Rayleigh variables [16], which are obtained from *Synthetic Aperture Radar*. This instrument allows one to get images of the Earth and near planets and satellites in centimetric wavelengths. Unfortunately these images are affected by the so-called *speckle noise*, due to the coherent lighting. The distribution of the amplitude ρ follows a Rayleigh law:

$$p(\rho) = \frac{\rho}{\sigma^2} e^{-\frac{\rho^2}{2\sigma^2}} \tag{3.7}$$

where σ is a parameter which is characteristic of the ground rugosity. The PDF of the wavelet coefficients is derived from numerical experiments.

We have shown [16] that it is better to process the energy (ρ^2) instead of the amplitude, which leads us to examine the PDF of the wavelet coefficients of an exponential process. This PDF is also determined by numerical simulations.

It is easy to estimate the PDF of the wavelet coefficients in the case of Gaussian noise. This can be done analytically but simulation is easier. We compute the image of a simulated Gaussian noise with a variance 1. Then we compute its discrete wavelet transform and estimate the standard deviation $\sigma_n(i)$ at each scale. For the image to be processed, the standard deviation $\sigma(1)$ is estimated from the histogram of the wavelet coefficients $W(1, k, l)$. At this scale, the wavelet coefficient values essentially result from the noise. Knowing the variation of the noise with the scale from the simulation we deduce the $\sigma(i)$ set.

3.3.2.4 Denoising by using the significant coefficients

In a first approach with Starck [92], we have carefully examined this problem applying the *à trous* (with holes) algorithm [49] [15] for which the wavelet transform is computed scale by scale without decimation (appendix A). We have proposed four methods:

- a Wiener-like technique, which corresponds to a stationary Wiener filter done in the wavelet space;
- a hierarchical Wiener filtering, for which we take into account the correlation between the coefficients from one scale to the following one;
- an adaptive filtering which corresponds to Donoho's method, but with a redundant transform;
- hierarchical adaptive filtering for which the threshold depends on the wavelet coefficient in the previous plane.

Using a multiresolution quality criterion, the results pleaded for the hierarchical Wiener filtering, which led to the minimum of artifacts.

Due to the redundancy of the wavelet transforms, all the developed algorithms were iterative leading to an image $F^{(n)}$ at iteration n. In order to reduce the residual artifacts we assume that the difference $R^{(n)}$ between the observed image F and $F^{(n)}$ must be similar to noise [90] [95]. Consequently, a necessary, but not sufficient, condition holds in the lack of detection of significant coefficients in the $R^{(n)}$ wavelet transform.

We build a Boolean image $M(i, k, l)$ such that $M(i, k, l) = 1$ if the wavelet coefficient $W(i, k, l)$ is significant; and $M(i, k, l) = 0$ in the opposite case. This mask corresponds also to a volume V of the wavelet transform, the one where the wavelet coefficients are significant. We can reconstruct the image from the significant set. If $F(I, k, l)$ designs the smoothed image at the largest scale I, we can restore the image by the expression:

$$\tilde{F}(k, l) = F(I, k, l) + \sum_{1, I} W(i, k, l) \tag{3.8}$$

If we do not threshold, the reconstruction is exact. In the opposite case, we write:

$$F^{(0)}(k, l) = F(I, k, l) + \sum_{1, I} M(i, j, k) W(i, k, l) \tag{3.9}$$

Then if we get the wavelet transform of $F^{(0)}(k, l)$ we get another set $V^{(0)}(i, k, l)$ which is different from the set $W(i, k, l)$ even for the significant coefficients. This deviation is due to the redundancy of the *à trous* wavelet transform. By consequence, $F(k, l) - F^{(0)}(k, l)$ may show significant differences. This remark leads to the algorithm described in appendix C [23].

Bendjoya *et al.* [8] have applied also thresholding in a redundant wavelet transform space for denoising the profile of Saturn's rings, from data taken by Voyager 2. In their approach a coefficient statistic was done on the data, scale by scale, without taking into account white Gaussian noise.

Oosterloo [72] has developed a similar adaptive wavelet smoothing in a data cube resulting from synthetic observations obtained with a radio interferometer.

3.3.3 Multiscale adaptive deconvolution

This classical problem has played an important part recently in enhancing the observations taken with the Hubble Space Telescope (HST), before its refurbishment. We write:

$$F = O \star P + N \tag{3.10}$$

where O designs the true object distribution, P is the response function which is called *the point spread function* (PSF) in image processing and N the noise. A large number of methods were proposed to get O knowing F. This operation called *deconvolution* depends upon:

- the PSF P: if P has frequency holes, regularization techniques are necessary;
- the noise N: the methods are not the same for Gaussian, Poisson, or another kind of noise;
- the statistical properties of the object O. We do not use the same method to deconvolve remote sensing images, with building edges, or diffuse galaxies.

The application of multiresolution techniques for deconvolution is widely connected to the nature of astronomical objects. Many authors have developed different techniques, with or without redundancy. Bendinelli [7] first introduced a multiscale approach using Gabor expansions in series of elementary Gaussians. With Starck [92] [93] we have derived a deconvolution technique based on a wavelet transform associated with the PSF. The resolution increased but this algorithm was only adapted to a quite regular PSF, which was not the case for the HST. So we have developed a simpler idea connected to the denoising method described above. Our algorithm [15, 58] is based on the formulation of classical restoration algorithms:

$$O^{(n+1)}(k, l) = G(O^{(n}(k, l), R^{(n)}(k, l) \tag{3.11}$$

where $G(O, R)$ may be:

Van Cittert [114] $G(O, R) = O + R$;
Fixed step gradient [54] $G(O, R) = O + \alpha \tilde{P} \star R$, where $\tilde{P}$ is the joint operator related to the PSF ($\tilde{P}(k, l) = P(-k, -l)$), and α a parameter which is easily estimated;
Lucy [62] $G(O, R) = O + O(\frac{R}{O \star P}) \star \tilde{P}$.

The algorithm is similar to the smoothing one [15], and it is described in appendix D.

In Figure 3.1 a simulated astronomical image is plotted. It was convolved by a given PSF and a Gaussian noise was added. The bottom images show the restoration by the Lucy algorithm and by the proposed modified one. Automated analysis software was applied to the two images, allowing us to compare the results to the initial image:

- the noise suppression helps the object detection;
- the position accuracy is improved by this technique;
- the Signal to Noise ratio is only reduced of 3.5 dB for our method, while it is reduced by 10 dB using Lucy.

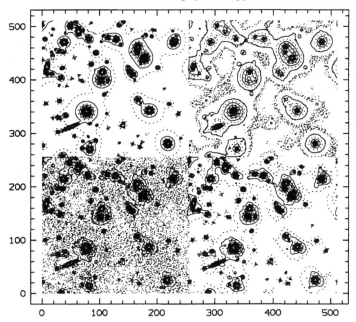

Fig. 3.1. Results of the simulation (isophotal display). At the top: left, the simulated image; and right, this image after the convolution with the PSF and the noise addition. At the bottom: left, the restored image by Lucy's method; and right, the one obtained with the proposed method (from [15]).

Some processing was done for HST images [68], for CCD ground based observations [5], and for deconvolution of X-ray images [105]. Other authors have developed methods based on the multiresolution analysis [80]. A multiscale support constraint is introduced interactively, by the user, in a first approach. An automated determination of the optimal mask is then introduced.

3.3.4 *The restoration of aperture synthesis observations*

With an interferometer with two telescopes we measure directly the amplitude of the Fourier transform of the image at a given $2D$ spatial frequency (u, v). By combining the information from three telescopes the phase can also be obtained, so that the image is directly sampled in the Fourier space. But the measurements need specific positions of the telescope, and the sampling in the Fourier space is necessarily irregular. The image, called the dirty map, is obtained by a simple inverse Fourier transform of the data, and the PSF, named the dirty beam, by an inverse Fourier transform of the frequency coverage. The presence of secondary lobes in the dirty beam creates large artifacts in the dirty map and deconvolution is necessary. This image restoration problem has led to the development of many different methods. Generally, radio astronomers apply the CLEAN algorithm [48] for which detected point-like sources are iteratively subtracted.

Wakker and Schwarz [110] have introduced the concept of Multi-Resolution Clean (MRC) in order to alleviate the difficulties with CLEAN for extended sources. The MRC approach consists of building two intermediate images, the first (called the smooth map) by smoothing the data to a lower resolution with a Gaussian function, and the second (called the difference map) by subtracting the smoothed images from the original data. Both images are then processed separately. By using a standard CLEAN on them, a smoothed clean map and difference clean map are obtained. The recombination of these two maps gives the clean map at full resolution.

Let us consider an image characterized by its intensity distribution $F(x, y)$. If $w_j^{(F)}$ are the wavelet coefficients of the image F at the scale j, we get:

$$\hat{w}_j^{(F)}(u, v) = \hat{w}_j^{(P)}\hat{O}(u, v) \qquad (3.12)$$

where $w_j^{(P)}$ are the wavelet coefficients of the PSF at the scale j. We deconvolve each wavelet plane of the image by the wavelet plane of the PSF by using the classical algorithm CLEAN to obtain the clean wavelet map. If B is the ideal PSF(clean beam) and L_j is the list of peaks found by CLEAN at the resolution j, the estimation of the wavelet coefficients of the object is:

$$w_j^{(E)}(x, y) = L_j * w_j^{(B)}(x, y) \qquad (3.13)$$

The clean map at full resolution is found by the reconstruction algorithm. We apply CLEAN to each plane of the wavelet transform. This allows us to detect at each scale the significant structures. An optimization of the height of the CLEAN peaks is further obtained, in order to get an image which is fully compatible with the observations.

Some astronomical applications were done for restoring stellar images from observations done by speckle interferometry [97]. Yan and Peng [114] have also applied this wavelet approach from radio observations, and they have shown its power.

3.3.5 *Applications to data fusion*

Today the same source is often observed at different wavelengths, and at different epochs. The comparison of these observations requires adapted image fusion tools. The fusion is usually based on the pixel values. But many phenomena may reduce the quality of this approach, and specifically the existence of a background, and a difference in resolution.

The wavelet transform splits the information along the scale axis, so that the data fusion may be done scale by scale. Consequently the variations in resolution are easy to take into account, scale by scale. The wavelet function has a null mean, and the background variations are automatically removed.

Data fusion in the wavelet transform space presents many advantages. We have given an application for optimal image addition [13]. The image registration was done by taking into account the information in the wavelet transform space. For each scale and for each image we estimated weights which account for the signal level and the standard deviation due to the noise. This method was applied also for registering remote sensed images [29]. A comparison of wavelet coefficients allows us to detect faint variations at a given scale for a set of images.

This approach could be applied to many other fields, such as biology, medicine, surveillance or industrial imaging, for geometrical registration and for detection of variable phenomena at a given scale.

3.4 Multiscale vision

3.4.1 Astronomical surveys and vision models

Astronomical images contain typically a large set of point-like sources (stars), some quasi point-like objects (faint galaxies, double stars,...) and some complex and diffuse structures (galaxies, nebulous, planetary nebulae, clusters, etc.). A *vision model* is defined by the sequence of operations required for automated image analysis. Astronomical images need specific analyses which take into account the scientific purpose, the characteristics of the objects and the existence of hierarchical structures.

The classical vision model for robotic and industrial images is based on edge detection. We have applied this concept to astronomical imagery [14]. We chose the Laplacian of the intensity as the edge line. The results are independent of large scale spatial variations, such as those due to sky background. The main disadvantage of the resulting model lies in the difficulty of getting a correct object classification: astronomical sources cannot be accurately recognized from their edges.

Many reduction procedures were built using a model for which the image is the sum of a slowly varying background with superimposed small scale objects [98] [88] [104] [53]. We created first a background mapping [10].

For that purpose we need to introduce a scale: the background is defined in a given area. Each pixel with a value significantly greater than the background is considered to belong to a real object. The same label is given to each significant pixel belonging to the same connected field. For each field we determine the area, the position, the flux and some pattern parameters. Generally, this procedure leads to quite accurate measurements, with correct detection and recognition. The model works very well for poor fields. If it is not the case, a labelled field may correspond to many objects. The background map is done at a given scale: larger objects are removed. Smoothing is only adapted to star detection, not to larger objects.

The classical vision models fail to yield a complete analysis of astronomical images because they are based on a single spatial scale for the adapted smoothing and background mapping. They are only suited to stars or quasi stellar sources with a slowly varying background. The multiscale analysis has allowed us to get a background adapted to a given object and to optimize the detection of objects with different size.

3.4.2 A multiscale vision model for astronomical images

3.4.2.1 Object definition in the wavelet transform space

An object has to be defined in the wavelet transform space (WTS). In the image, an object occupies a physically connected region and each pixel of this region can be linked to the others. The connectivity in the direct space has to be transported to the WTS. All structures form a $3D$ connected set which is hierarchically organized: the structures at a given scale are linked to smaller structures of the previous scale. This set gives the description of an object in the WTS. The steps of the multiscale model can now be defined.

After applying the wavelet transform to the image, a thresholding in the WTS is performed in order to identify the statistically significant pixels. These pixels are regrouped in connected fields by a scale by scale segmentation procedure, in order to define the object structures. Then, an interscale connectivity graph is established. The object identification procedure extracts each connected sub-graph that corresponds to $3D$ connected sets of pixels in the WTS and, by referring to the object definition, the sub-graph which can be associated with the objects. From each set of pixels an image of the object can be reconstructed using reconstruction algorithms. Finally, measurement and classification operations can be carried out.

3.4.2.2 Scale by scale segmentation and the interscale relation

After thresholding, the region labelling is done by a classical growing technique. At each scale, neighbouring significant pixels are grouped together to form a segmented field. A label $n > 0$ is assigned to each field pixel. If a pixel is not significant, it does not belong to a field and its label is 0. We denote by $L(i, k, l)$ the label corresponding to the pixel (k, l) at scale i and by $D(i, n)$ a segmented field of label n at the same scale.

Now we have to link the fields labelled at a given scale to the ones belonging to the following scale, in order to construct a graph from which we can extract the objects. Let us consider the fields $D(i, n)$ at scale i and $D(i + 1, m)$ at scale $i + 1$. The pixel coordinates of the maximum coefficient $W(i, k_{i,n}, l_{i,n})$ of $D(i, n)$ are $(k_{i,n}, l_{i,n})$. $D(i, n)$ is said to be connected to $D(i + 1, m)$ if the maximum position belongs to the field $D(i + 1, m)$, i.e $L(i + 1, k_{i,n}, l_{i,n}) = m$. With this *criterion of interscale neighbourhood*, a field of a given scale is linked to at most one field of the upper scale. So we have a set of fields $D(i, n)$ and a relation $\mathcal{R}$:

$$D(i, n) \; \mathcal{R} \; D(i + 1, m) \qquad \text{if} \qquad L(i + 1, k_{i,n}, l_{i,n}) = m \qquad (3.14)$$

This relation leads us to build the interscale connectivity graph whose summits correspond to the labelled fields. Statistically, some significant structures can be due to the noise. They contain very few pixels and are generally isolated, i.e they are connected to no field at upper and lower scales. So, to avoid false detection, the isolated fields are removed from the initial interscale connection graph.

3.4.2.3 The object identification

An object is associated with each local maximum of the image wavelet transform. For each field $D(i, n)$ of the interscale connection graph, its highest coefficient $W(i, k_{i,n}, l_{i,n})$ is compared with the corresponding coefficients of the connected fields of the upper scale, $W(i + 1, k_+, l_+)$ and lower scale, $W(i - 1, k_-, l_-)$.

If $W(i - 1, k_-, l_-) < W(i, k_{i,n}, l_{i,n})$ and $W(i, k_{i,n}, l_{i,n}) > W(i + 1, k_+, l_+)$, $D(i, n)$ corresponds to a local maximum of the wavelet coefficients. It defines an object. No other fields of the scale i are attributed to the object; $D(i, n)$ concentrates the main information which permits the object image to be reconstructed. Only the fields of the lower scales connected to $D(i, n)$ are kept. So the object is extracted from larger objects that may contain it. On the other hand, some of these fields may define other objects. They are sub-objects of the object. To get an accurate representation of the object cleaned of its components, the fields associated with the sub-objects cannot be directly removed; as experiments show, their images will have to be restored and subtracted from the reconstructed global image of the object. By construction, $D(i, n)$ is the root of a sub-graph which defines a tree noted T. T expresses the hierarchical overlapping of the object structures.

3.4.2.4 The object image reconstruction

Let us consider an object (or a sub-object) $\mathcal{O}$ previously defined and its associated tree T. The object corresponds to a set of wavelet coefficients $\mathcal{V}$ defined on a 3D support S in WTS:

$$\mathcal{O} \Longleftrightarrow \{\mathcal{V}(i, k, l), \text{ for } (i, k, l) \in S\} \tag{3.15}$$

where

$$S = \{(i, k, l) \text{ such that } W(i, k, l) \in D(i, n) \text{ element of } T\} \tag{3.16}$$

F is an image and W is its corresponding wavelet transform. F can be considered as a correct restored image of the object $\mathcal{O}$ if:

$$\mathcal{V}(i, k, l) = W(i, k, l) \qquad \forall (i, k, l) \in S \tag{3.17}$$

Let us denote by P_S the projection operator in the subspace S and by WT the operator associated with the wavelet transform. We can write:

$$\mathcal{V} = (P_S \circ WT)(F) = A(F) \tag{3.18}$$

We have to solve the inverse problem which consists of determining F knowing A and $\mathcal{V}$. We minimize the distance $\|\mathcal{V} - A(F)\|$ leading to:

$$\tilde{A}(\mathcal{V}) = (\tilde{A} \circ A)(F) \tag{3.19}$$

The initial equation (3.18) is modified with the introduction of $\tilde{A}$, the adjoint operator associated with A. $\tilde{A}$ is applied to a wavelet transform W and gives an image $\tilde{F}$. The equation (3.19) is solved either by *the gradient algorithm* [12] or by *the conjugate gradient algorithm* [82] which improves the restoration quality and the convergence speed.

The previous vision scheme has also been applied to a pyramidal wavelet transform. The interscale connectivity graph is determined taking into account the decimation from one scale to the following one. The restoration algorithm is derived from the conjugate gradient algorithm.

3.4.2.5 *Applications to astronomical images*

We test the multiscale models on the image L384-350 (see Figure 3.2) corresponding to the galaxy 384350 of the *Surface photometry catalogue of ESO-Uppsala galaxies.*

We performed a 7-scales wavelet transform of L384-350, and 58 objects are detected. The restored image with the *à trous* algorithm, made of the reconstructed images of each object, is given in Figure 3.2. The restored image of the central galaxy is plotted in Figure 3.3. A sub-object of the galaxy, which corresponds to a spiral arm, has been extracted; its image is shown in the same figure.

In the case of simple objects of small size, usual astronomical imagery methods and the multiscale model give very close results [82]. But, the multiscale model permits not only point-like objects to be identified but also objects which are much more complex (for instance the central galaxy of L384-350). Such objects with their structure hierarchy can be decomposed by our model thanks to the notion of sub-object.

3.4.3 *Applications to the analysis of astrophysical sources*

Many applications of the wavelet transform have been done on the analysis of astrophysical sources by different approaches. Our vision model is too recent to have been currently applied, but similar ideas were partly implemented.

The fractal behaviour of interstellar clouds was one of the specific applications of the wavelet transform in astrophysics. Gil and Henriksen [39] first

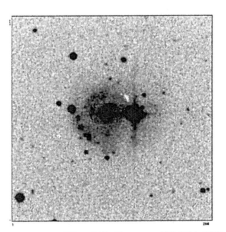
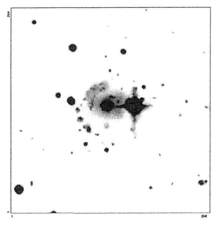

Fig. 3.2. Image of L384-350 and the restored image (from [82]).

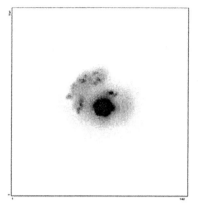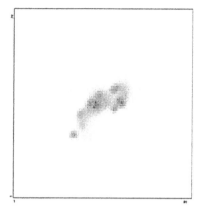

Fig. 3.3. Restored images of the galaxy object and one of its sub-object (from [82]).

analysed the ^{13}CO spectral data of the outflow region of a molecular cloud. The fractal Hausdorff dimension was obtained for some specific regions. The wavelet transform was applied with a Mexican hat. A deeper analysis led them to believe that a substantial quantity of gas is present between the emission peaks.

From observations obtained with the infrared satellite IRAS, a similar analysis was done on the 60 and 100 μm emissions at high galactic latitudes [1]. A local spectral index was obtained for each pixel. Its map is more homogeneous than the brightness map.

Langer *et al.* [55] have used a Laplacian pyramid transform for analysing molecular clouds. This transform can be considered as a discrete wavelet transform with four wavelets. They identified all the components, with positive coefficients for fragments or clumps, and with negative ones for cavities or bubbles. A hierarchical structure was shown. The studied cloud seems to be more chaotic than that predicted for incompressible turbulence, probably because of the importance of long-range gravitational forces, the compressibility of clumps and the presence of turbulent dissipation.

A 1D application was done by Lepine *et al.* [60] [65] for analysing discrete stochastic components found on emission lines in Wolf-Rayet stars. This allowed them to identify a dominant scale. The wavelet power spectrum was used to verify the consistency of the data with a model based on scaling laws.

Coupinot *et al.* [26] have developed a multiscale method for analysing complex objects. They showed by numerical simulations that they obtained an available photometric accuracy. They applied their method on different sources. On high angular resolution images of the galaxy M31 they succeeded in detecting new faint globular clusters [3]. For a clumpy irregular galaxy they extracted about twenty clumps [46]. Using a model, they deduced the age and the mass of these structures. Evidence from the self-propagating star formation mechanism was derived.

Bendjoya *et al.* [8] [89] applied a 1D wavelet transform for the analysis of rings of Saturn and Uranus [74]. They showed their hierarchies, and they identified the components. The nuclei of comets have also been studied using a multiscale analysis [69].

Clusters of galaxies include three components: galaxies, an intracluster medium made of hot gas radiating in the X-ray window, and an amount of dark matter. The baryonic matter is mainly in the form of ionized hydrogen, which appears in images obtained in the X-ray bands by satellites like Einstein or ROSAT. A subclustering has been found in many clusters [32] [36] and its analysis is of great importance in establishing the dynamical state of the cluster of galaxies. Multiscale approaches based on the wavelet transform have contributed to quantify this subclustering [35]. The correlation between the subclusterings and the gravitational potential wells must be stated precisely.

In Figure 3.4 the raw image of ABCG 2256 taken with the ROSAT PSPC detector is plotted. In Figure 3.5 its reconstructed image from its significant components shows the matter organization. The substructures found with the wavelet transform at different scales are superimposed in Figure 3.6. This study has been done on a set of clusters of galaxies and it raised the question of the validity of the hydrodynamic equilibrium hypothesis of the gas in the potential well [87].

Today, structural features are identified and measured on X-ray images by a multiscale analysis [42] [109] [30]. The wavelet transform is also used for detecting new X-ray clusters [81] [103].

3.4.4 Applications to galaxy counts

The complexity of the distribution of galaxies and of clusters of galaxies is now clearly established up to scales of 50 Mpc[†] (with a Hubble constant $= 100$ km s^{-1} Mpc^{-1}) [4]. Valuable information on the three-dimensional clustering of galaxies is provided by wide-angle redshift surveys, such as the Center for Astrophysics (hereafter CfA) redshift survey slices [51]. The

---

[†] The parsec (pc) is the current astronomical unit of distance. Its value is approximatively 3.26 light-years.

Albert Bijaoui

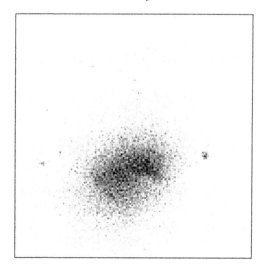

Fig. 3.4. Raw image of ABCG 2256 taken with the ROSAT PSPC (from [87]).

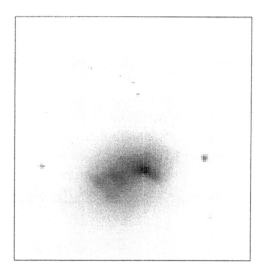

Fig. 3.5. Reconstructed image of ABCG 2256 after a wavelet analysis (from [87]).

main feature of the galaxy distribution is the departure from homogeneity at all scales within reach. The topology of the distribution is characterized by a complex network of structures, $1D$ filaments [40] or $2D$ sheets [28]. The high-density structures appear to connect clusters of galaxies and to delineate large spherical regions which are devoid of bright galaxies: voids are frequent events of the distribution.

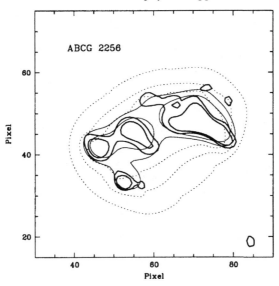

Fig. 3.6. Isocontours of two wavelet images of ABCG 2256. Full thin line, at 4 pixel scale, full thick line at 2 pixel scale, dashed line the restored image (from [87]).

Various statistical methods have been used to detect local structures and to discriminate among theoretical models. Generally the statistical indicators provide an objective way to compare observational data with numerical simulations, but they measure only an average value of the parameter used to characterize the distribution of galaxies. The wavelet transform provides a space-scale analysis in which both over- and under-dense structures are detected according to their typical size.

In Figure 3.7 we have superimposed the positions of a part of the CfA catalogue with the limit curves of the significant positive wavelet coefficients at different scales. For each scale we compute a wavelet coefficient image, then we threshold it taking into account a threshold statistical level, here equal to 0.001. The superimposed curves show clearly the hierarchy of structures formed by the galaxies. In Figure 3.8 we show the same plot where the negative significant pixels and the voids appear distinctly. The wavelet analysis allowed us for the first time to objectively detect and to locate voids in the CfA slice and to replace the often used subjective visual criteria with quantitative parameters [86].

This approach was applied to other catalogues of cosmological interest and always showed a hierarchically structured distribution of objects [23]. A full description of the large structure was done around the South Galactic Pole, leading to the identification of clusters and superclusters of galaxies [33]

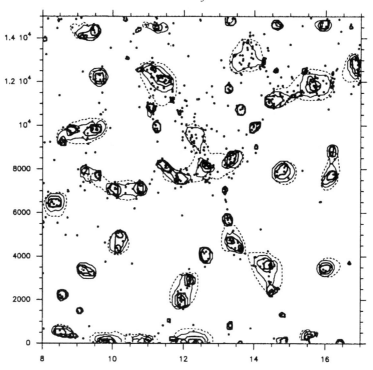

Fig. 3.7. Map of the significant over-dense structures. The lines define areas corresponding to a statistical level 0.005. Four scales are plotted. For a given structure, the contours enclosing the larger areas correspond to the larger wavelet scales (from [86]).

[34]. Larger structures were also obtained by analysing the distribution of the clusters of galaxies [22].

This method for identifying hierarchical groups has also been adapted to the analysis of the distribution of asteroids in the space of dynamical parameters [9]. A new decomposition of asteroids in dynamical families was done and compared successfully to a classical cluster analysis [25].

A continuous wavelet transform, with a so-called *Mexican hat* wavelet, was applied on the radial velocity of galaxies in order to determine an irrotational velocity field [77]. The application of this method to cosmic velocity fields allows one to derive the potential or, similarly, any linear function on the vectorial field.

3.4.5 *Statistics on the large-scale structure of the Universe*

Obviously the analysis of the scalogram obtained on the the galaxy distribution may disclose information on the fractality of the Universe. Many diffi-

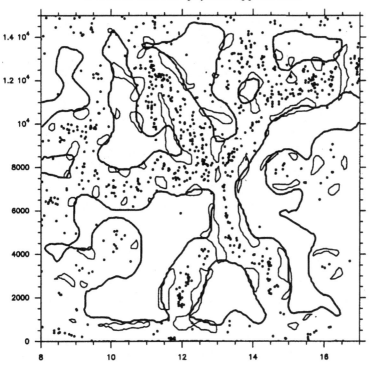

Fig. 3.8. Map of the significant underdense structures. The statistical threshold of detection is 0.005. The two plotted contours correspond to wavelet scales $350\,\mathrm{km\,s^{-1}}$ and $700\,\mathrm{km\,s^{-1}}$ (from [86]).

culties remain in interpreting the results: bias due to the Poisson noise, velocity dispersion in bound gravitational structures, homogeneity in the object selection, etc. Applying such an analysis, Martinez *et al.* [64] found the existence of different exponents showing the multifractal nature of the distribution of galaxies.

The spectrograms of the quasi stellar objects (QSO) exhibit a large set of fine absorption lines which have been identified as the redshifted Lyα absorptions of hydrogen galactic halos along the sight line. This Lyα forest became a fundamental marker of the large-scale structures. Pando and Fang [73] applied a Daubechies wavelet transform for analysing these data. Clustering was identified on scales similar to the one of the cluster of galaxies. The intensity of this clustering seems to decrease with the distance.

The correlation function is not sufficient to separate between cosmological scenarios. Bromley [21] has proposed an approach based on the histogram of the wavelet coefficients. We derived a new statistical indicator for that purpose, based on a morphological parameter of the structures detected by the wavelet transform. We implemented on a Connection Machine CM-200 an *n*-body simulation program corresponding to different classical scenarios [58]. In Figure 3.9 we show a simulation with a *Cold Dark Matter* (CDM) scenario while the *Hot Dark Matter* (HDM) is given in Figure 3.10. It can be seen clearly that the CDM scenario favours clustered structures while the HDM favours filamentary ones. An objective method is necessary to characterize the structures.

We applied our vision model to 3*D* images resulting from simulations. After the structure identification, we quantify the morphological properties by the shape parameter:

$$L(a) = 36\pi \frac{V^2(a)}{S^3(a)} \tag{3.20}$$

where $V(a)$ and $S(a)$ are the volume and the surface of a structure at scale *a*. The mean of $L(a)$ gives a description of the deviation from sphericity at the scale *a*. In Figure 3.11 the results obtained for the two scenarios are plotted. Error bars are given by the variance on $< L(a) >$ obtained with five simulations for each scenario. It appears clearly that the CDM scenario is made of structures of almost spherical shape with a slight variation towards the elon-

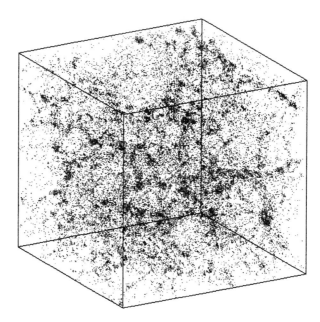

Fig. 3.9. Numerical simulation of the CDM universe on a 128^3 grid (the physical size of the box is 192Mpc) (from [58]).

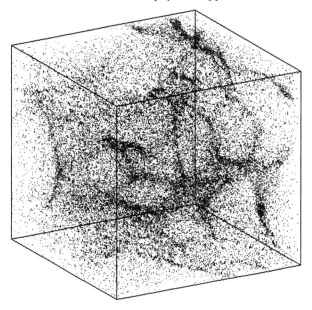

Fig. 3.10. Numerical simulation of the HDM universe (from [58]).

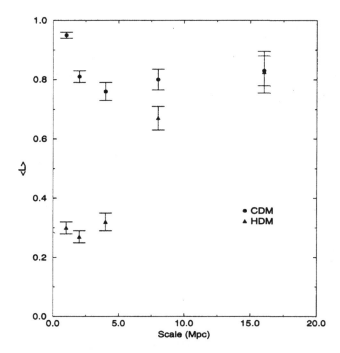

Fig. 3.11. Variation of the sphericity indicator with the scale for the structures generated from the CDM and HDM cosmological scenarios (from [58]).

gated ones, while the HDM scenario is made of elongated structures even at very small scales.

This new indicator, based on the morphology of structures in the wavelet space, was applied on observational data, the CfA catalogue [57], or the Abell cluster of galaxies [22] and tends to favour an intermediate scenario.

3.5 Conclusion

The wavelet transform is used more and more often for astrophysical data analysis. Its main domain of applications consists essentially of nonstationary processes which exhibit long range correlations. We often observe this situation in astrophysical data, either for time series (solar granulation, distribution of the magnetic fields, quasar flux, GRB, etc.), for images (molecular clouds, galaxies, comets, X-ray cluster of galaxies, etc.) and also for astronomical catalogues (spatial distribution of galaxies, distribution of asteroids in a dynamical parameter space, etc.).

As we have shown in this chapter, the astrophysicists have applied different discrete wavelet transforms: Morlet's transform, for time-frequency analysis, the *à trous* algorithm and the pyramidal transform for image restoration and analysis, pyramidal with Fourier transform for synthesis aperture imaging. The choice is determined by the problem. For example image compression needs a compact representation, consequently the multiresolution analysis is *a priori* the most suitable tool for this application. Aliasing must be avoided for analysis and if we have no problem with memory or computing time, the *à trous* algorithm is the best, and if that is not the case, a pyramidal transform would be better than the multiresolution analysis. For times series, often the phase is required, so a Morlet or a Gabor transform is indicated.

Nevertheless, the sampling, the compactness in the direct space and the regularity of the wavelet are the main elements that the astrophysicist has to take care of in applying a given discrete wavelet transform. With my collaborators I did not apply the multiresolution analysis which provides an unredundant image representation, but I used instead the wavelet transform as a set of pass-band filters. So our use of this transform was characterized by the following.

- The use of the *à trous* algorithm, in order to keep the same sampling for each scale.
- The use of a quasi isotropic wavelet, so that no direction is privileged by this analysis.

- A wavelet with no positive bumps, of course outside the central peak. Since the wavelet function has a null mean, it has necessarily at least one negative bump: significant negative coefficients may result of significant positive structures, but no other positive structures must be detected around a positive peak.
- A decision rule based on the PDF of the wavelet coefficient for a uniform image is applied. We can estimate the probability of a given structure to be due to the noise.
- The groups and the voids are detected by the same objective procedure.
- We detect a group or a void superimposed on a background.
- The image of each structure can be restored by application of partial restoration.
- The structures are characterized by an indicator based on the morphology in the wavelet transform space.

Many wavelet packages are now available, some of them are accessible by anonymous FTP. A large part of our work is now integrated in the SADAM (*Système d'Analyse des Données Astronomiques Multiéchelles*) package developed in the framework of a collaboration between our group and Starck's one at the CEA [91]. The methods and the algorithms implemented in this package are described in a book which will allow the physicists to apply our multiscale methods [94].

Appendices to Chapter 3

A. The à trous *algorithm*

Let us consider the 1D algorithm. The sampled data are considered as the scalar product of the image function $f(x)$ with the translated scaling functions $\phi(x - k)$:

$$c(0, k) = \langle f(x), \phi(x - k) \rangle \tag{3.21}$$

Let us consider the scalar products at the scale i:

$$c(i, k) = \frac{1}{2^i} \left\langle f(x), \phi(\frac{x - k}{2^i}) \right\rangle \tag{3.22}$$

If $\phi(x)$ satisfies the dilation equation [99]:

$$\frac{1}{2}\phi\left(\frac{x}{2}\right) = \sum_n h(n)\phi(x - n) \tag{3.23}$$

$c(i, k)$ can be iteratively computed according to the relation:

$$c(i, k) = \sum_n h(n)c(i - 1, k + 2^{i-1}n) \tag{3.24}$$

We choose the cubic central B-spline $B_3(x)$ with $h(n) = \{\frac{1}{16}; \frac{1}{4}; \frac{3}{8}; \frac{1}{4}; \frac{1}{16}\}$ $(-2 \leq n \leq 2)$ and we use the wavelet resulting from the difference between two successive approximations:

$$w(i, k) = c(i - 1, k) - c(i, k) \tag{3.25}$$

Hence we get an easy reconstruction algorithm by adding all the wavelet images with the smoothest one. The n-dimensional algorithm works with separable B-splines for the successive approximations. The $B_3(x)$ function is close to a Gaussian function and the results are quasi isotropic.

We notice that this wavelet function has no positive bumps, i.e. the origin is the maximum. This point is essential since bumps lead to rings around each field.

B. The pyramidal algorithm

Let us consider again the 1D algorithm. We start with the same sampled data $c(0, k)$ but the scalar products at the scale i are:

$$c(i, k) = \frac{1}{2^i} \left\langle f(x), \phi\left(\frac{x}{2^i} - k\right) \right\rangle \tag{3.26}$$

The sampling step is then 2^i, and the number of coefficients are reduced by a factor 2 from one step to the following one. Taking into account the dilation equation $c(i, k)$ can be iteratively computed according to the relation:

$$c(i, k) = \sum_n h(n)c(i - 1, 2k + n) \tag{3.27}$$

We choose again the cubic central B-spline $B_3(x)$ and the wavelet results also from the difference between two successive approximations, but we have to take into account the decimation, i.e. the reduction by a factor 2 of the sampling. So we introduce the approximation before this operation by:

$$\tilde{c}(i, k) = \sum_n h(n)c(i - 1, k + n) \tag{3.28}$$

and the wavelet coefficients are:

$$w(i, k) = c(i - 1, k) - \tilde{c}(i, k) \tag{3.29}$$

The reconstruction algorithm, based on an iterative scheme, is less trivial. The n-dimensional algorithm works also with separable B-splines for the successive approximations.

C. The denoising algorithm

(i) Set $n = 0$;

(ii) determine the significant coefficients, and consequently the mask $M(i, k, l)$;

(iii) restore the image $F^{(n)}$ for the significant coefficients and the last smoothed image;

(iv) compute the residue $R^{(n)} = F - F^{(n)}$;

(v) compute the wavelet transform $V^{(n)}$ of the residue;

(vi) keep the significant values and reconstruct the significant residue $\tilde{R}^{(n)}$ from this set;

(vii) if the significant residue is negligible, stop;

(viii) otherwise, add the significant residue to the previous restored image in order to get $F^{(n+1)}$;

(ix) increment n and return to point (iv).

At the end, no significant difference may be detected between the original image and the restored one. That needs between 6–10 iteration steps. Some variants are possible. We see that the set of significant wavelet coefficients is modified during the iterative process. We can process differently, by modifying the wavelet coefficients outside the mask $M(i, k, l)$, in order to get the correct solution. In another method, we reconstruct the image by applying a fixed step or a conjugate gradient. The application of a given variant depends on the purpose.

D. The deconvolution algorithm

(i) set $n = 0$;

(ii) set $O^{(0)} = 0$;

(iii) restore the image $F^{(n)} = O^{(n)} \otimes P$;

(iv) compute the residue $R^{(n)} = F - F^{(n)}$;

(v) compute the significant residue $\tilde{R}^{(n)}$;

(vi) if the significant residue is negligible, stop;

(vii) otherwise, get $O^{(n+1)} = G(O^{(n)}, \tilde{R}^{(n)})$;

(viii) increment n and return to point (iii).

Some variants are also possible, by taking into account only the significant coefficients of the original image $F(k, l)$.

Acknowledgements

My work on the application of the wavelet transform to astrophysical data was done in collaboration with many astronomers, engineers and students. I thank them and also A. Arnéodo, A. Grossmann, Y. Meyer and S. Mallat who helped me when I was starting in this field.

References

[1] A. Abergel, F. Boulanger, J. M. Delouis, G. Dudziak, and S. Steindling. Local correlations of the fluctuations at different scales of the 60 and 100mum emissions of a high galactic latitude complex. *Astron. Astroph.*, **309**: 245–257, (1996)

[2] F.J. Anscombe. The transformation of Poisson, binomial and negative-binomial data. *Biometrika*, **15**: 246–254, (1948)

[3] M. Aurière, G. Coupinot, and J. Hecquet. New globular clusters in the bulge of M31. *Astron. Astroph.*, **256**: 95–103, (1992)

[4] N. Bahcall. Large-scale structure in the universe indicated by galaxy clusters. *An. Rev. Astron. Astroph.*, **26**: 631–686, (1988)

[5] T. Bauer, H. Weghorn, E. K. Grebel, and D. J. Bomans. The young star cluster R64 in the OB association LH9 resolved with ground-based CCD observations. *Astron. Astroph.*, **305**: 135–139, (1996)

[6] C. W. Baumgart. Taxonomy of gamma ray burster data using a self-organizing neural network. In *Presented at the 16th SPIE Thermosense: International Conference on Thermal Sensing and Imaging and Diagnostic Applications, Orlando, FL, 4-8 Apr. 1994*, 1993.

[7] O. Bendinelli. Abel integral equation inversion and deconvolution by multi-gaussian approximation. *Astroph. Journ.*, **366**: 599–604, (1991)

[8] Ph. Bendjoya, J.M. Petit, and F. Spahn. Wavelet analysis of the Voyager data on planetary rings. I. Description of the method. *Icarus*, **105**: 385–399, (1993)

[9] Ph. Bendjoya, E. Slezak, and C. Froeschlè. The wavelet transform - a new tool for asteroid family determination. *Astron. Astroph.*, **251**: 312–330, (1991)

[10] A. Bijaoui. Skybackground estimation and applications. *Astron. Astroph.*, **84**: 81–84, (1980)

[11] A. Bijaoui. Algorithmes de la transformation en ondelettes. Application à l'imagerie astronomique. *Ondelettes et paquet d'ondes*. Cours CEA/EDF/INRIA. Ed. INRIA, Rocquencourt France, 1991.

[12] A. Bijaoui and F. Rué. A multiscale vision model adapted to the astronomical images. *Sign. Proc.*, **46**: 345–362, (1995)

[13] A. Bijaoui and M. Giudicelli. Optimal image addition using the wavelet transform. *Experimental Astronomy*, **1**: 347–363, (1991)

[14] A. Bijaoui, G. Lago, J. Marchal, and C. Ounnas. Le traitement automatique des images en astronomie. In INRIA, editor, *Traitement des Images et Reconnaissance des Formes*, pp. 848–854, 1978.

[15] A. Bijaoui, J.L. Starck, and F. Murtagh. Restauration des images multi-échelles par l'algorithme à trous. *Traitement du Signal*, **11**: 229–243, (1994)

[16] A. Bijaoui, Y. Bobichon, Y. Fang, and F. Rué. Méthodes multiéchelles appliquées à l'analyse des images SAR. *Traitement du Signal*, **14**: 179-193, (1997)

[17] Y. Bobichon and A. Bijaoui. A regularized image restoration algorithm for lossy compression in astronomy. *Experimental Astronomy*, **7**: 239–255, (1996)

[18] K. Bocchialini and F. Baudin. Wavelet analysis of chromospheric solar oscillations. *Astron. Astroph.*, **299**: 893–896, (1995)

[19] P. T. Boyd, P. H. Carter, R. Gilmore, and J. F. Dolan. Investigating the frequency evolution of roAp star HD 60435. *Bull. Am. Astr. Soc.*, **185**: 8001, (1994)

[20] P.T. Boyd, P.H. Carter, R. Gilmore, and J.F. Dolan. Nonperiodic variations in astrophysical systems: Investigating frequency evolution. *Astroph. Journ.*, **445**: 861–871, (1995)

[21] B. C. Bromley. A wavelet analysis of large-scale structure. *Bull. Am. Astr. Soc.*, **181**: 7903, (1992)

[22] P. Bury. *De la distribution de matière à grande échelle à partir des amas d'Abell.* PhD thesis, University of Nice Sophia Antipolis, 1995.

[23] P.Bury and A.Bijaoui. Multiscale analysis of the Abell-ACO catalogue. In *Cosmological Aspects of X-Ray Clusters of Galaxies.* NATO ASI workshop, 1993.

[24] M. Cappacioli, E.V. Held, H. Lorenz, G.M. Richter, and R. Ziener. Application of an adaptive filtering technique to surface photometry of galaxies. The method tested on NGC 3379. *Astron. Nachrichten*, **309**: 69–80, (1988)

[25] A. Cellino and V. Zappala. Asteroid 'clans': Super-families or multiple events? *Celestial Mechanics and Dynamical Astronomy*, **57**: 37–47, (1993)

[26] G. Coupinot, J. Hecquet, M. Auriere, and R. Futaully. Photometric analysis of astronomical images by the wavelet transform. *Astron. Astroph.*, **259**: 701–710, (1992)

[27] I. Daubechies. Orthogonal bases of compactly supported wavelets. *Com. Pure and Appl. Math.*, **41**: 909–996, (1988)

[28] V. de Lapparent, M.J. Geller, and J.P. Huchra. Measures of large-scale structure in the CfA redshift survey slices. *Astroph. Journ.*, **369**: 273, (1991)

[29] J.P. Djamdji, A. Bijaoui, and R. Manière. Geometrical registration of images. The multiresolution approach. *Photog. Eng. Rem. Sens.*, **59**: 645–653, (1993)

[30] M. Donahue and J.T. Stocke. Rosat observations of distant clusters of galaxies. *Astroph. Journ.*, **449**: 554–566, (1995)

[31] D. Donoho. Non linear wavelet methods for recovery of signals, densities and spectra from indirect and noisy data. *Symposia in Applied Mathematics*, **47**: 173–205, (1993)

[32] E. Escalera, A. Biviano, M. Girard, G. Giuricin, F. Mardorissian, A. Mazure, and M. Mezzetti. Structures in the galaxy clusters. *Astroph. Journ.*, **423**: 539–552, (1994)

[33] E. Escalera and H. T. MacGillivray. Topology in galaxy distributions: method for a multi-scale analysis. a use of the wavelet transform. *Astron. Astroph.*, **298**: 1–21, (1995)

[34] E. Escalera and H. T. Macgillivray. Detection of structures on multiple scales around the south galactic pole. a two-dimensional analysis of the COSMOS/UKST southern sky galaxy catalogue. *Astronomy and Astrophysics Supplement Series*, **117**: 519–555, (1996)

[35] E. Escalera and A. Mazure. Wavelet analysis of subclustering – an illustration, Abell 754. *Astroph. Journ.*, **388**: 23–32, (1992)

[36] E. Escalera, E. Slezak, and A. Mazure. New evidence for subclustering in the coma cluster using the wavelet analysis. *Astron. Astroph.*, **264**: 379–384, (1992)

[37] K. Fritze, M. Lange, H. Oleak, and G.M Richter. A scanning microphotometer with an on-line data reduction for large field schmidt plates. *Astron. Nachrichten*, **298**: 189–196, (1977)

[38] D. Gabor. Theory of communication. *Jour. IEE*, **93**: 429–457, (1946)

[39] A. G. Gill and R. N. Henriksen. A first use of wavelet analysis for molecular clouds. *Astroph. Jour. Lett.*, **365**: L27–L30, (1990)

[40] R. Giovanelli, M. Haynes, S.T. Myers, and J. Roth. A 21cm survey of the Pisces Perseus supercluster. ii. the declination from +21.5 to 27.5 degrees. *Astron. J.*, **92**: 250, (1986)

[41] M. J. Goupil, M. Auvergne, and A. Baglin. Wavelet analysis of pulsating white dwarfs. *Astron. Astroph.*, **250**: 89–98, (1991)

[42] S. A. Grebenev, W. Forman, C. Jones, and S. Murray. Wavelet transform analysis of the small-scale X-ray structure of the cluster Abell 1367. *Astroph. Journ.*, **445**: 607–623, (1995)

[43] Ph. Guillemain and R. Kronland-Martinet. Characterization of acoustic signals through continuous linear Time-Frequency representations. *Proc. IEEE.*, **84**: 561–585, (1996)

[44] W.W. Harman. *Principles of the Statistical Theory of Communication*, chapter 11, page 217. McGraw-Hill, New York, 1963.

[45] A. Haar. Theorie der orthogonalen Funktionensysteme. *Math. Ann.*, **69**: 331–371, (1910)

[46] J. Hecquet, R. Augarde, G. Coupinot, and M. Auriere. Starbursts in the clumpy irregular galaxy VV 523. *Astron. Astroph.*, **298**: 726–736, (1995)

[47] P. G. Hjorth, L. F. Villemoes, J. Teuber, and R. Florentin-Nielsen. Wavelet analysis of 'double quasar' flux data. *Astron. Astroph.*, **255**: L20–L23, (1992)

[48] J.A. Högbom. Aperture synthesis with a non-regular distribution of interferometer baselines. *Astron. Astroph. Sup. Ser.*, **15**: 417–426, (1974)

[49] M. Holschneider, R. Kronland-Martinet, J. Morlet, and P. Tchamitchian. A Real-Time Algorithm for Signal Analysis with the Help of the Wavelet Transform. In *Wavelets: Time-Frequency Methods and Phase-Space*, pp. 286–297. Springer, Berlin, 1989.

[50] L. Huang and A. Bijaoui. Astronomical image data compression by morphological skeleton transform. *Experimental Astronomy*, **1**: 311–327, (1991)

[51] J.P. Huchra, M. Davis, D. Latham, and J. Tonry. A survey of galaxy redshifts. IV. The data. *Astrophys. J. Sup Ser.*, **52**: 89, (1983)

[52] R. W. Komm. Wavelet analysis of a magnetogram. *Solar Phys.*, **157**: 45–50, (1995)

[53] A. Kruszewski. Inventory-searching, photometric and classifying package. In *1ˢᵗ ESO/ST-ECF Data Analysis*. Warsaw University Observatory, 1989.

[54] L. Landweber. An iteration formula for Fredholm integral equations of the first kind. *Am. J. Math.*, **73**: 615–624, (1951)

[55] W.D. Langer, R.W. Wilson, and C.H. Anderson. Hierarchical structure analysis of interstellar clouds using nonorthogonal wavelets. *Astroph. Jour. Lett.*, **408**: L45–L48, (1993)

[56] J. K. Lawrence, A. C. Cadavid, and A. A. Ruzmaikin. Turbulent and chaotic dynamics underlying solar magnetic variability. *Astroph. Journ.*, **455**: 366–375, (1995)

[57] E. Lega, A. Bijaoui, J. M. Alimi, and H. Scholl. A morphological indicator for comparing simulated cosmological scenarios with observations. *Astron. Astroph.*, **309**: 23–29, (1996)

[58] E. Lega, H. Scholl, J.M. Alimi, A. Bijaoui, and P. Bury. A parallel algorithm for structure detection based on wavelet and segmentation algorithm. *Parallel Comp.*, **21**: 265–285, (1995)

[59] R. B. Leighton, R. W. Noyes, and G. W. Simon. Velocity fields in the solar atmosphere: I. Preliminary report. *Astroph. Jour.*, **135**: 474–499, (1962)

[60] S. Lepine. Wavelet analysis of Wolf-Rayet emission line variability: Evidence for clumping. *Astroph. and Space Sci.*, **221**: 371–382, (1994)

[61] H. Lorenz, G.M. Richter, M. Cappacioli, and G. Longo. Adaptative filtering in astronomical image processing. *Astron. Astroph.*, **277**: 321–330, (1993)

[62] L.B. Lucy. An iteration technique for the rectification of observed distributions. *Astron. Journal*, **79**: 745–754, (1974)

[63] S. Mallat. A theory for multiresolution signal decomposition: the wavelet representation. *IEEE Trans. Pattern Anal. Mach. Intelligence*, **11(7)**: 674–693, (1989)

[64] V.J. Martinez, S. Paredes, and E. Saar. Wavelet analysis of the multifractal character of the galaxy distribution. *Mon. Not. Royal Astron. Soc.*, **260**: 365–375, (1993)

[65] A. F. J. Moffat, S. Lepine, R. N. Henriksen, and C. Robert. First wavelet analysis of emission line variations in wolf-rayet stars. *Astroph. and Space Sci.*, **216**: 55–65, (1994)

[66] J. Morlet, G. Arens, E. Fourgeau, and D. Giard. Wave propagation and sampling theory. *Geophysics*, **47**: 203–236, (1982)

[67] B. Mosser, F.X. Schmieder, Ph. Delache, and D. Gautier. A tentative identification of Jovian global oscillations. *Astron. Astroph.*, **251**: 356–364, (1991)

[68] F. Murtagh, J.L. Starck, and A. Bijaoui. Image restoration with noise suppression using the wavelet transform ii. *AA Sup Ser*, **112**: 179–189, (1995)

[69] F. Murtagh, W. Zeilinger, J.L. Starck, and A. Bijaoui. Object detection using multi-resolution analysis. *Astronomical Data Analysis Software and Systems IV, ASP Conference Series, Vol. 77, 1995, R.A. Shaw, H.E. Payne, and J.J.E. Hayes, eds.*, **4**: 260, (1995)

[70] J. P. Norris, R. J. Nemiroff, J. D. Scargle, C. Kouveliotou, G. J. Fishman, C. A. Meegan, W. S. Paciesas, and J. T. Bonnel. Detection of signature consistent with cosmological time dilation in gamma-ray bursts. *Astroph. Journ.*, **424**: 540–545, (1994)

[71] A.R. Ochadlick, Jr., H.N. Kritikos, and R. Giegengack. Variations in the period of the sunspot cycle. *Geoph. Res. Let.*, **20**: 1471–1474, (1993)

[72] T. Oosterloo. Adaptive filtering and masking of HI data cubes. *Vistas in Astron.*, **40**: 571–577, (1996)

[73] J. Pando and L.Z. Fang. A wavelet space-scale decomposition analysis of structures and evolution of QSO Ly alpha absorption lines. *Astroph. Journ.*, **459**: 1–11, (1996)

[74] J.M. Petit and Ph. Bendjoya. A new insight in Uranus rings: a wavelet analysis of Voyager 2 data, in *Proc. of the Pacific Astronomical Society* (in the press 1997)

[75] W.K. Pratt. Digital Image Processing, chapter 10, p. 255. (John Wiley and Sons, New York, 1978)

[76] W. L. Press. Wavalet-based compression software for FITS images. *Astronomical Data Analysis Software and Systems I, ASP Conference Series*, **25**: 1, (1992)

[77] S. Rauzy, M. Lachièze-Rey, and R.N. Henriksen. Wavelet analysis of cosmic velocity fields. *Astron. Astroph.*, **273**: 357–366, (1993)

[78] G.M Richter. Zur auswertung astronomischer aufnahmen mit dem automatischen flächenphotometer. *Astron. Nachrichten*, **299**: 283–303, (1978)

[79] G.M. Richter, P. Böhm, H. Lorenz, A. Priebe, and M. Cappacioli. Adaptative filtering in astronomical image processing. *Astron. Nachrichten.*, **312**: 345–349, (1991)

[80] S. Roques, K. Bouyoucef, L. Touzillier, and J. Vigneau. Prior knowledge and multiscaling in statistical estimation of signal-to-noise ratio. Application to deconvolution regularization. *Sign. Proc.*, **41**: 395–401, (1995)

[81] P. Rosati, R. Della Ceca, R. Burg, C. Norman, and R. Giacconi. A first determination of the surface density of galaxy clusters at very low X-ray fluxes. *Astroph. Jour. Lett.*, **445**: L11–L14, (1995)

[82] F. Rué and A. Bijaoui. A multiscale vision model to analyse field astronomical images. *Experimental Astronomy*, **7**: 129-160, (1997)

[83] J.D. Scargle, T. Steiman-Cameron, K. Young, D.L. Donoho, J.P. Crutchfield, and J. Imamura. The quasi-periodic oscillations and very low frequency noise of Scorpius X-1 as transient chaos – a dripping handrail? *Astroph. Jour. Lett.*, **411**: L91–L94, (1993)

[84] K. G. Schaefer, H. E. Bond, and G. Chanmugam. High speed UV photometry of AM Herculis binaries. *Bull. Am. Astr. Soc.*, **185**: 1603, (1994)

[85] E. Slezak, A. Bijaoui, and G. Mars. Identification of structures from galaxy counts – use of the wavelet transform. *Astron. Astroph.*, **227**: 301–316, (1990)

[86] E. Slezak, V. De Lapparent, and A. Bijaoui. Objective detection of voids and high-density structures in the first CfA redshift survey slice. *Astroph. Journ.*, **409**: 517–529, (1993)

[87] E. Slezak, F. Durret, and D. Gerbal. A wavelet analysis search for substructures in eleven X-ray clusters of galaxies. *Astron. Journ.*, **108**: 1996–2008, (1994)

[88] E. Slezak, G. Mars, A. Bijaoui, C. Balkowski, and P. Fontanelli. Galaxy counts in the coma supercluster field: automated image detection and classification. *Astron. Astrophys. Sup. Ser.*, **74**: 83–106, (1988)

[89] F. Spahn, J.M. Petit, and Ph. Bendjoya. The gravitational influence of satellite pan on the radial distribution of ring-particles in the region of the Encke-division in Saturn's a ring. *Celestial Mechanics and Dynamical Astronomy*, **57**: 391–402, (1993)

[90] J. L. Starck and F. Murtagh. Image restoration with noise suppression using the wavelet transform. *Astron. Astroph.*, **288**: 342–348, (1994)

[91] J.L. Starck. SADAM: Système d'Analyse des Données Astronomiques en Multiéchelles. CEN Saclay, DAPNIA/SEI, CEA, Gif/Yvette, France, 1996.

[92] J.L. Starck and A. Bijaoui. Filtering and deconvolution by the wavelet transform ii. *Sign. Proc.*, **35**: 195–211, (1994)

[93] J.L. Starck and A. Bijaoui. Multiresolution deconvolution. *J. Opt. Soc. Am. A*, **11**: 1580–1588, (1994)

[94] J.L. Starck, F. Murtagh, and A. Bijaoui. *Image processing and data analysis : the multiscale approach.* (Cambridge University Press, 1998.)

[95] J.L. Starck, F. Murtagh, and M. Louys. Astronomical image compression using the pyramidal median transform. *Astronomical Data Analysis Software and Systems IV, ASP Conference Series, Vol. 77, 1995, R.A. Shaw, H.E. Payne, and J.J.E. Hayes, eds.*, **4**: 268, (1995)

[96] J.L. Starck, F. Murtagh, B. Pirenne, and M. Albrecht. Astronomical image compression based on noise suppression. *Pub. Astron. Soc. Pacific*, **108**: 446–455, (1996)

[97] J.L. Stark, A. Bijaoui, B. Lopez, and C. Perrier. Image reconstruction by the wavelet transform applied to aperture synthesis. *Astron. Astroph.*, **283**: 349–360, March 1994.

[98] R.S. Stobie. The COSMOS image analyzer. *Pattern Recognition Letters*, **4**: 317–324, (1986)

[99] G. Strang. Wavelets and dilation equations: a brief introduction. *SIAM Review*, **31**: 614–627, (1989)

[100] K. Szatmary, J. Gal, and L. L. Kiss. Application of wavelet analysis in variable star research. ii. the semiregular star V Bootis. *Astron. Astroph.*, **308**: 791–798, (1996)

[101] K. Szatmary and J. Vinko. Periodicities of the light curve of the semiregular variable star Y Lyncis. *Month. Not. Royal Astron. Soc.*, **256**: 321–328, (1992)

[102] K. Szatmary, J. Vinko, and J. Gal. Application of wavelet analysis in variable star research. i. Properties of the wavelet map of simulated variable star light curves. *Astronomy and Astrophysics Supplement Series*, **108**: 377–394, (1994)

[103] M. P. Ulmer, A. K. Romer, R. C. Nichol, B. Holden, C. Collins, and D. Burke. A progress report on serendipitous high-redshift archival ROSAT Cluster (SHARC) survey. *Bull. Am. Astr. Soc.*, **187**: 9503, (1995)

[104] F. Valdes. Faint object classification and analysis system standard test image. In *1ˢᵗ ESO/ST-ECF Data Analysis*. IRAF group, Tucson, Arizona, 1989.

[105] S. E. Vance and J. E. Grindlay. Applications of a wavelet-based filtering and deconvolution technique. *Bull. Am. Astr. Soc.*, **185**: 4008, (1994)

[106] P.H. Van Cittert. Zum einfluss der spaltbreite auf die intensitätsverteilung in spektrallinien ii. *Z. Physik*, **69**: 298, (1931)

[107] A. Vigouroux and Ph. Delache. Fourier versus wavelet analysis of solar diameter variability. *Astron. Astroph.*, **278**: 607–616, (1993)

[108] A. Vigouroux and Ph. Delache. Sunspot numbers uncertainties and parametric representations of solar activity variations. *Solar Phys.*, **152**: 267–274, (1994)

[109] A. Vikhlinin, W. Foreman, and C. Jones. Mass concentrations associated with extended X-ray sources in the core of the coma cluster. *Astroph. Journ.*, **435**: 162–170, (1994)

[110] B.P. Wakker and U.J. Schwarz. The multiresolution CLEAN and its application to short-spacing problem in interferometry. *Astron. Astroph.*, **200**: 312–322, (1988)

[111] R.L. White and J.W. Percival. Compression and progressive transmission of astronomical images. *Bull. Am. Astr. Soc.*, **185**: 1504, (1994)

[112] R.L. White. High-performance compression of astronomical images. Available via anonymous FTP from stsci.edu:/software/hcompress, 1992

[113] R.L. White, M. Postman, and M.G. Lattanzi. Compression of the Guide Star digitised Schmidt plates. In *Digitised sky surveys,*, pages 167–175. Kluwer Academic Publishers, 1992

[114] Y. Yan and B. Peng. Noise suppression with wavelets in image reconstruction for aperture synthesis. In *Astronomical Data Analysis Software and Systems V, ASP Conference Series.*, 1995

4

Turbulence analysis, modelling and computing using wavelets

M. FARGE, N.K.-R. KEVLAHAN, V. PERRIER

Laboratoire de Météorologie Dynamique du CNRS,
Ecole Normale Supérieure,
24, rue Lhomond, 75231 Paris Cedex 5, France

K. SCHNEIDER

Institut für Chemische Technik,
Universität Karlsruhe (TH),
Kaiserstrasse 12, 76128 Karlsruhe, Germany

Abstract

We have used wavelets to analyse, model and compute turbulent flows. The theory and open questions encountered in turbulence are presented. The wavelet-based techniques that we have developed to study turbulence are explained and the main results are summarized.

4.1 Introduction

In this chapter we will summarize the ten years of research we have done to try to better understand, model and compute fully developed turbulent flows using wavelets and wavelet packets. Fully developed turbulence is a highly nonlinear regime (very large Reynolds number tending to infinity) and is distinct from the transition to turbulence (low Reynolds number). We have chosen to present a personal point of view concerning the current state of our understanding of fully developed turbulence. It may not always coincide with the point of view of other researchers in this field because many issues we are addressing in this chapter are still undecided and highly controversial. This paper is a substantially revised and extended version of: *Wavelets and Turbulence* by Farge, Kevlahan, Perrier and Goirand which appeared in *Proceedings of the IEEE*, vol. 84, no. 4, April 1996, pp. 639–669.

After more than a century of turbulence study [30], [173], no convincing theoretical explanation has produced a consensus among physicists (for a historical review of various theories of turbulence see [160], [158], [72],

[91]). In fact, a large number of *ad hoc* 'phenomenological' models exist that are widely used by fluid mechanicians to interpret experiments and to compute many industrial applications (in aeronautics, combustion, meteorology . . .) where turbulence plays a role. For these models there is no need to suppose the universality of turbulence since they are not derived from first principles. Their predictions are compared with experiments, such as wind tunnel measurements, in order to tune the parameters necessary to match the model to the observations. This procedure is done case by case, for a given type of turbulent flow and for a given geometry of the internal or external boundaries. Actually, it is still not known whether fully developed turbulence has the universal behaviour (independence from initial and boundary conditions) which is generally assumed in the limit of small scales. Already in 1979 one of us (M.F. [69]) expressed reservations about our understanding of turbulence and thought that we did not yet know the pertinent questions to ask in order to guide research in this field. Nearly twenty years of work on the subject have persuaded her that we have not yet identified the appropriate objects, by which we mean the structures and elementary interactions, from which it will be possible to construct a satisfactory theory of turbulence. Turbulent flows are chaotic, i.e. sensitive to initial conditions, therefore we are looking for a statistical theory, but the classical averages used at present do not appear to be adequate. This point has been beautifully discussed in a conference given in 1956 by Kampé de Fériet [112], where he rightly concluded that:

In order to become really useful to research in turbulence theory, the statistical definition of the average still requires, we believe, that the theory of the integration of Navier–Stokes equations should have made substantial progresses.

This remark is as pertinent today as it was in 1956.

In our opinion, our present ignorance of the elementary physical mechanisms at work in turbulent flows arises in part from the fact that we perform averages using point measurements and also because we analyse them in terms of correlations or Fourier modes. This problem has already been pointed out by Zabusky [208] when he wrote:

In the last decade we have experienced a conceptual shift in our view of turbulence. For flows with strong velocity shear . . . or other organizing characteristics, many now feel that the spectral description has inhibited fundamental progress. The next 'El Dorado' lies in the mathematical understanding of coherent structures in weakly dissipative fluids: the formation, evolution and interaction of meta-stable vortex-like solutions of nonlinear partial differential equations.

By using point measurements or the Fourier representation, we probably miss the point, because these classical methods ignore the presence of the coherent vortices that one observes in physical space and whose dynamic role seems essential. As Hans Liepmann, successor to Von Karman as director of the Aeronautical Laboratory of Caltech, likes to comment [141], in turbulence research we are like the drunk man who has lost his keys in a dark alley, but who finds it easier to search for them under the street light. Everyone knows that turbulence has to do with vortex production and interaction. This is even embedded in the Latin etymology of the word 'turbulence': *turba* for crowd and *turbo* for vortex. Namely, a turbulent flow can be described as 'a crowd of vortices in nonlinear interaction'. However, because we do not have a good enough theoretical grasp of the structure of these vortices, on the mechanism of their production by nonlinear instabilities in shear-layers, on their long-range collective dynamics and their nonlinear interactions, we prefer to forget about them and content ourselves with studying turbulence as far as possible from regions where vortices are produced, in particular, as far as possible from solid walls.

This approach has led turbulence research for the last fifty years to explore the unphysical academic case of statistically stationary, homogeneous and isotropic turbulence, which, under those hypotheses and neglecting the essential effect of walls in considering periodic boundary conditions, represents turbulent fields in terms of Fourier modes and predict the scaling properties of ensemble averages. To construct this theory one needs to suppose that the injection of energy is confined to the low wavenumbers, and that the dissipation of energy is confined to the high wavenumbers. This assumption allows us to define an intermediate range of wavenumbers, called the inertial range, where the flow behaves in a conservative manner, which then enables us to predict the scaling of the energy spectrum in this range. Unfortunately these hypotheses are incompatible with the local production of vorticity in boundary layers or shear layers, due to the duality between physical localization and spectral localization: if you have one you cannot have the other and vice versa (Heisenberg's uncertainty principle). The same remark holds for the dissipation of energy. Incidentally, we are convinced that this lack of physical soundness of the statistical theory proposed in 1941 by Kolmogorov [117] [118], [119], and developed by Batchelor [16], explains why G. I. Taylor had never been convinced by this redirection of turbulence research, where the dynamics of individual turbulent flow realizations, resulting from vortex interactions, is not taken into account. In fact, as early as 1938 Taylor had already recognized the importance of vortices in turbulence when he wrote [190]:

The fact that small quantities of very high frequency disturbances appear, and increase as the speed increases, seems to confirm the view frequently put forward by the author that the dissipation of energy is due chiefly to the formation of very small regions where the vorticity is very high.

Nowadays if we want to refocus turbulence research towards a more physical and dynamical approach valid also for inhomogeneous flows, we should take up the challenge proposed by Hans Liepmann during a workshop we organized in February 1997 in Santa Barbara:

As long as we are not able to predict the drag on a sphere or the pressure drop in a pipe from first principles (namely from continuous, Newtonian, incompressible assumptions, without any other complications), we will not have made it!' [142].

As astonishing as it may seem, these two very 'simple' and basic problems are still open and should be taken as a serious challenge. Our conviction is that the wavelet representation, because it keeps track of both position and scale, can help us to address these problems, in combining dynamical and statistical approaches to improve our understanding of fully developed turbulence and propose new turbulence models.

As far as we know, we have been the first to introduce wavelets to analyse turbulence in two [81], [77] and then three dimensions [75], to design orthogonal wavelet algorithms to solve nonlinear PDEs [165], to use wavelets and wavelet packets to extract coherent vortices out of turbulent flows [74], to solve the Navier–Stokes equations in a wavelet basis [96], [41], and to locally force turbulent flows using wavelets [185]. We apply the wavelet transform to decompose the vorticity field onto a set of smooth functions with compact (or quasi-compact) support and thus permit a representation in both space and scale. The choice of vorticity for both two- and three-dimensional turbulent flows, rather than velocity, matters because vorticity is, from a dynamical point of view (considering Helmholtz's and Kelvin's theorems), the essential field which triggers the evolution of velocity. We share the views of Chorin [43], [44] who has been advocating for 25 years the importance of vorticity for the computation of turbulent flows.

We are convinced that the wavelet transform is an appropriate tool, not only for analysing and interpreting experimental results, but also for attempting to construct a more satisfactory statistical theory, design new turbulence models and define new numerical methods to compute fully developed turbulent flows. Moreover, the unconditional approximation property of the wavelet representation may help us to compute high Reynolds number flows presenting a strong intermittency, to replace periodic boundary conditions by more physical ones, and to simulate the local production of vortices at the walls or in shear layers, while controlling the quality (local resolution and smoothness) of the approximation. This is the program we will expose in this chapter. We will discuss the results we have obtained in the last ten years, but it is still very much work in progress and ten more years will be needed before its potential can be confirmed or denied.

Our chapter is organized as follows. We first state the problem of turbulence and the main open questions. We then focus on how wavelets can be used to answer these questions. We present fractal and multifractal analysis, turbulence analysis and turbulence modelling, and finally the use of wavelets to numerically solve the Navier–Stokes equations. In conclusion, we present several perspectives and point out where new methods need to be developed in order to improve our understanding of fully developed turbulence.

4.2 Open questions in turbulence

4.2.1 Definitions

Turbulence is a highly unstable state of fluid flows, where by fluids we mean continuously movable and deformable media. Liquids, gases and plasmas are considered to be fluids when the scale of observation is much larger than the molecular mean free path. Turbulence is characterized by the Reynolds number, which is the ratio of the nonlinear inertial forces, responsible for the flow instability, to the linear dissipative damping, which converts kinetic energy into thermal energy. We will focus on 'fully developed turbulence', namely the limit of very large Reynolds numbers, which corresponds to, either very large velocities (strong advection), and/or very small viscosity (weak dissipation, which tends to a constant as the Reynolds number tends to infinity), and/or very large turbulent scales. For flows encountered in hydraulics and naval engineering Reynolds numbers are of the order of 10^2 to 10^6, in aeronautics (engines, airplanes, shuttles) 10^6 to 10^8, in meteorology and oceanography 10^8 to 10^{12}, and in astrophysics larger than 10^{12}.

While the dissipation term is optimally represented in Fourier space because Fourier modes diagonalize the Laplacian operator (for periodic boundary conditions or unbounded domains), the nonlinear advective term is very complicated in Fourier space where it becomes a convolution, i.e. all Fourier modes are involved and coupled. As fully developed turbulence corresponds to flows where nonlinear advection is dominant, i.e. is larger than linear dissipation by a factor of the order of Reynolds number, it is obvious that the Fourier representation is inadequate for studying and computing flows in this large Reynolds limit. We need to find a mathematical tool to optimally solve the nonlinear advection term, in the same way as the Fourier transform is the most economical representation to solve the linear dissipation term for the rather unphysical case of periodic boundary conditions. Surprisingly, however, all classical methods in turbulence rely on the Fourier representation, which is inappropriate for the nonlinear advection term. For a review of these methods the best references are Monin and Yaglom [158] for the statistical theory of three-dimensional turbulence and Kraichnan and Montgomery [123] for the statistical theory of two-dimensional turbulence.

Turbulence remains an unsolved problem because our traditional conceptual and technical tools are inadequate. For instance, classical Hamiltonian mechanics describes steady states of conservative systems, but turbulent flows are non-stationary and dissipative. Classical dynamics only solves systems with a few degrees of freedom, while fully developed turbulent flows have a very large, perhaps even infinite, number of degrees of freedom. Classical statistical theories deal with closed reversible systems in thermal equilibrium, but turbulent flows are open irreversible systems out of thermal equilibrium. Classical mathematical methods solve linear differential equations, but cannot integrate analytically the nonlinear partial differential equations encountered in the study of turbulence (apart from a very few cases for which an appropriate transform allows to reformulate the problem as a linear one, such as Burgers' equation using the Hopf–Coles transform). In fact, even the existence and uniqueness of solutions of the Navier–Stokes equations describing the fluid motions is an unsolved problem when nonlinear advection becomes dominant, i.e. in the fully developed turbulent regime. We should mention here recent mathematical results which give, using multi-scale (Paley–Littlewood) decomposition, a global existence theorem [35] and a global unicity theorem [98] for Navier–Stokes equations in $\mathbb{R}^3$ if initial conditions are sufficiently oscillating (in a Besov norm sense). Some other mathematical attempts have been made using divergence free vector wavelets [86], [19], but in all cases these proofs are done in an unbounded space. However, physical fluid flows are bounded either internally or externally, and we still do not know what is the optimal functional space for describing real turbulent flows.

In summary, the theory of fully developed turbulence is in what we may call a pre-scientific phase, because we do not yet have an equation, nor a set of equations, that could be used to efficiently compute turbulent flows. The incompressible Navier–Stokes equations, which are the fundamental equations of fluid mechanics, are not the right ones for turbulence because their computational complexity becomes intractable for large Reynolds number flows. However, in this limit it should then be possible, as it is done in statistical mechanics, to define averaged quantities which would be the appropriate variables to describe turbulence and then find the corresponding transport equations to compute the evolution of these new quantities. Likewise, the Navier–Stokes equations can be derived from the Boltzmann equation by considering appropriate limits (Knudsen and Mach numbers tending to zero [11], [12]) and appropriate averaging procedures to define new coarse-grained variables (velocity and pressure) and associated transport coefficients (viscosity and density); the turbulence equations should be derived as a further step in this hierarchy of embedded approximations, but this scientific programme may be impaired by the possible non-universality of turbulence, which remains an essential question to address.

More precisely, it is easier to define the appropriate parameters to go from Boltzmann to Navier–Stokes than from Navier–Stokes to turbulence equations. In the first case only a linear averaging procedure is needed, while in the second case we have to find an appropriate nonlinear procedure, namely some conditional averaging which depends on each flow realization. For this we should first identify the dynamically active structures constituting turbulent flows, classify their elementary interactions and define the averaging procedures needed to construct appropriate statistical observables. Wavelet analysis is a good tool for exploring this conditional averaging and for seeking an atomic decomposition of phase space, defined in both space and scale. Tennekes and Lumley in 1972 [191] had already the intuition of such a phase-space decomposition when they proposed to consider a turbulent flow as a superposition of Gaussian-shaped wave packets, they were calling 'eddies'; but we know since Balian's theorem [10] that we cannot build orthogonal bases with such functions. This is why we propose to use instead wavelet or wavelet packet bases to study how phase-space 'atoms' exchange energy, or other important dynamical quantities, during the flow evolution and possibly combine to form phase-space 'molecules', such as coherent structures.

Wavelets may supply new functional bases better adapted to represent and compute turbulent flows, i.e. to extract their elementary dynamical entities, perform the appropriate averages on them, and predict the evolution of these statistical quantities. We still hope that there will be enough universality in the behaviour of these phase-space 'molecules' so that we can find a general theory and a set of equations to describe their evolution, but this may well be an unrealistic goal.

4.2.2 Navier–Stokes equations

The fundamental equations of the dynamics of an incompressible (constant density) and Newtonian (rate of strain proportional to velocity gradients) fluid are the Navier–Stokes equations:

$$\frac{\partial v}{\partial t} + (v \cdot \nabla)v + \frac{1}{\rho} \nabla P = \nu \nabla^2 v + F, \qquad (4.1)$$

$$\nabla \cdot v = 0, \qquad (4.2)$$

plus initial and boundary conditions,

where t is the time, v the velocity, P the pressure, F the resultant of the external forces per unit of mass, ρ a constant density and ν a constant kinematic viscosity.

The mathematical difficulty of the Navier–Stokes equations arises from the fact that the small parameter ν, which tends to zero in the limit of infinite Reynolds numbers, i.e. for fully developed turbulent flows, appears in the term containing the highest-order derivative, namely the dissipation term $\nu \nabla^2 v$. Thus the character of the equations changes as ν tends to zero, since in this limit it is the nonlinear advection term $(v \cdot \nabla)v$ which dominates. This singular limit seems similar to the semi-classical limit of quantum mechanics when the Planck's constant tends to zero; incidentally Planck's constant has the same dimension as kinematic viscosity. When $\nu = 0$, i.e. for infinite Reynolds numbers, the Navier–Stokes equations are called Euler's equations.

One of the physical difficulties of the Navier–Stokes equations comes from the incompressibility condition, namely the divergence-free requirement imposed by equation (4.2), which implies that the speed of sound is infinite. In this case any local perturbation is instantaneously transmitted throughout the whole domain. This requirement seems too drastic and quite unphysical because the speed of sound is large in real flows but never infinite. In the future we may prefer to consider instead weakly compressible Navier–Stokes equations to simplify the computation of turbulent flows and represent their local behaviour more accurately. Moreover, on physical grounds Euler's equations are unrealistic because the limit $\nu = 0$ contradicts the fluid hypothesis, which supposes that the system is locally close to thermodynamical equilibrium due to molecular collisions (which implies macroscopic dissipation).

Taking the curl of equations (4.1) and (4.2) gives the equation of vorticity ω, the curl of velocity,

$$\frac{\partial \omega}{\partial t} + (v \cdot \nabla)\omega = (\omega \cdot \nabla)v + \nu \nabla^2 \omega + \nabla \times F. \qquad (4.3)$$

In three dimensions this equation shows that vortex tubes may be stretched by velocity gradients, a mechanism which has been proposed to explain the transfer of energy towards the smallest scales of the flow. In two dimensions the vortex stretching term becomes zero, because the vorticity is then a pseudo-scalar $\omega = (0, 0, \omega)$ perpendicular to the velocity gradients. The vorticity, and its infinitely many moments, are therefore Lagrangian invariants of the flow (Helmholtz's theorem). In this case there is no vortex stretching and energy cannot cascade towards the smallest scales, but tends to accumulate into the largest scales, the so-called inverse energy cascade [121], [15], while enstrophy (vorticity squared) instead cascades towards the smallest scales where it accumulates.

4.2.3 Statistical theories of turbulence

The first statistical method to analyse turbulent flows was proposed in 1894 by Reynolds [174] who assumed that turbulent flows can be separated into mean fields and fluctuations. He decomposed the velocity field $v(x)$ into a mean contribution $\bar{v}_i$ plus fluctuations v_i' and rewrote the Navier–Stokes equations to predict the evolution of $\bar{v}_i$, which gives the Reynolds equations

$$\frac{\partial \bar{v}_i}{\partial t} + \bar{v}_j \frac{\partial \bar{v}_i}{\partial x_j} + \frac{1}{\rho} \frac{\partial \bar{P}}{\partial x_i} = \frac{\partial}{\partial x_j} \left(\nu \frac{\partial \bar{v}_i}{\partial x_j} - \overline{v_i' v_j'} \right) + \bar{F}_i. \tag{4.4}$$

To obtain the time evolution of the mean velocity $\bar{v}_i$ one should compute the second order moment of the velocity fluctuations $\overline{v_i' v_j'}$, called the Reynolds stress tensor, which in fact depends on the third order moment $\overline{v_i' v_j' v_k'}$ (i, j, and k are dummy indices), which depends on the fourth order moment, and so on *ad infinitum*. This is the closure problem: there are more unknowns than equations and, to solve the hierarchy of Reynolds equations, the traditional strategy is to introduce another equation, or system of equations, chosen from some *a priori* phenomenological hypotheses, to close the set of equations.

For instance, to close the hierarchy of Reynolds equations, Prandtl introduced a characteristic length scale for the velocity fluctuations, called the mixing length, which led him to rewrite the Reynolds stress tensor term as a turbulent diffusion. Following an hypothesis proposed by Boussinesq [30], and by analogy with molecular diffusion which smoothes velocity gradients for scales smaller than the molecular mean free path, Prandtl assumed that there exists a turbulent diffusion which regularizes the mean velocity gradients for scales smaller than the mixing length. Unfortunately this hypothesis is wrong because, contrary to molecular diffusion which is decoupled from the large scale motions and can then be modelled by a linear operator (Laplacian) with an appropriate transport coefficient (viscosity), turbulent motions interact nonlinearly at all scales and there is no spectral gap to decouple large scale motions from small scale motions. This is a major obstacle faced by all turbulence models and therefore the closure problem remains open. This is also the reason why renormalization group techniques [207], nonlinear Galerkin numerical methods [149] and Large Eddy Simulation (LES) [135] have not yet lived up to their promises. An important direction of research is to find a new representation of turbulent flows in which there is a gap, decoupling motions out of equilibrium from well thermalized motions, which can then be modelled. Such a separation seems only possible with a nonlinear closure, based on conditional averages which depend on the local behaviour of each flow realization. We have proposed to use nonlinear wavelet filters for this (see section 4.5.2).

Taylor [189], under the influence of Wiener with whom he was in correspondence [18] since his famous paper on turbulent diffusion [188], proposed in 1935 to characterize turbulent fields by their correlation functions, in particular by the Fourier transform of their two-point correlation function which gives their energy spectrum. This relies on Wiener–Khinchin's theorem, which states that the modulus of the Fourier transform of one realization of a stationary and ergodic random process in $\mathbb{R}^n$ is the same as the Fourier transform of the two-point correlation function of this process. Twenty years before, Einstein [62] had outlined the same method to characterize fluctuating data, but he was not followed at the time [206]. To simplify the computation of correlation functions, Taylor made the hypothesis of statistical homogeneity and isotropy of turbulent flows, supposing that the averages are invariant under both translation and rotation. In the 1930s Gebelein proposed applying the probability theory of Kolmogorov to hydrodynamics, a method later developed by Kolmogorov himself and his student Obukhov [161], who published in 1941 three key papers on the statistical theory of fully developed turbulence. Kolmogorov [117], [118], [119] studied the way in which the energy density of the two-point correlation of a turbulent flow in three dimensions is distributed among the different wavenumbers. This type of approach is common in statistical mechanics, but a difficulty arises here from the fact that turbulent flows are open thermodynamical systems, due to the injection of energy by external forces and its dissipation by viscous frictional forces. To resolve this difficulty Kolmogorov supposed that external forces act only on the largest scales while frictional forces act only on the smallest scales, which, in the limit of very large Reynolds numbers, leaves an intermediate range of scales, called the inertial range, in which energy is conserved and only transferred from large to small scales at a rate ϵ which is supposed to be constant. But this cascade of energy is supposed for ensemble averages and not for an individual flow realization; moreover, this cascade hypothesis is only phenomenological and has never been proved from first principles. Following Taylor [189], Kolmogorov supposed that turbulent flows are statistically homogeneous and isotropic; as a consequence of these two hypotheses and using Navier–Stokes equations, von Karman and Howarth [113] have shown that the skewness, namely the departure from Gaussianity of the velocity increment probability distribution, is a non-zero constant. All these assumptions lead Kolmogorov to propose the K41 model, which predicts the following energy spectrum scaling, known as the $k^{-5/3}$ law

$$E(k) = C\epsilon^{2/3}k^{-5/3} \qquad (4.5)$$

where k is the modulus of the wavenumber averaged over directions and C is called Kolmogorov's constant. Classically in turbulence k is interpreted as the inverse of a scale, but this is only true for averaged fields of statistically homogeneous and isotropic flows.

Landau criticized Kolmogorov's hypothesis of a constant rate of energy transfer ϵ independent of the scale, arguing that the dissipation field should also be considered random. Following this remark, and due to observational evidence of small-scale intermittency introduced by Townsend in 1951 ([194], [195]), Kolmogorov proposed to model the energy transfer as a multiplicative random process where only a fraction β of energy is transferred from one scale to another. Assuming that the probability density of the dissipation field varies randomly in space and time with a log-normal law, this led him to propose the K62 model which predicts the following energy spectrum scaling

$$E(k) = C\epsilon^{\frac{2}{3}}k^{-\frac{5}{3}}\ln\left(\frac{k}{k_I}\right)^{\beta} \qquad (4.6)$$

where k_I is the wavenumber at which energy is injected (inverse of the integral length scale).

Kolmogorov 1962's paper opened a debate, which is still very lively today, but which was already very well addressed 24 years ago by Kraichnan [122] when he wrote in 1974:

The 1941 theory is by no means logically disqualified merely because the dissipation rate fluctuates. On the contrary, we find that at the level of crude dimensional analysis and eddy-mitosis picture the 1941 theory is as sound a candidate as the 1962 theory. This does not imply that we espouse the 1941 theory. On the contrary, the theory is made implausible by the basic physics of vortex stretching. The point is that this question cannot be decided *a priori*; some kind of non-trivial use must be made of the Navier–Stokes equation.

Kraichnan claims that one needs to understand the generic dynamics of Navier–Stokes equation before constructing a statistical theory able to take into account intermittency:

If the Kolmogorov law $E(k) \propto k^{-5/3-\mu}$ is asymptotically valid, it is argued that the value μ depends on the details of the nonlinear interaction embodied in the Navier–Stokes equations and cannot be deduced from overall symmetries, invariances and dimensionality [122].

To his criticism of Kolmogorov 62's theory, Kraichnan added:

Once the 1941 theory is abandoned, a Pandora's box of possibilities is open. The 1962 theory of Kolmogorov seems arbitrary, from an a priori viewpoint [. . .]. We make the point that even in the general framework of some kind of self-similar cascade, and of intermittency which increases with the number of cascade steps, the 1962 theory is only one of many possibilities [122].

Kraichnan also commented on the fact that Kolmogorov 41's theory has proved to be valid even in cases where its hypotheses are not satisfied:

Kolmogorov's 1941 theory has achieved an embarrassment of success. The -5/3 spectrum has been found not only where it reasonably could be expected, but also at Reynolds numbers too small for a distinct inertial range to exist as in boundary layers and shear flows where there are substantial departures from isotropy, and such strong effects from the mean shearing motion that the stepwise cascade appealed to by Kolmogorov is dubious [122].

For two-dimensional turbulence there is a statistical theory similar to Kolmogorov's theory which has been proposed by Kraichnan in 1967 [121] and then developed by Batchelor in 1969 [15]. This theory takes into account, in addition to the conservation of energy in the inertial range, the conservation of enstrophy (integral of vorticity squared), which is true only in dimension two. Making the same kind of hypotheses as Kolmogorov, they predicted a direct enstrophy cascade, from large to small scales, giving a k^{-3} energy spectrum, and an inverse energy cascade, from small to large scales, giving a $k^{-5/3}$ energy spectrum. The problem is that the energy spectra obtained from numerical simulations are in most cases steeper than the predicted k^{-3}. There is another more recent statistical theory proposed by Polyakov [170] which takes into account, in addition to the energy conservation, the conservation of infinitely many moments of vorticity in two dimensions, which led him to predict different scaling laws depending on the way energy is injected; thus, Polyakov's theory is not universal. In fact the same non-universal behaviour of forced two-dimensional turbulence is also observed in numerical simulations [128].

Since the pioneering works of Onsager [163] and Joyce & Montgomery [111], there are several statistical theories for decaying two-dimensional turbulence [178], [152], [179], [54], [180], [67] which are not based on structure functions nor Fourier representation. These theories, unlike those of Kraichnan and Polyakov, do not discard the spatial flow structure. For a recent review of these theories a good reference is [148]. Onsager's theory assumes that all vorticity is concentrated into a finite number of point vortices and predicts that there exist negative temperature states; more precisely it predicts that high energy states can be favoured compared to low energy states, contrary to classical statistical physics. These negative temperature states correspond to the clustering of same-sign vortices characteristic of the inverse energy cascade of two-dimensional turbulence. But the extension of Onsager's approach to describe continuous vorticity fields, involving infinite number of degrees of freedom and therefore infinite Liouville measure, leads to a highly singular limit which has been overcome only recently using large deviation probabilities and maximum entropy techniques. This new theory, due independently to Robert [179], [180] and Miller [152], predicts for decaying 2D turbulent flows (i.e. in the absence of external forces) final stationary states characterized by a functional relation between coarse-grained vorticity and streamfunction. This relation is called the coherence function and it seems to be verified for strong mixing situations, such as two-dimensional shear layers or vortex merging [187].

In the case of 3D forced homogeneous turbulent flows Chorin proposed a new statistical theory [45], [46], which is a generalization to 3D of the 2D vortex equilibrium theory initiated by Onsager [163]. The small-scale structure is described as a perturbation of an ensemble of vortices in thermal equilibrium (by 'equilibrium' Chorin means 'Gibbsian equilibrium' and not 'statistical steady state'). This theory recovers the Kolmogorov spectrum and proposes an explanation for the origin of intermittency.

4.2.4 Coherent structures

Since the beginning of turbulence research there has been, alongside the statistical approach based on ensemble averages, a tendency to analyse each flow realization separately. This led to the recognition that turbulence contains coherent structures, even at very large Reynolds numbers [110]. Examples of coherent structures include the vortices observed by Roshko in 1961 at a Reynolds number of 10^7 [182], the horseshoe vortices observed in turbulent boundary layers and mixing layers [38], [181], and the vorticity tubes (often called filaments) [49], [32] observed in statistically homogeneous flows. Coherent structures are defined as local condensations of the vorticity field which survive for times much longer than the eddy turnover time characteristic of the turbulent fluctuations.

The vorticity field is easy to visualize in numerical experiments, but very difficult to visualize in laboratory experiments; therefore, one usually observes the pressure field instead. Indeed, if we take the divergence of equation (4.1) we obtain

$$2\nabla^2 P/\rho + s^2 - \omega^2 = 2\nabla \cdot \boldsymbol{F}, \tag{4.8}$$

where $s^2 = \frac{1}{2}(\partial_i v_j + \partial_j v_i)$ is the rate of strain which controls dissipation. This equation shows that vorticity concentrations, corresponding to coherent structures, are sources of low pressure, while strained regions, corresponding to dissipation, are sources of high pressure. Couder *et al.* [49], [32] recently measured the probability distribution function (estimated through a histogram) of pressure and showed that for the large negative pressures it is exponential, while for the pressures around zero it is Gaussian. In other words, the coherent structures, which are characterized by strong depressions, are responsible for the non-Gaussian behaviour of turbulent flows, which is consistent with observations made before by Van Atta and Antonia [197] from measurements of the spatial gradients of velocity. This has also been shown by Abry *et al.* [2], [3] using wavelet techniques to separate the coherent structures from the background flow in a one-dimensional cut of pressure signal.

The mere existence of finite (and quite small) number of coherent structures [203] may invalidate the ergodic hypothesis, which is an essential ingredient of any statistical theory, necessary to replace ensemble averages by space averages. Then, according to Taylor's hypothesis, which requires that fluctuating velocities should be much smaller than the mean velocity, space averages can be replaced by time averages, which are easier to obtain in laboratory experiments. As far as we know, almost all existing laboratory results measuring the turbulence energy spectrum rely on Taylor's hypothesis. We are therefore sceptical of their validity when the coherent structures produce rare but intense velocity fluctuations. In this case, even though the the velocity fluctuations remain in average small compared to the mean velocity, coherent structures produce bursts which exceed the mean value and it is dubious that time and space averages can then be interchanged.

Concerning numerical experiments, we interpret the energy spectrum, and its inertial range power-law form, as characteristic of the random processes responsible for turbulence. However in practice we analyse only one flow realization because in most simulations the correlation length is of the order of the size of the computational periodic domain. In this case a power-law behaviour could be interpreted as indicating the presence of some quasi-singular structures in the flow, and not as a proof of its random dynamics. This new point of view led Saffman [183] to interpret the energy power-law behaviour as resulting from the presence of vorticity fronts. Later Farge and Holschneider [76] proposed another interpretation based on the emergence of cusp-like coherent structures. In the limit of an infinite Reynolds number, these vorticity cusps will tend to point vortices, which correspond to the limit case of negative temperature states [34]. The wavelet transform, because it measures the local scaling of a field, is the appropriate tool for verifying these different interpretations in relating the power-law scaling of the energy spectrum to the shape of possible singularities.

Today we still do not have a complete theory to explain the formation and persistence of coherent structures, and we shall have to content ourselves with a qualitative description of their behaviour. This is more evidence that we may still be in a pre-scientific phase, having as yet only a limited grasp of the nature of turbulence. The new point of view is to consider that coherent structures are generic to turbulent flows, even at very high Reynolds numbers, and that they probably play an essential role in their intermittency. Indeed, several wind tunnel experiments [17], [5] have shown that the energy associated with the smallest scales of turbulent flows is not distributed densely in space and time. This has led various authors to conjecture that the support of the set on which dissipation occurs should be fractal [147], [92], or multifractal [164]. It is now thought, but not proven, that the time and space intermittency of turbulent flows is related to the presence of coherent structures [75]. This is still an open question and wavelet analysis seems to be one of the appropriate techniques to answer it.

The classical theory of turbulence is blind to the presence of coherent structures because their spatial support is small in the inertial range. Therefore low-order statistical moments are insensitive to them and characterize only the background flow whose spatial support is on the contrary dense in the inertial range. Moreover, in three-dimensional flows coherent structures (vorticity tubes often called filaments) are highly unstable [49] and therefore their temporal support is also small. Consequently, the presence of coherent structures only affects the high-order statistical moments of the velocity increments which are more sensitive to rare and extreme events (large deviations). The high-order structure functions have been measured only recently [5], because their calculation requires very long data sequences.

They do not follow Kolmogorov's theory which predicts a linear dependence of the scaling exponent of the velocity structure functions on their order. Van der Water [199] has observed that there are in fact two distinct nonlinear dependencies for odd and for even orders, which may be interpreted in terms of the multi-spiral model of Vassilicos [198].

It is important to provide statistical predictions based on coherent structure models. It has been shown by Min, Mezic and Leonard [153] that a system of singular vortex elements in two dimensions and three dimensions possesses statistics that deviate from Gaussian and that the probability density functions (PDFs) of velocity derivatives are non-Gaussian with a Cauchy distribution. The experimental evidence of similar findings is contained in the work of Goldburg and collaborators [193] in which the Cauchy distribution, predicted in [153] as a consequence of $1/r$ velocity decay of a singular vortex, is seen for the region of small velocity differences. The results of [153] also indicate that the tails of PDFs are determined by the structure of vortex cores.

In conclusion, we have shown [83] that the presence of coherent structures is responsible for the non-Gaussian statistics of fully developed turbulent flows in two dimensions, and we conjecture that this will still be valid in three dimensions. Due to the sensitivity to initial conditions of turbulent flows, any theory of turbulence should be statistical. But, before being able to construct a new statistical theory of turbulence, we need to find new types of averages able to preserve the information associated with coherent structures and therefore take into account the intermittency of turbulent flows. Wavelets can play a role there in separating the coherent (non-Gaussian) components from the incoherent (Gaussian) components of turbulent flows, in order to devise new conditional averages to replace the classical ensemble averages. This method will lead to new turbulence models based on the fact that the coherent components, namely the vortices, are out of statistical equilibrium, while on the contrary we can define a Gaussian equilibrium state for the incoherent components which correspond to the well-mixed background flow. Therefore this method to compute turbulent flows combines a deterministic approach, to solve the dynamical system describing the vortex motions, and a statistical approach, to model the effect of the background flow.

4.3 Fractals and singularities

4.3.1 Introduction

According to Kolmogorov's K41 model, turbulence in the inertial range has a power law energy spectrum (4.5), and thus does not have a characteristic length scale. Therefore turbulence in this range of length scales looks similar at any magnification and can be described as self-similar. According to experimental observations, however, turbulence is also characterized by quasi-singular structures such as vortices and is intermittent (quantities such as energy dissipation vary greatly in time and space). A quasi-singular structure is one that appears singular until the dissipation scale at which the smoothing effect of viscosity becomes important. In fact the theoretical $k^{-\frac{5}{3}}$ inertial range energy spectrum predicted by Kolmogorov's theory implies that some sort of quasi-singular distribution of velocity and vorticity must be present in turbulent flows [106], [154], [101]. This quasi-singular distribution could be the result of a set of quasi-singular structures (e.g. vortices), or due to a particular statistical distribution of structures (independently of their smoothness). One of the difficulties in turbulent flow analysis is how to disentangle these different contributions from the overall statistics.

It remains an open question whether this quasi-singular behaviour is due to the randomness of turbulent motions resulting from their chaotic dynamics or to the presence of localized quasi-singular structures resulting from an internal organization of the turbulent motions. Kolmogorov's theory is based on ensemble averages, but in using them we are unable to disentangle these two hypotheses. Ensemble averages should be replaced by an analysis of turbulence for each realization and be based on the local measurements and statistics of singularities for which we need effective ways of detecting and characterizing quasi-singularities in turbulent signals.

The types of possible singularities in the turbulent velocity or vorticity may be divided into two classes: *cusps* (i.e. non-oscillating singularities in which the function or one of its derivatives approaches infinity at a certain point, e.g. $1/x$) and *spirals* (i.e. oscillating singularities in which the frequency of oscillation approaches infinity at a certain point, e.g. $\sin(1/x)$). Figure 4.6 shows an example of a two-dimensional flow containing both cusps and a spiral (a cut through the spiral is an oscillating singularity over a certain range of length scales.) Likewise the distribution of singularities in turbulence may also be divided into two classes: *isolated* (singularities at a finite number of points) and *dense* (singularities at an infinite number of points in a finite area). Dense distributions of singularities are called fractals and are characterized by one (monofractal) or more (multifractal) fractal dimensions characterizing their scaling properties. Figure 4.1a shows a typical fractal signal. Note that fractals may contain both cusp and spiral type singularities. Turbulence might contain both fractal and isolated distributions of singularities, and spiral and cusp types of singularities. Figure 4.1b shows a spiral

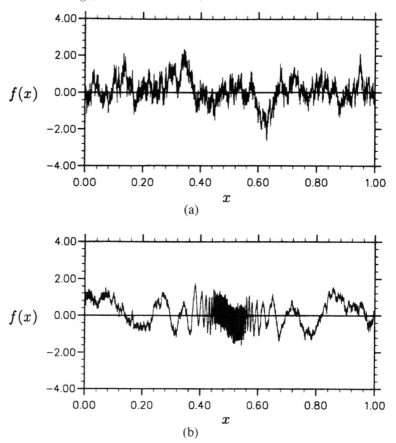

Fig. 4.1. Different types and combinations of singularities. (a) A fractal signal with energy spectrum $E(k) \propto k^{-\frac{5}{3}}$. (b) A spiral with fractal noise (both noise and spiral have the same energy spectrum $E(k) \propto k^{-\frac{5}{3}}$).

type singularity with fractal noise superimposed; both the noise and the spiral have the same energy spectrum scaling in $k^{-\frac{5}{3}}$.

This section is concerned with wavelet-based techniques for calculating quantities such as energy spectra, structure functions, singularity spectra and fractal dimensions. These subjects are connected by the fact that they all measure the local regularity of the signal (i.e. the strength of singularities in the signal). For example, the slope of the usual Fourier energy spectrum of a signal containing only isolated cusp singularities is determined by the strongest singularity [211]. The advantage of the wavelet transform is that it is able to analyse locally the singular behaviour of a signal. One can then use this local information to construct statistics describing the distribution and type of singularities (e.g. multifractals), and define local or conditionally averaged versions of traditional diagnostics such as the energy spectrum and structure functions. We are primarily concerned with cusp type singularities (either isolated or dense), although we also discuss methods for distinguishing between signals containing isolated spirals and dense fractal signals.

In subsection 4.3.2 we review the mathematical results on one of the key properties of wavelet transforms: their ability to detect and characterize singular structures. We then describe three related applications which rely on this property: calculation of local energy spectra, structure functions (subsection 4.3.3) and the singularity spectra which characterize multifractals (subsection 4.3.5). These wavelet methods generally require the assumption that the singularities of the signal are isolated. Because isolated spirals are likely to be present in turbulence (see on Figure 4.6 the production of spiraling vorticity filaments by vortex merging) it is essential to have a method of determining which sort of singularity a signal contains. In subsection 4.3.6 we review a different wavelet-based method for distinguishing between signals containing isolated spirals and purely fractal signals (the two types of signal most likely to be observed in a turbulent flow). Each section gives a practical review of the method and briefly summarizes some results that have been obtained for turbulence data. Formulating these techniques in terms of wavelet transforms brings out the connections between them as well as providing new information, and this point is emphasized throughout this section.

4.3.2 Detection and characterization of singularities

The most useful property of the wavelet transform is its ability to detect and accurately measure the strength (given by the Hölder exponent) of individual singularities in a signal. We will first give a definition of the Hölder exponent. A function $f(x)$, such that

$$f : \mathbb{R} \to \mathbb{R} \tag{4.8}$$

is said to belong to the Hölder space C^α for α a positive non-integer if there exists a constant C such that, for each x_0, there exists a polynomial P of order less than α such that

$$|f(x) - P(x - x_0)| \leq C|x - x_0|^\alpha. \tag{4.9}$$

f is said to have the Hölder exponent $\alpha(x_0)$ at point x_0 if $\alpha(x_0) = \sup\{\theta > 0 / f \in C^\theta(x_0)\}$. The exponent $\alpha(x_0)$ therefore measures the smoothness of the function $f(x)$ near x_0: the larger $\alpha(x_0)$ is, the smoother or more regular the function $f(x)$ is near x_0, while the smaller $\alpha(x_0)$ is, the rougher or more singular the function is. If the Hölder exponent is less than one, there is an actual singularity of the function at x_0 (or a quasi-singular behaviour near x_0 over a certain range of length scales if one is measuring a physical quantity like vorticity or velocity).

It is important to note that equation (4.9) does not hold for oscillating singularities because in this case the Hölder exponent increases by more than one when the function is integrated. This anomalous behaviour is due to the fact that there are an infinite number of accumulating oscillations in the neighbourhood of such a singularity.

Consider the L^1 norm wavelet transform (which conserves the L^1 norm of a function)

$$\tilde{f}_1(x, r) = \frac{1}{r} \int_{-\infty}^{\infty} f(x') \psi\left(\frac{x' - x}{r}\right) dx'. \tag{4.10}$$

The wavelet transform is thus a two-dimensional function in position x and scale $r > 0$. Mallat and Hwang [146] have shown that singularities in $f(x)$ produce a maximum in the modulus of the wavelet transform $|\tilde{f}_1(x, r)|$ and that following the position of a wavelet modulus maximum as $r \to 0$ gives the position x_0 of the singularity. Furthermore, each singularity has an associated 'influence cone' defined by

$$|x - x_0| \le Cr, \tag{4.11}$$

and, if the singularity is an isolated cusp, then the wavelet transform modulus for all points within the influence cone is

$$|\tilde{f}_1(x, r)| \le Ar^{\alpha(x_0)}, \tag{4.12}$$

provided that at least the first $n > \alpha(x_0)$ moments of the analysing wavelet $\psi(x)$ vanish, where the nth moment is defined by the integral

$$\int_{-\infty}^{+\infty} x^n \psi(x) \, dx. \tag{4.13}$$

Equation (4.12) shows that the Hölder regularity $\alpha(x_0)$ can be found from the slope of the graph of $\log |\tilde{f}_1(x, r)|$ versus $\log r$ at a position x satisfying inequality (4.11). When several singularities are present only the non-overlapping parts of the cones associated with each singularity satisfy (4.12). Intuitively, it is the self-similar scaling property of the wavelet which allows the wavelet transform to measure the rate of self-similar narrowing with decreasing scale, characterizing the strength of a cusp singularity.

If the singularity is not isolated and there is only one zero-crossing of the wavelet transform near x_0, one can find the regularity in the left and right neighbourhoods of x_0 by measuring the decay of the wavelet coefficient modulus along maxima lines of the wavelet transform to the left and right of the influence cone of x_0.

In practice, such graphs of $\log |\tilde{f}_1(x, r)|$ versus $\log r$ contain oscillations superimposed on the power-law behaviour which can make it difficult to determine the slope at larger scales. Vergassola and Frisch [200] showed that these oscillations are necessarily present for any self-similar random process whether or not the signal is multifractal (the lacunarity of multifractal signals should also produce oscillations). These oscillations can be reduced by finding the average decay of the wavelet coefficient modulus along many lines in the influence cone, or by averaging the decay along vertical lines at many different points (e.g. one may be interested in the conditionally averaged scaling of points in regions of irrotational straining, see Figure 4.2). Arnéodo, Bacry and Muzy [6] have suggested that the deviations from a strict power-law may be reduced by measuring the decay of the modulus of the wavelet transform along the line of maximum modulus within the influence cone.

The analysis of signals containing spiral singularities, either isolated (e.g. $\sin(1/|x - x_0|)$) or fractal (e.g. the Riemann–Weierstrass function), is more complicated because the worst singular behaviour of a spiral singularity appears outside the cone of influence. In this case one measures the decay as $r \to 0$ of the modulus of the wavelet transform along the set of points which are general maxima outside the cone of influence (i.e. maxima in both the position and scale directions). This gives an upper bound on the Hölder exponent, but in general one has to use lines of maximum modulus both inside and outside the cone of influence to fully determine the singular behaviour of an oscillating singularity.

Arnéodo, Bacry and Muzy [7] have recently carried out work defining two wavelet-based exponents that measure the strength of an oscillating singularity. They find that the faster the frequency increases, the more irregular its derivative. In general, oscillating behaviour appears in fractal objects that are self-similar under non-hyperbolic mappings, e.g. the Riemann–Weierstrass function or the Farey-tree partitioning of rationals.

4.3.3 Energy spectra

The Fourier energy spectrum has been one of the most popular techniques for turbulence analysis, indeed traditional turbulence theory was constructed in Fourier space [16]. The energy spectrum $E(k)$ of a one-dimensional function $f(x)$ is the Fourier transform of its two-point correlation, which is equal (Wiener–Khinchin's theorem) to

$$E(k) = \frac{1}{2\pi} |\hat{f}(k)|^2 \quad \text{for} \quad k \geq 0 \tag{4.14}$$

where $(\hat{\ })$ signifies Fourier transform. Note that when analysing turbulence velocity signals one should ensemble average the energy spectra from many realizations. In practice, one assumes ergodicity and averages only one flow realization split into many pieces whose lengths are larger than the integral scale (which is the largest correlated scale in a turbulent signal). In traditional turbulence theory only the modulus of the Fourier transform is used (e.g. the energy spectrum) and thus the phase information is lost. This is probably a major weakness of the traditional way of analysing turbulence since it neglects any spatial organization of the turbulent velocity field.

The wavelet transform extends the concept of energy spectrum so that one can define a local energy spectrum $\tilde{E}(x, k)$ using the L^2 norm wavelet transform (which conserves the L^2 norm of a function) rather than the L^1 norm used in subsection 4.3.2 (i.e. the wavelet transform is normalized by $1/r^{\frac{1}{2}}$ rather than by $1/r$ and the resulting function is designated by $\tilde{f}$ instead of $\tilde{f}_1$)

$$\tilde{E}(x, k) = \frac{1}{2c_\psi k_0} \left| \tilde{f}\left(x, \frac{k_0}{k}\right) \right|^2 \quad \text{for} \quad k \geq 0 \tag{4.15}$$

where k_0 is the peak wavenumber of the analysing wavelet ψ and

$$c_\psi = \int_0^{+\infty} \frac{|\hat{\psi}(k)|^2}{k^2} \, d^2k. \tag{4.16}$$

By measuring $\tilde{E}(x, k)$ at different places in a turbulent flow one might estimate what parts of the flow contribute most to the overall Fourier energy spectrum and how the energy spectrum depends on local flow conditions. For example, one can determine the type of energy spectrum contributed by coherent structures, such as isolated vortices, and the type of energy spectrum contributed by the unorganized part of the flow called background flow.

Since the wavelet transform analyses the flow into wavelets rather than sine waves it is possible that the mean wavelet energy spectrum may not always have the same slope as the Fourier energy spectrum. Perrier, Philipovitch and Basdevant [166] have shown, however, that the mean wavelet spectrum $\tilde{E}(k)$

$$\tilde{E}(k) = \int_0^{+\infty} \tilde{E}(x, k) \, dx \tag{4.17}$$

gives the correct Fourier exponent for a power-law Fourier energy spectrum $E(k) \propto k^{-\beta}$ provided that the analysing wavelet has at least $n > (\beta - 1)/2$

vanishing moments. This condition is obviously the same as that for detecting singularities derived in the previous section since $\beta = 1 + 2\alpha$ for isolated cusps. Thus, the steeper the energy spectrum the more vanishing moments of the wavelet we need. The inertial range in turbulence has a power-law form. The ability to correctly characterize power-law energy spectra is therefore a very important property of the wavelet transform (which is of course related to its ability to detect and characterize singularities).

Note that if the singularities are all isolated then the exponent of the Fourier energy spectrum is determined by the strongest singularity α of the signal [211]

$$E(k) = Ck^{-2(\alpha+1)}, \tag{4.18}$$

where C is a constant. If the singularities are spirals and/or are not isolated then the strongest singularity sets a lower bound on the exponent of the energy spectrum [211]

$$E(k) \leq Ck^{-2\alpha}. \tag{4.19}$$

The way the dense singularities accumulate can make the signal effectively more singular, decreasing the magnitude of the exponent of the energy spectrum by up to 2. Because they are both controlled in the same way by singularities, the wavelet energy spectrum can be thought of as a sort of local Fourier transform.

The mean wavelet energy spectrum $\tilde{E}(k)$ is a smoothed version of the Fourier energy spectrum $E(k)$. This can be seen from the following relation between the two spectra

$$\tilde{E}(k) = \frac{1}{2c_\psi k_0} \int_0^{+\infty} E(k') \left| \hat{\psi}\left(\frac{k_0 k'}{k}\right) \right|^2 dk' \tag{4.20}$$

which shows that the mean wavelet spectrum is an average of the Fourier spectrum weighted by the square of the Fourier transform of the analysing wavelet shifted at wavenumber k. Note that the larger k is, the larger the averaging interval, because wavelets are passband filters at $\frac{\Delta k}{k}$ constant. This property of the mean wavelet energy spectrum is particularly useful for turbulent flows. The Fourier energy spectrum of a single realization of a turbulent flow is too spiky to be useful, but one can measure a well-defined slope from the mean wavelet energy spectrum.

The Mexican hat wavelet

$$\hat{\psi}(k) = k^2 \exp(-k^2/2) \tag{4.21}$$

has only two vanishing moments and thus can correctly measure energy spectrum exponents up to $\beta < 5$. Only the zeroth order moment of the Morlet wavelet

$$\hat{\psi}(k) = \frac{1}{2\pi}\exp(-(k-k_\psi)^2/2) \quad \text{for} \quad k > 0$$

$$\hat{\psi}(k) = 0 \text{ for } k \leq 0 \tag{4.22}$$

is zero, but the higher n^{th} order moments are very small ($\propto k_\psi^n \exp(-k_\psi^2/2)$) provided that k_ψ is sufficiently large. Therefore the Morlet wavelet transform should give accurate estimates of the power-law exponent of the energy spectrum at least for approximately $\beta < 7$ (if $k_\psi = 6$).

Perrier, Philipovitch and Basdevant [166] present a family of new wavelets with an infinite number of cancellations

$$\hat{\pi}_n(k) = \alpha_n \exp\left(-\frac{1}{2}\left(k^2 + \frac{1}{k^{2n}}\right)\right), \quad n \geq 1, \tag{4.23}$$

where α_n is chosen for normalization. The wavelets defined in (4.23) can therefore correctly measure any power-law energy spectrum. Furthermore, these wavelets can detect the difference between a power-law energy spectrum and a Gaussian energy spectrum ($E(k) \propto \exp(-(k/k_0)^2)$). It is important to be able to determine at what wavenumber the power-law energy spectrum becomes exponential since this wavenumber defines the end of the inertial range of turbulence and the beginning of the dissipative range.

The first measurements of local energy spectra in three-dimensional turbulence were reported by Farge *et al.* [75] and Meneveau [150]. Meneveau used the discrete wavelet transform to measure local energy spectra in experimental and Direct Numerical Simulation (DNS) flows and found that the standard deviation of the local energy (a measure of the spatial fluctuation of energy) was approximately 100% throughout the inertial range. Meneveau also calculated the spatial fluctuation of $T(k)$ which measures the transfer of energy from all wavenumbers to wavenumber k. On average $T(k)$ is negative for the large scales and positive for the small scales, indicating that in three-dimensional turbulence energy is transferred from the large scales to the small scales where it is dissipated (in agreement with Kolmogorov's [117], [118], [119] model of turbulence). Meneveau found, however, that at many places in the flow the energy cascade actually operates in the opposite direction, from small to large scales, indicating a local inverse energy cascade (also called back-scattering). This local spectral information, which links the physical and Fourier space views of turbulence, can only be obtained using the wavelet transform but not with the Fourier transform.

4.3.4 Structure functions

Another fundamental quantity in the classical theory of turbulence [117] is the pth order structure function $S_p(r)$

$$S_p(r) = \frac{1}{L} \int_0^L |f(x) - f(x+r)|^p \, dx, \tag{4.24}$$

where $L \gg r$ is the length of the signal, and L must be long enough so that $S_p(r)$ does not change if L is increased (and thus the increments of f should be stationary in x). The velocity signal of a turbulent flow varies in both space and time and between different realizations of the flow. Thus the integral in (4.24) should, in general, be replaced by a suitably defined ensemble average in order to calculate the structure function of turbulent velocities. To justify the use of space or time averages instead of ensemble averages (over different realizations of the flow), one supposes that the turbulent flow motions are ergodic, which is an unvalidated hypothesis and is probably wrong for two-dimensional turbulence [203]. If the energy spectrum exponent β is in the range $1 < \beta < 3$ (as is usually the case for the inertial range of turbulence) the velocity increments are a stationary function even though the velocities themselves are not [51]; this is a good reason to work with velocity increments rather than the velocities themselves since stationarity is necessary in order to justify estimating a quantity by averaging. The larger p the more $S_p(r)$ is dominated by extreme events. Thus the pth order structure function charac-terizes more and more extreme events as p increases.

If $f(x)$ is self-similar then, just as in the case of the energy spectrum, the structure functions will have a power-law dependence on the scale r

$$S_p(r) = r^{\zeta(p)}. \tag{4.25}$$

The first order structure function $\zeta(1)$ provides a measure of the smoothness of $f(x)$, and in fact $\zeta(1)$ is related to the box dimension D_F of the graph of $f(x)$

$$D_F = 2 - \zeta(1) \tag{4.26}$$

where D_F measures the space-fillingness of $f(x)$. The second order structure function is related to the energy spectrum by

$$\beta = \zeta(2) + 1. \tag{4.27}$$

The Kolmogorov theory [117] showed that the inertial range of turbulence has $\beta = 5/3$, or equivalently that

$$\zeta(p) = p/3, \tag{4.28}$$

however experiments [5] have shown that the structure function exponents increase more slowly than linearly with p for $p > 5$, contradicting Kolmogorov's 1941 theory. The cause of this difference is generally thought to be the fact that the energy dissipation $\varepsilon(x) = (du(x)/dx)^2$ is intermittent in space, i.e. it varies greatly from place to place.

The velocity increment $\Delta f(x, r) = |f(x) - f(x + r)|$ is equivalent to a wavelet transform with DOD (difference of Diracs) wavelet $\psi_\Delta(x) = \delta(x + 1) - \delta(x)$. In fact Jaffard [107] has shown that the exponent $\eta(p)$ defined by

$$\tilde{S}_p(r) = \frac{1}{L} \int_0^L |\tilde{f}(x, r)|^p \, dx \sim r^{\eta(p)} \tag{4.29}$$

is the same as $\zeta(p)$ provided $p > 1$ and $\zeta(p) < p$, no matter what wavelet is used. The wavelet-based method of calculating the structure functions unifies the analysis of structure functions with the calculation of energy spectra and the strength of local singularities. If one uses a wavelet with a sufficient number of vanishing moments, then the wavelet-based structure function $\tilde{S}_p(r)$ should also be more sensitive to larger α singularities since the equivalent wavelet for the structure function, which is the Haar wavelet $\psi_\Delta(x)$, has only one vanishing moment. By changing from an integral to a sum over wavelet maxima we circumvent the divergence of the integral for negative p and thus one can extend the definition of structure functions to include negative ps (as in Arnéodo, Bacry and Muzy's [6] Wavelet Transform Modulus Maximum method discussed in the following section).

The wavelet-based version of the structure function allows us to see directly how the structure function is determined by the singular behaviour of $f(x)$. From equation (4.12) the wavelet transform modulus is proportional to $r^{\alpha(x_0)}$ and thus, since $r \ll L$, the stronger singularities contribute most to the higher order structure functions and least to the lower order structure functions. In other words, the value of $\zeta(p)$ is determined mostly by the stronger singularities for large ps and mostly by the weaker singularities for small ps.

Davis, Marshak and Wiscombe [51] point out that the 'dissipation' of a discrete function f_j, $\varepsilon_j = |f_j - f_{j-1}|$, is in fact a measure. Because ε_j is a measure, the generalized dimension $D(p)$ of $f(x)$ can be calculated from the exponent $K(p)$ of the structure function of $\varepsilon(x)$,

$$D(p) = 1 - \frac{K(p)}{p - 1}. \tag{4.30}$$

The generalized dimension $D(p)$ is the dimension of the set containing the singularities that contribute most to the pth order structure function. Because

$\varepsilon(x)$ is a homogeneous variable (for $1 < \beta < 3$) we have $0 < \beta_\varepsilon(x_0) < 1$ and thus $-1/2 < \alpha(x_0) < 0$. Because $\alpha(x_0) < 0$ the dissipation contains actual singular behaviour (the dissipation tends to infinity).

In general terms the exponents $\zeta(p)$ characterize the homogeneity of the field, while the exponents $K(p)$ characterize the singularity of the field. One can learn a great deal about the behaviour of a signal from the variability of $\zeta(p)$ and $K(p)$ and from the value of the first structure function exponents $\zeta(1)$ and $K(1)$. This information is summarized in Table 4.1.

Davis, Marshak and Wiscombe [51] introduced the 'mean multifractal plane' defined as the plane with coordinates given by the most informative exponents $0 < \zeta(1) = 2 - D_F < 1$ and $0 < \frac{dK}{dp}(1) = 1 - D(1) < 1$ (where D_F is the fractal dimension and $D(1)$ is by definition the information dimension). The position of a particular flow or model on the mean multifractal plane is a good indicator of its self-similar characteristics. The higher the flow's $\frac{dK}{dp}(1)$ component the more intermittent and multifractal it is, and the higher the flow's $\zeta(1)$ component, the smoother and less stationary it is. Experimental turbulent velocity fields lie in the centre of the mean multifractal plane. Turbulence models, however, tend to lie along the boundaries of the multifractal plane: purely multiplicative cascade models (such as δ-functions) lie on the $\frac{dK}{dp}(1)$ axis and purely additive models (such as fractional Brownian motion) lie on the $\zeta(1)$ axis! This clearly indicates that the current turbulence models do not represent correctly the self-similar structure of turbulent flows.

4.3.5 The singularity spectrum for multifractals

In order to characterize a multifractal function it is necessary to calculate its singularity spectrum. The singularity spectrum $D(\alpha)$ may be defined as the Hausdorff (or 'fractal') dimension of the set of points with Hölder exponent α:

$$D(\alpha) = D_F\{x, \alpha(x) = \alpha\}. \tag{4.31}$$

Note that this definition is equally valid for multifractal functions and measures. The singularity spectrum of a monofractal has only one point, e.g. the singularity spectrum of the fractional Brownian signal $B_{1/3}(x)$ which has a $k^{-\frac{5}{3}}$ energy spectrum is $D(\alpha = 1/3) = 1$ (the function $B_{1/3}(x)$ is singular everywhere with $\alpha = 1/3$), while the singularity spectrum of a multifractal is a curve.

Parisi and Frisch [164] found a way of estimating the singularity spectrum from the Legendre transform of the structure function exponents $\zeta(p)$

Table 4.1. *Properties of a signal from the behaviour of the exponents of its structure function $\zeta(p)$ and the structure function of the modulus of its derivative $K(p)$*

Value of structure function	Type of signal
$\zeta(1) = 0$	stationary, $D_F = 2$
$\zeta(1) = 1$	noiseless, $D_F = 1$
$\frac{dK}{dp}(1) = 0$	weak variability
$\frac{dK}{dp}(1) = 1$	δ-function
$\zeta(p)$ variable	non-stationary multifractal
$\zeta(p)$ constant	non-stationary monofractal
$K(p)$ variable	stationary multifractal
$K(p)$ constant	stationary monofractal

$$D(\alpha) = \inf_p(p\alpha - \zeta(p) + 1) \tag{4.32}$$

where, as explained in subsection 4.3.4, $\zeta(p)$ may be calculated using the wavelet transform.

Equation (4.32) can be derived heuristically by noticing that near a singularity of order α

$$|\tilde{f}(x, r)| \sim r^\alpha, \tag{4.33}$$

where we have used equation (4.12) and have written $\alpha = \alpha(x_0)$ for simplicity. Now, if the dimension of the points with singularity α is $D(\alpha)$ then there are about $r^{-D(\alpha)}$ 'boxes' (in this case wavelets) with the scaling (4.33) in each interval r, so that the total contribution to the integral (4.29) is $r^{\alpha p - D(\alpha)+1}$. To leading order the magnitude of the integral is given by the largest contribution so that

$$\zeta(p) = \inf_\alpha(\alpha p - D(\alpha) + 1). \tag{4.34}$$

Since $\zeta(p)$ is concave, formula (4.32) can be obtained by an inverse Legendre transform.

However, Jaffard [107] proved mathematically that structure function calculations of the singularity spectrum can, in general, only set an upper bound on $D(\alpha)$ and he gave some counterexamples where such calculations give completely misleading answers.

Arnéodo, Bacry and Muzy [6] have developed a method for calculating the singularity spectrum called the Wavelet Transform Modulus Maximum (WTMM) method. This method is closely related to the calculation of structure functions by wavelet transforms except that, instead of integrating (or summing in case of discretely defined functions) the wavelet transform over all positions, one only sums the wavelet transforms located at maxima, i.e.

$$\tilde{\Sigma}_p(r) = \sum_{l \in L(r)} \left(\sup_{(x,r')} |\tilde{f}(x, r')|^p \right), \tag{4.35}$$

where l is a maxima line of the wavelet transform modulus on $[0, r]$ and $\sup_{(x,r')}$ means that the supremum is taken for (x, r') on l (so that $r' \leq r$). The wavelets are in fact playing the role of 'generalized boxes' in a new form of the standard box-counting algorithm used to estimate fractal dimensions $D(\alpha)$. Summing only over the wavelet modulus maxima makes sense since, as Mallat and Hwang [146] showed, most of the information in the wavelet transform is carried by the wavelet maxima lines. Furthermore, because one does not sum over places where the wavelet modulus is zero, $\tilde{\Sigma}_p(r)$ is also defined for $p < 0$ as well as for $p \geq 0$. Note that the structure function methods are defined only for $p \geq 0$.

Arnéodo, Bacry and Muzy draw the analogy with statistical thermodynamics and interpret $\tilde{\Sigma}_p(r)$ as a 'partition function' (see Table 4.2).

If $f(x)$ is a self-similar function then $\tilde{\Sigma}_p(r) \propto r^{\tau(p)}$ and the singularity spectrum can be found by calculating the Legendre transform

$$D(\alpha) = \inf_p (p\alpha - \tau(p)). \tag{4.36}$$

To avoid technical problems associated with calculating the Legendre transform in (4.36) Arnéodo, Muzy and Bacry [6] recommend an alternative way of finding $D(\alpha)$ (see their paper for details).

Jaffard [107] proved mathematically that the WTMM method, unlike the structure function methods, gives the correct singularity spectrum for all p provided it is slightly modified. Indeed a problem might arise if the wavelet modulus maxima are too close together; in that case the sum in an interval of width r must be restricted to the largest maxima. Jaffard also showed that even the modified WTMM method fails if the function $f(x)$ contains too many oscillating singularities.

Arnéodo, Bacry and Muzy [6] found the relation between $\tau(p)$ and $\zeta(p)$ from their respective definitions in terms of $D(\alpha)$, but given the limitations of equation (4.32), it is perhaps better (and more intuitive) to find the connection directly through the structure functions. In terms of discrete signals, the wavelet transform-based calculation of the structure function (4.29) becomes

Table 4.2. *Analogies between statistical thermodynamics and the Wavelet Transform Modulus Maximum method for multifractals*

Thermodynamic parameter	Multifractal parameter
Temperature	p^{-1}
Partition function	$\tilde{\Sigma}_p(r)$
Free energy	$\tau(p)$
Entropy	$D(\alpha)$

$$\tilde{S}_p(r) = \frac{1}{N} \sum_{j=1,N} |\tilde{f}(x_j, r)|^p. \tag{4.37}$$

Each cone of influence of width r must contain only maxima lines with the same scaling (since the scaling $r^{\alpha(x_0)}$ is the same for all points within the influence cone of point x_0) and if the function is everywhere singular all intervals of size r must contain at least one maxima line. If one follows Jaffard's [107] refinement to WTMM, and only counts one maximum for each interval of length r, then the number of terms in the sum must be proportional to N/r. Therefore, if the wavelet moduli are only summed over their maxima the structure function becomes

$$\tilde{S}_p(r) = \frac{1}{N/r} \sum_{l \in L(r)} \left(\sup_{(x_j, r')} |\tilde{f}(x_j, r')|^p \right) = \frac{1}{N/r} \tilde{\Sigma}_p(r). \tag{4.38}$$

We thus find that the relation between the structure function exponents $\zeta(p)$ and the WTMM 'free energy' exponents $\tau(p)$ is

$$\zeta(p) = \tau(p) + 1. \tag{4.39}$$

Note that equation (4.39) only holds if the function $f(x)$ has singularities everywhere and WTMM is modified by only counting one wavelet modulus maximum for each interval of length r.

Arnéodo, Bacry & Muzy [6] applied the WTMM method to single point high Reynolds number (the Taylor scale based Reynolds number is $R_\lambda = 2720$) velocity data obtained by Gagne [99] from the wind tunnel of ONERA at Modane. The self-similar inertial range follows the Kolmogorov $E(k) \sim k^{-\frac{5}{3}}$ law for almost three decades. The WTMM analysis was carried out for this inertial range of scales on a section of data 100 integral (energy containing) scales long.

The histogram of singularities $\alpha(x_0)$ in the turbulence data was found to be quite wide and centred about the Kolmogorov value $\alpha = 1/3$. Surprisingly, at some places in the flow α is negative which implies actual singular behaviour (velocity tending towards infinity). These negative α values may be spurious or may indicate the (rare) presence of strong vortices. The function $\tau(p)$ is convex which suggests that the regularity of the flow varies greatly from place to place. The singularity spectrum is peaked at the Kolmogorov value $\alpha_{\max}(p = 0) = 0.335 \pm 0.005$ with $D(\alpha_{\max}) = 1.000 \pm 0.001$. This result indicates that the signal is fractal everywhere because the fractal support of $D(\alpha_{\max})$ is equal to its topological dimension (i.e. the dimension of the signal, which is 1).

4.3.6 Distinguishing between signals made up of isolated and dense singularities

Although the inertial range of turbulence has a self-similar structure, not all self-similar functions are fractal; in fact some of the most physically plausible turbulence structures, the spiral vortices, can generate self-similar oscillating singularities with a non-trivial box-counting dimension (a technique to estimate the Hausdorff or fractal dimension). The conclusion drawn by Arnéodo, Bacry and Muzy [6] that turbulence is everywhere singular with a multifractal structure may be invalid if the turbulent velocity signal they analysed contains oscillating singularities. Because the WTMM method is only valid for signals that contain dense distributions of cusp type singularities, one should first try to determine whether a signal has isolated oscillating singularities before attempting to use the WTMM method. Unfortunately, the difference between signals containing singularities everywhere ('fractals') and signals containing a large number of isolated oscillating singularities (isolated 'spirals' in multi-dimensions or isolated 'chirps' in one dimension, see Figure 4.1) is not obvious: both signals can have non-trivial box-counting dimensions.

Kevlahan and Vassilicos [115] developed two methods for distinguishing between isolated spiral and fractal signals based on the wavelet transform. In fact their method only distinguishes between isolated and dense singularities, however isolated cusp singularities have a trivial box-counting dimension and thus can be distinguished from fractal signals on the basis of box-counting dimension alone. The first method takes advantage of the fact that the singularities in a fractal are dense (there are singularities at an infinite number of points, see Figure 4.1), whereas the singularities in an isolated spiral signal are isolated (the signal contains oscillating singularities only at the centres of spirals). If one averages the wavelet transforms of many realizations, or different data segments (separated by more than one integral scale L in order to be decorrelated) together, one can prove that the average wavelet transform modulus $\langle |\tilde{f}(x, r)| \rangle$ decays differently for the two types of singularity

$$\text{as } \left\langle |\tilde{f}(x, r)| \right\rangle \propto N^{-1/2} |\tilde{f}(x_0, r)| \quad \text{for fractal signals,} \tag{4.40}$$

but

$$\text{as } \left\langle |\tilde{f}(x, r)| \right\rangle \propto |\tilde{f}(x_0, r)|, \quad r < L/N \quad \text{for spiral signals,} \tag{4.41}$$

where N is the number of realizations or of decorrelated segments averaged together and L is the length of each segment. Thus, the average wavelet transform of the random phase fractal signal is $N^{-1/2}$ times a single realization, while that of the spiral signal does not depend, below a certain scale, on the number of realizations. The difference in the behaviour of $\langle |\tilde{f}(x, r)| \rangle$ is striking, and provides a diagnostic for determining whether a signal contains spiral-type singularities or not. This method was applied to the Gagne [99] turbulence data. The results were inconclusive, perhaps due to insufficient resolution near expected spiral scales or rarity of spiral vortices passing near the pointwise velocity probe.

The second method for distinguishing between isolated spiral and fractal singularities derives from the observation that the spatial fluctuation of wavelet energy $\tilde{E}(x, k)$ (measured by the standard deviation $\tilde{\sigma}(k)$ of $\tilde{E}(x, k)$) is independent of wavenumber for a random phase fractal signal, but increases with wavenumber for a spiral signal with the same energy spectrum. Analysis of the turbulent signal shows that $\tilde{\sigma}(k)$ increases with wavenumber (although at a slower rate than for the purely spiral test signal), indicating that turbulence probably contains some sort of isolated oscillating singularities. This conclusion should be borne in mind when interpreting the results of multifractal analyses of turbulence.

4.4 Turbulence analysis

4.4.1 New diagnostics using wavelets

It is impossible to define a local Fourier spectrum, because Fourier modes are non-local, but it is possible to define a local wavelet spectrum, since wavelets are localized functions. Actually, due to the inherent limitation of the uncertainty principle stating that there is a duality between spectral and spatial selectivity, we should be aware that the spectral accuracy will be poor in the small scales and that the spatial accuracy will be poor in the large scales.

Since turbulent flows are either two-dimensional or three-dimensional, in the following section we will use the two-dimensional continuous wavelet transform. Let us consider a two-dimensional scalar field $f(\boldsymbol{x})$ and a two-dimensional real isotropic wavelet $\psi(\boldsymbol{x})$. We generate the family $\psi_{\boldsymbol{x},r}(\boldsymbol{x}')$ of wavelets, translated by position parameter $\boldsymbol{x} \in \mathbb{R}^2$, and dilated by scale parameter $r \in \mathbb{R}^+$, all having the same L^2 norm

$$\psi_{\boldsymbol{x},r}(\boldsymbol{x}') = r^{-1}\psi\left(\frac{\boldsymbol{x}' - \boldsymbol{x}}{r}\right). \tag{4.42}$$

The two-dimensional wavelet transform of $f(\boldsymbol{x})$ is

$$\tilde{f}(\boldsymbol{x}, r) = \int_{\mathbb{R}^2} f(\boldsymbol{x}')\,\psi_{\boldsymbol{x},r}(\boldsymbol{x}')\,d^2\boldsymbol{x}'. \tag{4.43}$$

The local wavelet spectrum of $f(\boldsymbol{x})$ is defined as

$$\tilde{E}(\boldsymbol{x}, r) = \frac{1}{2c_\psi k_0}|\tilde{f}(\boldsymbol{x}, r)|^2. \tag{4.44}$$

A characterization of the local 'activity' of $f(\boldsymbol{x})$ is given by its wavelet intermittency $\tilde{I}(\boldsymbol{x}, r)$, which measures the local deviation from the mean spectrum of f at each position $\boldsymbol{x}$ and scale r, defined as follows

$$\tilde{I}(\boldsymbol{x}, r) = \frac{|\tilde{f}(\boldsymbol{x}, r)|^2}{\int_{\mathbb{R}^2} |\tilde{f}(\boldsymbol{x}, r)|^2\, d^2\boldsymbol{x}}. \tag{4.45}$$

Another measure of interest for turbulence is the wavelet Reynolds number $\tilde{Re}(\boldsymbol{x}, r)$, given by

$$\tilde{Re}(\boldsymbol{x}, r) = \frac{\tilde{u}(\boldsymbol{x}, r)r}{\nu}, \tag{4.46}$$

where r is the scale parameter, ν the kinematic viscosity of the fluid, and $\tilde{u}$ the root mean square value of the velocity field contribution at position $\boldsymbol{x}$ and scale r defined as

$$\tilde{u}(\boldsymbol{x}, r) = \left(\frac{1}{C_\psi}\sum_{i=1}^{3}|\tilde{u}_i(\boldsymbol{x}, r)|^2\right)^{1/2}, \tag{4.47}$$

with the constant

$$C_\psi = \int_{\mathbb{R}^2} |\hat{\psi}(\boldsymbol{k})|^2 \frac{d^2 \boldsymbol{k}}{|\boldsymbol{k}|^2}. \tag{4.48}$$

The expectation is that at large scales $r \sim L$, the wavelet Reynolds number should coincide with the usual large-scale Reynolds number $Re = UL/\nu$, where U is the r.m.s. turbulent velocity and L is the integral scale, which is the energy containing scale of the flow. In the smallest scales (say $r \sim \eta$, where η is the Kolmogorov scale of the flow which characterizes the high wavenumber limit of the inertial range where dissipation becomes significant), one expects this wavelet Reynolds number to be close to unity when averaged spatially. The question we want to address here is the variability of such a wavelet Reynolds number defined in space and scale: are there locations where such a Reynolds number in the small scales is much larger than elsewhere, and how do such regions correlate with regions of small-scale activity in the flow? Actually $\tilde{Re}(\boldsymbol{x}, r)$ gives an unambiguous measure of the nonlinear activity at small scales (or at any desired scale), because regions of high wavelet Reynolds number correspond to regions of strong nonlinearity.

Concerning the computation of energy and enstrophy transfers and fluxes, we should be aware that the results depend on the functional basis we consider. Indeed, due to Heisenberg's uncertainty principle, each representation measures different types of transfers and fluxes. In Fourier space one computes transfers between different independent wavenumber bands, which detect the modulations and resonances excited under the flow dynamics. In wavelet space one computes exchanges between different locations and different scales, which detect instead advections and scalings. But one should never forget that in wavelet space spatial resolution is bad in the large scales and good in the small scales, while, by duality, space resolution is good in the small scales but bad in the large scales. In an orthogonal wavelet basis, although all wavelets are independent in space and scale, they are not necessarily independent in wavenumber. In an orthogonal wavelet packet basis all wavelet packets are independent in space, scale and wavenumber, but their Fourier spectrum may present several peaks at distant wavenumbers and they may be quite delocalized in wavenumber space; therefore wavelet packets are not appropriate to precisely measure transfers between different wavenumber bands. This is the reason why a comparison between transfers computed in wavelets, wavelet packets and Fourier modes is misleading: these three diagnostics do not measure the same quantities!

4.4.2 Two-dimensional turbulence analysis

Unlike the velocity field, the vorticity field is invariant with respect to uniform rectilinear translations of the inertial frame (Galilean invariance). The dependence of streamlines and streaklines on the reference frame causes considerable difficulties in the study of fluid flows, particularly in observing and defining vortices. In fact, due to its Galilean invariance, vorticity is the most suitable field for tracking the dynamics of turbulent flows, in both two and three dimensions. Moreover, due to Helmholtz's theorem stating the Lagrangian conservation of vorticity in 2D and of vortex tubes in 3D, we are convinced that vorticity is, for both 2D and 3D flows, the fundamental field whose evolution controls all other relevant fields; the importance of vorticity has been advocated for years by Saffman [184] and Chorin [44]. We think that turbulence analysis, modelling and computing should be based on a segmentation of the vorticity field into coherent vortices (or vortex tubes in 3D) and random background of vorticity filaments (namely 1D structures embedded in 2D or 3D) produced by the nonlinearly interactions between the coherent vortices. The vorticity field is directly accessible from numerical simulations, but is difficult to obtain from laboratory experiments. This is why we will now focus on vorticity fields obtained from direct numerical simulation (DNS) results. The drawback with DNS, i.e. the integration of Navier–Stokes equations without any *ad hoc* turbulence modelling, is that current supercomputers are only able to compute low Reynolds number flows (up to a few thousand).

Let us show an example of a wavelet analysis of an instantaneous vorticity field computed using the Navier–Stokes equations [168], [71]. We segment it into three regions using the Weiss criterion [202], [63], namely into rotational regions corresponding to the coherent structures, strongly strained regions corresponding to the shear layers surrounding the coherent structures, and weakly strained regions corresponding to the background flow made of vorticity filaments (these vorticity filaments encountered in two-dimensional turbulence are not the same dynamical objects as the vorticity tubes encountered in three-dimensional turbulence and often called filaments). We then decompose the vorticity field into a continuous wavelet representation using an isotropic (Hermite) wavelet to integrate in space the wavelet coefficients for each type of region. This decomposition is in fact a conditional statistical analysis because the energy spectrum is computed separately for each type of region. The energy spectrum of the coherent structure regions tends to scale as k^{-5}, the sheared regions as k^{-4} and the background regions as k^{-3} (Figure 4.2). We found [80], [168], [73] that each region has energy throughout the

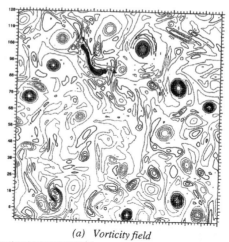

(a) *Vorticity field*

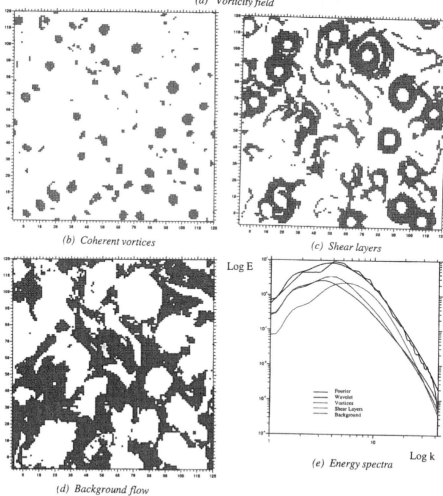

(b) *Coherent vortices*

(c) *Shear layers*

(d) *Background flow*

(e) *Energy spectra*

inertial range and therefore there is no scale separation. This is why the Fourier representation cannot disentangle these different regions.

The scaling of the coherent structures seems compatible with the cusp-like model for vortices proposed by Farge and Holschneider [76], the scaling of the shear layers seems compatible with the vorticity fronts model proposed by Saffman [183] and only the scaling of the homogeneous background regions seems to verify the Batchelor–Kraichnan prediction for 2D homogeneous isotropic turbulence [121]. From this analysis we confirm that there is no universal power-law scaling for two-dimensional turbulent flows; the slope of the Fourier energy spectrum varies with the density of coherent structures (their number per unit area in 2D and per unit volume in 3D), which depends on initial conditions and forcing (energy injection by external forces). We have then conjectured [80] that there may be a universal scaling for each region of the flow considered separately, but this has not yet been proven. Extensive wavelet analysis of very different types of turbulent flows would be necessary to check this conjecture.

A key question, which remains open, is the following: is there a generic shape (namely a typical vorticity distribution) for coherent structures? The answer to this question influences our analysis, in particular our interpretation in terms of scale, because the notion of scale is intrinsically linked to the generic shape we assume for the coherent structures. *A prioris* are as essential in statistical analysis as hypotheses are in modelling: we should state them clearly, otherwise our results would be nonsensical. For instance, without a definition of vortex shape the notion of vortex size and vortex circulation would be meaningless. A misunderstanding has persisted for years in the field of turbulence due to the identification of scale with the inverse wavenumber,

---

Fig. 4.2. Conditional wavelet spectra (this computation was done in collaboration with Thierry Philipovitch). (Colours referred to in this caption are shown at www.cambridge.org/resources/0521533538) (a) Vorticity field. In red: elliptic regions, dominated by rotation (antisymmetric part of the stress tensor ∇v), which correspond to the coherent vortices. In blue: hyperbolic regions, dominated by strain (symmetric part of the stress tensor ∇v), which correspond to the incoherent background flow. (b) Coherent vortices where rotation dominates. (c) Shear layers where strain and strong velocity dominate. (d) Background flow where strain and weak velocity dominate. (e) Energy spectra. In black: Fourier energy spectrum, which tends to scale as $k^{-4.5}$ in the inertial range. In dark blue: wavelet energy spectrum, which is a smooth approximation of the Fourier spectrum and tends to scale as $k^{-4.5}$. In red: wavelet energy spectrum of the coherent vortices, which tends to scale as k^{-5}. In green: wavelet energy spectrum of the shear layers, which tends to scale as k^{-4}. In light blue: wavelet energy spectrum of the background flow, which tends to scale as k^{-3}.

which is true only if one assumes a wave-like shape for the vorticity field. Conversely, in other papers one encounters different implicit models of coherent structures (point vortices [203], vortex patches [127], Gaussian vortices [144], or cusp-like vortices [76]), which indeed condition our statistical analysis. Therefore one first needs a method to extract coherent structures out of turbulent flows in order to study them individually. The classical method consists of thresholding the vorticity field and identifying as coherent vortices all regions where vorticity is larger than this threshold. However, the spectral information is then lost due to the discontinuity introduced by the threshold. We have proposed instead [74], [80] two new methods based on the continuous wavelet representation, which preserves the smoothness of the vorticity field and therefore its spectrum.

These methods depend on the choice of the analysing wavelet (although this dependance is weak) and ideally we should use a wavelet which is a local solution of the linearized Navier–Stokes equations, namely a solution of the heat equation, such as any isotropic and smooth distribution of vorticity. This is why we propose to use two-dimensional Hermite wavelets (derivatives of the Gaussian), which are solutions of the heat equation. The higher the derivative, the better the cancellations and the more sensitive the wavelet will be to quasi-singular vortices, however its spatial selectivity will not be as good as for low order derivative wavelets. In two examples shown in this chapter (Figures 4.2 and 4.5) we use Marr's wavelet which is the Laplacian of the Gaussian.

The new approach we have proposed is to decompose turbulent flows into coherent and inhomogeneous components versus incoherent and homogeneous components. This decomposition should be performed for each flow realization before averaging, because these two classes of components correspond to different statistical distributions and present different scaling laws. The first method to perform this decomposition consists of extracting the coherent structures by retaining only the wavelet coefficients inside the influence cones (namely the spatial support of the wavelets) attached to the local maxima of the vorticity field corresponding to the centres of the coherent structures; the wavelet coefficients outside the influence cones are discarded before reconstructing the coherent components of the vorticity field [80]. We can also extract just one coherent structure, analyse its shape, and compute its coherence function, namely the pointwise relation between vorticity and streamfunction, to check if it corresponds to the stationary states predicted by Montgomery's [111] or Robert's [178], [179], [180] statistical theories. The second method to split the flow into coherent and incoherent components consists of retaining only the wavelet coefficients which are larger than a given threshold and to discard all other coefficients before reconstructing the coherent vorticity field. We have thus extracted the coherent structures (corresponding to the wavelet coefficients larger than the threshold) from the background flow (corresponding to the weaker wavelet coefficients). By computing the Fourier spectrum of these two fields we have confirmed our previous analysis: the energy spectrum of coherent structures tends to scale as k^{-5} and that of the background field as k^{-3} [74], [80], thus recovering the scaling predicted by the statistical theory of homogeneous 2D turbulence [121]. This confirms the conjecture stating that the coherent structures are responsible for the intermittency of 2D turbulent flows. We think this conjecture is also true for 3D turbulent flows.

Inspired by Donoho's theorem for optimal denoising [58], we have recently proposed [83] the threshold $\widetilde{\omega_T} = (2 <\omega^2> \log_e N)^{1/2}$ to select the wavelet coefficients to be retained to extract coherent structures. This threshold depends only on the variance of the vorticity field $<\omega^2>$ and on the number of grid-point samples of the vorticity field N, without any adjustable parameter. For statistically steady turbulent flows, whose variance is by definition stationary, this threshold remains constant during the whole time evolution. Using this method we have analysed decaying [82], wavelet forced [83] and Fourier forced 2D turbulent flows [84] (see Figure 4.3). For these three different types of turbulent flows we have observed that the coherent components, obtained from the wavelet coefficients of vorticity larger than the threshold $\widetilde{\omega_T}$, have non-Gaussian vorticity and velocity PDFs, while the incoherent components, obtained from the wavelet coefficients of vorticity smaller than the threshold $\widetilde{\omega_T}$, have Gaussian PDFs [83], [82], [84] (see Figures 4.3f and 4.10d). There is still some hope of finding universal statistical distributions for each component taken separately, and we may be able to propose new turbulence models based on the Gaussianity of the background flow. Even if such a universal distribution exists for the coherent structures, we would still need to calculate the dynamics of these structures in detail because they remain out of equilibrium (unlike the background).

Using the wavelet segmentation technique we have just described, we analysed a 2D forced turbulent flow computed with 256^2 Fourier modes, and found that only 0.7% of the wavelet modes retain 94.3% of the total enstrophy and 99.2% of the total energy. These modes correspond precisely to the coherent structures as exhibited on the coherence scatter plot (Figure 4.3d). These coherent modes are responsible for the PDFs of the total vorticity and velocity fields, while the incoherent modes (corresponding to the 99.3% remaining wavelet coefficients) have a Gaussian PDF, with a flatness 3 and

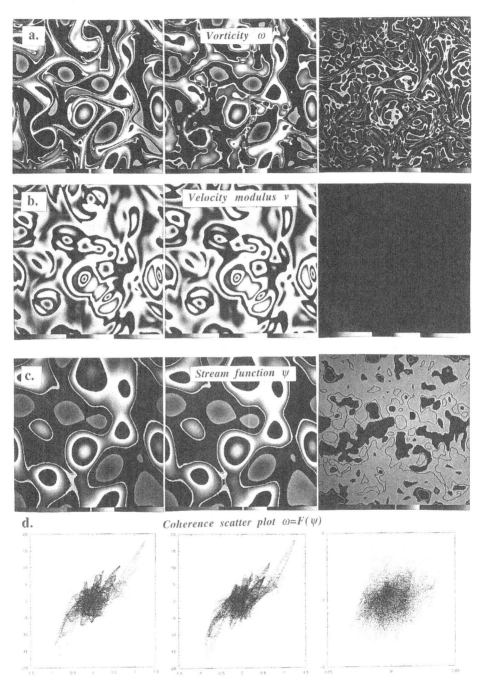

Fig. 4.3

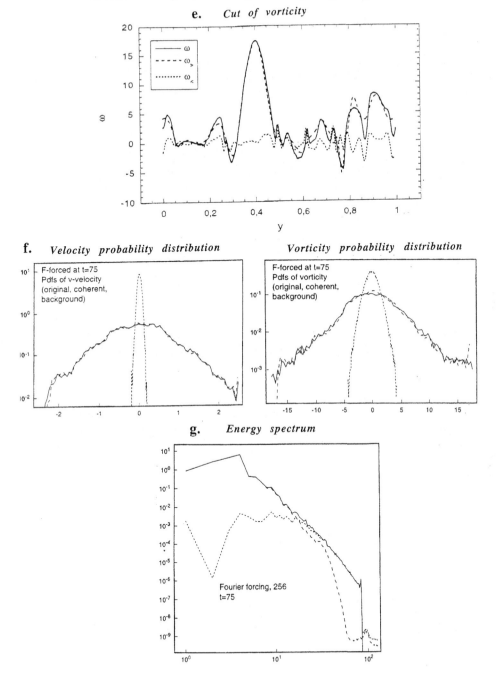

Fig. 4.3 (*continued*)

a much smaller variance than that of the total fields (Figure 4.3f). On the contrary the incoherent background flow is responsible for the energy spectrum scaling in the high-wavenumbers (Figure 4.3g), because the contribution of the coherent modes to the small scales are too localized to be detected by the two-point correlation (whose Fourier transform gives the energy spectrum). We have also shown that the coherent modes are responsible for the total flow dynamics because they trigger the total velocity field. The incoherent modes are passively advected by the coherent velocity, because the incoherent velocity field generated by the background flow is nearly zero (Figure 4.3b). For all these reasons we think that the coherent modes are essential, and that analysing of turbulent flows in terms of energy spectrum scaling in the high-wavenumbers alone is misleading, and one should also consider the PDFs of vorticity and velocity.

Another application of the wavelet representation in turbulence is to design new types of forcing for numerical simulations. The method, proposed by Schneider and Farge [185] consists of injecting energy and enstrophy at each time step, but only into the wavelet coefficients inside the influence cone corresponding to a given location. Depending on the type of forcing we want, we could either excite the same vortices or randomly select new vortices at each time step. Forcing is currently done in Fourier space and is rather unphysical, while wavelet-based forcing could simulate the production of vorticity in boundary layers or shear layers, which is a local process. This is another promising application of wavelet techniques for turbulent flow

---

Fig. 4.3. Wavelet compression of vorticity. (a) The vorticity. (b) The modulus of velocity. (c) The streamfunction. (d) The coherence scatter plot. (e) Cut of vorticity (f) PDFs of velocity and vorticity. (g) Energy spectrum.

The solid lines correspond to the total vorticity ω, the dashed lines to the coherent part $\omega_>$, and the dotted lines to the incoherent part $\omega_<$.

We observe that only 0.7% of the total number of wavelet coefficients are sufficient to represent all coherent structures, while the remaining 99.3% correspond to the incoherent background flow, which is much weaker and homogeneous. The coherent vorticity $\omega_>$ contains 94.3% of the total enstrophy. Moreover, the velocity associated with the coherent structures $v_>$ is quasi-identical to the total velocity v and contains 99.2% of the total energy. As for the coherent stream function, $\Psi_>$ is perfectly identical to the total stream function Ψ. The fact that the coherence scatter plot of the background $F_<$ is isotropic proves that our method has extracted all coherent structures. The PDFs of velocity and vorticity show that only 0.7% of the wavelet coefficients are sufficient to capture the non-Gaussian one-point statistical distribution of vorticity, while the remaining 99.3% have a Gaussian distribution. The energy spectrum, on the contrary, is dominated by the background at small scales and therefore is insensitive to coherent structures, because they are too localized in the small scales to affect the energy spectrum (which is the Fourier transform of the two-point correlation function) in the high wavenumbers. Parts (a) to (d) of this figure are also shown at www.cambridge.org/resources/0521533538.

simulation (the results obtained with this wavelet forcing method are discussed in section 4.6.4.3 and shown on Figure 4.10).

4.4.3 Three-dimensional turbulence analysis

We have analysed different flow fields resulting from direct numerical simulations of three-dimensional turbulent flows [75], using the complex-valued Morlet wavelet, which plays the role of a numerical polarizer due to its angular selectivity, and whose complex modulus directly measures the energy density. We have first studied the temperature, velocity and pressure fields of a channel flow near the wall and have used the wavelet intermittency to pinpoint the regions of the flow dominated by strong nonlinear dynamics, corresponding to locally stronger wavelet Reynolds numbers. It appears that the most intermittent regions are correlated with those of large vertical velocity, corresponding to ejections from the boundary layer. We have found that temperature behaves as a passive scalar almost everywhere, except in these very localized ejection regions. We have also observed that there is no return to isotropy in the small scales, contradicting one of the hypotheses of the statistical theory of turbulence, which supposes that turbulent flows become homogeneous and isotropic at small scales.

We have then analysed the vorticity, velocity and a passive scalar in a temporal mixing layer after the mixing transition. We have found that wavelet intermittency is very strong, up to 120, in the collapsing regions where the ribs (streamwise vorticity tubes produced by a three-dimensional instability) are stretched and engulfed into the primary spanwise vortex (produced by a two-dimensional Kelvin–Helmholtz instability). On the other hand, the wavelet intermittency in the braids, i.e. outside the spanwise vortex, remains very low, not exceeding 5. We have also noticed in this case and contrarily to the channel flow, a return to isotropy in the small scales. From the local spectrum of the vertical vorticity we have observed that the collapsing regions have a spectral slope much shallower than the one of the braid regions; this departure from the space average wavelet spectrum increases with the scale and confirms the strong intermittency of the mixing layer. If we extrapolate the observed slopes, we conjecture that intermittency should increase with Reynolds number. We have then visualized the iso-surfaces of the wavelet Reynolds number, which can be interpreted as surfaces of iso-nonlinearity in the flow. The peaks on these iso-surfaces, which are associated with the most unstable regions, are located in the primary vortex core; this confirms our previous conclusions concerning the concentration of small-scale nonlinear activity there, due to the stretching of the ribs rolled around the primary vortex. We have also shown that the Kolmogorov scale, corresponding to the iso-surface $Re(x, r) \simeq 1$ where linear dissipation balances nonlinear advection, varies with location, being at much smaller scales in the vortex core than in the braids, with a scale variation of four octaves. This means that there may also be some (spatially localized) dissipation at scales belonging to the inertial range. This observation contradicts Kolmogorov's hypothesis of non-dissipative energy transfers in the inertial range, but is in agreement with Castaing's theory of turbulence [36], [37], with Frisch and Vergassola's [93] multifractal model and with Benzi *et al.*'s [21] extended self-similarity model, which assume a weak dissipation in the inertial range.

For shear flows, such as the channel flow or the mixing layer we have studied, there is a clear correlation between large-scale events and small-scale activity, due to the presence of coherent structures. Wavelet analysis has been an essential tool for identifying them as wavelet phase-space regions correlated in both space and scale, where intermittency decreases with scale [75]. We conjecture that for large Reynolds numbers these regions may become more and more localized and very intense in small-scale enstrophy. Therefore they correspond to rare but strong events, which are susceptible to develop singularities at very large Reynolds numbers. For the mixing layer these quasi-singular regions correspond to collapsing events, where the ribs are stretched and accumulated inside the primary vortex core, while for the channel flow these regions correspond to the tip of the hairpin vortices ejected from the wall boundary layer. According to the Cafarelli–Kohn–Nirenberg theorem [33], singularities of Navier–Stokes equations, if they exist, should be at most a set of Hausdorff measure one in space-time for any Reynolds numbers, which confirms the fact that they could only be rare events to which standard statistical tools, such as two-point correlation and energy spectrum, remain insensitive. Incidentally if we want to look for quasi-singularities in three-dimensional turbulent flows it may be better to use a space-time continuous wavelet transform, whose theory has been initiated by Duval-Destin and Murenzi [61], but has not yet been sufficiently developed.

4.5 Turbulence modelling

We will now reconsider the closure problem mentioned in subsection 4.2.3, taking advantage of the new observations we have made of turbulent flows, and in particular the dynamical role of coherent structures, thanks to the wavelet analysis.

4.5.1 Two-dimensional turbulence modelling

To compute turbulent flows we must separate the active components, responsible for their chaotic behaviour (namely sensitivity to initial conditions), from the passive components, which are advected by the velocity field resulting from the overall coherent structure motion. The active components are not in thermal equilibrium, while the passive components are well thermalized. Therefore the active components should be computed explicitly, while the passive components can be modelled by some *ad hoc* parametrization.

Classical numerical techniques (Galerkin methods [103], Large Eddy Simulation [133], [175], [135] and Nonlinear Galerkin methods [149]) assume that the active components are the low-wavenumber Fourier modes, or the scales resolved by the computational grid, while the passive components are the high-wavenumber Fourier modes, or the sub-grid scales. This scale separability of the turbulent dynamics is assumed to be true in both two and three dimensions.

We have shown [204] that a compression in the wavelet or wavelet packet representation extracts the coherent structures out of the background flow, while the same amount of compression done in the adapted local cosine (Malvar) representation, which is a type of windowed Fourier basis, does not have this property (Figure 4.4). Indeed, the more you compress in Fourier or windowed Fourier representations, the more you smooth the coherent structures, and consequently lose their enstrophy, destroy their phase information, and introduce parasitic wiggles in the background. Indeed, the more you compress, the larger the effect of the analysing function. Therefore wavelets and wavelet packets, being localized functions, tend to separate coherent structures from the background flow (Figure 4.4a), while Fourier and windowed Fourier, being non-localized functions, tend to smear coherent structures into the background flow (Figure 4.4b).

We have shown [74], using nonlinear wavelet packet compression, that there is no scale separability in two-dimensional turbulence; we conjecture that this result is also true in 3D turbulence. To prove this we have computed the time evolution of a two-dimensional turbulent flow which we use as our high-resolution reference flow. We have then compressed the initial vorticity field in two ways: either by retaining only the lower wavenumber Fourier modes, or by selecting the strongest (in L^2-norm) wavelet packet coefficients. We found that for a compression ratio of 200 the wavelet packet representation preserves, in a statistical sense (namely the energy spectrum is well predicted), the reference flow evolution while the Fourier representation

(a) *Uncompressed vorticity field*

(b) *Vorticity compressed*
in a wavelet packet basis

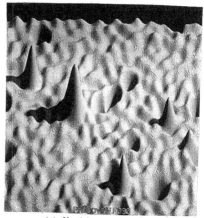

(c) *Vorticity compressed*
in an adapted local cosine basis

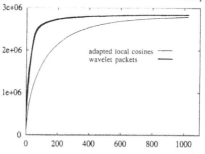

(d) *Amount of enstrophy retained*
for each compression

Figure 4
Comparison between
wavelet packet and
adapted local cosine
compression

leads to a statistically different solution. This conclusion is not surprising, considering the existence of an inverse energy cascade in two-dimensional turbulence which implies that the high-wavenumber Fourier modes remain active and affect the evolution of the low-wavenumber modes. The implication of this behaviour should be implemented in turbulence models, because we now have wavelet-based numerical methods to replace grid-point or Fourier representations and integrate Navier–Stokes equations (see 4.6).

In the same paper [74] we showed that there is a possible separability between active modes, namely the coherent structures corresponding to the strong wavelet packet coefficients, and passive modes, namely the vorticity filaments of the background flow corresponding to the weak wavelet packet coefficients. Both components are multi-scale, which is why the Fourier representation is not able to disentangle them and *a fortiori* to model them. According to Weiss analysis [202] the coherent structures correspond to elliptic regions (nearby fluid trajectories remain nearby) where rotation ω^2 dominates strain s^2, while the background flow corresponds to hyperbolic regions (two nearby fluid trajectories separate exponentially) where strain s^2 dominates rotation ω^2. In the elliptic regions the wavelet Reynolds number $\tilde{Re}(x, r)$ is larger than one, while in the hyperbolic regions it is smaller than one, which indicates that the background flow is actually laminar (Figure 4.5).

We have shown ([83], [82], [84]) that probability distribution functions (PDF) of the vorticity and velocity fields associated to the coherent structures are non-Gaussian, while they are Gaussian for the background flow (Figure 4.3). Therefore the coherent structures are out of thermal equilibrium, while the background flow has already thermalized due to the very strong mixing resulting from the strain imposed by the coherent structures. Therefore the probability distributions of the background flow are stationary and do not depend on the spatial configuration of the coherent structures. We should then be able to model the incoherent background flow by

---

Fig. 4.4. Comparison between wavelet packet and adapted local cosine compression (this computation was done in collaboration with Echeyde Cubillo). (a) The uncompressed vorticity field computed with 128^2 modes. (b) The vorticity field reconstructed from the 70 strongest wavelet packet coefficients, which contain 90% of the enstrophy. (c) The vorticity field reconstructed from the 425 strongest adapted local cosine coefficients, which contain 90% of the total enstrophy. (d) Enstrophy contained in the retained coefficients versus their number. We observe, for instance, that 70 wavelet packet coefficients retain 90% of the total enstrophy, while 70 adapted local cosine coefficients retain only 50% of the total enstrophy. This figure is also shown at www.cambridge.org/resources/0521533538.

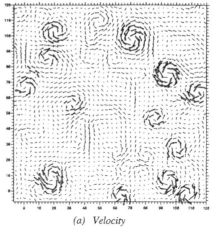

(a) *Velocity*

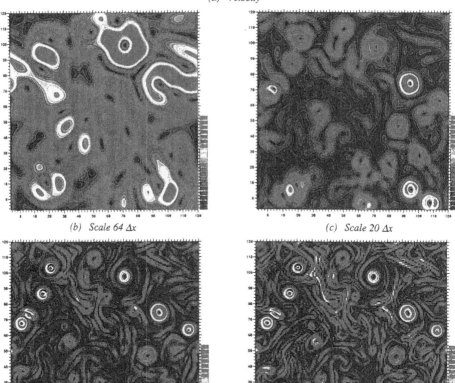

(b) *Scale 64 Δx*   (c) *Scale 20 Δx*

(d) *Scale 8 Δx*   (e) *Scale 2 Δx*

an *ad hoc* stochastic process having the same enstrophy and the same statistics, in particular the same spectral slope, or using simple turbulence models (Boussinesq, Smagorinsky or $k - \epsilon$ [155]), whereas the coherent structures should be explicitly computed in wavelet phase-space. A possible direction would be to construct a wavelet or wave packet frame (namely a quasi-orthogonal basis) made of local solutions of the linearized Navier–Stokes equations (namely any isotropic smooth function, such as the Mexican hat). We do not yet know how to construct it, nor to compute Navier–Stokes equations in it, although we know how to compute Navier–Stokes equations in an orthogonal wavelet basis (see 4.6), which is a promising first step in the same direction.

We have also shown [114] that the presence of coherent structures inhibits the nonlinear instability of the background flow, namely the formation of new coherent structures. Using the wavelet packet representation to extract the coherent structures we then computed the evolution of the remaining background flow, in the absence of coherent structures, and observed the emergence of new ones out of it (Figure 4.6). Actually when coherent structures are present, they impose a strain on the background flow, which then inhibits the formation of new coherent structures, and therefore there is no energy or enstrophy backscatter from the incoherent to the coherent components of two-dimensional flows. The next step to validate this observation will be to compute the different transfers between coherent and incoherent components of the flow (namely from coherent structures to coherent structures, from coherent structures to background, from background to coherent structures and from background to background) and check that there is no transfer from background to coherent structures. If this is confirmed, there will be a possible wavelet separability between the coherent and incoherent flow components and we may then be able to propose new turbulence models based on this property.

Fig. 4.5. Wavelet Reynolds number (this computation was done in collaboration with Thierry Philipovitch). (a) Velocity field computed with resolution 128^2 ($\Delta x = 1$ unit length between two grid-points). (b) Wavelet Reynolds number at scale $64\Delta x$, which fluctuates between 148 and 2700 with a mean value of 1713. (c) Wavelet Reynolds number at scale $20\Delta x$, which fluctuates between 31 and 578 with a mean value of 365. (d) Wavelet Reynolds number at scale $8\Delta x$, which fluctuates between 1 and 27 with a mean value of 17. (e) Wavelet Reynolds number at scale $2\Delta x$, which fluctuates between 0 and 3 with a mean value of 2. This figure is also shown at www.cambridge.org/resources/0521533538.

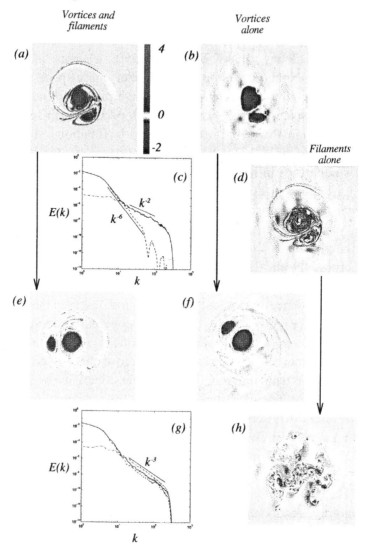

Fig. 4.6. Dynamical analysis of coherent structures and incoherent background flow. (Colours referred to in this caption are shown at www.cambridge.org/resources/ 0521533538) (a) Total vorticity at $t = 30$ computed with a resolution 1024^2. (b) Vorticity corresponding to the coherent vortices alone at $t = 30$. They are made up of 31 strong wavelet packet coefficients which contain 83% of the total enstrophy. (c) Energy spectra at $t = 30$. In green: the total energy spectrum. In red: the coherent vortices energy spectrum. In blue: the filament energy spectrum. (d) Vorticity corresponding to the filaments alone at $t = 30$. They are made up of 1 048 545 weak wavelet packet coefficients which contain 17% of the total enstrophy. (e) Integration of the total vorticity until $t = 120$. (f) Integration of the coherent vortices alone until $t = 120$. (g) Energy spectra at $t = 120$. In green: the total energy spectrum. In red: the coherent vortices energy spectrum. In blue: the filament energy spectrum. (h) Integration of the filaments alone until $t = 120$.

4.5.2 Three-dimensional turbulence modelling

The assumption that the high-wavenumber Fourier modes are slaved to the active low-wavenumber Fourier modes, is probably also wrong for three-dimensional turbulence due to the evidence of energy backscattering [55], [54], [56], [129], [169], i.e. inverse energy transfer from small to large scales, resulting from the presence of organized structures which locally interact and transfer energy to larger scales. We should take this observation with caution knowing that the amount of backscattering observed depends sensitively on the sharpness of the spectral filter used. There are two other reasons to explain why the assumption that the high-wavenumber modes are slaved to the low-wavenumber modes is not valid and should be revised.

The first reason comes from the fact that we do not have any universal theory of turbulence aside from the statistical theory which deals with homogeneous and isotropic ensemble averages, while numerical simulations compute one flow realization at the time (at the highest resolution possible with present supercomputers) and not ensemble averages (which will require too many computations to obtain several realizations of the same turbulent flow). Actually each flow realization is, unlike an ensemble average, highly inhomogeneous due to the presence of coherent structures. As we have shown in performing wavelet analyses of two- and three-dimensional turbulent flows, coherent structures are multi-scale and, through their mutual nonlinear interactions, are responsible for inverse energy transfers. If the computational grid is too coarse, its resolution is insufficient to accurately compute these transfers. Likewise subgrid-scale parametrization is only able to model direct transfers (from resolved to unresolved scales) and inverse transfers (from unresolved to resolved scales) in a statistical sense, assuming homogeneity, but cannot exactly compute the tranfers for the given inhomogeneous flow realization one integrates. In fact backscattering is a major unresolved drawback of current numerical methods, which will last as long as they will be unable to separate the coherent structures from the background flow and take into account the parametrization of homogeneous turbulent components separately from the inhomogeneous components. This difficulty comes from the fact that both components are multi-scale and therefore low-pass filters are inadequate here.

The second reason comes from the fact that our current numerical methods are defined, either in grid-point, finite element or Fourier representation, and are unable to compute multi-scale objects with a small number of coefficients. This would be possible using either adapted multi-grid or wavelet numerical methods. Multi-grid techniques were proposed 20 years ago by Achi Brandt

[31] for solving elliptic problems, such as the diffusion equation; they were then adapted to quasi-stationary problems, but do not seem yet optimal to solve time-dependent problems. Actually the multi-grid approach is similar to a wavelet approach using a Haar wavelet, which is very well localized in physical space and corresponds to a set of embedded grids, but which is too delocalized in spectral space and tends to produce large errors in the higher order derivatives of the solution. As far as we know, locally refined multi-grid techniques have been tried for the Navier–Stokes equations, but not yet in the turbulent regime.

One possible approach is to use the wavelet Reynolds number to split the Navier–Stokes equations at each time step into advection and diffusion operators, which will be solved separately using the most appropriate numerical method and turbulence parametrization for each operator. Namely, the advection term should be computed only where $\tilde{Re}(x, r) > O(1)$, and the diffusion term where $\tilde{Re}(x, r) \leq O(1)$. This method makes sense only if the flow is computed either in a multi-grid or in a wavelet representation (see section 4.6).

Actually, as we have already said, the Navier–Stokes equations are computationally intractable for the large Reynolds number limit which corresponds to fully developed turbulent flows. Although the use of wavelets may improve current numerical methods of solving the Navier–Stokes equations (see section 4.6), a more promising direction may be to look for a new set of equations specific to the turbulent regime. Such equations would be written in terms of a small number of new variables corresponding to the degrees of freedom attached to the coherent structures. As a consequence of this drastic reduction of degrees of freedom to compute, these new equations may break some of the symmetries of Euler or Navier–Stokes equations. This is analogous to the way in which Boltzmann's equation, describing the macroscopic level, breaks the time reversibility of Newton's equation, describing the microscopic level. For modelling turbulent flows we ought to go one step further in the hierarchy of embedded equations going from Boltzmann's to Navier–Stoke's and define a new 'organized' level emerging out of the thermalized background flow.

4.5.3 Stochastic models

The idea is to find stochastic models of turbulence that mimic the behaviour of Navier–Stokes equations at high Reynolds numbers, but which would be easier to solve numerically, and perhaps even analytically. These models could then be used to study some properties of turbulent flows, such as energy spectrum, probability density functions, intermittency and departure from Kolmogorov's scaling.

The first attempt was done in 1974 by Desjanski and Novikov [52] who devised a so-called shell model where the Navier–Stokes equations were represented on a discrete set of wavenumbers in Fourier space, each Fourier shell corresponding to one octave. The coupling between different octaves was supposed to be local in Fourier space and energy was transferred only from large to small scales. Such shell models, sometimes also called cascade models, are still popular because with them it is easy to obtain very large inertial range, up to Reynolds numbers 10^{10}, at a limited computational cost. The number of degrees of freedom needed to compute three-dimensional Navier–Stokes equations by standard direct simulations scale as $Re^{9/4}$, whereas they scale as Re for shell models. The weak point of shell models is that the vectorial structure of Navier–Stokes equations is lost, the incompressibility condition is not satisfied and they do not give accurate information on the spatial structure of the flow.

In 1981 Zimin [209], [90], [210] proposed another model, called the hierarchical model, defined in both space and scale. He projected the three-dimensional Navier–Stokes equations onto Littlewood–Paley basis and discretized them by octaves, considering a limited number of vortices for each octave, few in the large scales and more in the small scales in accordance to Heisenberg's uncertainty principle. He then assumed that each vortex is advected by the velocity field of the larger vortices, which lead him to propose a set of semi-Lagrangian wavelets to compute the flow evolution. This impressive work foreshadowed the wavelet decomposition, and has since been developed by Frick [89], [88], [8]. Hierarchical models are more physical than shell models because they also take into account the vortex motions, but they are still not very realistic from a physical standpoint because they neglect the vortex deformation which is responsible for energy transfers and subsequent dissipation. Recently Eyink [68] has criticized this approach in showing that semi-Lagrangian wavelets do not remove the effect of large-scale convection to the energy transfers and therefore do not guarantee their locality in wavenumber space. This is again due to Heisenberg's uncertainty principle and is related to the fact that it is impossible to compare transfers between wavenumbers and transfers between scales (this point has already been discussed in section 4.4).

Ideas on turbulence evolve at a very slow pace. As an example of this, let us quote what Liepmann wrote in the proceedings of the turbulence conference held in Marseille in 1961 [140]:

The success of the spectral representation of turbulent fields is due, after all, not to the belief in the existence of definite waves but to the possibility of representing quite general functions as Fourier integrals. In the application to stochastic problems the usefulness of the Fourier representation stems essentially from their translational invariance. Consequently, really successful models for representing turbulent shear flows will require far broader invariance considerations. It is clear that the essence of turbulent motion is vortex interaction. In the particular case of homogeneous isotropic turbulence this fact is largely masked, since the vorticity fluctuations appear as simple derivatives of the velocity fluctuations. In general this is not the case, and a Fourier representation is probably not the ultimate answer. The proposed detailed models of an eddy structure represent, I believe, a groping for an eventual representation of a stochastic rotational field, but none of the models proposed so far has proven useful except in the description of a single process.

These remarks, written 37 years ago, are still very pertinent and define the direction we should take for future research in turbulence.

Nowadays, using wavelets we can construct more elaborate stochastic processes. As Liepmann has perceived we should be able to synthesize stochastic rotational fields, built from a set of randomly translated, rotated and dilated elementary vortices, which should have the same non-Gaussian statistics as those observed for two- and three-dimensional turbulent flows (see Figure 4.3). Recently Elliott and Majda [64], [65] have used wavelets to build a Gaussian, stationary and self-similar stochastic process for synthesizing turbulent velocity fields satisfying Taylor's hypothesis and displaying Kolmogorov's energy spectrum. Using these synthetic velocity fields they recover Richardson's law for scalar pair dispersion [66]. It is well-known that the Gaussian hypothesis is incompatible with the turbulence cascade (Skewness being non zero [113]), but their method may be useful to model the background flow, which, contrary to coherent structures, may present Gaussian statistics, although this point is still very controversial [192].

4.6 Turbulence computation

4.6.1 Direct numerical simulations

The direct numerical simulation of turbulent flows, based on the integration of the Navier–Stokes equations at high Reynolds number without any sub-grid turbulence model, requires a very large number of degrees of freedom. This number increases like Re in two dimensions and like $Re^{9/4}$ in three dimensions. Among the numerous Eulerian and Lagrangian numerical schemes, one may identify two different points of view: spectral and physical.

The first long-time simulations of two-dimensional turbulent flows [14], [143], [22] were based on spectral methods, i.e. Fourier decomposition, and had a resolution of 512^2. More recently, resolutions of 4096^2 have been calculated [42], but even these high-resolution simulations cannot attain realistic Reynolds numbers which are several orders of magnitude larger. The observation of the formation of coherent structures in both laboratory and numerical experiments lead to the recognition of the important dynamical role played by vortices in turbulent flows and resulted in the development of Lagrangian methods ([1], e.g. vortex methods [134], [47] or contour dynamics methods [127]) which follow the motion of each vortex, but which are imprecise concerning the background flow between the vortices. Finite-element, -difference or -volume methods allow mesh refinement in regions of the flow where small structures appear, for instance in the boundary layer of an obstacle; unfortunately automatic adaptive refinement requires post-processing to follow these small structures.

Wavelet bases, in the context of the numerical simulation of PDEs (partial differential equations), appear to be a good compromise between spectral methods (precise, but expensive), contour dynamics (which automatically follow coherent structures, but not the background flow) and finite element or finite difference methods (local in space, of low order and therefore not precise). Wavelet methods have already been used to solve Burgers' equation in one [9], [104] and two dimensions [25], Stokes' equation in two dimensions [196], the Kuramoto–Sivashinsky equation [159], Benjamin–Davis–Ono–Burgers' equation [87], the heat equation in two dimensions [40], some reaction-diffusion equations in one and two dimensions [94], [29], [28], the nonlinear Schrödinger equation [100], Euler's equation [172] and Navier–Stokes' equation in two dimensions [41], [96].

4.6.2 Wavelet-based numerical schemes

The localization of wavelet bases, both in space and scale, leads to an effective nonlinear compression of the solution as well as of the operators involved in equations (4.3). Such a sparse representation is obtained by performing nonlinear thresholding of the wavelet coefficients of the solution and of the matrices representing the operators, i.e. those coefficients with absolute value below a given threshold are set to zero. This thresholding can be justified by theoretical results [53] and verified by numerical experiments.

The sparsity of the wavelet expansion of a given function is linked to its local smoothness: where the function is smooth, the corresponding wavelet coefficients decrease with scale. This fact is related to the characterization of point-wise Hölder spaces [108], [105] (see subsection 4.3). Recall that for the Fourier decomposition, the decay of the coefficients depends on the global regularity of the function [211]. Another important property of wavelets is the nonlinear approximation of functions: the approximation error between a function and its wavelet series taken as the N largest coefficients (in a given norm) can be estimated, in some Lebesgue space, by a (negative) power of N which depends on the smoothness, or non-smoothness, of this function. This result follows from the characterization of Sobolev and Besov spaces by means of wavelet coefficients [151], [53], [58]. Note that the nonlinear wavelet approximation of a given function is associated with a grid in physical space which is refined where there are singularities of this function. A comparison of Fourier versus wavelet and wavelet packet nonlinear compression for a turbulent vorticity field is shown in Figure 4.7. We observe that the wavelet packet compression is the most efficient, both in terms of the minimal number of coefficients used and the quality of the approximation.

Another important consequence of the simultaneous localization in space and scale of wavelet bases is that many pseudo-differential operators and their inverse have a sparse representation, i.e. are almost diagonal or have a typical finger structure, depending on the employed (i.e. non-standard or standard) form [24]. This is the case for the gradient operators and the heat kernel. For a theoretical justification in the general context of Calderon–Zygmund operators we refer the reader to [151]. As an example, the discretized heat kernel (on a 1024^2 grid) is projected onto a wavelet basis and we observe that only 9.5 % of the coefficients are greater than 10^{-8}, absolute value to be compared to the largest eigenvalue which is order 1, instead of 21 % for a finite difference projection.

These two fundamental properties (compression of the solution and of the operator) allow us to define adaptive wavelet-based numerical schemes for solving nonlinear PDEs. By neglecting small coefficients in the solution and/ or in the operator's wavelet representation, each step of the algorithm is based on approximate but fast matrix-vector products computed in wavelet space. Note that the schemes based on scaling functions (often deliberately confused with wavelets) [102], [124], [87] instead of wavelet functions are no more efficient than classical finite element methods on a regular grid! Theoretical error and stability estimates for some particular wavelet schemes

---

Fig. 4.7. Nonlinear compression of a vorticity field. In each case the reconstructions using the strong coefficients (containing 95% of the total enstrophy) are displayed on the left, and using the weak coefficients (containing 5% of the total enstrophy) are displayed on the right. (a) Uncompressed vorticity field computed with a resolution of 512^2. (b) Compression in a Fourier basis (813 strong coefficients). (c) Compression in a wavelet basis (338 strong coefficients). (d) Compression in a wavelet packet basis (156 strong coefficients). This figure is also shown at www.cambridge.org/resources/0521533538.

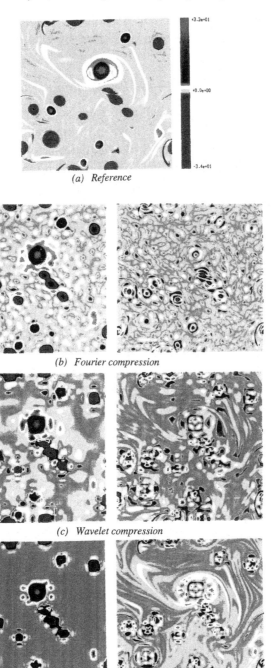

(a) *Reference*

(b) *Fourier compression*

(c) *Wavelet compression*

(d) *Wavelet packet compression*

may also be derived [24], [50], [26]. A scaling function scheme for solving the Euler equations has already been developed by Qian and Weiss [172].

4.6.3 Solving Navier–Stokes equations in wavelet bases

Before presenting wavelet-based numerical schemes to solve the Navier–Stokes equations, we should mention a very interesting direction which consists of simplifying the Navier–Stokes equations by re-writing them in an appropriate wavelet basis. Jacques Lewalle has shown that some continuous wavelets, namely the Hermitian wavelets (derivatives of the Gaussian), simplify the resolution of the linear term and allow a simpler convolution formula for the nonlinear term [136], [137]. He has found that the first derivative of the Gaussian gives a Hamiltonian form of the diffusion equation, where dissipation is replaced by spectral transport, namely Hermitian wavelets are propagators for the diffusion equation [138].

The first adaptive wavelet schemes for the Navier–Stokes equations have been derived by Charton and Perrier [39] and Fröhlich and Schneider [96]. Different approaches can be used to solve the two-dimensional Navier–Stokes equations. We will focus here on the two recently developed wavelet schemes for solving Navier–Stokes equations: the algebraic wavelet method of Charton and Perrier [41] and the Petrov–Galerkin scheme of Fröhlich and Schneider [96, 97]. Both methods are based on the discrete wavelet transform and take advantage of the nonlinear compression of the operators and the solution.

Apart from the above Eulerian schemes another possible approach would be to develop Lagrangian-type wavelet methods, based on the continuous wavelet transform. The travelling wavelet method in which wavelets behave like particles evolving in phase-space coordinates has been proposed in 1990 by Basdevant, Holschneider and Perrier [13]. The travelling wavelet method looks for an approximate solution of equation (4.50) (see below), which is a finite sum of N wavelets evolving in phase-space:

$$\omega(x, t) \approx \sum_{i=1}^{N} c_i(t) \psi\left(\frac{x - b_i(t)}{a_i(t)}\right), \, a_i > 0, \tag{4.49}$$

where ψ is the wavelet and c_i, a_i, b_i, are respectively the time dependent amplitude, scale and position parameters.

This method works well for linear equations, such as the convection-diffusion equation, and also for the Korteweg–de-Vries equation. It has also been applied to the study of the formation of galaxies [20]. However, in the nonlinear case the method encounters technical difficulties which have not yet been completely overcome. These difficulties arise when two wavelets approach each other in phase-space which leads to a 'phase-space atom collision'.

Now let us consider the two-dimensional Navier–Stokes equations written in terms of vorticity and stream function, which are scalars

$$
\begin{cases}
\dfrac{\partial \omega}{\partial t} + \boldsymbol{v} \cdot \nabla \omega = \nu \nabla^2 \omega + F, \; x \in [0, 1]^2, \; t > 0 \\
\nabla^2 \Psi = \omega, \; \boldsymbol{v} = \left(-\dfrac{\partial \Psi}{\partial y}, +\dfrac{\partial \Psi}{\partial x} \right).
\end{cases} \tag{4.50}
$$

We complete the problem with periodic or Dirichlet or Neumann boundary conditions and a suitable initial condition.

By introducing a classical semi-implicit time discretization and a time step δt, and setting $\omega^n(x) \approx \omega(x, n\delta t)$ to be the approximate solution at time $n\delta t$, equation (4.50) is replaced, for example (for notational ease we take here the simplest, but unstable, time scheme), by

$$
\begin{cases}
(1 - \nu \delta t \nabla^2) \omega^{n+1} = \omega^n + \delta t (f^n - \boldsymbol{v}^n \cdot \nabla \omega^n) \\
\nabla^2 \Psi^{n+1} = \omega^{n+1}, \; \boldsymbol{v}^{n+1} = (-\partial_y \Psi^{n+1}, +\partial_x \Psi^{n+1}), \\
\text{with } f = \nabla \times \boldsymbol{F}.
\end{cases} \tag{4.51}
$$

The spatial discretization is then performed by approximating, at time $n\delta t$, ω^n by a function ω_J^n belonging to a finite dimensional subspace V_J obtained from a multiresolution analysis $(V_j)_{j \geq 0}$ of the space $L^2([0, 1]^2)$.

At this point the algebraic method of Charton and Perrier differs significantly from the Petrov–Galerkin scheme of Fröhlich and Schneider. The method proposed by Charton and Perrier [41] starts with a finite difference scheme on a regular Cartesian grid. Wavelets are then used to speed up the solution procedure by compression of the discrete inverse operator and the actual solution during the time advancement. Furthermore, operator splitting by means of an ADI (Alternating Direction Implicit) technique is introduced. The two-dimensional wavelet basis employed relies on a tensor product of two one-dimensional multiresolution analyses. The method proposed by Fröhlich and Schneider [97, 96] uses a two-dimensional multiresolution analysis as the projection basis. In this case the inverse operator is applied during the time advancement, using special test functions.

We will attempt to clarify the principle of these wavelet methods. The spatial approximation can be either of collocation type, i.e. grid-point values, or of Galerkin type, i.e. a projection onto a basis. The transformation between the single level representation of a function, i.e. its values at regular collocation points, and a multi-level wavelet Galerkin representation uses an orthogonal wavelet transform. However, problems arise with adaptive schemes because it is difficult to take advantage of the sparsity of the wavelet decomposition when going back and forth between grid point and wavelet representations. Let us be more precise, and consider the one-dimensional case. Suppose that dim $V_J = 2^J$. Then the function ω_J^n can be expanded onto the scaling function basis (single level representation) $(\varphi_{J,k})_{k=0,2^J-1}$ of V_J

$$\omega_J^n(x) = \sum_{k=0}^{2^J-1} c_{J,k}^n \varphi_{J,k}(x), \tag{4.52}$$

or onto a wavelet basis $(\psi_{j,k})_{0\leq j<J, k=0,2^j-1}$ of V_J

$$\omega_J^n(x) = \sum_{j=0}^{J-1}\sum_{k=0}^{2^j-1} d_{j,k}^n \psi_{j,k}(x) + c_{0,0}^n. \tag{4.53}$$

The transition between both representations is done by the orthogonal wavelet transform (Mallat's algorithm).

In the collocation method, the function ω_J^n is naturally associated with a regular grid $(x_k = k2^{-J})_{k=0,2^J-1}$ of $[0, 1]$ and its corresponding collocation values $\omega_J^n(x_k)$. Often, by using properties of scaling functions $\varphi_{J,k}$ one can identify

$$\omega_J^n(x_k) \approx 2^{-J/2} c_{J,k}^n. \tag{4.54}$$

The wavelet Galerkin method is based on the wavelet coefficients $d_{j,k}^n$, and in practice uses only a few (non-negligible) coefficients larger than a given threshold ε: $\{d_{j,k}^n; |d_{j,k}^n| > \varepsilon\}$. Mallat's fast wavelet algorithm works well for regular grids, but is not efficient for irregular grids made up of irregularly spaced grid points x_k corresponding to the 'centres' of wavelets $\psi_{j,k}$, for which the coefficients of $\omega_J^n(x_k)$ satisfy $|d_{j,k}^n| > \varepsilon$. To avoid this problem, one can introduce in many cases, an interpolating function of V_J [201] and adapt Mallat's fast wavelet algorithm [95], [97]. Another way to overcome this problem is to directly construct the interpolating scaling functions $\varphi_{J,k}$ and the corresponding interpolating wavelet basis $\psi_{j,k}$ [57], [25]. Finally, one can also construct an adaptive multiresolution analysis [171], [4].

The algorithm (4.51) for solving the two-dimensional Navier–Stokes equations can now be split into four steps which we will discuss below: (1) time-stepping of the heat equation, (2) solving a Poisson equation, (3) computing the nonlinear term, (4) imposing the boundary conditions.

4.6.3.1 The heat equation solution

Let us consider the discretized heat equation

$$(1 - \nu\delta t\nabla^2)\omega^{n+1} = \omega^n + \delta t f^n. \tag{4.55}$$

The biorthogonal approach introduced in [139], [126], [94], [97] consists of building a biorthogonal system from a classical wavelet basis $\psi_{j,k}$, first setting

$$\theta_{j,k} = (1 - \nu\delta t\nabla^2)^{-1}\psi_{j,k}, \tag{4.56}$$

with suitable hypotheses on ψ. Then a system $\tilde{\theta}_{j,k}$ biorthogonal to $\theta_{j,k}$ is constructed, and solving equation (4.55) reduces to the change of basis

$$\langle\omega^{n+1}|\psi_{j,k}\rangle = \langle\omega^n|\theta_{j,k}\rangle + \delta t\langle f^n|\theta_{j,k}\rangle, \tag{4.57}$$

where the notation $\langle|\rangle$ denotes scalar product. The functions $\theta_{j,k}$ and $\tilde{\theta}_{j,k}$ are called vaguelettes and have localization properties similar to those of wavelets [151]. This approach avoids assembling and solving a linear system. For the collocation projection operator-adapted cardinal functions [97] have been constructed which allow the construction of efficient interpolatory quadrature formulas. The decomposition of the right hand side of equation (4.57) can then be calculated using the fast adaptive vaguelette decomposition of [97] based on a hierarchical subtraction strategy. This approach has been used for one- and two-dimensional problems.

The Galerkin approach is to project (4.55) onto a classical, orthogonal or biorthogonal, wavelet basis $(\psi_{j,k})$ of the space V_J. We can write

$$\left(\langle\omega_J^{n+1}|\psi_{j,k}\rangle\right)_{j,k} = K\left(\langle\omega_J^n + \delta t f|\psi_{j,k}\rangle\right)_{j,k} \tag{4.58}$$

where

$$K_{(j,k),(j',k')} = \langle(1 - \nu\delta t\nabla^2)^{-1}\psi_{j,k}|\psi_{j',k'}\rangle \tag{4.59}$$

is the heat kernel, which is almost diagonal, as explained in section 4.6.2. This step is based on approximated, but fast, matrix-vector products. An easy way to reduce the previous two-dimensional system to several one-dimensional systems is to use a tensor wavelet basis $(\psi_{j,k}(x).\psi_{j',k'}(y))$ and to split the two-dimensional heat kernel into two one-dimensional operators

$$(1 - \nu\delta t\nabla^2)^{-1} \approx (1 - \nu\delta t\frac{\partial^2}{\partial x^2})^{-1}(1 - \nu\delta t\frac{\partial^2}{\partial y^2})^{-1} \tag{4.60}$$

as in ADI methods. Such a method is applied in [40], [41].

4.6.3.2 The Poisson equation

The solution of the Poisson equation

$$\nabla^2\Psi^{n+1} = \omega^{n+1} \tag{4.61}$$

can be obtained using a pseudo-transient technique, i.e. calculating the steady state solution of the heat equation, which, as in ADI methods, is reached in only a few iterations by considering iterated powers K^n of the heat kernel K (4.59) which become sparser with n [40].

An alternative approach, proposed by Jaffard [108], is to consider the well-conditioned system

$$\mathbf{P A P P}^{-1} \big(\langle \Psi_J^{n+1} | \psi_{j,k} \rangle \big)_{(j,k)} = P \big(\langle \omega_J^{n+1} | \psi_{j,k} \rangle \big)_{(j,k)} \tag{4.62}$$

where A is the Galerkin matrix of the Laplacian in a wavelet basis: $A_{(j,k),(j',k')} = \langle \nabla^2 \psi_{j,k} | \psi_{j',k'} \rangle$ and P is the diagonal preconditioning matrix: $P_{(j,k),(j',k')} = 2^{-j} \delta_{j,j'} \delta_{k,k'}$, in one dimension (in two dimensions this should be modified according to the chosen 2D wavelet basis). Jaffard proved that the condition number of PAP does not depend on the dimension of the system. Then the solution of (4.62) can be reached in a few iterations by a classical conjugate gradient method.

The biorthogonal approach is also possible using operator-adapted biorthogonal vaguelettes for homogeneous operators, i.e. $\theta_{j,k} = (\nabla^2)^{-1} \psi_{j,k}$ and $\tilde{\theta}_{j,k} = \nabla^2 \psi_{j,k}$. The solution of the Poisson equation then reduces to a change of basis, analogously to the case of the heat equation.

4.6.3.3 The nonlinear term

The nonlinear term $v^n \cdot \nabla \omega^n$ can be computed either by a collocation or by a Galerkin method. The collocation (also called pseudo-wavelet by analogy with pseudo-spectral) method can be sketched as follows: starting from the wavelet coefficients of ω^n we obtain the values of ω^n on a locally refined grid through an inverse wavelet transform. Solving the Poisson equation with one of the methods described in the previous section, we get the wavelet coefficients of the stream function Ψ^n. Applying an inverse wavelet transform, the stream function is reconstructed on a locally refined grid. Subsequently, the velocity v^n and $\nabla \omega^n$ are calculated using finite differences on an adaptive grid. Then the scalar product $v^n \cdot \nabla \omega^n$ is calculated at each grid point. Finally, the wavelet coefficients of the nonlinear term are obtained by a wavelet transform. However, in the bi-orthogonal approach the right hand side of the first equation of (4.51) is summed up on the adaptive grid in physical space and then the wavelet coefficients of the vorticity ω^{n+1} are calculated using the adaptive vaguelette decomposition. This collocation method requires a fast wavelet transform between grid points and sparse coefficients sets. This problem was mentioned in the previous section. Fröhlich and Schneider [97], [95] have developed such a wavelet transform for lacunary bases which enables the adaptive evaluation of terms of the form $f(\omega)$ without derivatives. This method has been applied for the full adaptive discretization of reaction-diffusion problems [28]. The algorithm described above will enable the adaptive evaluation of the convective term.

On the other hand, a Galerkin method works only in the wavelet coefficient space, avoiding transforms between physical and wavelet space [23]. The nonlinear term is then written as a convolution of the wavelet coefficients of v^n and $\nabla \omega^n$; these convolutions involve triple wavelet connection coefficients of the form $\langle \psi_{j_1,k_1} \psi'_{j_2,k_2} | \psi_{j_3,k_3} \rangle$. *A priori* the complexity of such a calculation is very large, but the method can be competitive for two reasons. First, since the wavelets are localized both in space and scale, connection coefficients vanish when two of the three wavelets are separated either in scale or space. Hence, only a limited number of terms in the convolution are significant. Secondly, the method can handle adaptive description of the fields, i.e. the convolution can be restricted to the significant components of the flow [167].

Let us mention that at the moment the nonlinear term is computed by a collocation method either on a regular grid [41], [96], or on an adapted grid [28].

4.6.3.4 The boundary conditions

Boundary conditions are in general included in the definition of the spaces $(V_j)_{j \in Z}$ when constructing the multiresolution analysis. The simplest and most popular (due to the development of Fourier spectral methods) are periodic boundary conditions for which periodic wavelets, in one or several dimensions, can be easily constructed [165]. For Dirichlet or Neumann boundary conditions, compactly supported bases have recently been constructed in one dimension [48], [156], [157], and these bases are also associated to fast orthogonal wavelet transforms, like for the periodic case. They can easily be included in some of the previous algorithms, since the extension to cubic domains in several dimensions is trivial using tensor products of wavelets (in practice all two-dimensional orthogonal wavelet bases are tensor products, which raises the problem of the lack of isotropy).

One should also mention the existence of divergence-free wavelet bases [131], [130], which can be used for the velocity-pressure formulation of Navier–Stokes equation (4.1) and automatically take into account the incompressibility condition [196].

4.6.4 Numerical results

To illustrate the previously described adaptive wavelet methods we present some numerical results for two different cases, i.e. a strong nonlinear interaction of three vortices and a decaying turbulent flow. For comparison a classical pseudo-spectral method serves as a reference. Furthermore in order to study statistically stationary turbulent flows we discuss results computed with a recently developed wavelet based forcing method [185]. In all computations presented below the method of Frölich and Schneider [96] using cubic spline wavelets of Battle–Lemarié type have been used.

4.6.4.1 Three vortex interaction

As a prototype for vortex merging we consider the strong nonlinear interaction of three Gaussian vortices [186]. This is an important test case, because the flow dynamics is highly nonlinear, but not yet chaotic (although the motion of four vortices would be). This allows us to compare in a deterministic manner the time evolution of the solution computed with different numerical schemes, presenting different truncation errors (here we will compare a pseudo-Fourier scheme and a pseudo-wavelet scheme) with the same number of nodes. As soon as the dynamics of the system one computes becomes chaotic, namely sensitive to initial conditions and therefore to numerical errors, it becomes tricky to compare the predictions of different numerical schemes. A 'deterministic comparison' (based on the L^2-norm of the difference between two solutions computed with two different schemes) works well for laminar flows, but it should be replaced by a 'statistical comparison' as soon as the dynamics becomes chaotic (namely beyond the onset of the transitory regime). The choice of the appropriate statistical diagnostics has been addressed for several years by Farge and Wickerhauser [74], [205], but is still an open issue, not yet sufficiently discussed in the numerical analysis literature.

For details on the numerical simulation we refer the reader to Schneider, Kevlahan and Farge [186]. The initial condition is given by the superposition of two positive and one negative Gaussian vortices, $\omega(x, y) = \sum_{i=1}^{3} A_i \exp\left(-((x - x_i)^2 + (y - y_i)^2)/\sigma_i^2\right)$ with amplitudes $A_1 = A_2 = -2A_3 = \pi$ and $\sigma_i = 1/\pi$. The maximum resolution of the computation corresponds to a finest scale $J = 8$ which is equivalent to 256^2 possible degrees of freedom. As threshold for the adaptive method [97] we used $\varepsilon = 10^{-6}$, i.e. only wavelet coefficients with absolute value larger than ε have been computed.

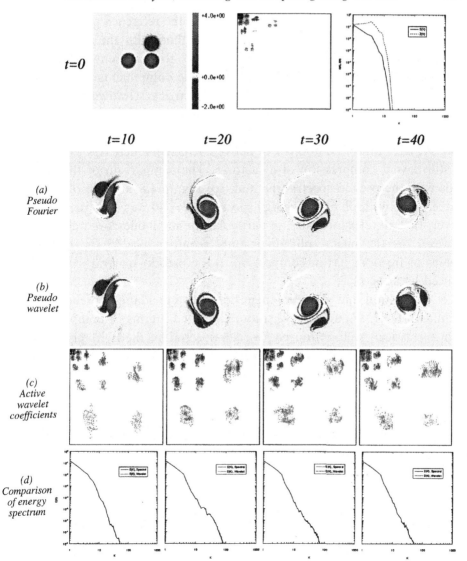

Fig. 4.8. Simulation of the merging of three vortices at times $t = 10, 20, 30, 40$. (a) Vorticity field, reference pseudo-spectral method. (b) Vorticity field, adaptive wavelet method. (c) Wavelet coefficients used in the adaptive wavelet method. (d) Comparison of Fourier energy spectra for the pseudo-spectral and adaptive pseudo-wavelet methods (note that the two curves are identical). This figure is also shown at www.cambridge.org/resources/0521533538.

In Figure 4.8 we show the vorticity field for the reference pseudo-spectral method and the adaptive pseudo-wavelet method with the corresponding computed wavelet coefficients (dark entries) for different instants. We observe that during the interaction small scale components are produced, which is directly reflected in the active (i.e. the strongest) wavelet coefficients. At $t = 0$ the vorticity field is highly regular and the strongest wavelet coefficients (namely those larger than a given threshold) represent only 3 % of the total. At later times the number of active coefficients increases to 20 %, i.e. we still have a compression of a factor 5. The comparison of the vorticity fields with the pseudo-spectral method, see Figure 4.8, shows no significant difference. If we look at the energy spectra at $t = 40$ we can observe quantitatively that all relevant scales, in particular the small ones, are well resolved. However, as the fine resolution is only required locally, the number of degrees of freedom has in comparison to the pseudo-spectral method been reduced by a factor 5.

Let us mention that at the moment both existing adaptive pseudo-wavelet methods [41, 97, 96] are not yet more efficient in terms of computing time than a classical, well-optimized, pseudo-spectral method. In principle the adaptive wavelet methods have a computational complexity of order $O(N_{ad})$, where N_{ad} denotes the number of the degrees of freedom adapted to the solution. In comparison the pseudo-spectral methods are of order $O(N_{reg}log_2N_{reg})$ complexity, where N_{reg} denotes the number of degrees of freedom on the regular grid. The actual numerical cost depends directly on the constant multiplying the order term. At the moment this factor is rather high for the adaptive wavelet methods. Therefore for simulations at moderate resolutions, such as $N = 128^2$ or 256^2, the adaptive pseudo-wavelet methods cannot yet outperform the classical spectral methods, although their operation count scales slower, as $O(N)$ instead of as $O(N \, log_2N)$. But we have some hope to be able to significantly reduce the time step needed with the adaptive wavelet code, due to the fact that the retained coefficients are attached to vortices, namely locations of strong vorticity but weak translational velocity. Therefore the CFL (Courant–Friedrich–Lewy) criterion, defining the largest time step to guarantee stability for an explicit time scheme, can be based on a much larger spatial step than the smallest scale computed by the adaptive wavelet scheme [85].

4.6.4.2 Freely decaying turbulence

For the computation of freely decaying turbulence one typically uses a statistical initial condition, generated by means of a Gaussian random distribution and imposing a given energy spectrum. Here we used a broad band spectrum of the form $E(k) = ck^2/(k_0^6 + k^6) \exp(-k^2/k_v^2)$ with $k_0 = 10$ and $k_v = 80$. The constant c has been chosen such that the total kinetic energy was equal to 1/2. The maximal resolution was 256^2 numbers of degrees of freedom, with $v = 10^{-3}$. Using a classical pseudo-spectral method we calculated the solution up to $t = 4$ corresponding to 12 initial eddy turnover times. The resulting vorticity field, exhibiting coherent structures and a smooth spectrum with an inertial range, was then taken as initial condition for the adaptive wavelet calculation and therefore we assigned the time $t = 0$. The threshold for the wavelet coefficients was $\varepsilon = 5 \cdot 10^{-5}$. In Figure 4.9 we give an example of the vorticity field at $t = 2$ for the pseudo-spectral method and the adaptive wavelet method with the corresponding wavelet coefficients which have been computed. As observed in the case of the three vortices, the wavelet solution does not exhibit a visible difference with respect to the spectral method. However, out of the total 256^2 wavelet coefficients, only about 20 % have been used during the calculation of the solution. The energy spectrum also does not deviate significantly from the reference, thus we may conclude that all scales are well-resolved with only 1/5th of the possible degrees of freedom. We should mention that the resolution of the present calculations with 256^2 is fairly small. Since for higher resolutions larger Reynolds number flows can be computed, the compression rate of the wavelet representation will increase due to the greater intermittency of the flow. Therefore the impact of adaptive wavelet methods will become particularly attractive for high Reynolds number flows.

4.6.4.3 Wavelet-forced turbulence

The numerical simulation of turbulent flows has been performed considering two different regimes: either the freely decaying regime, where the flow is excited initially and its evolution is computed without any forcing, or the forced regime, where the flow is excited in such a way that it reaches a statistically steady state for which the dissipation must be compensated by the forcing. The advantage of the freely decaying regime is that it depends only on the flow's intrinsic nonlinear dynamics, with the hope of thus observing a universal behaviour. The problem with this method is that it never reaches a statistically steady state because energy or enstrophy tends to decay in time. The advantage of the forced regime is that the turbulent flow reaches a statistically steady state, but this state depends on the kind of forcing performed [14], which precludes a universal turbulent behaviour.

Classically, two forcing schemes are used which both operate in Fourier space. Either a negative dissipation within a given wavenumber band, with a complex amplification coefficient which depends on the wavenumber, or a

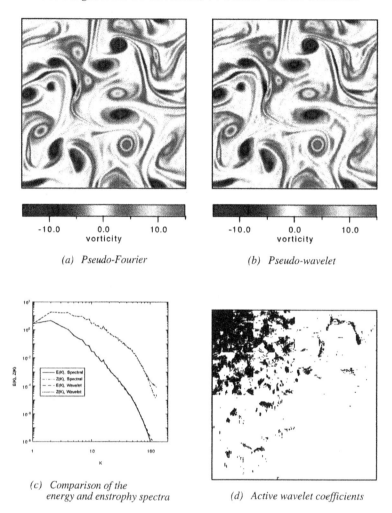

(a) *Pseudo-Fourier*   (b) *Pseudo-wavelet*

(c) *Comparison of the*
     energy and enstrophy spectra   (d) *Active wavelet coefficients*

Fig. 4.9. (a) Vorticity field of the freely decaying turbulence simulation at time $t = 2$ for the reference pseudo-spectral method. (b) The vorticity field at $t = 2$ for the adaptive wavelet method. (c) The Fourier energy and enstrophy spectra for the two methods. (d) The corresponding wavelet coefficients used by the adaptive wavelet method. This figure is also shown at www.cambridge.org/resources/0521533538.

white or coloured noise in time with a prescribed isotropic spectral distribution, strongly peaked in the vicinity of a given wavenumber, with random phases. Neither of the two schemes is a satisfactory model because they inject energy and enstrophy locally in Fourier space and therefore non-locally in physical space. This forcing mechanism is neither intrinsically related to the flow's chaotic dynamics, nor simulates the production of enstrophy on walls and in shear layers, which is local in physical space and therefore broad-band in Fourier space. Another drawback of such a forcing is that the scale of the coherent vortices produced by the nonlinear dynamics of the flow is imposed by the scale at which the forcing is done.

To overcome these drawbacks of the Fourier forcing, a wavelet forcing scheme has been proposed by Schneider and Farge [185], which excites vortices locally in physical space and as smoothly as possible (in order to avoid creating any unphysical discontinuities in the vorticity field), without affecting the background. This wavelet forcing is based on the fact that vortices produced in two-dimensional turbulent flows correspond to the strongest wavelet coefficients of the vorticity fields, while the remaining weaker coefficients correspond to the residual background flow [81], [70], [71], [79]. Therefore it injects enstrophy only into the strongest wavelet coefficients, hence in an inhomogeneous way, in order to excite the vortices without affecting the background flow. This procedure does not interfere with the emergence of vortices and does not impose a scale on them, contrary to the Fourier forcing. The distribution and size of the vortices depend only on the intrinsic nonlinear dynamics of the flow.

For the numerical results presented here both energy and enstrophy are kept steady during more than 60 eddy turn over times. Figure 4.10a displays the vorticity field in a stationary regime at different instants showing that neither the energy spectrum (Figure 4.10c) nor the PDF of vorticity (Figure 4.10d) change significantly in time. The vortices present in the initial condition become more circular and well isolated during the flow evolution because they are better able to withstand the mutual strain due to the additional enstrophy injected into them. We observe that the slopes of the spectra (see Figure 4.10c) are much steeper (close to k^{-6}) than the k^{-3} law predicted by the statistical theory of homogeneous turbulence. This discrepancy, as observed for other types of forcing [14], confirms the fact that the spectral behaviour of two-dimensional turbulent flows is not universal, but instead depends on the forcing. In Figure 4.10b we observe that the spatial support of the active wavelet coefficients decreases with the scale, which reveals a strong intermittency of the flow. Consequently the vorticity field is efficiently compressed in a wavelet basis, because only about 20% of the 128^2 coefficients are needed to represent the flow dynamics. We also show that the PDF of vorticity (Figure 4.10d) is Gaussian for the weak values, corresponding to the background flow, and presents non-Gaussian tails for the strong values, corresponding to the vortices.

In the work presented here, we only excite the vortices produced by the flow's nonlinear dynamics. We can also use the same wavelet forcing to create new vortices by injecting enstrophy locally in the regions of the background flow where the strain (imposed by the coherent structures to the background

$t=60$

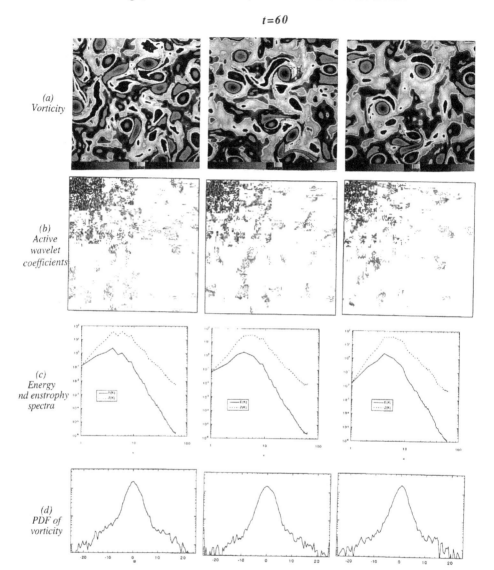

Fig. 4.10. Temporal evolution of the wavelet-forced turbulence simulation. (a) The vorticity field at $t=0$, 60, 120. (b) The wavelet coefficients used at $t=0$, 60, 120. (c) The Fourier energy and enstrophy spectra at $t=0$, 60, 120. (d) The PDF of vorticity at $t=0$, 60, 120. This figure is also shown at www.cambridge.org/resources/0521533538.

flow) becomes weaker than the background vorticity, this in order to simulate the formation of new vortices by instabilities, such as the Kelvin–Helmholtz instability.

4.7 Conclusion

The main factor limiting our understanding of turbulent flows is that we have not yet identified the structures responsible for its chaotic and therefore unpredictable behaviour. Based on laboratory and numerical experiments, we think that coherent vortices are these elementary objects, from which we may be able to construct a new statistical mechanics and define new equations appropriate for computing fully developed turbulent flows.

The quasi-singular vortices encountered in turbulent flows are, by their nature, very rare. In fact, the Cafarelli, Kohn and Nirenberg theorem [33] shows that singular structures, if they exist, must be of Hausdorff measure one in space-time. Most of the statistical diagnostics presently used to analyse turbulent flows are low order statistics and thus insensitive to rare events, while the effect of coherent structures appears only in the higher order statistics. An example of this is the fact that the two-point structure function follows Kolmogorov's 1941 law (which assumes a homogeneous structureless and non-intermittent flow), while the higher order structure functions depart strongly from this law. This deviation is due to the fact that turbulent flows are highly intermittent, and we think that this intermittency is due to the nonlinear interaction of coherent vortices, which correspond to strong but rare events. To efficiently analyse the role of coherent structures in turbulent flows one requires either a high order statistical method or some conditional averaging.

Using a wavelet representation instead of a Fourier representation minimizes the restrictions on the basis functions enlarging them from Sobolev (measuring global smoothness) to Hölder and Besov (measuring local smoothness) spaces. Moreover, the Fourier spectrum used by the present statistical theory of turbulence is not the appropriate way to analyse the physical structure of a turbulent flow, because it loses all spatial information which is present only in the phase of the Fourier coefficients. Since the Fourier spectrum is by definition the Fourier transform of the two-point velocity correlation (which is by Wiener–Khinchin's theorem the modulus of the Fourier transform of velocity), the phase is lost. Furthermore, the Fourier energy spectrum is sensitive to only the strongest isolated singularity in the flow, and even then can give no information about the form or location of this singularity. In short, Fourier space analysis is unable to disentangle coherent vortices from the rest of the flow.

The complementary simultaneous space and scale information provided by the wavelet representation makes it an appropriate tool for identifying and analysing coherent vortices in turbulent flows. The wavelet transform can be used to segment the vorticity field into coherent and incoherent components as the first stage in a conditional sampling algorithm. Such a segmentation method respects Galilean invariance because it is performed on the vorticity field and not on the velocity field. A local wavelet analysis can also give the strength and form of all quasi-singular isolated vortices and separate them from the background flow.

Different wavelet techniques must be used depending on whether the flow contains oscillating (e.g. spiral) or non-oscillating (e.g. cusp) type singularities, and whether it contains isolated (e.g. a single cusp or spiral) or dense (e.g. fractal) distributions of singularities. For example, the current wavelet-based methods for determining the singularity spectrum of a multifractal work only if the signal does not contain oscillating singularities. Turbulence may contain both types of singularities in either dense or isolated distributions. It is therefore important to determine from the beginning whether a given turbulence signal contains oscillating singularities and how these singularities are distributed. This classification is possible using a wavelet-based diagnostic.

In section 4.3 we reviewed the wavelet-based methods for detecting and analysing the singular structure of a signal. We saw that these methods are useful, not only because they provide new information which cannot be obtained using other methods, but also because they formally unify a wide range of previously disparate approaches. For instance the wavelet-based method of calculating the structure functions unifies their analysis with the calculation of energy spectra and the strength of local singularities. Furthermore, wavelets play the role of 'generalized boxes' in a new form of the standard box-counting algorithm used to estimate fractal dimensions. This algorithm brings out the intimate relationship between structure functions and multifractals. These techniques have been applied to analyse turbulent signals.

In section 4.4 we showed that wavelet analysis has been an essential tool for identifying coherent structures as phase-space regions correlated in both space and scale, and for studying their scaling properties. This method has helped to relate the intermittency of turbulent flows to the presence of organized coherent vortices, and explained why the predictions of the statistical theory of turbulence are not verified for high-order statistics. The wavelet representation has also been used to compute the transfers of energy and of enstrophy between coherent and incoherent components of turbulent flows. Wavelet extraction of coherent structures has shown that they have non-Gaussian one-point PDFs, while the background has Gaussian one-point PDFs.

In section 4.5 we reviewed several applications of wavelets for turbulence modelling. In particular, we showed that the wavelet representation, associated with nonlinear filtering, extracts the coherent vortices in a computationally efficient way. Turbulent motions are non-separable in the Fourier representation, while a wavelet representation is able to provide such separability. Based on the analysis mentioned above, we expect a separation in wavelet coordinates between organized vortices (having non-Gaussian statistics) to be explicitly computed, and background flow (having Gaussian statistics) to be modelled by an appropriate stochastic process. This decomposition is the basis of a new way of numerically simulating turbulent flows called Coherent Vortex Simulation (CVS) [84], and possibly other kinds of intermittent phenomena having similar statistics.

In section 4.6 we summarized the progress that has been made in actually computing partial differential equations in wavelet space. Numerous promising experiments have been carried out using wavelets on Burgers' equation in one or two dimensions, heat equation or Stokes equation in two dimensions and Navier–Stokes equations in two dimensions. All these experiments have shown that wavelet approaches are valid, although they are still computationally expensive.

In conclusion, we think that the wavelet functional representation may be the proper tool for building a statistical mechanics of turbulence based on the identification of elementary dynamical structures from the observational data we have. This theory may replace the present Fourier-based statistical theory of turbulence which relies on the symmetries of the Navier–Stokes equations, but is unable to treat near-wall regions where turbulence is produced by instabilities and symmetries are broken. We are now convinced that the Navier–Stokes equations are not the practical model equations to compute large Reynolds number flows. Indeed in this limit, there is probably some symmetry breaking associated with the production of coherent structures out of the random background flow in shear layers.

Turbulence research is a kind of tragi-comedy – tragic due to its military (atomic bomb, missiles, fighter airplanes) applications – and comic because at each generation we seem fated to rediscover old ideas. For instance, our understanding of dissipation and turbulence modelling is the same as what Richardson was suggesting 68 years ago when he wrote

Diffusion is a compensation for neglect of detail. By an arbitrary choice we try to divide motions into two classes: (a) Those which we treat in detail. (b) Those which we smooth away by some process of averaging [177],

and the program we develop corresponds to the prescription for turbulence research proposed 48 years ago by Dryden when he wrote: 'It is necessary to separate the random processes from the non-random element' [60]. This corresponds precisely to what we do: we split each flow realization between rare events out of statistical equilibrium (the coherent vortices) that we compute as a nonlinear dynamical system, and random events in Gaussian statistical equilibrium (the incoherent background flow produced by the nonlinear interactions between coherent vortices) which can be modelled by a Gaussian stochastic process.

Wavelets, as a new mathematical tool, bring new insights to evaluate current methods and we hope that they will lead to a better understanding of turbulent flows. But, knowing the past difficulties encountered in this field, we should not be overly optimistic, nor should we oversell wavelets. As Robert Sadourny likes to say ironically: 'Wavelets? You mean this new approach which will waste another 20 years of turbulence research!'.

Acknowledgements

This work has been supported by the NATO 'Collaborative Research' on 'Wavelet Methods in Computational Turbulence' (contract CRG–930456), by the Training and Mobility of Researchers program of the EC (contract ERBFMBICT950365) and by the French–German DFG/CNRS/MENRT program on Computational Fluid Dynamics (contract GR 1144/7-1). These grants are gratefully acknowledged. We thank Echeyde Cubillo, Eric Goirand and Thierry Philipovitch who computed some of the results presented here and Jean-François Colonna who did the visualizations for figures 3, 4 and 10.

References

[1] Abgrall, R. and Basdevant, C. (1987). Un schéma numérique semi-Lagrangien pour la turbulence bidimensionelle, *C. R. Acad. Sci. Paris*, **305**, série I, 315–318.

[2] Abry, P., Fauve, S., Flandrin, P. and Laroche, C. (1994). Analysis of pressure fluctuations in swirling turbulent flows, *J. Physique II*, **4**, 725–733.

[3] Abry, P. (1997). *Ondelettes et turbulences*, (Diderot).

[4] Amellaoui, A. (1996). Ondelettes splines à support compact sur une grille irrégulière et approximations numériques stables, *Thèse de doctorat, Université Paris 11*.

[5] Anselmet, F., Gagne, Y., Hopfinger, E. J., and Antonia, R. A. (1984). High-order velocity structure function in turbulent shear flows, *J. Fluid Mech.*, **140**, 63–89.

[6] Arnéodo, A., Bacry, E. and Muzy, J. F. (1995). The thermodynamics of fractals revisited with wavelets, *Physica A.*, **213**, 232–275.

[7] Arnéodo, A., Bacry, E. and Muzy, J. F. (1995). Oscillating singularities in locally self-similar functions, *Phys. Rev. Lett.*, **74**(24), 4823–4826.

[8] Aurell, E., Frick, P. and Shaidurov, V. (1994). Hierarchical tree-model of 2D-turbulence, *Physica D*, **72**, 95–109.

[9] Bacry, E., Mallat, S. and Papanicolaou, G. (1992). A wavelet based space-time adaptive numerical method for partial differential equations, *Math. Mod. Num. Anal.*, **26**, 793.

[10] Balian, R. (1981). Un principe d'incertitude fort en théorie du signal ou en mécanique quantique, *C. R. Acad. Sci. Paris*, **292**, II, 1357–1361.

[11] Bardos, C., Golse, F., and Levermore, D. (1991). Fluid dynamical limit of kinetic equations: 1. formal derivations, *J. Stat. Phys.*, **63**, 323–344.

[12] Bardos, C., Golse, F., and Levermore, D. (1993). Fluid dynamical limit of kinetic equations: 2. convergence proofs for the Boltzmann equation, *Comm. Pure Appl. Math.*, **46**(5), 667–754.

[13] Basdevant, C., Holschneider, M. and Perrier, V. (1990). Méthode des ondelettes mobiles, *C. R. Acad. Sci. Paris*, série I, **310**, 647–652.

[14] Basdevant, C., Legras, B., Sadourny, R. and Béland, M. (1981). A study of barotropic model flows: intermittency waves and predictability, *J. Atmos. Sci.*, **38**, 2305–2326.

[15] Batchelor, G. K. (1969). Computation of the energy spectrum in homogeneous two-dimensional turbulence, *Phys. Fluid, suppl. II*, **12**, 233–239.

[16] Batchelor, G.K. (1953). *Homogeneous turbulence*, (Cambridge University Press).

[17] Batchelor, G. K., and Townsend, A. A. (1949). The nature of turbulent motion at large wave-numbers, *Proc. Roy. Soc. Lond.* A, **199**, 238–255.

[18] Battimelli, G. (1984). The mathematician and the engineer: the statistical theories of turbulence in the 20's, *Riv. Stor. Sci.*, **1**(1), 73–94.

[19] Battle, G., and Federbush, P. (1993). Divergence-free vector wavelets, *Michigan Math. J.*, **40**, 181–195.

[20] Benhamidouche, N. (1995). Ondelettes mobiles et l'instabilité gravitationnelle, *Thèse de doctorat de l'Université d'Aix-Marseille II*.

[21] Benzi, R., Ciliberto, S., Baudet, C., Ruiz-Chavarria, G. and Tripiccione, R. (1993). Extended self-similarity in the dissipation range of fully developed turbulence, *Eur. Phys. Let.*, **24**(4), 275–279.

[22] Benzi, R., Paladin, G., Parisi, G. and Vulpiani, A. (1990). Power spectra in two-dimensional turbulence, *Phys. Rev. (Rap. Comm.)* A42, 3564.

[23] Beylkin, G. (1991). Wavelets, MRA and fast numerical algorithms, *INRIA lectures on Wavelets*.

[24] Beylkin, G., Coifman, R. and V. Rokhlin, (1991). Fast wavelet transforms and numerical algorithms I, *Comm. Pure Appl. Math.*, **43**, 141–183.

[25] Bertoluzza, S. (1996). Adaptive wavelet collocation method for the solution of Burgers equation, *Trans. Th. Stat. Phys.*, **25**.

[26] Bertoluzza, S., Maday, Y. and Ravel, J. C. (1994). A dynamically adaptive wavelet method for solving partial differential equations, *Comput. Meth. Appl. Mech. Eng.*, **116**, 293–299.

[27] Bertoluzza, S., Naldi, G. and Ravel, J. C. (1994). Wavelet methods for the numerical solution of boundary value problems on the interval, *Wavelets: Theory Algorithms and Applications*, eds. Chui, Montefusco and Puccio, (Academic Press).

[28] Bockhorn, H., Fröhlich, J. and Schneider, K. (1997). An adaptive two-dimensional wavelet-vaguelette algorithm for the computation of flame balls, *Preprint CPT-97/P.3565, Centre de Physique Théorique, CNRS-Luminy, Marseille*.

[29] Bockhorn, H., Fröhlich, J. and Schneider, K. (1998). Direkte numerische Simulation turbulenter reaktiver Strömungen, *Forschungszentrum Karlsruhe, Wissenschaftliche Berichte* FZKA **6084**, 3–19.

[30] Boussinesq, J. (1877). Essai sur la théorie des eaux courantes, *Mémoire de l'Acad. Sci. Paris*, **23**(1), 1–680.

[31] Brandt, A. (1984). Multigrid techniques with applications to fluid dynamics: 1984 guide, *VKI Lecture Series*, 1–176.

[32] Cadot, O., Douady, S., and Couder, Y. (1995). Characterization of the low-pressure filaments in a three-dimensional shear flow, *Phys. Fluids*, **7**(3), 630–646.

[33] Cafarelli, L., Kohn, R. and Nirenberg, L. (1982). Partial regularity of suitable weak solutions of the Navier–Stokes equations, *Comm. Pure Appl. Math.*, **35**, 771–831.

[34] Caglioti, E., Lions, P. L., Marchioro, C. and Pulvirenti, M. (1992). A special case of stationary flows for two-dimensional Euler equations: a statistical mechanics description, *Comm. Math. Phys.*, **143**, 501–525.

[35] Cannone, M. (1995). *Ondelettes, paraproduits et Navier–Stokes*, (Diderot Editeur).

[36] Castaing, B. (1989). Consequences d'un principe d'extremum en turbulence, *J. Physique*, **59**, 147–156.

[37] Castaing, B., Gagne, Y. and Hopfinger, E. (1990). Velocity probability density functions of high Reynolds number turbulence, *Physica D*, **46**, 177–200.

[38] Cantwell, B. J. (1981). Organized motion in turbulent flow, *Ann. Rev. Fluid Mech.*, **13**, 457–515.

[39] Charton, P. (1996). Produits de matrices rapides en bases d'ondelettes : application à la résolution numérique d'équations aux dérivées partielles, *Thèse de doctorat de l'Université Paris 13*.

[40] Charton, P. and Perrier, V. (1995). Factorisation sur bases d'ondelettes du noyau de la chaleur et algorithmes matriciels rapides associés, *C. R. Acad. Sci. Paris*, série I, **320**, 1013–1018.

[41] Charton, P. and Perrier, V. (1996). A pseudo-wavelet scheme for the two-dimensional Navier–Stokes equation, *Comp. Appl. Math.* **15**, 139–160.

[42] Chasnov, J. K. (1997). On the decay of two-dimensional homogeneous turbulence, *Phys. Fluids*, **9**(1), 171–180.

[43] Chorin, A. J. (1973). Numerical study of slightly viscous flow, *J. Fluid Mech.*, **57**, 785–796.

[44] Chorin, A. J. (1994). *Vorticity and turbulence*, (Springer).

[45] Chorin, A. J. (1996). Turbulence cascades across equilibrium spectra, *Phys. Rev. E*, **54/3**, 2616–2619.

[46] Chorin, A. J. (1996). Turbulence as a near-equilibrium process, *Lecture Notes in Appl. Math.*, **31**, 235–249.

[47] Chorin A. J. (1996). Vortex methods, *Computational Fluid Mechanics*, eds. Lesieur, Comte and Zinn-Justin, (Elsevier).

[48] Cohen, A., Daubechies, I. and Vial, P. (1993). Wavelets on the interval and fast wavelet transforms, *Appl. Comp. Harm. Anal.*, **1**, 157–188.

[49] Couder, Y., Douady, S. and Cadot, O. (1994). Vorticity filaments, *Turbulence: a Tentative Dictionary, NATO ASI Series*, eds. Tabeling and Cardoso, (Plenum Press).

[50] Dahmen, W., Prössdorf, S. and Schneider, R. (1994). Multiscale methods for pseudo-differential equations on smooth closed manifolds, *Wavelets: Theory Algorithms and Applications*, eds. Chui, Montefusco and Puccio, (Academic Press).

[51] Davis, A., Marshak, A. and Wiscombe, W. (1994). Wavelet-based multifractal analysis of non-stationary and/or intermittent geophysical signals, *Wavelet Transforms in Geophysics*, eds. Foufoula-Georgiou and Kumar, 249–298.

[52] Desnjanski, V. and Novikov, E.A. (1974). Model of cascade processes in turbulent flows, *Appl. Math. Mech.*, **38**(3), 507–513.

[53] DeVore, R., Jawerth, B. and Popov, V. (1992). Compression of wavelet decomposition, *Am. J. Math.*, **114**, 737–785.

[54] Domaradzki, J. A. (1992). Nonlocal triad interactions and the dissipation range of isotropic turbulence, *Phys. Fluids* A, **4**, 2037.

[55] Domaradzki, J. A. and Rogallo, R. S. (1990). Local energy transfer and nonlocal interactions in homogeneous, isotropic turbulence, *Phys. Fluids* A, **2**, 413.

[56] Domaradzki, J. A., Rogallo, R. S. and Wray, A. A. (1990). Interscale energy transfer in numerically simulated turbulence, *CTR, Proceedings of the Summer Program*.

[57] Donoho, D. L. (1992). Interpolating Wavelet Transforms. *Preprint Stanford University.*

[58] Donoho, D. (1993). Unconditional bases are optimal bases for data compression and statistical estimation, *Appl. Comp. Harm. Anal.*, **1**, n. 1, 100–115.

[59] Dupree, T. (1992). Coarse-grain entropy in two-dimensional turbulence, *Phys. Fluid B*, **10**, 3101–3114

[60] Dryden, H. (1948). Recent advances in the mechanics of boundary layer flow, *Advances in Applied Mechanics*, **1**, 1–40, (Academic Press).

[61] Duval-Destin, M. and Murenzi, R. (1993). Spatio-temporal wavelets: application to the analysis of moving patterns, *Progress in Wavelet Analysis and Applications*, eds. Meyer and Roques, (Editions Frontières), 399–408.

[62] Einstein, A. (1914). Méthode pour la détermination de valeurs statistiques d'observations concernant des grandeurs soumises à des fluctuations irrégulières, *Archive des Sciences Physiques et Naturelles*, **37**, 254–255.

[63] Elhmaidi, D., Provenzale, A., and Babiano, A. (1993). Elementary topology of two-dimensional turbulence from a Lagrangian viewpoint and single-particle dispersion, *J. Fluid Mech.*, **257**, 533–558.

[64] Elliott, F. W. and Majda, A. J. (1994). A wavelet Monte-Carlo method for turbulent diffusion with many spatial scales, *J. Comput. Phys.*, **113**(1), 82–111.

[65] Elliott, F. W. and Majda, A. J. (1995). A new algorithm with plane waves and wavelets for random velocity fields with many spatial scales, *J. Comput. Phys.*, **117**(1), 146–162.

[66] Elliott, F. W. and Majda, A. J. 1996 Pair dispersion over an inertial range spanning many decades, *Phys. Fluids* **8**(4), 1052–1060.

[67] Eyink, G.L. and Spohn, H. (1993). Negative-temperature states and large-scale, long-lived vortices in two-dimensional turbulence, *J. Stat. Phys.*, **70**, n. 3/4, 833–887.

[68] Eyink, G.L. (1995). Space-scale locality and semi-Lagrangian wavelets (energy dissipation without viscosity in ideal hydrodynamics), *Preprint, Physics Department, University of Illinois at Urbana-Champain.*

[69] Farge, M. (1979). Notes sur la turbulence, *Rapport pour le cours de J. M. Levy-Leblond, Université Paris VII.*

[70] Farge, M. (1992). Wavelet transforms and their applications to turbulence, *Ann. Rev. Fluid Mech.*, **24**, 395–457.

[71] Farge, M. (1992). The continuous wavelet transform of two-dimensional turbulent flows, *Wavelets and their Applications*, eds. Ruskai *et al.*, (Jones and Bartlett), 275–302.

[72] Farge, M. (1992). Evolution des théories sur la turbulence développée, *Chaos et déterminisme*, eds. Dahan *et al.*, (Le Seuil), 212–245.

[73] Farge, M. (1994). Wavelets and two-dimensional turbulence, *Computational Fluid Dynamics*, **1**, (John Wiley & Sons), 1–23.

[74] Farge, M., Goirand, E., Meyer, Y., Pascal, F. and Wickerhauser, M. V. (1992). Improved predictability of two-dimensional turbulent flows using wavelet packet compression, *Fluid Dyn. Res.*, **10**, 229–250.

[75] Farge, M., Guezennec, Y., Ho, C. M. and Meneveau, C. (1990). Continuous wavelet analysis of coherent structures, *Proceedings of the Summer Program of the CTR, NASA-Ames and Stanford University*, 331–348.

[76] Farge, M. and Holschneider, M. (1991). Interpretation of two-dimensional turbulence spectrum in terms of singularity in the vortex cores, *Europhys. Lett.*, **15**(7), 737–743.

[77] Farge, M., Holschneider, M. and Colonna J. F. (1990). Wavelet analysis of coherent structures in two-dimensional turbulent flows, *Topological Fluid Dynamics*, eds. Moffatt and Tsinober, (Cambridge University Press), 765–776.

[78] Farge, M., Hunt, J. C. R. and Vassilicos, J. C. (1993). *Wavelets, fractals and Fourier transforms*, (Clarendon Press).

[79] Farge M., Kevlahan N., Perrier V. and Goirand E., (1996). Wavelets and Turbulence, *Proc. IEEE*, **84**, No. 4, p. 639–669.

[80] Farge M. and Philipovitch, T. (1993). Coherent structure analysis and extraction using wavelets, in *Progress in Wavelet Analysis and Applications*, eds. Meyer and Roques, (Editions Frontières), 477–481.

[81] Farge, M. and Rabreau, G. (1988). Transformée en ondelettes pour détecter et analyser les structures cohérentes dans les écoulements turbulents bidimensionnels, *C. R. Acad. Sci. Paris Ser. II*, **307**, 433–462.

[82] Farge, M., Schneider, K., Tieng, Q. and Anh V. (1997). Non-Gaussian statistics, intermittency and long-range dependence in two-dimensional freely decaying 2D turbulence. *Preprint LMD-CNRS, Paris.*

[83] Farge, M., Schneider, K. and Kevlahan N. (1998). Coherent structure eduction in wavelet-based two-dimensional turbulent flows, *Dynamics of slender vortices*, ed. Krause, Kluwer 65–83.

[84] Farge, M., Schneider, K. and Kevlahan N. (1999). Non-Gaussianity and Coherent Vortex Simulation in two-dimensional turbulence using an orthonormal wavelet basis. *Phys. Fluids*, **11**(8), 2187–2201.

[85] Farge, M. and Schneider, K. (1998). CFL criterion for adapted wavelet scheme. *Preprint LMD-CNRS, Paris*.

[86] Federbush, P. (1993). Navier and Stokes meet the wavelet, *Comm. Math. Phys.*, **155**, 219–248.

[87] Fournier, A. (1995). Wavelet representation of lower atmospheric long nonlinear wave dynamics governed by the Benjamin-Davis-Ono-Burgers equation, *Proceedings SPIE: Wavelet Applications II*, ed. H. H. Szu, **2491**, 672–687.

[88] Frick, P. (1993). Choix des ondelettes pour les modèles hiérarchiques de la turbulence, *Progress in Wavelet Analysis and Applications*, eds. Meyer and Roques, (Editions Frontières), 483–490.

[89] Frick, P., Dubrulle, P. and Babiano, A. (1995). Scale invariance in a class of shell models, *Phys. Rev. E*, **51**(5).

[90] Frick, P. and Zimin, V. (1993). Hierarchical models of turbulence, *Wavelets, Fractals and Fourier Transforms*, eds. Farge, Hunt and Vassilicos, (Clarendon Press), 265–283.

[91] Frisch, U. (1995). *Turbulence*, (Cambridge University Press).

[92] Frisch, U., Sulem, P. L. and Nelkin, M. (1978). A simple dynamical model of intermittent fully developed turbulence, *J. Fluid Mech.*, **87**, 719–736.

[93] Frisch, U. and Vergassola, M. (1990). A prediction of the multifractal model: the intermediate dissipation range, *Europhys. Lett.* **14**(5), 439–444.

[94] Fröhlich, J. and Schneider, K. (1994). An adaptive wavelet Galerkin algorithm for one and two dimensional flame computation, *Eur. J. Mech. B/Fluids*, **13**(4), 439–471.

[95] Fröhlich, J. and Schneider, K. (1995). A fast algorithm for lacunary wavelet bases related to the solution of PDEs, *C. R. Math. Rep. Acad. Sci. Canada*, Vol. XVII, no. 6, 283–286.

[96] Fröhlich, J. and Schneider, K. (1996). Numerical simulation of decaying turbulence in an adaptive wavelet basis, *Appl. Comput. Harm. Anal.*, **3**, 393–397.

[97] Fröhlich, J. and Schneider, K. (1997). An adaptive wavelet-vaguelette algorithm for the Solution of PDEs, *J. Comput. Phys.*, **130**, 174–190.

[98] Furioli G., Lemarié-Rieusset P. G. and Terrano E. (1997). Sur l'unicité des solutions "mild" des équations de Navier-Stokes, *C. R. Acad. Sci. Paris, Série I*, **235**, 1253–1256.

[99] Gagne, Y. (1987). Etude expérimentale de l'intermittence et des singularités dans le plan complexe en turbulence développée, *Thèse de doctorat, Université de Grenoble, France*.

[100] Gagnon, L. and Lina, J. M. (1994). Symmetric Daubechies' wavelets and numerical solution of the nonlinear Schrödinger's equation, *Rapport techniaue, Laboratoire de Physique Nucléaire, Université de Montréal*, n. 16, 1994.

[101] Gilbert, A. D. (1988). Spiral structures and spectra in 2-D turbulence, *J. Fluid Mech.*, **193**, 475–497.

[102] Glowinski, R., Lawton, W. M., Ravachol, M. and Tenenbaum, E. (1990). Wavelet solution of linear and nonlinear elliptic, parabolic and hyperbolic problems in one space dimension, *Proceedings of the 9th International*

Conference on Numerical Methods in Applied Sciences and Engineering, SIAM, Philadelphia.

[103] Gottlieb, D. and Orszag, S.A. (1977). Numerical analysis of spectral methods: theory and applications, *Regional Conference Series, SIAM-CBMS, Philadelphia.*

[104] Harten, A. (1994). Adaptive multiresolution schemes for shock computations, *J. Comput. Phys.*, **115**, 319–338.

[105] Holschneider, M. and Tchamitchian, P. (1991). Pointwise analysis of Riemann's 'non-differentiable' function, *Inventiones Mathematicae*, **105**, 157–176.

[106] Hunt, J. C. R. and Vassilicos, J. C. (1991). Kolmogorov's contributions to the physical and geometrical understanding of small-scale turbulence and recent developments, *Proc. R. Soc. Lond.* A, **434**, 183–210.

[107] Jaffard, S. (1994). Some mathematical results about multifractal formalism for functions, *Wavelets: Theory, Algorithms, and Applications*, eds. Chui, Montefusco and Puccio, 1–37.

[108] Jaffard, S. (1992). Wavelet methods for fast resolution of elliptic problems, *SIAM J. Numer. Anal.*, **29**, 965–987.

[109] Jaffard, S. (1991). Pointwise smoothness, two-microlocalization and wavelet coefficients, *Publicacions Matematiques*, vol. 35, pp. 155–168, 1991.

[110] Jimenez, J. (1981). *The role of coherent structures in modelling turbulence and mixing*, (Springer).

[111] Joyce, G. and Montgomery, D. (1973). Negative temperature states for the two-dimensional guiding-center plasma, *J. Plasma Phys.*, **10**, 107.

[112] Kampé de Fériet, J. (1956). La notion de moyenne en théorie de la turbulence, *Seminario Matematico e Fisico di Milano*, **XXVII**.

[113] Karman T. von and Howarth L. (1938). On the statistical theory of isotropic turbulence *Proc. Roy. Soc. London Ser. A*, **164**, 192–215.

[114] Kevlahan, N. K.-R. and Farge, M. (1997). Vorticity filaments in two-dimensional turbulence: creation, stability and effect, *J. Fluid Mech.*, **346**, 49–76.

[115] Kevlahan, N. K.-R. and Vassilicos, J. C. (1994). The space and scale dependencies of the self-similar structure of turbulence, *Proc. R. Soc. Lond. A.*, **447**, 341–363.

[116] Kline, S. J., Reynolds, W. C. Schraub, F. A. and Rundstadler, P.W. (1967). The structure of turbulent boundary layer, *J. Fluid Mech.*, **30**, 741.

[117] Kolmogorov, A. N. (1941). The local structure of turbulence in incompressible viscous fluid for very large Reynolds numbers, *C. R. Acad. Sci. USSR*, **30**, 301–305.

[118] Kolmogorov, A. N. (1941). On degeneration of isotropic turbulence in an incompressible viscous liquid, *C. R. Acad. Sci. USSR*, **31**, 538–540.

[119] Kolmogorov, A. N. (1941). Dissipation of energy in the locally isotropic turbulence, *C. R. Acad. Sci. USSR*, **32**, 16–18.

[120] Kolmogorov, A. N. (1962). A refinement of previous hypotheses concerning the local structure of turbulence in a viscous incompressible fluid at high Reynolds number, *J. Fluid Mech.*, **13**, 82–85.

[121] Kraichnan, R. H. (1967). Inertial ranges in two-dimensional turbulence, *Phys. Fluids*, **10**, 1417–1423.

[122] Kraichnan, R. H. (1974). On Kolmogorov's inertial-range theories, *J. Fluid Mech.*, **62**, 305–330.

[123] Kraichnan, R. H. and Montgomery, D. (1982). Two-dimensional turbulence, *Reports on Progress in Physics*, **45**, 547.

[124] Latto, A. and Tenenbaum, E. (1990). Compactly supported wavelets and the numerical solution of Burgers' equation, *C. R. Acad. Sci. Paris*, Série I, **311**, 903–909.

[125] Lazaar, S., Liandrat, J. and Tchamitchian, P. (1994). Algorithme à base d'ondelettes pour la résolution numérique d'équations aux dérivées partielles à coefficients variables, *C. R. Acad. Sci. Paris*, série I, **319**, 1101–1107.

[126] Lazaar, S., Ponenti, P. J. Liandrat, J. and Tchamitchian, P. (1994). Wavelet algorithms for numerical resolution of partial differential equations, *Comput. Methods Appl. Mech. Eng.*, **116**, 309–314.

[127] Legras, B. and Dritschel, D. G. (1993). A comparison of the contour surgery and pseudo-spectral methods, *J. Comput. Phys.*, **104**(2), 287–302.

[128] Legras, B., Santangelo, P. and R. Benzi, R. (1988). High-resolution numerical experiments for forced two-dimensional turbulence, *Europhys. Lett.*, **5**, 37–42.

[129] Leith, C. E. (1990). Stochastic backscatter in a subgrid-scale model: plane shear mixing layer, *Phys. Fluids* A **3**, 297–299.

[130] Lemarié-Rieusset, P. G. (1998). Some remarks on divergence-free vector wavelets, *Communication at the International Congress of Mathematics, Berlin*, 19–24 August.

[131] Lemarié-Rieusset, P. G. (1992). Analyses multirésolutions non-orthogonales, commutation entre projecteur et derivation, et ondelettes à divergence nulle, *Revista Mat. Iberoamerica*, **8**, 221–236.

[132] Lemarié-Rieusset, P. G. (1997). Un théorème d'inexistence pour les ondelettes vecteur à divergence nulle, *CRAS Série I*, **310**, 811–813.

[133] Leonard, A. (1974). On the energy cascade in large eddy simulations of turbulent fluid flows, *Adv. Geophys.*, A **18**, 237.

[134] Leonard, A. (1980). Vortex methods for flow simulation, *J. Comp. Phys.*, **37**, 289–335.

[135] Lesieur, M. and Métais, O. (1996). New trends in large-eddy simulations of turbulence, *Ann. Rev. Fluid Mech.*, **28**, 45–82.

[136] Lewalle, J. (1993). Wavelet transforms of the Navier–Stokes equations and the generalized dimensions of turbulence, *Appl. Sci. Res.* **51**, 109–113.

[137] Lewalle, J. (1993). Energy dissipation in the wavelet-transformed Navier–Stokes equations, *Phys. Fluids* A **5**, 1512–1513.

[138] Lewalle, J. (1997). Hamiltonian formulation for the diffusion equation, *Phys. Rev. Lett.* E, **55**(2), 1590–1599.

[139] Liandrat, L., Perrier, V. and Tchamitchian, P. (1992). Numerical resolution of nonlinear partial differential equations using the wavelet approach, in *Wavelets and their applications*, Ruskai et al. (ed.), Jones and Barlett, 227–238.

[140] Liepmann, H. W. (1962). Free turbulent flows, *Mécanique de la turbulence*, ed. Favre, 211–227.

[141] Liepmann, H. (1979). The rise and fall of ideas in turbulence, *Scientific American*, March-April.

[142] Liepmann, H. W. (1997). *Private communication*.

[143] Mc Williams, J. C. (1984). The emergence of isolated coherent vortices in turbulent flows, *J. Fluid Mech.*, **146**, 21–43.

[144] Mc Williams, J. C. (1984). The vortices of two-dimensional turbulence, *J. Fluid Mech.*, **219**, 361–385.

[145] Maday, Y., Perrier, V. and Ravel, J. C. (1991). Adaptativité dynamique sur base d'ondelettes pour l'approximation d'équations aux dérivées partielles, *C. R. Acad. Sci. Paris*, série I, **312**, 405–410.

[146] Mallat, S. and Hwang, W. H. (1992). Singularity detection and processing with wavelets, *IEEE Trans. Inform. Theory*, **38**, 617–643.

[147] Mandelbrot, B. (1975). Intermittent turbulence in self-similar cascades: divergence of high moments and dimension of carrier, *J. Fluid Mech.*, **62**, 331–358.

[148] Marchioro, C. and Pulvirenti, M. (1994). *Mathematical theory of incompressible nonviscous fluids*, (Springer).

[149] Marion, M. and R. Temam, R. (1989). Nonlinear Galerkin Methods, *SIAM J. Num. Anal.*, **26**, 1139–1157.

[150] Meneveau, C. (1991). Analysis of turbulence in the orthonormal wavelet representation, *J. Fluid Mech.*, **232**, 469–520.

[151] Meyer, Y. (1990). *Ondelettes et opérateurs*, (Hermann).

[152] Miller, J. (1990). Statistical mechanics of Euler equations in two dimensions, *Phys. Rev. Lett.*, **65**, n. 17, 2137–2140.

[153] Min, I. A., Mezić, I. and Leonard, A. (1996). Lévy stable distributions for velocity and velocity difference in systems of vortex elements, *Phys. Fluids* **8**(5), 1169–1180.

[154] Moffatt, H. K. (1984). Simple topological aspects of turbulent velocity dynamics, *Proc. IUTAM Symp. on turbulence and chaotic phenomena in fluids*, ed. T. Tatsumi, 223, (Elsevier).

[155] Mohammadi, B. and Pirroneau, O. (1994). *Analysis of the k-ε Turbulence Model*, (Wiley-Masson).

[156] Monasse, P. and Perrier, V. (1995). Ondelettes sur l'intervalle pour la prise en compte de conditions aux limites, *C. R. Acad. Sci. Paris*, **321**, Série I, 1163–1169.

[157] Monasse, P. and Perrier, V. (1998). Orthonormal wavelet bases adapted for partial differential equations with boundary conditions, *SIAM Journal of Mathematical Analysis*, **29**, 4, 1040.

[158] Monin, A. S. and Yaglom, A. M. (1975). *Statistical Fluid Mechanics: Mechanics of Turbulence*, (The M.I.T. Press).

[159] Myers, M., Holmes, P., Elezgaray, J. and Berkooz, G. (1995). Wavelet projections of the Kuramoto-Sivashinsky equation I: heteroclinic cycles and modulated waves for short systems, *Physica D*, **86**, 396–427.

[160] Neumann, J. von (1949) Recent theories of turbulence. In *Collected works (1949–1963)* **6**, 437–472, (ed. A. H. Taub). Pergamon Press, (1963).

[161] Obukhov, A. M. (1941). Energy distribution in the spectrum of a turbulent flow, *Izv. AN URSS*, **4**, 453–466.

[162] Onsager, L. (1945). The distribution of energy in turbulence, *Phys. Rev.*, **68**, 286.

[163] Onsager, L. (1949). Statistical hydrodynamics, *Suppl. Nuovo Cimento*, suppl. **6**, 279–287.

[164] Parisi, G. and Frisch, U. (1985). Fully developed turbulence and intermittency, *Turbulence and Predictability in Geophysical Fluid Dynamics and Climate Dynamics*, eds. Ghil, Benzi and Parisi, (North-Holland), 71–88.

[165] Perrier, V. and Basdevant, C. (1989). Periodical wavelet analysis, a tool for inhomogeneous field investigation. Theory and algorithms, *Rech. Aérosp. - n 1989-3*.

[166] Perrier, V., Philipovitch, T. and Basdevant, C. (1995). Wavelet spectra compared to Fourier spectra, *J. Math. Phys.*, **36**(3), 1506–1519.

[167] Perrier, V. and Wickerhauser, M. V. (1996). Multiplication of short wavelet series using connection coefficients, *Proc. workshop on Wavelets and their Applications, 1175 Chinese University*, Hong-Kong, Springer.

[168] Philipovitch, T. (1994). Applications de la transformée en ondelettes continue à la turbulence homogène isotrope bidimensionnelle, *Thèse de Doctorat, Université Paris VI*.

[169] Piomelli, U. Cabot, W. H. Moin, P. and Lee, S. (1990). Subgrid-scale backscatter in transitional and turbulent flows, *CTR, Proceedings of the Summer Program*, 19–30.

[170] Polyakov, A. M. (1993). Conformal Turbulence, *Nucl. Phys. B*, **396**, 367.

[171] Ponenti, P. (1994). Algorithmes en ondelettes pour la résolution d'équations aux dérivées partielles, *Thèse de Doctorat, Université d'Aix-Marseille I*.

[172] Qian, S. and Weiss, J. (1993). Wavelets and the numerical solution of Partial Differential Equations, *J. Comp. Phys.*, **106**, 155–175.

[173] Reynolds, O. (1883). An experimental investigation of the circumstances which determine whether the motion of water shall be direct and sinuous, and the law of resistance in parallel channels. *Phil. Trans. Roy. Soc. Lond.*, 51–105.

[174] Reynolds, O. (1894). On the dynamical theory of incompressible viscous fluids and the determination of the criterion, *Phil. Trans. Roy. Soc. Lond.*, **186**, 123–164.

[175] Reynolds, W. C. (1976). Computation of turbulent flows, *Ann. Rev. Fluid Mech.*, **8**.

[176] Richardson, L. F. (1922). *Weather prediction by numerical process*, (Cambridge University Press).

[177] Richardson, L. F. and Gaunt, J. A. (1993). Diffusion regarded as a compensation for smoothing, *Memoirs of the Royal Meteorological Society*, **3**, no. 30, 171–175, 1930 (Reprinted in the collected papers of Lewis Fry Richardson, **1**, (Cambridge University Press), 773–777).

[178] Robert, R. (1990). Etat d'équilibre statistique pour l'écoulement bidimensionnel d'un fluide parfait, *C. R. Acad. Sci. Paris, Série I*, **311**, 575–578.

[179] Robert, R. (1991). A maximum-entropy principle for two-dimensional perfect fluid dynamics, *J. Stat. Phys.*, **65**, 531–553.

[180] Robert, R. and Sommeria, J. (1991). Statistical equilibrium states for two-dimensional flows, *J. Fluid Mech.*, **229**, 291–310.

[181] Robinson S. K. (1991). Coherent motions in the turbulent boundary layer, *Ann. Rev. Fluid Mech.*, **23**, 601–639.

[182] Roshko, A. (1961). Experiments on the flow past a circular cylinder at very high Reynolds number. *J. Fluid Mech.*, **10**, 345.

[183] Saffman, P. G. (1971). A note on the spectrum and decay of random two-dimensional vorticity distribution, *Stud. Appl. Math.*, **50**, 377–383.

[184] Saffman, P. G. (1972). *Vortex Dynamics*, (Cambridge University Press).

[185] Schneider, K. and Farge, M.(1997). Wavelet forcing for numerical simulation of two-dimensional turbulence, *C. R. Acad. Sci. Paris*, **325**, Série II, 263–270.

[186] Schneider, K., Kevlahan, N. and Farge, M., (1997). Comparison of an adaptive wavelet method and nonlinearly filtered pseudo-spectral methods for two-dimensional turbulence, *Theoret. Comput. Fluid Dynamics*, **9(3/4)**, 191–206.

[187] Sommeria, J. Staquet, C. and Robert, R. (1991). Final equilibrium state of a two-dimensional shear layer, *J. Fluid Mech.*, **223**, 661–689.

[188] Taylor, G. I. (1921). Diffusion by continuous movements, *Proc. Lond. Math. Soc. Ser. 2*, **20**, 196–211.

[189] Taylor, G. I. (1935). Statistical theory of turbulence, *Proc. Roy. Soc. Lond.* A, **151**, 421–478.

[190] Taylor, G. I. (1938). The spectrum of turbulence. *Proc. Roy. Soc. Lond.* A **164**, 476–490.

[191] Tennekes, H. and Lumley, J. L. (1972). *A first course in turbulence*, (MIT Press).

[192] Tsinober A., Shtilman L. and Vaisburd H. (1997). A study of properties of vortex stretching and enstrophy generation in numerical and laboratory experiments, *Fluid Dyn. Res.*, **21**, 477–494.

[193] Tong, P. and Goldburg, W. I. (1988). Experimental study of relative velocity fluctuations in turbulence. *Phys. Lett.* A, **127**, 147.

[194] Townsend, A. A. (1951). On the fine-scale structure of turbulence *Proc. Roy. Soc. Lond.*, **208**, 534–542.

[195] Townsend, A. A. (1956). *The structure of turbulent shear flow*, (Cambridge University Press).

[196] Urban, K. (1994). A wavelet-Galerkin algorithm for the driven-cavity-Stokes-problem in two space dimensions, *Numerical Modelling in Continuous Mechanics*, eds. Feistauer, Rannacher and Kozel, Karls Universität Prag, 278–289.

[197] Van Atta, C. W. and Antonia, R. A. (1980). Reynolds dependence of sknewness and flatness factors of turbulent velocity derivatives, *Phys. Fluids*, **23**, 252–257.

[198] Vassilicos, J. C. (1992). The multi-spiral model of turbulence and intermittency, *Topological Aspects of the Dynamics of Fluids and Plasmas*, Kluwer, 427–442.

[199] van der Water, W. (1993). Experimental study of scaling in fully developed turbulence, *Turbulence in Spatially Extended Systems*, eds. Benzi, Basdevant and Ciliberto, 189-213.

[200] Vergassola, M. and Frisch, U. (1991). Wavelet transforms of self-similar processes, *Physica D*. **54**, 58–64.

[201] Walter, G. (1992). A sampling theorem for wavelet subspaces. *IEEE Trans. Inform. Theory*, **38**, 881.

[202] Weiss, J. (1992). The dynamics of enstrophy transfer in two-dimensional hydrodynamics, *Physica D*, **48**, 273.

[203] Weiss, J. B. and McWilliams, J. C. (1991). Non-ergodicity of point vortices. *Phys. Fluids* A, **3**, 835.

[204] Wickerhauser, M. V., Farge, M., Goirand, E. Wesfreid, E. and Cubillo, E. (1994). Efficiency comparison of wavelet packet and adapted local cosine bases for compression of a two-dimensional turbulent flow, *Wavelets: Theory, Algorithms and Applications*, eds. Chui, Montefusco and Puccio, (Academic Press), 509–531.

[205] Wickerhauser, M. V., Farge, M. and Goirand, E. (1997). Theoretical dimension and the complexity of simulated turbulence, *Multiscale Wavelet Methods for Partial Differential Equations*, eds. Dahmen, Kurdila and Oswald, (Academic Press), 473–492.

[206] Yaglom, A. M. (1986). Einstein's work on methods of processing fluctuating series of observations and the role of these methods in meteorology, *Izv. Atmos. Ocean. Phys.*, **22**, 1.

5

Wavelets and detection of coherent structures in fluid turbulence

LONNIE HUDGINS

Acoustics & Signal Processing Group
Northrop Grumman Corporation
Hawthorne, California 90251, USA

JON HARALD KASPERSEN

Dept. Applied Mechanics, Thermo- and Fluiddynamics
The Norwegian Institute of Technology
University of Trondheim
7034 Trondheim, Norway

Abstract

The energy cascade found in fully developed fluid turbulence is believed to originate as large-scale organized motions called coherent structures. The process of detecting, locating, and tracking these coherent structures is therefore of central importance to the continued study of turbulence. A number of researchers have applied wavelet-based methods to the problem of coherent structure detection, and significant performance improvements over other existing methods have already been reported.

In this paper, we compare the performance of various conventional as well as wavelet-based detector algorithms for cylinder wake flow data. The resulting ROC curves quantitatively demonstrate the effectiveness of wavelet methods. The detections are then used to form conditional averages of the velocity time-series, revealing their underlying physical structure.

5.1 Introduction

Recently, advances in the theoretical understanding and implementation of wavelets have led to their increased use in analysis and signal processing. Wavelet methods can be very effective in the study of non-stationary phenomena [7], and have thus sparked a concerted interest in applying them to

201

the analysis of turbulent flows in general, and to the detection of coherent structures in particular [8].

It is possible that some coherent structure detectors are better suited for certain types of turbulent flows, or operate most effectively under specific conditions. As the numbers and types of detectors grow, it becomes increasingly important to measure their relative performance in quantitatively meaningful ways. In this chapter, we describe an approach to the comparison of detector algorithms by means of the Receiver Operating Characteristic (ROC) curves, and demonstrate the utility of this method for the case of cylinder wake flow. The resulting detections are then used to form conditional averages from the velocity time-series data, so that the precise morphology of the coherent structures can be studied in detail. For additional discussion regarding the theoretical basis for such conditional averages, see Chapter 4 by Farge *et al.* in this volume.

In the early decades of the twentieth century turbulence was considered to be a purely stochastic process, and well defined statistical quantities for turbulent flows were measured. In 1895, Reynolds [21] became the first to divide the flow into its mean and fluctuating parts. By doing this he was able to derive what is called the Reynolds averaged Navier Stokes equations. These equations show that the fluctuating part of the velocity gives rise to convective stress terms which are of great importance in turbulent flows. Based on the assumption that the turbulent scales interact randomly, Prandtl [20] and Taylor [23] introduced the eddy viscosity model.

Probably the first investigator to realize that turbulent flows contain non-random structures was Townsend [25]. He found that turbulent motion includes a system of large convecting eddies with sizes comparable to the outer scales of the flow. Townsend further discovered that these eddies are much larger than those which contain most of the turbulent energy, and observed that they had a deterministic form. These findings were probably the main trigger for the present great interest in the nature of coherent structures. While there is no consensus on exactly how to define a coherent structure, flow visualizations made by Kline *et al.* [16] as well as others have shown that coherent structures physically exist, and that they underlie the random three-dimensional vorticity that characterizes turbulence. This means for example that a turbulent shear flow can be decomposed into a sum of coherent structures and incoherent turbulence. Consider the Reynolds momentum equations as discussed by Tennekes and Lumley [24]:

$$\frac{\partial U_i}{\partial t} + U_j \frac{\partial U_i}{\partial x_j} = \frac{1}{\rho} \frac{\partial}{\partial x_j} \left[\sigma_{ij} - \rho \overline{u_i u_j} \right], \text{ where} \qquad (5.1)$$

$$\sigma_{ij} = -P \delta_{ij} + 2\mu S_{ij}, \text{ and}$$

$$S_{ij} = \frac{1}{2} \left(\frac{\partial U_i}{\partial x_j} + \frac{\partial U_j}{\partial x_i} \right).$$

This equation relates the mean flow U_i to the pressure P. The Reynolds shear stress, which appears in the fluctuating terms $\frac{1}{\rho} \frac{\partial}{\partial x_j} \left(-\rho \overline{u_i u_j} \right)$ on the right-hand side of (5.1) would identically vanish if the velocity components were uncorrelated random variables. However, it is known that the shear stress $-\rho \overline{uv}$ is always negative.

It is generally accepted that large scale coherent structures are responsible for the transport of significant amounts of mass, heat and momentum. On the other hand, they do not necessarily possess high levels of kinetic energy [12]. Although there remains some disagreement about this, turbulent kinetic energy is primarily associated with incoherent turbulence. Coherent structures in the wake of a solid body have been shown by Zhou [26] to measurably contribute to both $\overline{u^2}$ (10–20%) and $\overline{v^2}$ (40–60%), where we use an overline, e.g., $\overline{u}$, to indicate time averaging. They are also responsible for most of the turbulent shear stress, $\overline{uv}$: indeed, recent studies by Krogstad and Kaspersen [17] have shown that they contribute as much as 70% to the total. Advances in the physical understanding of these processes will surely be aided by experimental measurements of coherent structures, which in turn depend on effective methods for detection.

Drag is a force which occurs in all moving fluids. This frictional, dissipative force is due to viscosity, and converts kinetic energy into heat. The potential for reducing its effects would be of great value in many industrial problems. In the last few years there has been a growing interest in trying to control and manipulate coherent structures for the purpose of reducing drag and/or heat transfer. Direct numerical simulations of channel flow at low Reynolds numbers have shown that active control of coherent structures could reduce drag by as much as 50% [19]. If regions of high fluid friction can be reliably identified in real-time, they might be controlled either by injection or suction of fluid. Active control of coherent structures is of course strongly dependent on the algorithm used to detect them. Therefore, it is of great interest to find a method to evaluate the relative performance of various detection algorithms.

Several authors have tried to define coherent structures, and some definitions are more nebulous than others. In Robinson [22], a coherent structure is defined as a 'three-dimensional region of the flow over which at least one fundamental flow variable (velocity component, density, temperature, etc.) exhibits significant correlation with itself or with another variable over a range of space and/or time that is significantly larger than the smallest local scales of the flow.' A more specific definition is given by Hussain [13]: 'A coherent structure is a connected turbulent fluid mass with instantaneously phase-correlated vorticity over its spatial extent.' Coherent structures can be further classified as either *bursts* or *sweeps* [18] by their location in the $u - v$ plane. While the correct definitions for bursts and sweeps are still being debated in the literature, the reader may wish to think of a burst as a near-wall 'streak' of fluid which moves slowly outward into the boundary layer. Typically, when this streak reaches a distance of $y^+ \approx 40$ (the quantity y^+ is a dimensionless distance, normalized by the ratio of wall shear velocity to the kinematic viscosity) it suddenly ejects into the outer portion of the boundary layer. On the other hand, a sweep is a volume of fluid that moves inward from the outer portion of the boundary layer toward the wall. For a more complete description of bursts and sweeps, see Bogard and Tiederman (1983) [4] in which the authors used flow visualization techniques to study their structure in detail. These fluid entities are physically coherent structures, and in our analysis we will always seek to detect bursts. In this study we will adopt the point of view that coherent structures are distributions of fluid having a large-scale energetic deterministic form, with statistically random fluctuations superimposed on them. The use of this definition provides us with a simple mechanism to treat our coherent structures in a mathematically precise way.

Equally difficult as defining the coherent structures themselves is the problem of constructing an algorithm to detect them. Many such detector algorithms have been proposed, and it is not surprising that some perform better than others. Based on this fact there is no doubt that conclusions made about the dynamics and significance of coherent structures will depend in part on which detection algorithm is used. This provides additional motivation to test and compare the various detection algorithms.

The conventional approach to studying coherent structures begins with a *detector* – an algorithm for determining when and where a coherent structure exists in the flow. Because of the relative difficulty in following a given coherent structure as it convects downstream, we are typically forced to collect data from specific locations in space, and employ statistical methods to infer the nature and motion of the structures. Such statistical methods generally amount to forming ensemble averages conditioned on the detections, and examining their structure in the mean. In our analysis we start with standard ensemble averages and combine them with computer-generated noise based on the measured statistics of the data. We thus obtain synthetic data suitable for comparing the statistical performance of each of the detectors.

This chapter is organized as follows: section 5.2 summarizes the main advantages that wavelets bring to the detection of coherent structures in fluid turbulence. Section 5.3 describes the experimental details for the data used in this study. Section 5.4 outlines the overall approach used to quantitatively measure the relative performance of the various detectors, which are defined in sections 5.5 and 5.6. Our results are discussed in section 5.7, and summarized in section 5.8.

5.2 Advantages of wavelets

There are several advantages that wavelets bring to the detection of coherent structures in fluid turbulence. The aim of this paper is to compare and illustrate several conventional and wavelet methods in a physically meaningful and intuitive way. It will be seen that the general performance of the wavelet detectors, as measured by their ROC curves, is superior to the non-wavelet conventional detectors. In fact, we show that the best conventional method examined in this study is actually a wavelet method, devised long before the term 'wavelet' came into use.

Unlike the Fourier basis, wavelets are analysing functions that are localized in space but have a variable width. For turbulence studies where the event scales aren't necessarily known ahead of time, wavelet analysis can lead to rapid determination of the relevant scales. Wavelets are short enough to be able to detect individual events, and do not require periodicity or stationarity. Finally, with the modifications that we will describe, the *bivariate wavelet cross-transform* can be made to optimally detect individual pairs of signal events that are in *quadrature*, i.e., whose temporal inner product vanishes,[†] in an optimal fashion. This is in contrast to Fourier methods such as the cross-spectrogram, which provide optimal detection only for one scale.

5.3 Experimental details

The data used in this study were collected in the turbulent wake of a circular cylinder, where coherent structure generation, often referred to as *vortex shedding*, occurs with great regularity. Furthermore, coherent structures within the wake flow are very distinct, and their existence has been abun-

[†]Physically, events in quadrature belong to distinct modes of the system.

dantly confirmed by many researchers (see for example Bisset, Antonia and Browne [2]). Fluctuating velocity data are relatively easy to measure under these circumstances, and can even be used for flow visualization, *ibid*. These features make cylinder wake flow data particularly suitable to our investigation.

The measurements were made in a $0.7 \, \text{m} \times 1.0 \, \text{m}$ closed return wind tunnel. A sketch of the experimental setup is shown in Figure 5.1. The cylinder, which spanned the test section, had a diameter of $d = 23.6 \, \text{mm}$, which gave it a length-to-width ratio of 42.6. All measurements were made at a free stream velocity of $U_\infty = 13.4 \, \text{m/s}$, and thus a Reynolds number of ≈ 2200 at room temperature. The turbulence level of the undisturbed flow, i.e., the standard deviation of the fluctuating part of the velocity divided by the local mean velocity, was less than 0.3%.

Instantaneous velocity signals were measured in both the streamwise and transverse-vertical directions. For the purposes of this analysis, we decompose each component into its mean and fluctuating parts by writing $u(t) + U_0$ for the streamwise and $v(t) + V_0$ for the transverse velocities. An array of hot-wire anemometers (X-wire probes) was fabricated in-house, and permitted data collection at eight simultaneous equally spaced positions along the transverse y-direction. The frequency response of each probe was approximately 15 kHz. The signals were low pass filtered at 5 kHz and then digitized at 7874 Hz. Each record contained 315 392 samples per channel for a total sampling time of 40.05 s. A 12-bit 16 channel digitizer board from R.C. Electronics was employed in conjunction with a 486 Compaq PC. Raw

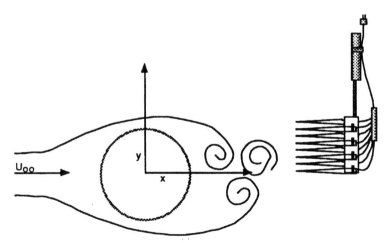

Fig. 5.1. Sketch of the experimental setup. The flow came from the left, passed around the circular cylinder, and was measured by the array of anemometers.

data were then transferred to a Digital Equipment Alpha workstation for further processing and subsequent tape storage.

Although measurements were made at several positions, we chose to use data that was taken at a distance of $x/d = 8$ from the centre of the cylinder. At this downstream position, near-wake effects like backflow are negligible while the structures are still very distinct. To better understand the nature of the fluid flow at this position, it is helpful to visualize the average Reynolds stresses, $\rho \overline{u_i u_j}$, as functions of their transverse displacement from the centre-line. To display these as dimensionless quantities, we have divided by the density and mean flow speed to obtain $\overline{u}^2/U_0^2$, $\overline{v}^2/U_0^2$, and $\overline{uv}/U_0^2$ (see Figure 5.2). As part of the validation process for our experimental setup, our measurements were compared with other published sources (see [26]) and were found to be in close agreement. Figure 5.3 shows the mean flow velocity, also as a function of distance from the centreline.

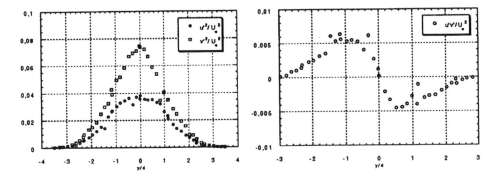

Fig. 5.2. Reynolds stress profiles at $x/d = 8$.

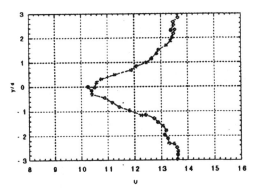

Fig. 5.3. Mean velocity profile at $x/d = 8$.

5.4 Approach

The task of a detector algorithm is to identify the presence or absence of some specified deterministic signal that may be corrupted by noise. Furthermore, if that deterministic signal is present, the aim will be to provide an estimate of its time-of-arrival. In much of the existing literature, such an algorithm is called a *receiver* and its performance is measured by a set of statistics called the Receiver Operating Characteristics (ROC). For any given detector functioning under a fixed set of conditions, its ROC statistics are the probability of detection and the false-alarm rate. The probability of detection, or P_D, is defined as the likelihood that an event will be properly detected when presented to the algorithm. The false-alarm rate, P_{FA}, is the probability that a detection will be reported when in fact the event was not present.

The probability of detection and the false-alarm rate are interdependent quantities. Generally speaking, the detector must make a trade-off between maximizing P_D while minimizing P_{FA}. When this trade-off can be parametrized, the result may be displayed as a ROC curve: a plot of P_D vs. P_{FA} for various values of the parameter. Typically, a receiver consists of some linear filter followed by a threshold detector. In that case, both P_D and P_{FA} are parametrized by the threshold level.

5.4.1 Methodology

We will assume that the coherent structures we wish to detect are realizations of some fixed deterministic function $g(t)$. A velocity signal that we have available from the instrumentation might be represented by $f(t)$. This signal will contain coherent structures located at some random set of times $\{\tau_n\}$, along with high levels of additive random noise. If the noise signal is a mean-zero process $\epsilon(t)$, our signal can be written as

$$f(t) = \epsilon(t) + \sum_n g(t - \tau_n).$$

We have at hand several detectors which have been specifically designed to detect occurrences of g and to provide us with estimates for the τ_ns. The purpose of this study is to quantitatively compare their performance, and we would like to apply the method of ROC curves to accomplish this. Our task then becomes to parametrically measure both the probability of detection and the false-alarm rate for each detector. Unfortunately, we do not know *a priori* the form of g, the times $\{\tau_n\}$, nor even the precise statistics of $\epsilon(t)$.

The first step is to analyse the noise so that we can synthesize some of our own. Then, using synthetic noise (containing no coherent structures) we can experimentally determine the false-alarm rate function $P_{FA}(\theta)$ for each of the detectors. Next, we obtain an estimate of the function g using the best conditional averages that each detector can provide. We then synthesize as many data points as we need by generating synthetic noise and placing copies of g at *known* locations. Using such data, we then estimate the probability of detection $P_D(\theta)$ for each candidate detector.

5.4.2 Estimation of the false-alarm rate

Figure 5.4 shows the measured power spectra and frequency distributions for the u and v components in the data record. All curves have been normalized: power spectra by the sample variance, and probability densities by the standard deviation. The spectral 'line' that appears at $\approx 125\,\mathrm{Hz}$ corresponds to the rate at which coherent structures are generated in the wake of the cylinder. This rate is in agreement with theoretical predictions based on the fluid

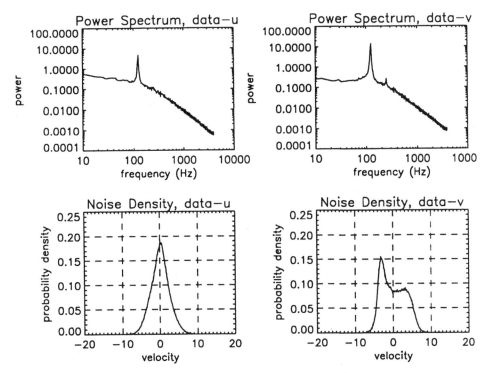

Fig. 5.4. Measured power spectra and frequency distributions for the u and v components in the data.

speed and the dimensions of the cylinder. Its second harmonic is also visible in the spectrum of v (at $\approx 250\,\text{Hz}$). These harmonic components are primarily associated with the deterministic function $g(t)$, i.e., the coherent structures. However, since the signal-to-noise ratio of the data is very low, the distribution and spectrum of the noise $\epsilon(t)$ was simply taken to be that of the data itself. Therefore, to estimate the spectrum of ϵ we ignored these harmonic lines and assumed that its spectrum was smooth. Having done that, we then synthesized pseudo-random noise data that approximated the characteristics of ϵ while at the same time contained *no coherent structures*. Figure 5.5 shows the measured power spectrum and frequency distribution of the synthetically generated noise.

We assume that the deterministic function $g(t)$ has support on an interval of length T, so that N consecutive realizations of data can be written as

$$f(t) = \epsilon(t) + \sum_{n=0}^{N-1} g(t - nT). \tag{5.2}$$

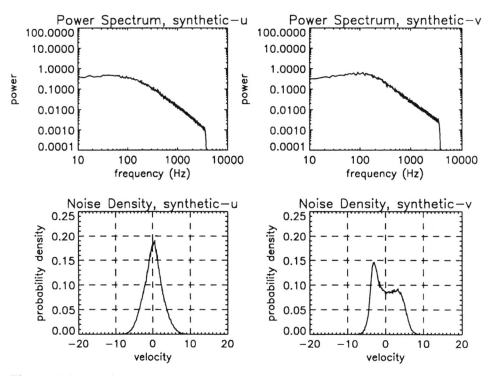

Fig. 5.5. Measured power spectra and frequency distributions for the u and v components in the synthetic noise.

Setting $g(t) = 0$ in equation (5.2) above, we then process $f(t)$ with each detector and determine the maximum value achieved at the filter output for each interval. Thus, at threshold θ,

$$P_{FA}(\theta) \approx \frac{\# \text{ intervals a detection} \geq \theta \text{ was observed}}{N}$$

is an estimate of the false-alarm rate for that detector.

5.4.3 Estimation of the probability of detection

Next, we will need an estimate for $g(t)$. We assume $\epsilon(t)$ to be stationary and uncorrelated over the time T, and that the occurrences of $g(t)$ in the signal aren't too close together, i.e.,

$$m \neq n \quad \Rightarrow \quad |\tau_m - \tau_n| \geq T. \tag{5.3}$$

Each detector returns a set of times $D = \{\hat{\tau}_n\}$ which we use together with the noisy time series data $f(t)$ to estimate $g(t)$ in the following way. Portions of the signal $f(t)$, each having length T and centred at one of the times $\hat{\tau}_n$ are excerpted and averaged. This technique is called *conditional averaging* because the average is 'conditioned' on the detection times. It produces an estimate of $g(t)$ which converges for large numbers of detections:

$$\hat{g}(t) = \frac{1}{N} \sum_n f(t + \hat{\tau}_n) \tag{5.4}$$

$$= \frac{1}{N} \sum_n \{\epsilon(t + \hat{\tau}_n) + g(t)\}$$

$$= g(t) + \frac{1}{N} \sum_n \epsilon(t + \hat{\tau}_n)$$

$$\rightarrow g(t).$$

By using this estimate of $g(t)$ in equation (5.2), we generate portions of data that contain coherent structures. Again, we process $f(t)$ with each detector and determine the maximum for each interval. Finally,

$$P_D(\theta) \approx \frac{\# \text{ intervals a detection} \geq \theta \text{ was observed}}{N}$$

is an estimate of the probability of detection for that detector.

5.5 Conventional coherent structure detectors

In this section we summarize some of the methods that have classically been used for detection and location of coherent structures. A more detailed description of these algorithms can be found in Kaspersen and Hudgins [15]. Each such method is equivalent to a definition of coherent structure. Different detectors use different criteria, and therefore will report different sets of times. This leads to different conditional averages and consequently different estimates for the underlying coherent structures.

5.5.1 *Quadrant analysis (Q2)*

We begin our description of methods for detecting coherent structures with a simple bivariate example. A *burst event* is physically associated with an out-rush of fluid from the wall, during which the transverse velocity is positive while the streamwise velocity temporarily falls below its mean value. In general, the quadrant algorithm can detect large $|uv|$ products in any specific quadrant of the u–v plane. However, to qualify a coherent event as a burst, we restrict our attention to the second quadrant. The method is therefore called 'Q2'. Define

$$Q_{fg}(t) = \begin{cases} f(t)g(t) & : \quad f > 0, g > 0 \text{ and} \\ 0 & : \qquad \text{otherwise.} \end{cases} \tag{5.5}$$

Let M represent the set of local maxima of $Q_{fg}(t)$, and for each $\theta > 0$ define the detection set as

$$D_Q(\theta) = \{\tau \in M : Q_{fg}(\tau) \geq \theta\}. \tag{5.6}$$

In practice, the Q2 algorithm is employed with $f = -u$ and $g = +v$, which forces $Q_{-uv}(t)$ to take positive values in the second quadrant of the u–v plane.

5.5.2 *Variable Interval Time Average (VITA)*

The Variable Interval Time Average (VITA) algorithm of Blackwelder and Kaplan [3] is an example of a univariate detector. This method looks for large and/or abrupt changes in the streamwise velocity by calculating its short-time variance over a fixed time interval.

For $\Delta > 0$, define the short-time average of f as

$$\widehat{f}(t) = \frac{1}{\Delta} \int_{t-\Delta/2}^{t+\Delta/2} f(x)\,dx. \tag{5.7}$$

The short-time variance may then be expressed as

$$\widetilde{f}(t) = \widehat{f^2}(t) - [\widehat{f}(t)]^2. \tag{5.8}$$

In our analysis, we let M represent the set of local maxima of $\widetilde{f}(t)$. Then for each θ we define the detection set as

$$D_{\text{VITA}}(\theta) = \{\tau \in M : \widetilde{f}(\tau) \geq \theta\}. \tag{5.9}$$

In terms of wavelets, the above filter can be understood as the lowpass branch of a perfect reconstruction filter bank for the Haar system. In practice, the algorithm is applied to the u-component of velocity using $\Delta = 10$–20 wall time units, which for our data was approximately 2.5 ms.

The first author has demonstrated [9] for an atmospheric boundary layer over the open ocean that a bivariate wavelet method based on both the u- and v-velocity components outperforms the corresponding univariate algorithm. In this study we wish to compare both univariate and bivariate detectors whenever both methods can reasonably be defined. We therefore offer the following bivariate version of the VITA algorithm.

Write the short-time covariance as

$$V_{fg}(t) = \widehat{fg}(t) - \widehat{f}(t)\widehat{g}(t). \tag{5.10}$$

Let M represent the set of local maxima of $V_{fg}(t)$, and for each θ define the detection set as

$$D_{\text{Bi-VITA}}(\theta) = \{\tau \in M : V_{fg}(\tau) \geq \theta\}. \tag{5.11}$$

For the same reasons as before, this algorithm is employed with $f = -u$ and $g = +v$ which forces $V_{-uv}(t)$ to take positive values in the second and fourth quadrants of the u–v plane.

The resulting coherent structures can be further discriminated as bursts by restricting our attention to the second quadrant. In this way, additional algorithms may be defined in terms of the existing ones. Let

$$Q = \{\tau : f(\tau) > 0 \text{ and } g(\tau) > 0\}. \tag{5.12}$$

Then the VITA + Q detection set is given by

$$D_{\text{VITA}}^{Q}(\theta) = D_{\text{VITA}}(\theta) \cap Q, \tag{5.13}$$

in other words, those VITA detections which occur in the specified quadrant. Bursts are detected when we let $f = -u$ and $g = +v$. Similarly,

$$D_{\text{Bi-VITA}}^{Q}(\theta) = D_{\text{Bi-VITA}}(\theta) \cap Q \tag{5.14}$$

gives the Bi-VITA detections which qualify according to their quadrant.

5.5.3 Window Average Gradient (WAG)

The Window Average Gradient (WAG) algorithm was introduced by Antonia and Fulachier [1] in 1989 as an alternative to VITA. The abrupt velocity changes associated with coherent structures can be detected by measuring the average gradient over some appropriate window width. Loosely speaking, Taylor's hypothesis states that turbulent fluid velocities are practically constant over any interval of time less than or equal to their size divided by their convection speed. This assumption allows us to infer the 'instantaneous' shape of a coherent structure by observing it at a single point in space as it moves under the influence of fluid convection. We can also use this to estimate the streamwise spatial gradient by observing its temporal gradient.

Filter the signal f according to

$$\widetilde{f}(t) = \frac{1}{\Delta}\left[\int_{t+0}^{t+\Delta/2} f(x)\,dx - \int_{t-\Delta/2}^{t-0} f(x)\,dx\right], \tag{5.15}$$

and let M be the set of local maxima of $\widetilde{f}$. For each θ, the set of detection points is given by

$$\mathrm{D_{WAG}}(\theta) = \{\tau \in \mathrm{M} : \widetilde{f}(\tau) \geq \theta\}. \tag{5.16}$$

The reader will recognize that the filter employed by the WAG algorithm is simply the continuous wavelet transform of f using the *Haar wavelet* at fixed scale Δ. As such, it identifies those places where the signal changes over the characteristic time scale of Δ. In practice, the algorithm is applied with $f = -u$, and with $\Delta = \frac{0.9\delta}{U_\infty}$, where δ is the boundary layer thickness, and U_∞ is the free-stream velocity.

The algorithm is easily extended to a bivariate version (Bi-WAG) by defining $\widetilde{f}$ and $\widetilde{g}$ according to equation (5.15), and letting M represent the set of local maxima of $\widetilde{f}(t)\widetilde{g}(t)$. For each θ define

$$\mathrm{D_{Bi-WAG}}(\theta) = \{\tau \in \mathrm{M} : \widetilde{f}(\tau)\widetilde{g}(\tau) \geq \theta\}. \tag{5.17}$$

To further qualify the detected events as bursts, we can restrict them to quadrant II by defining the WAG + Q detection set as

$$\mathrm{D_{WAG}^Q}(\theta) = \mathrm{D_{WAG}}(\theta) \cap \mathrm{Q}. \tag{5.18}$$

Similarly, in the bivariate (Bi-WAG + Q) case,

$$\mathrm{D_{Bi-WAG}^Q}(\theta) = \mathrm{D_{Bi-WAG}}(\theta) \cap \mathrm{Q}. \tag{5.19}$$

5.6 Wavelet-based coherent structure detectors

In the previous section it was observed that the WAG detector and all of its variants employ the Haar wavelet. Specifically, the continuous wavelet transform is used at a single fixed scale. We are now prepared to generalize this approach. A comprehensive treatment of the subject of wavelets can be found in [5], [6], [14], and others. Further details regarding wavelet spectra and their application to detection of coherent structures may be found in [10], [11], and [8].

Conventional Fourier methods identified the repetition rate to be approximately 125 Hz. In this analysis we employ a cubic spline wavelet at fixed scale $a = 1/125$ Hz. The choice of the cubic spline wavelet provides an excellent trade-off between short length (which gives good time localization) and high peak-to-sidelobe ratio (which minimizes filter leakage) while maintaining phase linearity (by virtue of its symmetry).

5.6.1 *Typical wavelet method (psi)*

Except for the wavelet itself and the method of selecting the analysis scale, the wavelet-based detectors described in this section are identical to the WAG algorithms above. Filter the signal f according to

$$W_f(t) = \int f(x)\psi_a(x - t)dx, \tag{5.20}$$

where a is the scale parameter, and the functions

$$\psi_a(t) = \frac{1}{\sqrt{|a|}}\,\psi\!\left(\frac{t}{a}\right)$$

are defined in terms of an admissible wavelet $\psi(t)$. Let M be the set of local maxima of $W_f(t)$. For each θ the ψ-set of detection points is given by

$$D_\psi(\theta) = \{\tau \in M : W_f(\tau) \geq \theta\}. \tag{5.21}$$

The algorithm is applied with $f = -u$. The bivariate version (Bi-ψ) defines W_f and W_g according to equation (5.20). Let M represent the set of local maxima of $W_f(t)W_g(t)$. For each θ define

$$D_{\mathrm{Bi}-\psi}(\theta) = \{\tau \in M : W_f(\tau)W_g(\tau) \geq \theta\}. \tag{5.22}$$

As before the $\psi + Q$ and Bi-$\psi + Q$ detection sets are given by

$$D_\psi^Q(\theta) = D_\psi(\theta) \cap Q, \text{ and} \qquad (5.23)$$

$$D_{\text{Bi}-\psi}^Q(\theta) = D_{\text{Bi}-\psi}(\theta) \cap Q. \qquad (5.24)$$

5.6.2 Wavelet quadrature method (Quad)

Each of the aforementioned detection methods was applied to the data record, with their thresholds individually adjusted so that the average detection frequency was equal to the peak in Figure 5.4. Conditional averages were then formed for both the u and v time series according to equation (5.4), producing estimates for the deterministic parts of these signals. The resulting averages are shown in Figure 5.6. The u and v averages are very nearly periodic, and essentially 90° out-of-phase with each other. This observation led us to make a refinement of the bivariate wavelet method which we call *wavelet quadrature* (Quad).

Both Fourier- and complex wavelet-based methods have been devised to recover the phase relationship between pairs of signals. (For a detailed treatment of the wavelet cross-transform and its relationship to the Fourier cross-transform see [9].) But the present problem merely requires a real-valued signal that responds well to events that are in quadrature.

Like the Haar wavelet, the cubic spline tends to detect the rising edges of a signal as it changes over a characteristic time scale. These are examples of anti-symmetric spline wavelets. On the other hand, the quadratic and quartic splines display even symmetry: they respond maximally to signal events that are fully symmetric. By employing one symmetric wavelet and one anti-symmetric wavelet in a bivariate wavelet detector, we can optimize its performance for events that are in quadrature. Additional details regarding the use of symmetric vs. anti-symmetric wavelets in the context of intermittent turbulence have been discussed by Hagelberg and Gamage in [8].

Let ψ and ψ' represent admissible wavelets possessing odd and even symmetry, respectively. Define

$$\mathcal{W}_f(t) = \int f(x)\psi_a(x - t)dx, \text{ and} \qquad (5.25)$$

$$\mathcal{W}_g'(t) = \int g(x)\psi_a'(x - t)dx.$$

Let M represent the set of local maxima of $\mathcal{W}_f(t)\mathcal{W}_g'(t)$ and for each θ define

$$D_{\text{quadrant}}(\theta) = \{\tau \in M : \mathcal{W}_f(\tau)\mathcal{W}_g'(\tau) \geq \theta\}. \qquad (5.26)$$

In our analysis, we have used the cubic and quartic spline wavelets, respectively. Because of the way these wavelets were chosen, the algorithm no

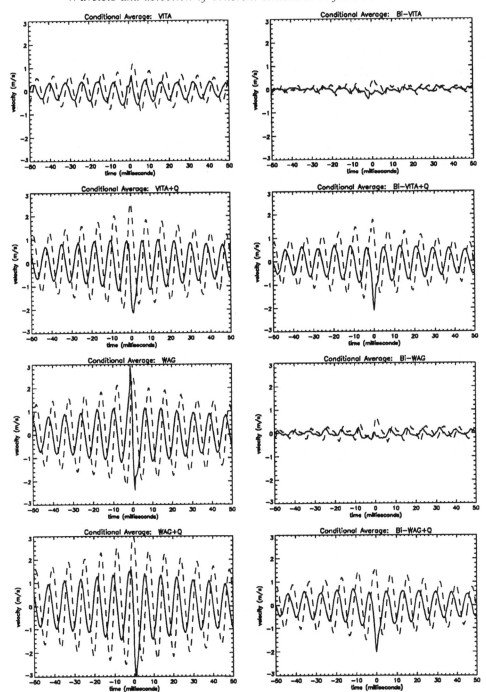

Fig. 5.6. (*continues on next page*)

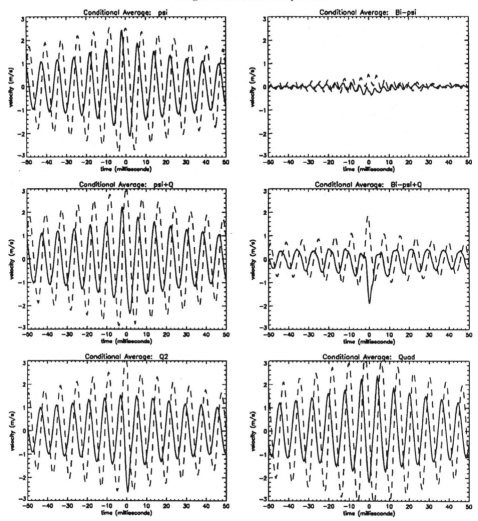

Fig. 5.6. (*continued*) Conditional averages of the *u* (solid line) and *v* (dashed line) velocity signals for each of the detection methods tested.

longer provides clear discrimination to quadrants in the $u-v$ plane: the response has been rotated 45° clockwise. This means that we now tend to detect structures that lie near the v-axis; precisely what is needed to detect events that are in quadrature. From Figure 5.6, it can be seen that when the v-velocity is at its positive extremum, the u-velocity is centred on its falling edge. Therefore, we again apply the algorithm with $f = -u$ and $g = +v$.

It is possible to proceed with the mathematical development of the wavelet quadrature method as a wavelet cross-transform of two signals, using a pair of wavelets that are Hilbert transforms of each other [15]. In this sense, it is a natural extension of the wavelet cross-transform which selectively detects signals that are 90° out-of-phase with each other.

5.7 Results

Each detector tends to pick out a certain component in the data. All of the univariate detectors have returned relatively strong harmonic conditional averages, while their bivariate counterparts appear to have much weaker responses. The Q2 and wavelet quadrature algorithms (both of which are inherently bivariate) have also returned large averages. But an algorithm attempts to identify segments of data that match its criteria, whether or not they arise from physically meaningful events. The probability that the algorithm will report a detection in the absence of any coherent structures determines the false-alarm rate (see section 5.4.1). It is therefore instructive to examine the response of each of the detectors to an input of random noise. In Figure 5.7 we display conditional averages of each algorithm's response to the synthetic noise data. These control samples were derived in a similar manner to the averages in Figure 5.6, by adjusting each detector threshold so that the mean frequency of detection was approximately equal to the vortex shedding rate in the original data record. The control samples provide us with clues about precisely what type of structure each method is detecting – at least in the mean. Not surprisingly, those methods with the largest control responses also returned the greatest conditional averages. Furthermore, only the WAG + Q, psi + Q, and Quadrature methods exhibit any tendency to selectively detect events that are in quadrature.

The measured $P_{FA}(\theta)$ and $P_D(\theta)$ for each detector were combined to form the respective ROC curves shown in Figure 5.8. Clearly, some of the detectors performed better than others. The wavelet based methods – WAG, psi, and Quad, are generally superior to the non-wavelet methods – VITA and Q2. However, each of the univariate algorithms, VITA, WAG, and psi, benefited by an additional qualification on quadrant II events. The bivariate versions that we offered for VITA, WAG, and psi, all gave disappointingly poor results, even when qualified on quadrant II. It should be observed that the nearly 'flat' control averages of VITA, WAG, and psi can be explained by noting that they respond equally to events in either quadrant II or IV. However, the fact that quadrant qualification did not appreciably help their performance indicates an erroneous physical basis.

Turning our attention toward the high detection rates, it can be seen (cf. bottom right panel of Figure 5.8) that the wavelet quadrature method

220   *L. Hudgins and J.H. Kaspersen*

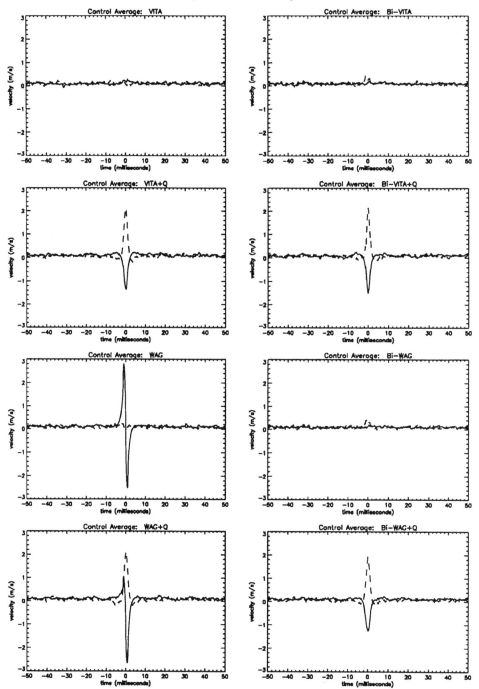

Fig. 5.7. (*continues on next page*)

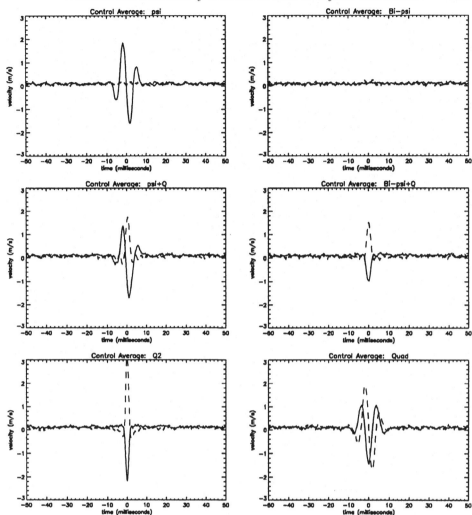

Fig. 5.7. (*continued*) Control averages of the synthetic signals for each of the detection methods tested.

outperformed all of the others. To better understand this behaviour, we computed some new Quadrature conditional averages using only detection thresholds that fell within a narrow range of values. Specifically, the possible threshold values were sorted into eight 'octiles', and conditional averages formed for each octile. Figure 5.9 demonstrates that the morphology of the coherent structures depends on the threshold level used to detect them. First of all, this means that the mathematical model in equation (5.2) is wrong: the deterministic function is not fixed. Indeed, the

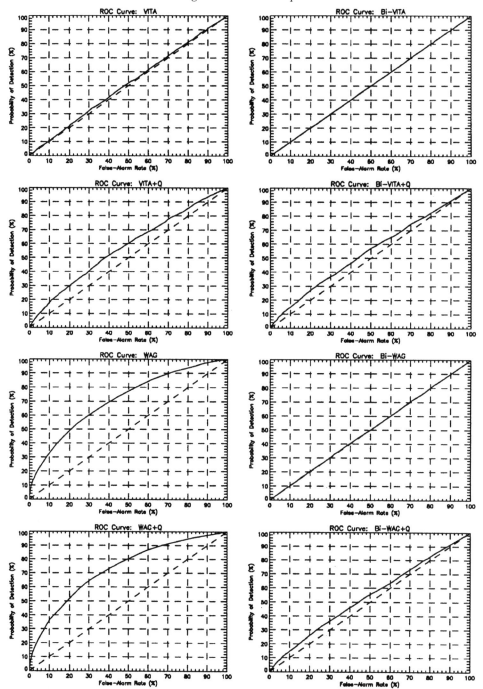

Fig. 5.8. (*continues on next page*)

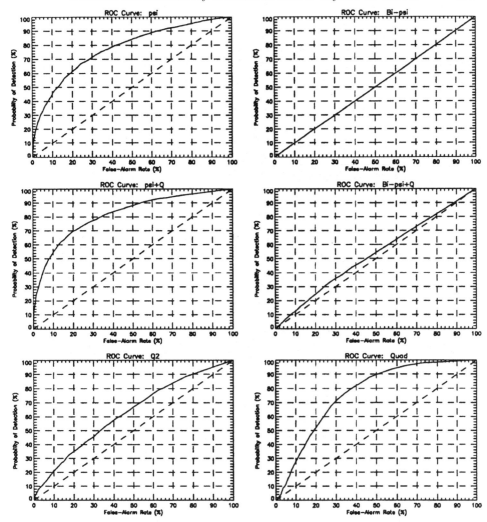

Fig. 5.8. (*continued*) Receiver Operating Characteristic (ROC) curves for each of the detection methods tested.

shape of the coherent structures detected at high threshold levels can be very different from those at low thresholds, especially in the u-velocity component. The shape of the low threshold averages (high detection rates) is nearly the same as the Quadrature global mean in Figure 5.6, but the relatively rare high threshold events (low detection rates) are many times more energetic. However, while the absolute value of the peak v-velocity depended somewhat on the threshold level, its overall shape was virtually constant. This would indicate that the rare, energetic events are

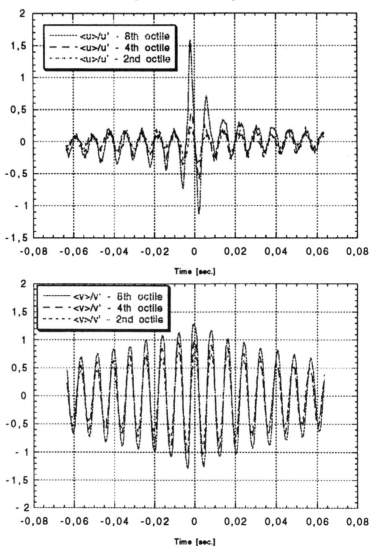

Fig. 5.9. Special conditional averages taken from Quadrature detections near the centreline of the wake. These demonstrate that the morphology of the coherent structures depends on the threshold level used to detect them.

much more closely associated with the streamwise component of motion than with the transverse component, but that the more common, low-energy events are essentially a mix of the *u*- and *v*-velocity components in quadrature. This explains why the Quadrature method works best at the highest detection rates, while the univariate wavelet method acting only on the *u*-component was superior at the lowest detection rates.

5.8 Conclusions

Several of the most commonly used algorithms for detecting coherent structures in turbulent flows have been tested in an objective manner by the use of ROC curves. Wavelet methods have been found to perform better than the conventional detection methods tested. It was also found that additional qualification on Q2 events improves the performance of each of the relevant methods. Finally, at high detection rates the wavelet quadrature method outperforms all of the others.

Conditional averages from the algorithms tested in this study showed that the coherent structures were in agreement with a quadrature model. In other regions of the wake or for other types of flow the quadrature detector may not be optimal; cf. [9] in which the bivariate wavelet method was superior to the univariate algorithm for an atmospheric boundary layer. Nevertheless, for any kind of flow conditional averaging methods which take into account the phase relationship between the velocity components would be strongly preferred. Complex-valued wavelets also retain phase information, and represent another way to implement a detection algorithm. This is in agreement with the findings of Hussain [12] who found that averaging based on constant phase is preferable.

Finally, the morphology of the coherent structures can be very different at different threshold levels, even for the fixed test geometry described in this report. At low threshold levels, the coherent structures are very harmonic, possess relatively low energy, and their effect is more or less evenly split between the u- and v-velocity components. At high thresholds, they are quite peaked in shape, highly energetic, and their effect is mainly coupled to the u-velocity.

Acknowledgements

The first author gratefully acknowledges the many thoughtful comments and contributions received from Dr Paul Neiswander at Northrop Grumman, and the insightful and constructive criticism from the referees.

References

[1] R.A.Antonia and L. Fulachier. Topology of the turbulent boundary layer with and without wall suction. *J. Fluid Mech.*, **198**: 429–451, (1989)

[2] D.K. Bisset, R.A. Antonia and L.W.B. Browne. Spatial organization of large structures in the turbulent far wake of a cylinder. *J. Fluid Mech.*, **218**: 439–461, (1990)

[3] R.F. Blackwelder and R.E. Kaplan. On the wall structure of the turbulent boundary layer. *J. Fluid Mech.*, **76**: 89–112, (1976)

[4] D.G. Bogard and W.G. Tiederman. Investigation of flow visualization techniques for detecting turbulent bursts. In Reed, Patterson and Zaki (eds.), *Symposium on Turbulence*, p. 289, 1983

[5] C.K. Chui. *An Introduction to Wavelets*, volume 1 of *Wavelet Analysis and Its Applications*. (Academic Press, San Diego, CA, 1992)

[6] I. Daubechies. *Ten Lectures on Wavelets*. (SIAM, Philadelphia, PA, 1992)

[7] P. Flandrin. Some aspects of non-stationary signal processing with emphasis on time-frequency and time-scale methods. In J.M. Combes, A. Grossmann and Ph. Tchamitchian (eds.), *Wavelets: Time-Frequency Methods and Phase Space*. (Springer-Verlag, Berlin, 1989)

[8] E. Foufoula-Georgiou and P. Kumar (eds.), *Wavelets in Geophysics*, volume 4 of *Wavelet Analysis and Its Applications*. (Academic Press, San Diego, CA, 1994)

[9] L. Hudgins. *Wavelet Analysis of Atmospheric Turbulence*. PhD thesis, University of California, Irvine, March 1992

[10] L. Hudgins. Wavelet transforms and spectral estimation. American Mathematical Society, Eastern Section, Washington, D.C., April 1993

[11] L. Hudgins, M.E. Mayer and C.A. Friehe. Fourier and wavelet analysis of atmospheric turbulence. In Y. Meyer and S. Roques (eds.), *Progress in Wavelet Analysis and Applications*, pages 491–498. (Editions Frontieres, 1993)

[12] A.K.M.F. Hussain. Coherent structures – reality and myth. *Physics of Fluids*, **26**: 2816–2850, (1983)

[13] A.K.M.F. Hussain. Coherent structures and turbulence. *Journal of Fluid Mechanics*, **173**: 303–356, (1993)

[14] G. Kaiser. *A Friendly Guide to Wavelets*. (Birkhäuser, Boston, Basel, 1994)

[15] J.H. Kaspersen and L. Hudgins. Wavelet quadrature methods for detecting coherent structures in fluid turbulence. In Unser (ed.), *Conference on Wavelets in Mathematical Imaging*. SPIE, 1996

[16] S.J. Kline, W.C. Reynolds, F.A. Schraub and P.W. Rundstandler. The structure of turbulent boundary layers. *Journal of Fluid Mechanics*, **30**: 741–773, (1967)

[17] P.A. Krogstad and J.H. Kaspersen. Methods to detect coherent structures – a comparison. In *Proc. 11th Australasian Fluid Mechanics Conference*, 1992

[18] T.S. Luchik and W.G. Tiederman. Timescale and structure of ejections and bursts in turbulent channel flows. *J. Fluid Mech.*, **174**: 529–552, (1987)

[19] P. Moin and T. Bewley. Application of control theory to turbulence. In *Proc. 12th Australasian Fluid Mechanics Conference*, 1995

[20] L. Prandtl. Bericht uber untersuchungen zur ausgebildeten turbulenz. *Z. Angew. Math. Mech.*, **5**: 136–139, (1925)

[21] O. Reynolds. *Phil. Trans. R. Soc.*, **186**: 123–164, (1895)

[22] S. Robinson. Coherent motion in the turbulent boundary layer. *Annual Review of Fluid Mechanics*, **23**: 601–639, (1991)

[23] G.I. Taylor. Note on the distribution of turbulent velocities in a fluid near a solid wall. In *Proc. R. Soc. London Ser. A* **135**: 678–684, (1932)

[24] H. Tennekes and J.L. Lumley. *A First Course in Turbulence*. (The MIT press, Cambridge, MA, 1987)

[25] A.A. Townsend. *The structure of turbulent shear flow*. (Cambridge Univ. Press, 2nd edition, 1956)

[26] Y. Zhou. *Organised motion in a turbulent wake*. PhD thesis, University of Newcastle, 1993

6

Wavelets, non-linearity and turbulence in fusion plasmas

B.Ph. VAN MILLIGEN

Asociación EURATOM-CIEMAT,
Avda. Complutense 22, 28040 Madrid, Spain

Abstract

Two fundamental properties of turbulence are intermittency and non-linearity. They imply that the standard Fourier spectral techniques are inadequate for its analysis. Spectral analysis based on wavelets provides a means to handle intermittency. New tools are required to handle non-linearity.

In this chapter, we redesign spectral analysis in terms of wavelet methods, paying particular attention to statistical stability, error estimates and non-linearity. The application to both computer simulations and measurements carried out in fusion plasmas provide some illustrative examples.

6.1 Introduction

Although the phenomenon of turbulence is only partially understood, there seems to be consensus on several aspects. First, that *intermittency* is a basic property of turbulence. This means that the characteristics of the turbulence (spectral distribution, amplitude etc.) vary on a short time scale. Analysis techniques that rely on the accumulation of data over time scales larger than this characteristic time scale will then average out much of the dynamics and obliterate relevant information (as may occur with Fourier analyses). Wavelet analysis provides an interesting starting point for redesigning the standard analysis techniques in order to tackle this problem. In this chapter we shall redefine some basic Fourier analysis techniques in terms of wavelets, such as cross coherence. We shall emphasize the need for statistical stability and provide noise level estimates. Finally, we provide some examples of these techniques.

Second, it is generally accepted that turbulence only arises in *non-linear* systems. Therefore, to understand the nature of turbulence, it is essential to employ analysis tools that are capable of handling this non-linearity. The usual analyses based on (cross-) spectra and (cross-) correlations, essentially linear analysis techniques, are not adequate. Non-linear analysis tools can be obtained by generalizing the common spectral analysis methods to higher order, which then are sensitive to non-linear interactions. In this chapter we shall focus on the so-called bispectral analysis, a method for the detection of quadratic interactions.

A definition of the bispectrum and the bicoherence based on wavelet analysis is given. Statistical stability of the bispectrum – a third-order spectrum – is again an important point, so we shall provide noise level estimates of this quantity. The main application of the bispectrum is the detection of phase coupling. It is shown how this phase coupling bears relation to the existence of 'structure' in turbulent or chaotic time series. Reasonable time resolution, relevant to intermittency in some turbulent phenomena, can be achieved.

The meaning of bicoherence is clarified analysing computer-generated chaotic time series. Then, some results obtained from measurements of turbulence in fusion plasmas are presented. Intermittency is detected, and an analysis of L/H transitions (Low- to High-confinement mode, a bifurcation in the behaviour of thermonuclear plasmas) is presented.

6.2 Linear spectral analysis tools

In this section, we shall redefine the traditional spectral analysis tools in terms of wavelets. In doing so, we shall mostly avoid reference to a specific type of wavelet, since the definitions are equally valid for all types. Although the definitions are given for continuous wavelets, extension to discrete wavelets is mostly self-evident. Such an extension is not possible, however, for the higher-order spectra (bicoherence), which will be introduced in section 6.3. The wavelet analysis is set up in such a way that it forms a natural extension to Fourier analysis, which will help the interpretation of the wavelet analysis results.

6.2.1 Wavelet analysis

The Fourier transform of a function $f(t)$ and its power spectrum are given by:

$$\hat{f}(\omega) = \int_{-\infty}^{\infty} f(t)\, e^{-i\omega t}\, dt \quad \text{and} \quad P_f(\omega) = \left|\hat{f}(\omega)\right|^2 \qquad (6.1)$$

A wavelet can be any function $\Psi(t)$ that satisfies the *wavelet admissibility condition*:

$$c_\Psi = \int_{-\infty}^{\infty} \left|\hat{\Psi}(\omega)\right|^2 |\omega|^{-1} \, d\omega \; < \; \infty \tag{6.2}$$

The corresponding wavelet family is obtained by means of the scale length parameter a:

$$\Psi_a(t) = \frac{1}{a^p} \, \Psi(t/a) \tag{6.3}$$

Some authors prefer to use the scale number s instead, where $s = 1/a$. The scale number s is proportional to the frequency of the wavelet. The factor p is the normalization choice. In the literature, values of p of 0, 1/2 and 1 are encountered [1, 2]. In the present work, we choose $p = 1/2$. Other choices of p may be motivated by computational efficiency or by wavelet power spectrum visualization demands, but the choice $p = 1/2$ implies that the L^2-norm of the wavelet is independent of a, and thus the wavelet analysis forms a natural extension of Fourier analysis.

As mentioned earlier, most of the definitions we shall be giving are independent of the actual wavelet choice. Nevertheless, for concreteness we select a specific wavelet (the Morlet wavelet) which has the benefit of conceptual closeness to the Fourier analysis base functions $e^{-i\omega t}$:

$$\Psi(t) = \left(d\sqrt{\pi}\right)^{1/2}(\exp[-i2\pi t] - c_0)\exp\left[-\tfrac{1}{2}(t/d)^2\right] \tag{6.4}$$

The factor $c_0 = \exp(-2\pi^2 d^2)$ is included to guarantee that Eq. (6.2) is satisfied; even so, due to its numerical smallness for values of d of the order of 1 it is usually omitted in practice. The normalization is such that the L^2-norm of this wavelet is equal to 1. The parameter d determines the exponential decay of the wavelet and thus permits a suitable combination of time- and frequency resolution to be selected. Comparing Eqs. (6.1), (6.3) and (6.4) we assign a frequency $\omega = 2\pi/a$ to each scale a. The frequency resolution of the wavelet $\Psi_a(t)$ is approximately $\Delta\omega = \omega/4d$ (FWHM – Full Width at Half Maximum). The time resolution is $\Delta t = 2ad$ (twice the e-folding length), given by the decay of the exponential part of the wavelet. Note that $\Delta\omega\Delta t = \pi$, independent of either a or d.

The *wavelet transform* of a function $f(t)$ is defined by:

$$W_f(a, \tau) = \int f(t)\,\Psi_a(t - \tau)\,dt \tag{6.5}$$

As d increases, frequency resolution improves but time resolution deteriorates. Interestingly, for $d \gg 1$ the wavelet analysis essentially becomes a Fourier analysis, for which the frequency resolution is optimal but there is no time resolution. At the other extreme, $d \downarrow 0$, the wavelet becomes a δ-function, and the wavelet transform yields the original signal – we have optimal time resolution but no frequency resolution. We set $d = 1$ in the following, which is, we believe, a reasonable compromise between frequency and time resolution, although for specific purposes other choices may be better. Note that the selected wavelet family is not orthogonal. This implies a certain redundancy in the wavelet transform coefficients which must be taken into account upon interpreting the results. This disadvantage is compensated by the mentioned conceptual closeness to the Fourier transform. The redundancy is actually a necessity when calculating higher-order spectra, as will become clear later. Even though the wavelets are not orthogonal, the inverse wavelet transform (for $p = 1/2$) can be calculated (for almost all t) by:

$$f(t) = \frac{1}{c_\Psi} \iint W_f(a, \tau) \Psi_a^*(\tau - t) \frac{da \, d\tau}{a^2} \tag{6.6}$$

which completes the analogy with the Fourier transform.

The wavelet transform $W_f(a, \tau)$ at any given a can be interpreted as a filtered version of $f(t)$, bandpassed by the filter Ψ_a. Usually $|W_f(a, \tau)|^2$ is plotted in the (a, τ)-plane for visualization purposes (scalogram). In many instances, the scalogram may be very instructive qualitatively, but provides no indication as to the statistical significance of the observed features. This problem will be discussed in section 6.2.2.

As we have tried to make plausible by rather hand-waving arguments, the wavelet transform can be regarded as a generalization of the Fourier transform. The main advantage over the Fourier transform is that, with a suitable choice of wavelet Ψ, time-resolved spectra can be calculated. Of course, the Short-Time (or Windowed) Fourier Transform (basically, chopping the signal $f(t)$ in short time intervals and calculating the Fourier transform for each section [3]) also achieves this. The particular advantage of the wavelet analysis lies in the fact that the time resolution is variable with frequency, so that high frequencies have a better time resolution. In other words, we abandon the rather mathematical idea of considering signals $f(t)$ to be composed of 'everlasting' monochromatic oscillations, and replace it by the more physical idea that the elementary oscillations composing the signal must die out in time, and more rapidly so the higher their frequency, which seems quite a natural state of affairs. Thus, one may expect that wavelet analysis is better adapted than Fourier analysis to the examination of systems with dissipation, or more generally non-linear systems, since the analysis functions used (wavelets) are more like the actual (short-lived) oscillations occurring in such systems than Fourier modes. Seen from another perspective, one may consider the problem of decomposing the oscillations of a system into modes. When the system is linear and the modes used are eigenmodes of that same system, the decomposition is likely to reveal much information. However, when the modes are not eigenmodes, the information is likely to be poorly represented on the basis of these non-eigenmode functions and each mode coeficient is the result of a number of eigenmodes – the information is 'scrambled'. In the non-linear systems with dissipation we are concerned with here, we know that in general no eigenmodes exist that allow such a decomposition. Nevertheless, *locally* one can often perform a mode decomposition, since in a local environment of a given point the equations describing the system can be linearized, provided the non-linearity of the system is not too strong. This *local* decomposition is precisely what the wavelet analysis pretends to do, and this explains the success of the wavelet analysis in turbulence analysis.

6.2.2 *Wavelet spectra and coherence*

We shall now proceed to reformulate the usual spectral analysis tools in terms of wavelets. In practice, the analysis usually starts with a signal $f(t)$ that has been digitally sampled. Therefore, the integral appearing in Eq. (6.5) should be replaced by a summation. Due to the fact that the wavelet decays rapidly, it is sufficient to evaluate the integral (sum) from, say, $-4ad$ to $+4ad$. This interval depends, of course, on the type of wavelet used. The digitally sampled signal has a finite record length, implying that the wavelet coefficients cannot be calculated correctly when τ is too close to the record boundaries. We simply set the wavelet coefficient to zero when τ is less than $4ad$ from a record boundary. It will be noted that this distance depends on the frequency.

One could, of course, take the usual definitions of power spectra, cross correlations, etc., based on Fourier analysis and simply replace all occurrences of Fourier coefficients with wavelet coefficients. That would lead, however, to highly unstable quantities, varying wildly with time, the practical value of which is limited. In order to obtain statistical stability while maintaining time resolution, we integrate (sum) the appropriate combinations of wavelet coefficients over a (small) finite time interval $T : \{T_0 - T/2 \le \tau \le T_0 + T/2\}$. As a bonus, this procedure allows the estimation of a noise level which will tell us the statistical significance of the obtained results. Apart from that, the definitions are completely analogous to the usual definitions used in Fourier analysis.

Thus, for example, the *wavelet cross spectrum* is given by:

$$C_{fg}^{w}(a, T_0) = \int_T W_f^*(a, \tau)\, W_g(a, \tau)\, d\tau \tag{6.7}$$

where $f(t)$ and $g(t)$ are two time series. We also introduce the *delayed wavelet cross spectrum*:

$$C_{fg}^{w}(a, T_0, \Delta\tau) = \int_T W_f^*(a, \tau)\, W_g(a, \tau + \Delta\tau)\, d\tau \tag{6.8}$$

which is a useful quantity for detecting e.g. structures flowing past two separated observation points. Note that C_{fg}^{w} is complex and both its phase and amplitude provide information. The *normalized delayed wavelet cross coherence* is:

$$\gamma_{fg}^{w}(a, T_0, \Delta\tau) = \frac{\left| \int_T W_f^*(a, \tau)\, W_g(a, \tau + \Delta\tau)\, d\tau \right|}{\left(P_f^w(a, T_0) P_g^w(a, T_0 + \Delta\tau) \right)^{1/2}} \tag{6.9}$$

which can take on values between 0 and 1. Usually either T_0 or $\Delta\tau$ is held fixed for visualization purposes. Here the *wavelet auto-power spectrum* is given by:

$$P_f^w(a, T_0) = C_{ff}^{w}(a, T_0) \tag{6.10}$$

Note that the wavelet power spectrum can also be written in terms of the Fourier power spectra of the wavelet and $f(t)$ when $T \to \infty$:

$$P_f^w(a) = \frac{1}{2\pi} \int P_{\Psi_a}(\omega)\, P_f(\omega)\, d\omega \tag{6.11}$$

Thus, in this limit the wavelet power spectrum is the Fourier power spectrum averaged by the power spectrum of the wavelet filter [4].

We shall provide an estimate of the statistical noise level of γ_{fg}^{w} (Eq. (6.9)). For that purpose, consider Eq. (6.8). In principle, the integration is over all samples in the interval T (although in practice one may devise a more efficient algorithm based on the considerations that follow). The frequency with which $f(t)$ is sampled is $\omega_{samp} = 2\pi f_{samp}$. Thus, theoretically the wavelet coefficients are determined for each of $N = T \cdot f_{samp}$ samples in the interval T and summed. However, these wavelet coefficients are not all statistically independent, since the chosen wavelet family is not orthogonal. Each coefficient is calculated by evaluating Eq. (6.5). Due to the periodicity a of the wavelets, two statistically independent estimates of the wavelet coefficients are separated by a time $a/2$, or by a number of points $M(a) = a\omega_{samp}/4\pi$. Thus, the integral appearing in Eq. (6.8) is carried out over $N/M(a)$ *independent* estimates of wavelet coefficients. The relative statistical error in the result is therefore $\sqrt{M(a)/N}$. Using $\omega = 2\pi/a$, we find:

$$\varepsilon\left(C_{fg}^w\right) \approx \left|C_{fg}^w\right| \cdot \left[\frac{\omega_{samp}}{2\omega}\frac{1}{N}\right]^{1/2} \tag{6.12}$$

Applying similar estimates to the denominator in Eq. (9), one arrives at:

$$\varepsilon\left(\gamma_{fg}^w\right) \approx \left[\frac{\omega_{samp}}{\omega}\frac{1}{N}\right]^{1/2} \tag{6.13}$$

It is found that the values of γ_{fg}^w obtained when analysing Gaussian noise (random data with a Gaussian probability distribution function, P.D.F.) conform quite well to this prediction. We consider that significant values of C_{fg}^w or γ_{fg}^w should at least be a factor 2 above noise level. Now the advantage of introducing the integration time interval T becomes clear: it provides us with a method to distinguish significant data from noise, something which is not at all evident for the individual wavelet coefficients, and which has been a severe point of criticism to the usual wavelet spectral analysis ever since its introduction. In fact, in the early wavelet papers the statistical fluctuation of the wavelet coefficients has often been mistaken for significant information.

6.2.3 Joint wavelet phase-frequency spectra

Another way of doing statistics with the wavelet coefficients that is very informative from a physical point of view is calculating the *joint wavelet phase-frequency probability distribution function* (for a motivation of this technique and a definition in terms of Fourier analysis see [5]). It consists in calculating the quantity $c = W_f^*(a, \tau)\,W_g(a, \tau + \Delta\tau)$ (being the argument appearing in the definition of C_{fg}^w) for a number of values of a and τ, with fixed $\Delta\tau$. As usual, we define $\omega = 2\pi/a$ and in addition we define ϕ to be the phase of c. A plot in the (ω, ϕ)-plane is then made by dividing the frequency range $0 \leq \omega \leq \omega_{samp}/2$ and the phase range $-\pi \leq \phi \leq \pi$ into bins and scoring how often each bin is 'hit'. This graph, when normalized so that the sum over all bins is 1, is referred to as $P(\phi, \omega)$, the joint wavelet phase-frequency probability distribution function. Another way of calculating consists of not summing 1 to each bin that is hit, but rather the instantaneous average wavelet transform value $\frac{1}{2}\left(|W_f|^2 + |W_g|^2\right)$. The resulting graph is $S(\phi, \omega)$, the *joint wavelet phase-frequency spectrum*. These graphs provide a marvellous insight into the frequency-dependent phase relations that may exist between two signals f and g (usually taken to be two spatially separated measurements of the same quantity). Moreover, in many cases, depending on the physical nature of the signals f and g, and under some additional assumptions (homogeneity of turbulence, see [5]) a relation can be made with the dispersion relation $\omega(k)$ for the processes driving the turbulence. For some examples of this technique, see section 6.4.2.

6.3 Non-linear spectral analysis tools

For the investigation of non-linear systems, proper non-linear spectral analysis tools are required. Given the importance of non-linearity in chaos and turbulence, it is surprising how little attention has been paid to the development of such tools. The focus has mostly been on statistical tools (referring to the determination of fractal dimensions, Lyapunov exponents etc.; for a recent review see [6]), while more recently time- and space-domain analysis (rather than frequency-domain analysis) has received a surge of interest, mostly due to wavelet analysis. Nevertheless, non-linear spectral analysis tools have been around for some time [7, 8, 9, 10] – although their practical use has always been hampered by the necessity for long time series in order to obtain statistical stability.

6.3.1 Wavelet bispectra and bicoherence

Basically, the (cross-) bispectrum is a third-order spectrum (in this terminology, the usual power spectrum would be a second-order spectrum). Its definition in terms of Fourier coefficients is $B_{fg}(\omega_1, \omega_2) = \left\langle \hat{f}^*(\omega)\hat{g}(\omega_1)\hat{g}(\omega_2) \right\rangle$, where $\omega = \omega_1 + \omega_2$ and $\langle \cdot \rangle$ signifies taking an ensemble average (averaging over many similar realizations). The interpretation of this quantity will be explained below in the discussion of the wavelet bispectrum. The point we wish to make here is that to obtain a statistically significant value of the bispectrum, the ensemble should consist of at least about 100, and preferably more, independent realizations, while each realization must consist of, say, at least 128 points to be able to obtain a spectrum with reasonable resolution. That means that experimental series must be well over 10^4 points long[†], during which time *the experimental conditions must not change appreciably.* Obviously, this is a very severe demand when analysing turbulence that may

---

[†]The demand that the realizations be strictly non-overlapping may be relaxed somewhat.

be intermittent, and this probably is the reason that the bispectral analysis has not been very popular or successful until recently.

Wavelet analysis now provides a second opportunity for these higher-order spectral methods, mainly by reducing the need for long time series. We replace the idea of the 'ensemble average' in the calculation of the bispectrum by the analogous concept of the time integral, along the lines of section 6.2.2. We define the *wavelet cross bispectrum* as in [11, 12]:

$$B_{fg}^{w}(a_1, a_2, T_0) = \int_T W_f^*(a, \tau) \, W_g(a_1, \tau) \, W_g(a_2, \tau) \, d\tau \qquad (6.14)$$

where T is again a short time interval centred at T_0 (see section 6.2.2), and

$$\frac{1}{a} = \frac{1}{a_1} + \frac{1}{a_2} \qquad (6.15)$$

(frequency sum-rule). Both the amplitude and the phase of B_{fg}^{w} contain significant information.

Now it becomes clear why we have chosen continuous wavelets as the basis for our analysis. Discrete wavelets provide wavelet coefficients for a set of scales $a_n \in \{2^n\}$ [2, 4, 13] (apart from constants), which cannot generally be combined in the manner of Eq. (6.15).

Likewise, we define the *wavelet auto bispectrum*

$$B^{w}(a_1, a_2, T_0) = B_{ff}^{w}(a_1, a_2, T_0) \qquad (6.16)$$

The squared *wavelet cross bicoherence* is the normalized squared cross bispectrum:

$$\left(b_{fg}^{w}(a_1, a_2, T_0)\right)^2 = \frac{\left|B_{fg}^{w}(a_1, a_2, T_0)\right|^2}{\left(\int_T \left|W_g(a_1, \tau) W_g(a_2, \tau)\right|^2 d\tau\right) P_f^{w}(a, T_0)} \qquad (6.17)$$

which can attain values between 0 and 1. Similarly, the squared *wavelet auto bicoherence* (henceforth simply referred to as bicoherence) is

$$\left(b^{w}(a_1, a_2, T_0)\right)^2 = \left(b_{ff}^{w}(a_1, a_2, T_0)\right)^2 \qquad (6.18)$$

The bicoherence is a measure of the amount of phase coupling that occurs in a signal or between two signals. Again, we assume that we are justified in setting $\omega = 2\pi/a$, i.e. we are using a 'well-behaved' wavelet whose Fourier transform has one well-defined peak frequency, ω. Phase coupling is defined to occur when two frequencies, ω_1 and ω_2, are simultaneously present in the signal(s) along with their sum (or difference) frequencies, and the sum of the phases ϕ of these frequency components remains constant in time. The bicoherence measures this quantity and is a function of two frequencies ω_1 and ω_2 which is close to 1 when the signal contains *three* frequencies ω_1, ω_2 and ω that satisfy the relation $\omega_1 + \omega_2 = \omega$ and $\phi_1 + \phi_2 = \phi + const$; if no such relation is satisfied, it is close to 0. Whereas the bicoherence measures the fraction of spectral power that is involved in the coupling process, the bispectrum measures the total power and phase. The concept of bicoherence is further explained in section 6.3.2.

When the analysed signal exhibits *structure* of any kind whatever, it may be expected that some phase coupling occurs – for example, to describe a non-sinusoidally shaped pulse (or pulse train) one needs several Fourier coefficients with definite phase relations. The definition of bicoherence in terms of wavelets, Eq. (6.17), is based on an integration over a (short) time interval T. Thus, one may be expected to be able to detect temporal variations in phase coupling (intermittent behaviour) or short-lived structures with a time resolution T, provided the calculation can be shown to be statistically significant [14]. This point will be discussed in the following.

Note that, in analogy with the wavelet cross coherence, one may also define the *delayed* cross bispectrum (and bicoherence) by replacing $W_g(a, \tau)$ by $W_g(a, \tau + \Delta\tau)$ in Eq. (6.14). Further, it is also possible to calculate the bispectrum from *three* time series f, g, and h instead of just two. For the sake of simplicity, we shall not include these possibilities explicitly in the definitions.

It is convenient to introduce the *summed bicoherence*, which is defined as $(b^w(a))^2 = \frac{1}{s(a)} \sum (b^w(a_1, a_2))^2$ (here explicit reference to T_0 has been omitted for convenience), where the sum is taken over all a_1 and a_2 such that Eq. (6.15) is satisfied and $s(a)$ is the number of summands in the summation. Similarly, the *total bicoherence* is defined as $(b^w)^2 = \frac{1}{S} \sum \sum (b^w(a_1, a_2))^2$ where the sum is taken over all a_1 and a_2 and S is again the number of terms in the summation. The factors $s(a)$ and S guarantee that the summed and total bicoherence are bounded between 0 and 1. These quantities summarize the information conveniently, as will be seen later.

The squared bicoherence $(b^w(a_1, a_2))^2$ is usually plotted in the (ω_1, ω_2)-plane rather than the (a_1, a_2)-plane for ease of interpretation. We allow ω_1, ω_2 and ω to take on negative values in order to be able to represent all sum and difference combinations of ω_1 and ω_2. There is no need to represent the whole plane; firstly, both ω_1, ω_2 and their sum ω must be smaller than the Nyquist frequency (half the sampling frequency); secondly, because ω_1 and ω_2 are interchangeable, we may restrict the plot to $\omega_1 \geq \omega_2$; and finally, the case (ω_1, ω_2) is identical to the case $(-\omega_1, -\omega_2)$ which is therefore not represented.

We note that Eq. (6.17) does not provide the only way of normalizing the bispectrum. For example, a symmetrical definition is possible by replacing the denominator by $P_g^w(a_1, T_0)P_g^w(a_2, T_0)P_f^w(a, T_0)$. With such a definition, it would be sufficient to plot only values $\omega_1, \omega_2 \geq 0$, which (taking into account the restrictions mentioned in the previous paragraph) is a triangular region. There is no strong objection to this; however, we shall see that some important information is lost (see section 6.3.2).

We proceed to estimate the error of the bicoherence (Eq. (6.17)) similar to the way we derived Eq. (6.13). Again we point out that the wavelet bispectrum (Eq. (6.14)) is calculated by integrating over a time interval T, corresponding to $N = T \cdot f_{samp}$ samples. As before, we observe that two statistically independent estimates of the wavelet coefficients are separated by a time $a/2$, or by a number of points $M(a) = a\omega_{samp}/4\pi$. To be on the safe side in the estimate of the error in the bispectrum, we say that the number of independent estimates of wavelet coefficient combinations appearing in Eq. (6.14) is at least $N/\max(M(a))$, where the maximum is taken over the values of a that come into play for the evaluation of a specific value of the squared bicoherence: $\{a, a_1, a_2\}$. An estimate for the statistical noise level in the bispectrum is, therefore:

$$\varepsilon\left(B_{fg}^w\right) \approx \left|B_{fg}^w\right| \cdot \left[\frac{\omega_{samp}}{2 \cdot \min(|\omega_1|, |\omega_2|, |\omega_1 + \omega_2|)} \frac{1}{N}\right]^{1/2} \tag{6.19}$$

From which one finds the statistical noise level in the bicoherence, using Eq. (6.17):

$$\varepsilon\left((b^w)^2\right) \approx \left[\frac{\omega_{samp}}{2 \cdot \min(|\omega_1|, |\omega_2|, |\omega_1 + \omega_2|)} \frac{1}{N}\right] \tag{6.20}$$

Observe that at low frequencies the statistical noise may dominate the bicoherence, and a significant interpretation must limit itself to (relatively) high frequencies. Again it is possible to confirm this theoretical estimate by analysing computer-generated Gaussian noise [12].

6.3.2 Interpretation of the bicoherence

The bicoherence is a complex quantity that contains much information but is not easy to interpret. We will provide some basic examples here meant to provide a 'feeling' and a guide to interpretation. In sections 6.4 and 6.5 we shall analyse data from numerical models and measurements. Together, these examples will hopefully clarify the significance and usefulness of the bicoherence.

We generate a test signal $f(t)$:

$$f(t) = A_p \sin(\omega_p t) + A_q \sin(\omega_q t) + A_r \sin(\omega_r t) \qquad (6.21)$$

such that the coupling condition $\omega_p = \omega_q + \omega_r$ is satisfied and A_p, A_q and A_r are constants. Three peaks with amplitude 1 will appear in the (ω_1, ω_2)-plane: one at $\omega_1 = \omega_q$, $\omega_2 = \omega_r$, one at $\omega_1 = \omega_p$, $\omega_2 = -\omega_r$ and one at $\omega_1 = \omega_p$, $\omega_2 = -\omega_q$. This is as true for the bicoherence based on Fourier analysis [9, 10] as it is for wavelet analysis. However, the coupling condition $\omega_p = \omega_q + \omega_r$ need only be satisfied to within the frequency resolution to produce a high value of the bicoherence, which in some cases is a significantly less strict requirement with wavelet than with Fourier analysis – thus, generally the wavelet bicoherence graph will show larger 'blobs' than the Fourier bicoherence graphs due to its lower frequency resolution, though the amplitude should be approximately the same (for an example see [12]).

One should be aware that the accuracy of the determination of the phases of the frequency components decreases rapidly as d gets smaller, since the smaller d is, the fewer oscillations are sampled (cf. Eq. (6.4)). A reliable phase determination is essential for a proper determination of the bicoherence. With the value of d we have chosen to use, $d = 1$, the wavelet transform samples about 5 oscillations of each frequency at any given time. Even so, the phase determination becomes unreliable for frequencies above roughly 95% of the Nyquist frequency. Therefore, the high values of the bicoherence that are often seen just below the Nyquist frequency can usually be ascribed to numerical phantoms.

A test signal that is mostly random noise, except for a short time period in which phase coupling is generated in the manner of Eq. (6.21), may cause the Fourier-based bicoherence not to detect the coupling due to the large time window used in its averaging process, whereas the wavelet bicoherence will detect the coupling during the relevant time window (provided it has a minimum duration of the order of T). This feature allows *intermittent coupling* to be detected; further, the time when the coupling occurs can be identified along with the *scales of the coupling interaction*. An interesting practical application of the wavelet bicoherence is therefore its use as a detector of intermittent non-linear behaviour.

Proceeding to a slightly more complex situation, imagine a test signal that is periodic but non-sinusoidal (e.g. a square wave or a sawtooth). Very strong coupling will be detected for a wide range of frequencies; wide areas of the bicoherence graph will show high values. This is as expected, since the non-sinusoidal wave can be built up from a number of Fourier components that are phase-locked with respect to each other. Compare this to a situation where a *non-periodic* pulse-train is generated; the pulses 'arrive' at intervals that are randomly distributed around some average. Then the bicoherence will show 'ridges': a wide range of frequencies couples to a single frequency (which corresponds to the average pulse frequency). Both of these situations are encountered when analysing turbulent data, and these considerations may serve as a guide to their interpretation.

It was noted before that the normalization of the bicoherence we have chosen (Eq. (6.17)) is asymmetric. Nevertheless, usually there is a triple symmetry of the three subregions of the bicoherence graph (i.e. $\{\omega_2 > 0\}$, $\{\omega_2 < 0 \wedge \omega_1 > -2\omega_2\}$ and $\{\omega_2 < 0 \wedge \omega_1 < -2\omega_2\}$), which can, however, be broken. It is observed often that one of the three peaks is significantly higher than the other two. Due to the symmetry of the numerator in Eq. (6.17), this asymmetry can only be due to its denominator. In fact, we can conclude that, during the time interval T, on average $\left|W_g(a_1, \tau)\right|$ and $\left|W_g(a_2, \tau)\right|$ do not simultaneously achieve significant values, causing the time integral of their squared product to be small (the *a*s are specified with reference to the high peak). The other two peaks, which are significantly smaller, indicate that $\left|W_g(a_1, \tau)\right|$ and $\left|W_g(a, \tau)\right|$, on the one hand, and $\left|W_g(a_2, \tau)\right|$ and $\left|W_g(a, \tau)\right|$, on the other hand, *do* show such a temporal correlation. We interpret these observations by saying that $\left|W_g(a_1, \tau)\right|$ and $\left|W_g(a_2, \tau)\right|$ *do not have a direct causal connection*, while $\left|W_g(a, \tau)\right|$ does show such a connection with $\left|W_g(a_1, \tau)\right|$ and $\left|W_g(a_2, \tau)\right|$, respectively. Thus, while the scales a_1 and a_2 are linearly independent, they interact (non-linearly) through scale a; one may conjecture that scale a *drives* scales a_1 and a_2. It should be noted that the observation of asymmetry in the bicoherence graph does not constitute proof for this conjecture. Summarizing: if the threefold symmetry in the bicoherence graph is broken and one peak is significantly higher than the other two, then the sum frequency belonging to the highest peak is most likely driving the coupling process. We shall encounter a beautiful example of this in section 6.4.1. The reverse may also occur: one peak is significantly lower than the other two, which are of similar height. By an argument analogous but opposite to the one given above, we conclude that it is likely that in this case the sum frequency belonging to the lowest peak is a consequence of the interaction of the other two frequencies. An example of this will be encountered in section 6.4.3. One must be careful applying this reasoning when harmonics are involved, leading to peaks on the lines $\omega_2 = -\omega_1/2$ (e.g. due to beat wave phenomena), since in this case two peaks coalesce, making their individual identification impossible.

The interpretation of the bicoherence in terms of the underlying physics is, in general, not straightforward. A simple interpretation is offered by a quadratic coupling model [7, 10, 15], which bears relevance to some elementary turbulence models. In this model, the coupling between three 'modes' or 'scales' a, a_1 and a_2 is expressed by means of a *coupling constant* $A(a_1, a_2, \tau)$:

$$W_f(a, \tau) = W_f^0(a, \tau) + A(a_1, a_2, \tau)W_g(a_1, \tau)W_g(a_2, \tau) \qquad (6.22)$$

where the frequency sum-rule (Eq. (6.15)) is satisfied. The component $W_f^0(a, \tau)$ is statistically independent of any other scales a_i. This equation expresses the existence of a (quadratic) relation between the three wavelet components. When the coupling constant $A(a_1, a_2, \tau)$ changes little during the time T (cf. Eq. (18)), the following equivalence holds:

$$|A(a_1, a_2, T_0)|^2 \approx \frac{|b_{fg}^w(a_1, a_2, T_0)|^2 P_f^w(a, T_0)}{\int_T |W_g(a_1, \tau)W_g(a_2, \tau)|^2 \, d\tau} \qquad (6.23)$$

Thus, the coupling constant in this simple quadratic phase-coupling model can be determined by evaluating the bicoherence, provided the averaging time T is smaller than the rate of change of the coupling constant.

In summary, we have two possible interpretations of the bicoherence: one in terms of 'coherent structures' (pulses or pulse trains, non-sinusoidal waves) passing by the observation point and one in terms of a coupling constant in a dynamical quadratic wave-interaction model. It is not possible to decide from the bicoherence alone which is the most appropriate interpretation.

More detailed interpretations are possible when data from more than one observation point are available. For example, two closely spaced observation points in a turbulent field allow the calculation of the cross bicoherence. This analysis has two advantages over the single-point measurement: first, any random noise present in the measurements will be more effectively suppressed provided the two measurements may be considered statistically independent; and second, the cross bicoherence decreases when the two points are separated, such that a determination of the average or typical size of each of the structures is possible (of course, the fluid velocity along the line connecting the two measurement points has to be taken into account – in some cases the cross correlation can give an estimate of this quantity).

6.4 Analysis of computer-generated data

The application of new analysis tools to computer-generated data has the important advantage that all the parameters of the studied system are known, so that the results of the interpretation based on the analysis can be checked against this knowledge and its informative value can be assessed. In the case of the bicoherence this type of analysis is particularly important, since its use has been very limited in the past so that little guidance as to its interpretation is available.

In section 6.4.1 we present an analysis of a chaotic system of coupled oscillators. Our main interest is of course the application of the analysis tools we have introduced to turbulence, but we consider chaotic systems to be illustrative simplifications of full-blown turbulence, possessing some of its most important characteristics, and therefore a good testing ground for our analysis tools.

In the following two sections we shall analyse data from numerical experiments that have bearing on turbulence in thermonuclear fusion plasmas. In a laboratory environment (as opposed to stellar environments), these plasmas are confined by magnetic fields or inertia, since the temperatures required for nuclear fusion are higher than any material vessel would withstand. The most succesful magnetic confinement scheme is the toroidal configuration (which can be of two general types: tokamak or stellarator), in which a magnetic field is induced with ring-like magnetic field lines. The ionized particles of the plasma are bound to these field lines by the Lorentz force, and thus circulate without colliding with material surfaces. To compensate for drifts in the necessarily inhomogeneous magnetic field, a second – poloidal, i.e. perpendicular to the main toroidal direction along the ring – field component is required. Thus the total magnetic field is helical, winds around a 'magnetic axis' and the average helical pitch varies with distance from this axis. The field lines are embedded in topologically toroidal (doughnut-shaped) surfaces. At certain radial intervals the ratio of toroidal to poloidal turns of a given field line (commonly referred to as 'safety factor', q) is a rational number, meaning that the field line connects with itself after a finite number of turns around the magnetic axis. Such magnetic surfaces are less stable to radial displacements of field lines than irrational surfaces, and magnetohydrodynamic instabilities may occur, leading to field-line reconnection and the formation of 'magnetic islands' – zones of plasma topologically isolated from the rest of the plasma by a separatrix enclosing the zone of reconnected field lines. Various such island chains may develop on various rational surfaces. The non-linear interaction between these island chains then leads to field-line stochastization in the intermediate zones. These zones of stochastic field line behaviour are expected to have a much higher radial heat transport than the zones that have their magnetic surfaces still intact.

The situation described above is known as 'magnetic turbulence', and is one of the main candidates for explaining the anomalously high radial heat transport found experimentally in toroidal fusion devices (i.e. higher than expected from a 'neoclassical' theory based on the assumption of the existence of unperturbed nested magnetic surfaces [16]). Another candidate is 'electrostatic' turbulence, driven by fluctuations of the plasma electric potential. The abundance of free energy available from the balance between the high pressure and temperature gradient forces and the confining magnetic field pressure may drive many other instabilities as well, drift waves being one of the more important examples. With drift waves, low-frequency ion motion perpendicular to the magnetic field is accompanied by electron motion along the field lines to preserve charge neutrality [17]. It will be clear that although the magnetic field introduces a strong anisotropy, it does not in general lead to quasi-two-dimensionality of the turbulence. The fundamental three-dimensional nature of the turbulence, the invalidity of isotropy assumptions and the large amount of instability drives available make numerical simulation of turbulence in thermonuclear plasmas a very difficult enterprise. The experimental identification of the main turbulence drive(s) is one of the most important and unresolved problems in thermonuclear plasma physics.

In sections 6.4.2 and 6.4.3 we analyse two plasma drift wave models, whose claim to a realistic description of plasma turbulence is limited since they focus on a single turbulence drive, but still may provide important clues as to how drift wave turbulence, if and when it occurs, may be recognized.

6.4.1 Coupled van der Pol oscillators

A system of two coupled van der Pol oscillators is one of the simplest numerical models that exhibits chaos in a self-sustaining way, i.e. without external driving [18]. The system is fully described by the equations:

$$\frac{\partial x_i}{\partial t} = y_i$$

$$\frac{\partial y_i}{\partial t} = \left[\varepsilon_i - \left(x_i + \alpha_j x_j \right)^2 \right] y_i - \left(x_i + \alpha_j x_j \right) \qquad (6.24)$$

The system $\{i = 1, j = 2\}$ describes the first oscillator, whereas $\{i = 2, j = 1\}$ describes the second. When $\alpha_i = 0$, the limit cycles of the uncoupled oscillators are determined completely by ε_i $(i = 1, 2)$. The parameters α_i describe the non-linear coupling between the oscillators.

The examination of this system presented in [18] proceeds along the standard lines of chaos analysis. Use is made of spectral analysis, bifurcation

Table 6.1. *System parameters of the coupled van der Pol oscillators in a periodic and a chaotic state*

System state	ε_1	ε_2	α_1	α_2
Periodic	1.0	1.0	0.5	-1.75
Chaotic	1.0	1.0	0.5	1.75

diagrams, fractal dimension estimates (a value of around 1.5 was found), etc. We shall not repeat any of this analysis here, but shall be asking whether wavelet analysis can provide additional insight into a chaotic system. For comparison, we also analyse the system in a periodic state. Table 6.1 lists the choice of control parameters for these two system states.

The Fourier spectrum of the signal $x_2(t)$ in the periodic case is shown in Figure 6.1a. The very clean spectrum shows only a few peaks with their harmonics. A section of 153 data points, sampled every $\Delta t = 0.2$ (units see [18]) and covering about 5 periods of the x_2 coordinate, was analysed using the wavelet bicoherence method. The calculation can also be done using a longer time series as input, but the results would be essentially the same. The result of the calculation is shown in Figure 6.1b. The strong, straight horizontal, diagonal and vertical ridges correspond to a frequency of roughly 0.34, which can be identified in Figure 6.1a with the second peak. Thus, the two dominant peaks in the Fourier spectrum at frequencies of 0.17 and 0.51 couple with their difference frequency at 0.34. Likewise, the difference in frequency between the second and the fourth peak, between the fourth and the sixth peak, etc., is always 0.34, and the same holds for the odd series of peaks. The difference frequencies between even and odd peaks (i.e. 0.17) are not reflected in the bicoherence plot, however (except when coupled to 0.34). It may therefore be conjectured that the odd peaks are the harmonics of the limit cyle of the oscillators, whereas the even peaks are due to the coupling interaction between the two oscillators. This interpretation is reaffirmed by the knowledge that with the combination of control parameters as given the limit cycle is asymmetric, which means that, were the coupling constants zero, only odd harmonics would appear [18]. These conjectures receive strong support from the cross bicoherence calculated from x_1 and x_2. Figure 6.1c shows a diagonal ridge at the same frequency as in Figure 6.1b (and a second one at a frequency of $0.68 = 2 \times 0.34$, a harmonic barely visible in Figure 6.1b), but more impor-

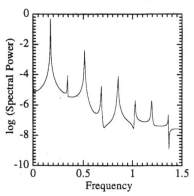

Fig. 6.1. System of two coupled van der Pol oscillators in a periodic state. For a description of the system and its parameters see text. a) Fourier spectrum of $x_2(t)$.

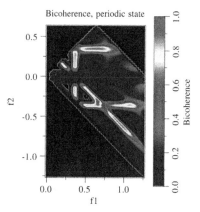

Fig. 6.1. b) Bicoherence graph of x_2. The bicoherence is calculated from 153 points covering about five periods. The horizontal and diagonal ridges are due to the coupling occurring at a frequency of 0.34. This figure is also shown at www.cambridge.org/resources/0521533538.

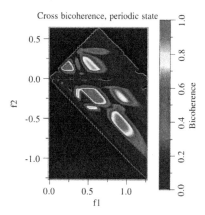

Fig. 6.1. c) Cross-bicoherence graph of x_1 and x_2. The graph is similar to b) except for its lack of symmetry. The asymmetry pinpoints 0.34 as the driving frequency (see text). This figure is also shown at www.cambridge.org/resources/0521533538.

tantly it shows a strong asymmetry. Applying the reasoning developed in section 6.3.2, we conclude that 0.34 is actually the driving frequency responsible for the coupling. This is an important conclusion, and we stress that while the previous results only permitted this thesis to be put forward as a conjecture, now it receives a firm basis. No other techniques are known to us that allow the identification of the driving frequency in such a simple and straightforward manner.

The Fourier spectrum of the signal $x_2(t)$ in the chaotic case is shown in Figure 6.2a. Several peaks are still visible, and some of these are related to the peaks in the periodic case through a frequency shift. New peaks have also appeared due to the process of period doubling in the transition to chaos. The chaos is apparent in the increase of the noisy (broad-band) part of the spectrum. Figure 6.2b shows the bicoherence as calculated using a section of 303 data points, sampled every $\Delta t = 0.2$, that covers about 8 pseudo-periods of $x_2(t)$. At first view, there is a striking similarity to Figure 6.1b. The main horizontal and diagonal ridges occur at a frequency of about 0.25, corresponding to the fourth major peak in Figure 6.2a. This frequency must therefore be identified with the frequency of 0.34 in Figure 6.1a – the change of control parameters of the coupled system, apart from introducing chaos, causes an overall frequency downshift with a factor of 0.73. Further it is observed that, although the high-frequency aspect of the graph has changed little – apart from a reduction in the value of the squared bicoherence – the low-frequency part is much more complex. The vertical line indicating the simple coupling at 0.34 of Figure 6.1b has split into several distinct coherent points at slightly shifted frequencies; observe the similarity in shape of these three points with the three points at double the frequency, which is obviously related to the period doubling process. New couplings have appeared at even lower frequencies (below 0.2), the biggest of which, at around 0.13, is easily identified as half the main coupling frequency of 0.25, and which is due to the period doubling effect also apparent in Figure 6.2a. The cross bicoherence shown in Figure 6.2c again confirms, by its asymmetry, the correct interpretation of 0.25 as the main coupling frequency.

6.4.2 A large eddy simulation model for two-fluid plasma turbulence

Direct numerical simulations of two-fluid plasma turbulence were carried out with the CUTIE code [19, 20, 21, 22]. The code is used to simulate low-frequency, relatively long wavelength drift-like fluctuations. It was developed to simulate tokamak turbulence, but in order to simplify the calculations the

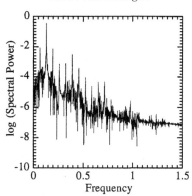

Fig. 6.2. System of two coupled van der Pol oscillators in a chaotic state. a) Fourier spectrum of $x_2(t)$.

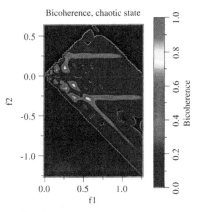

Fig. 6.2. b) Bicoherence graph of x_2, chaotic. The bicoherence is calculated from 303 points covering about eight pseudoperiods. The structure seen bears some similarity to the one seen in Fig. 6.1b, although it is less intense and more complex at lower frequencies due to the period doubling that has occurred in the transition to chaos. The main coupling frequency has been downshifted to 0.25. This figure is also shown at www.cambridge.org/resources/0521533538.

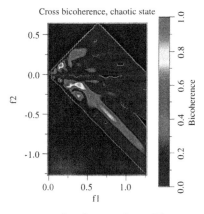

Fig. 6.2. c) Cross bicoherence graph of x_1 and x_2. The asymmetry seen in Fig. 6.1c survives in the chaotic régime, permitting the identification of the driving frequency even in chaos. This figure is also shown at www.cambridge.org/resources/0521533538.

geometry used is that of a periodic cylinder rather than that of a torus, and advantage is taken of the so-called tokamak ordering: $a/R \ll 1$, where a is the minor radius – (cylinder radius) – and R is the major radius – (of the corresponding torus) – $B_\theta \ll B_z$, i.e. the poloidal field is much smaller than the toroidal (longitudinal) field, and $\beta \ll 1$, where β is the pressure normalized to the magnetic field pressure. Quasi-neutrality is assumed and standard two-fluid/Maxwell equations are solved for the seven variables n_e (electron density), T_e, T_i (electron and ion temperature), $V_{//}$ (parallel plasma flow), Φ, Ψ and Ω (the electrostatic potential, the poloidal flux function and the parallel vorticity), taking account of the appropriate sources and relevant transport coefficients. The system is fully non-linear.

The code was run under the following conditions (typical of the COMPASS-D tokamak): $R = 55$ cm, $a = 23$ cm, $B_z = 2$ T, plasma current: 269 kA, initial central safety factor (q): 1.6 (related to the field line helicity), initial central density: 10^{14} cm^{-3}, initial central ion and electron temperatures: 500 eV and $Z_{eff} = 1.73$ (effective charge number).

In the following we analyse the simulated density fluctuations dn/n_0. Data are taken at various radial positions.

Figure 6.3a shows the wavelet spectrum of the fluctuating density at $r = 21.16$ cm. The U-shaped edge profile indicates the region beyond which the calculation of the wavelet transform is not possible due to the proximity of the data boundaries. Several modes can be distinguished in this figure: steady modes at 15 kHz and 100 kHz; and a mode at about 200–250 kHz that shows a beat phenomenon. Analysis of similar data obtained at a higher sampling rate indicates that this phenomenon is possibly due to the aliasing of a mode around 750 kHz. A weak and apparently not stationary mode is visible around 60 kHz. Analysis of the fluctuating density signal at other radii reveal very similar features.

Figure 6.3b shows the joint wavelet phase-frequency probability distribution function, calculated from the fluctuating density signals at two radii: $r = 18.4$ and $r = 22.08$ cm. The first striking feature that can be observed is that both steady modes (15 kHz and 100 kHz) show definite radial phase relations (the phase differences are -2.6 and 0.7 rad, respectively). This implies a very strong linear radial correlation for these modes. The mode above 200 kHz shows slightly less clear behaviour. The fact that the phase shift converges to 0 or π at frequencies close to the Nyquist frequency is a consequence of the impossibility to obtain accurate phase determinations at those frequencies, as explained in section 6.3.2, and not of any physical effect. The most interesting feature is seen in the intervals between 20 and 80 kHz, where a partial phase randomization occurs. The phase of the Fourier auto spectra of either of the

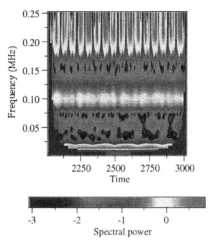

Fig. 6.3. Analysis of data from the CUTIE numerical turbulence model. a) The wavelet spectrum of the fluctuating density at $r = 21.16$ cm. This figure is also shown at www.cambridge.org/resources/0521533538.

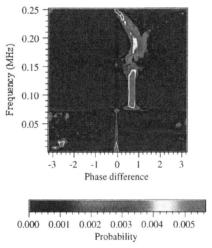

Fig. 6.3. b) The joint wavelet phase-frequency probability distribution function, calculated from the fluctuating density signals at two radii: $r = 18.4$ and $r = 22.08$ cm. This figure is also shown at www.cambridge.org/resources/0521533538.

two signals used here (not shown) show a π phase jump at 17.6 kHz and 103.5 kHz, indicative of the fact that these frequencies are probably driving the turbulence. Thus, the feature at 60 kHz is probably due to a non-linear interaction. The graph as a whole seems indicative of the existence of a low-dimensional attractor in the strongly non-linear dynamics.

6.4.3 A long wavelength plasma drift wave model

Another numerical model of drift wave turbulence in plasmas is studied in this section. This model simplifies the geometry even further to a slab (i.e. a box-shaped region). The x-coordinate is identified with the radial coordinate in a torus, the y-coordinate with the poloidal direction and the z-coordinate with the toroidal direction. The magnetic field does not have a radial (x-) component and its z-component is fixed while its y-component varies radially. Thus, a sheared magnetic field is created similar to that in a tokamak experiment. The numerical experiment is characterized by the shear length of the magnetic field, as well as the assumed electron density and temperature profiles.

The model studies the evolution of the ion density, which is separated in an average (n_i) and a fluctuating $(\tilde{n}_i)$ part. With the help of simplifications, such as the assumption of long wavelengths, an equation is derived for $\tilde{n}_i$ which is advanced in time. To do so, $\tilde{n}_i$ is Fourier-expanded in the toroidal and poloidal directions and a large number of modes is used to achieve sufficient accuracy for turbulence studies. In the radial direction, finite differences are used.

The detailed setup of these calculations is described elsewhere [23]. For the analysis that follows here, it is sufficient to mention just a few points: distance and time units are normalized to ρ_s, the ion gyroradius, and $1/\Omega_i$, the inverse ion gyrofrequency, respectively. The values of the model parameters have been chosen to provide a range of unstable modes with $6 \leq m \leq 76$, where m is the poloidal mode number. The safety factor, q, is equal to $\frac{3}{2}$ at the centre of the computational box. The standard box size (x-direction) is $60\rho_s$, and the number of unstable modes with resonant surfaces inside the computational box is about 250. In the calculation, we have included 439 Fourier components. The averaged density gradient is fixed, so saturation is caused by turbulence effects. The numerical data are for the saturated state.

Figure 6.4a shows raw data from this simulation at $r = 30.0\rho_s$ for the time interval $30.2 \times 10^4 < \Omega_i t < 40.4 \times 10^4$. In these units, the time step is 100. Figure 6.4b shows the corresponding wavelet power spectrum P^w. The spectrum is nearly featureless except for two small peaks at low frequencies, the frequencies being of the order of the linear mode frequency. Figure 6.4c shows the wavelet power spectrum for the whole range of radii available. The spectra are calculated over the interval $31.1 \times 10^4 < \Omega_i t < 39.4 \times 10^4$. The whole data time window is not used because the continuous wavelet transform cannot be evaluated near the data edges. The frequency is given in inverse time units. The radii are given in units of ρ_s.

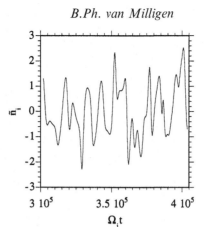

Fig. 6.4. Ion density perturbation of the drift wave model discussed in the text. a) Raw data of the ion density versus time at $r = 30.0\rho_s$.

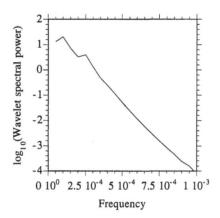

Fig. 6.4. b) Wavelet spectrum of the data shown in a). The relatively featureless turbulent spectrum shows two small peaks at low frequency.

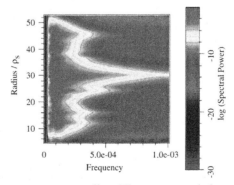

Fig. 6.4. c) Wavelet spectrum vs. radius. The spectrum is broadest at $r = 30.0\rho_s$. This figure is also shown at www.cambridge.org/resources/0521533538.

Figure 6.5a shows the RMS fluctuation level for this simulation, versus radius. The RMS level peaks at the position of the position of the $q = \frac{3}{2}$ rational surface. The wavelet spectrum (Figure 6.4c) broadens where the RMS level is high. At radial positions where the spectrum is narrow, the calculation of the wavelet transform suffers from numerical errors at high frequency. This is important to keep in mind when viewing the results presented below.

Figure 6.5b shows the total bicoherence vs. radial position for this simulation. Figure 6.5c shows the corresponding summed bicoherence as a function of both sum frequency and radial position. From Figure 6.4c it is apparent that e.g. the peak in Figure 6.5b at $r = 47.5\rho_s$ is due mainly to the numerical problems mentioned above and does not correspond to anything physical.

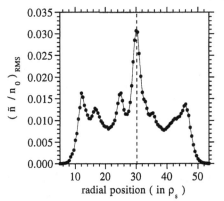

Fig. 6.5. a) Analysis of the ion density perturbation of the drift wave model. a) RMS fluctuation level of the ion density vs. radius.

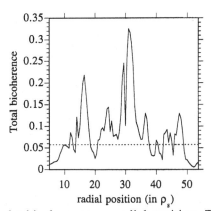

Fig. 6.5. b) Total wavelet bicoherence vs. radial position. The peaks are associated to, but do not coincide with, the peaks in a). Note the sharp drop in bicoherence at $r = 30.0\rho_s$.

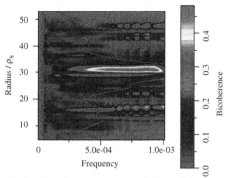

Fig. 6.5. c) Summed wavelet bicoherence vs. radial position and sum frequency. The
drawn line (a contour taken from Fig. 6.4c at log(wavelet power) $= -10$) indicates
roughly up to what frequency the bicoherence may be considered reliable. This figure
is also shown at www.cambridge.org/resources/0521533538.

Such numerical problems do not occur if the wavelet power transform coefficients are sufficiently large. We take the level -10 in Figure 6.4c to delimit
the zone with numerical problems; this level is indicated by the line in Figure
6.5c. Examining the remainder of Figs. 6.5b and c, one can make the following observations. (1) The bicoherence drops sharply at the position of the
$q = \frac{3}{2}$ rational surface, located at $30\rho_s$ (whereas neither the RMS value, nor
the spectrum exhibit such a local drop). (2) The maximum of the bicoherence
is at $31\rho_s$. (3) Secondary maxima occur, apart from some minor peaks, at
16.5, 24.5, 29.5 and $36.5\rho_s$. These positions coincide roughly (but *not* exactly)
with maxima in the RMS value (see Figure 6.5a). From this, we conclude that
the bicoherence provides information that pertains to an aspect of the turbulence (non-linear, or rather quadratic behaviour) that is not captured by
either of the other methods.

Having established the general interest of this analysis, the next question
must be: how does this information help in understanding turbulence? In the
following, we compare the results from the bicoherence calculation with a
more conventional approach. Figure 6.5d shows the cross correlation, the
weighted average cross coherence and the weighted average cross phase
between one radial position and the next; the weighting being done by the
spectral power (the cross spectra are calculated with the normal FFT). It is
observed that the cross correlation and cross coherence between adjacent
radial positions is generally high, but that around $30\rho_s$ these quantities
drop. Further, it is observed that the cross phase exhibits a peak around
$30.5\rho_s$, possibly indicating shear flow. The numerical results also show the
existence of a shear flow layer in this location. Although this analysis is by no

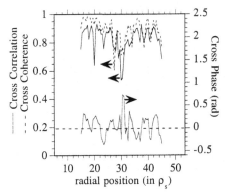

Fig. 6.5. d) Cross correlation, cross coherence, and cross phase between adjacent channels. Results are shown at position $x\rho_s$ for the cross analysis between $x\rho_s$ and $(x + 0.5)\rho_s$. The cross coherence and phase are computed from FFT (cross) spectra and are averaged over all frequencies by weighing with the spectral power.

means conclusive, it seems to suggest an explanation of the drop in bicoherence at $30\rho_s$ in terms of a decorrelation of the turbulence, possibly linked to a shear flow.

The maximum in the bicoherence at $31\rho_s$ is related to the presence of a long-living structure that is highly localized poloidally and radially. This structure has the (3,2) periodicity and lies close to the $q = \frac{3}{2}$ surface; such a structure is visible in a two-dimensional plot of the ion density (cf. Figure 6.5e), but because of its high spatial localization, it was only discovered after this analysis indicated the persistence of non-linear couplings over a time period of many decorrelation times (in [23] it is explicitly stated that coherent structures were not seen).

For a more detailed analysis we refer to Figure 6.5f, where the full bidimensional bicoherence is shown for a few selected radial positions. First, we draw attention to the graph corresponding to $30\rho_s$. Here the typical behaviour of a single mode coupling to broad-band turbulence is visible (horizontal band-like structure), with the main mode frequency around 1.4×10^{-4}. Turning now to the graph taken at the maximum of the bicoherence (at $31\rho_s$), one observes that each point of this band-like structure couples, in its turn, to a range of frequencies, thus nearly filling the two-dimensional plane of the bicoherence.

To summarize, we have been able to perform a rather detailed spectral analysis on computer-generated data of a turbulence simulation, for which (due to CPU-time limitations) only short data series were available, thus rendering Fourier analysis impracticable, or (in the case of the bicoherence)

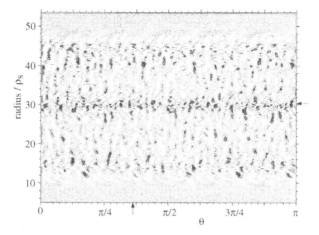

Fig. 6.5. e) Two-dimensional graph of $\tilde{n}_i$ in the $\rho - \theta$ plane. The location of a small coherent structure is indicated by the arrows. This figure is also shown at www.cambridge.org/resources/0521533538.

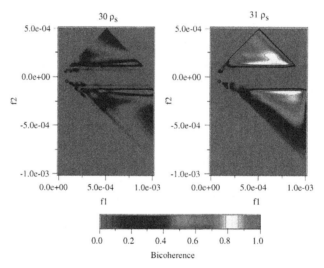

Fig. 6.5. f) Bicoherence for $r = 30.0\rho_s$ and $r = 31.0\rho_s$. This figure is also shown at www.cambridge.org/resources/0521533538.

even impossible. The analysis revealed a rather surprising narrow drop in non-linear coupling (bicoherence), precisely at the location where the radial correlation decreases locally and where the radial cross phase shows a peak. From these observations, we deduced that most likely a local shear flow is responsible for the observed decorrelation. Finally, the peak in bicoherence was associated with a small structure in the flow, which was not detected by other methods.

6.5 Analysis of plasma edge turbulence from Langmuir probe data

6.5.1 Radial coherence observed on the TJ-IU torsatron

In the present section we will analyse data from the TJ-IU Torsatron [24]. This is a toroidal device in which a hot plasma is confined by magnetic fields in the manner described at the beginning of section 6.4. The plasmas were heated by Electron Cyclotron Resonance Heating (ECRH), and have major radius $R = 0.6$ m, central rotational transform of the magnetic field $\iota(0) = 0.21$, minor radius $\langle a \rangle = 0.1$ m, toroidal field $B_T = 0.6$ T, and electron density $n_e = 5 \times 10^{18}$ m^{-3}.

Two Langmuir probes were inserted into the plasma edge region, where temperatures are sufficiently low to allow this without damaging the probes. The two probes are separated radially by 1 cm. Each of the two probes has three tips, aligned perpendicular to the magnetic field and separated poloidally by $\Delta = 0.2$ cm. The probes were designed and positioned to avoid the shadowing of one probe by another [25] ('shadow' referring to the influence cone of the probe along the direction of the magnetic field). The two extreme tips of each probe were configured to measure the floating potential, Φ_f, whereas the central tip was set up to measure the ion saturation current, I_{sat}. Thus it is possible to estimate the instantaneous radial turbulent flux for both probes: $\Gamma_T = \tilde{n}\tilde{E}_\theta / B_T$, using $\tilde{E}_\theta = (\tilde{\Phi}_f(1) - \tilde{\Phi}_f(2))/\Delta$ and $\tilde{n} \propto \tilde{I}_{sat}$ (where $\tilde{x}$ is the fluctuating part of x and the influence of temperature fluctuations is neglected). The signals were sampled at 1 MHz.

Figure 6.6 shows the cross spectrum and radial cross coherence between the $\tilde{I}_{sat}$ signals of the two radially separated probes. The influence of a MHD

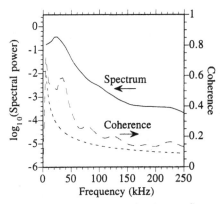

Fig. 6.6. Cross spectrum and cross coherence of $\tilde{I}_{sat}$. Calculations are made on measurements from two radially separated probes in the edge zone of a TJ-IU discharge. Continuous line: average wavelet spectrum; long dashes: wavelet coherence; short dashes: the noise level of the coherence.

mode is recognized in the peak of the spectrum at about 20 kHz. The cross coherence peaks at slightly higher frequency (35 kHz).

Figure 6.7 shows the same graph, now calculated for Γ_T. Features similar to the ones observed in Figure 6.6 can be seen, although at higher frequencies. This frequency shift is easily explained by the fact that Γ_T is a quadratic signal. What is most interesting is that the flux does not show a larger radial correlation than $\tilde{I}_{sat}$ (or $\tilde{\Phi}_f$ either, not shown). This seems to indicate that, for the present type of plasmas at least, the non-linear interactions in the turbulence, if present, do not generate a stronger coherence in the flux than in the fluctuating density and electric field, which is a hypothesis invoked by some turbulence models in order to explain the heat losses from thermonuclear plasmas [26].

Figure 6.8 shows the temporally resolved coherence of $\tilde{I}_{sat}$. The time resolution is 0.5 ms. The noise level is the same as in Figs. 6.6 and 6.7. The coherence is highly intermittent and occasionally very high values are achieved (much higher than the time-average value). This figure illustrates the necessity of using wavelet techniques for turbulence analysis; whereas results similar to Figs. 6.6 and 6.7 can be obtained using Fourier techniques, the intermittent character shown here is only evident from a wavelet analysis.

6.5.2 Bicoherence profile at the L/H transition on CCT

The data analysed in this section are from the Continuous Current Tokamak (CCT). It was operated with major radius $R = 1.5$ m, minor radius $a = 0.35$ m, toroidal field $B_T = 0.25$ T, plasma current $I_p \approx 40$ kA, electron density

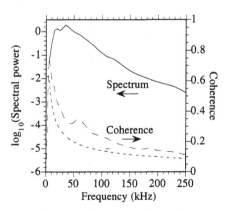

Fig. 6.7. Cross spectrum and cross coherence of Γ_T. Same as Fig. 6.6 for the instantaneous particle flux derived from the probe data (see text). The value of the coherence is smaller than for $\tilde{I}_{sat}$ (Fig. 6.6).

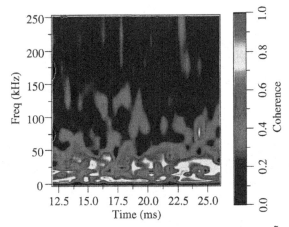

Fig. 6.8. Time-resolved wavelet coherence-versus-time graph of $\tilde{I}_{sat}$. The coherence is between two radially separated probes (see text). The noise level is the same as in Fig. 6.6. In several time intervals and at various frequencies the coherence obtains values far above the time-average value. This figure is also shown at www.cambridge.org/resources/0521533538.

$n_e = 2 \times 10^{18}$ m^{-3}, loop voltage $V_{loop} = 1.2 - 1.4$ V, central electron temperature $T_e(0) > 150$ eV, and central ion temperature $T_i(0) > 100$ eV. A transition from L-mode to H-mode confinement was induced by biasing a small electrode located about 0.1 m inside the limiter radius with respect to the vessel wall [27].

The H- or High confinement mode is a plasma state characterized by reduced global heat losses, and is the object of intense study by the fusion community. A full understanding of the reasons for the transition from the L- or Low confinement mode to the H-mode is not available. It is believed that a strong shear in the plasma rotation velocity near the edge (in the present case caused by the artificially generated radial electric field) may lead to suppression of turbulence in the plasma edge zone and thus to less heat losses. This belief is strengthened by the observation of strong density gradients in the edge zone during the H-mode in many devices. A detailed understanding of this process seems very important, since it may help to understand the general problem of confinement and may lead to methods for controlling the turbulence and thus the heat transport, which eventually may lead to smaller and cheaper thermonuclear fusion reactors.

For the present study we focus on a single probe from a poloidal Langmuir probe array [28, 29]. It was configured such that one of the probe tips was recording the floating potential locally. The sampling rate was 2.5 MHz. The probe was initially located just outside the last-closed-flux-surface (LCFS). During H-mode the increasing plasma pressure causes a slow movement of the plasma column out towards the low-field side. Thus, when the H-mode electrode bias is suddenly turned on, the outside midplane probe records a slow increase in the negative DC floating potential. This enabled a reconstruction of the radial profile in the H-mode of the quantities measured by the probe using an estimate of the instantaneous probe position relative to the LCFS. Knowing the value of the radial electric field E_r, the radial position r of the probe can be estimated as $r = \Phi_f/E_r$, where Φ_f is the floating potential measured by the probe. From Doppler shift measurements, we estimated the electric field to be $E_r \approx -100$ Vcm^{-1}. Uncertainties in this estimate translate into an uncertainty of 30–50% in the absolute reconstructed position, although the relative position is much more accurate since E_r does not vary significantly during the measurements.

Figure 6.9 shows the RMS values and the wavelet bicoherence of the measured ion saturation current (I_{sat}) for the outside midplane probe. The H-mode period (grey area) shows a slight reduction of the RMS and a gradual increase of the bicoherence as the plasma moves outward. A broad range of frequencies is involved in the production of the high bicoherence around $t = 80$ ms, with predominance of frequencies around 250 and 500 kHz.

Observations reported earlier for L/H transitions (using reflectometry) [12] showed an abrupt increase of the bicoherence and decrease of the RMS at the transition. The difference with the present apparently smooth transition may be explained by the fact that here the probe location is initially at the LCFS and moves gradually inward, whereas in the measurements reported earlier

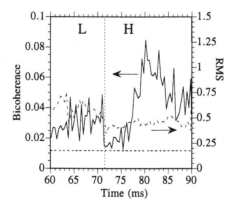

Fig. 6.9. Analysis of the saturation current I_{sat} measured by a Langmuir probe positioned on the outside midplane at CCT. Long dashes: RMS of I_{sat}. Drawn line: bicoherence of I_{sat}. Short dashes: noise level of the bicoherence. The grey area indicates the H-mode period, induced by probe biasing.

the measurements were taken well inside the LCFS. This would imply that the gradual change observed is due rather to the existence of a bicoherence profile than to a slow temporal change.

Performing standard statistical analysis on the outside midplane I_{sat} signal, we calculate the Probability Distribution Function (PDF) before and after the transition. Calculations are performed on records of 12500 samples after high-pass digital filtering with a cutoff frequency 1 kHz to remove drifts. We find that the L-mode PDF is Poisson-like (with skewness $S > 1$, and kurtosis $K > 5$), whereas in the H-mode it is more like a Gaussian ($0 < S < 1$, $3 < K < 4$) (Figure 6.10). The deviation from Gaussianity in the L-mode as contrasted with the near-Gaussianity in the H-mode is consistent with earlier studies of the relation between the PDF shape of turbulent signals and plasma conditions [26].

Using the above-mentioned estimate of the probe position, the profile of wavelet bicoherence was reconstructed during the H-mode phase from the signal of the outside midplane probe. The result is shown in Figure 6.11, along with a similar profile obtained from a different but similar discharge. The reproducibility is surprising. We recall that a high value of the bicoherence may either indicate the presence of non-linear interactions or of (quasi-static) structure [14]. It is interesting that these features should occur a small distance inside and not at the LCFS, which is possibly related to the fluid velocity shear layer at or near the LCFS, which may be decorrelating the turbulence or modifying the size of coherent structures. The precise meaning

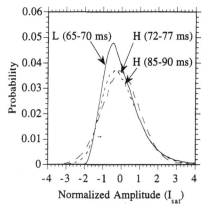

Fig. 6.10. PDF of the outside midplane I_{sat} signal in the L-mode and the H-mode. L-mode (65–70 ms): Skewness $S = 1.32 \pm 0.07$, Kurtosis $K = 5.94 \pm 0.22$; H-mode (72–77 ms): $S = 0.15 \pm 0.06$, $K = 3.27 \pm 0.11$; H-mode (85–90 ms): $S = 0.67 \pm 0.06$, $K = 3.83 \pm 0.13$.

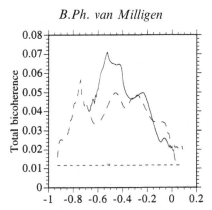

Fig. 6.11. H-mode bicoherence profile. Bicoherence profile in the H-mode deduced from the outside midplane probe signal shown in Fig. 6.9 and an estimate of the probe position relative to the LCFS (continuous line). Also included is a similar profile for another, similar discharge (long dashes). The noise level of the bicoherence is indicated by the short-dashed line.

of this maximum of bicoherence about half a centimetre inside the LCFS is as yet unclear but it seems relevant to H-mode physics.

6.6. Conclusions

The use of wavelets in the analysis of turbulence is a significant step forward with respect to the traditional spectral analysis. On the one hand, as was explained in section 6.2.1, the advance is fundamental in the sense that we liberate ourselves from the obligatory decomposition of signals in modes which are not eigenmodes of the system (Fourier modes) and which therefore lead to a scrambling of significant information. Wavelets can be seen as a local linear decomposition, and this procedure is justified provided the turbulence is not too strong, so that the non-linear equations describing the turbulence can be linearized locally. On the other hand, wavelets provide also a practical advance in the sense that they reduce the need for obtaining long stationary time series in order to obtain sufficient statistics, which is often not possible experimentally and difficult or expensive numerically. Thus, as we have illustrated through numerous examples, we have been able to perform rather complex analyses (e.g. bispectral analysis, which is sensitive to non-linear interactions) on rather short time series, something which was not possible before the advent of wavelets.

Acknowledgements

I am indebted to B. Carreras for discussions which inspired me to introduce the bicoherence based on wavelet analysis for the first time. I am grateful to I. Pastor for lending me his data on the system of van der Pol oscillators, to B. Carreras and L. García for permitting me to analyse their data on the drift-wave turbulence simulations and for many helpful thoughts on the problem of interpreting the bicoherence. I have been very lucky to be able to work with the Turbulence Group at the Asociación EURATOM-CIEMAT para Fusión who, apart from being excellent colleagues, provided the data analysed in section 6.5.1. In particular, I thank C. Hidalgo and M.-A. Pedrosa for their collaboration and help. I thank A. Thyagaraja and W. E. Han for the data they kindly made available (section 6.4.2). Finally, I am grateful to G.R. Tynan for his data and help in analysing the CCT data presented in section 6.5.2.

References

[1] C. Meneveau. Analysis of turbulence in the orthonormal wavelet representation. *J. Fluid Mech.*, **232**: 469, 1991.

[2] D.E. Newland. *An introduction to random vibrations, spectral & wavelet analysis.* (Longman Scientific & Technical, New York, 3rd edition, ISBN 0582 21584 6, page 295, 1993).

[3] G. Kaiser. *A friendly guide to wavelets.* (Birkhäuser, Boston, Basel, Berlin, 1994).

[4] L. Hudgins, C.A. Friehe and M.E. Mayer. Wavelet transforms and atmospheric turbulence. *Phys. Rev. Lett.*, **71**: 3279, 1993.

[5] J.M. Beall, Y.C. Kim and E.J. Powers. Estimation of wavenumber and frequency spectra using fixed probe pairs. *J. Appl. Phys.*, **53**: 3933, 1982.

[6] H.D.I. Abarbanel, R. Brown, J.J. Sidorovich and L. Sh. Tsimring. The analysis of observed chaotic data in physical systems. *Rev. Mod. Phys.*, **65**: 1331, 1993.

[7] Ch. P. Ritz, E.J. Powers and R.D. Bengtson. Experimental measurement of three-wave coupling and energy cascading. *Phys. Fluids*, B **1**: 153, 1989.

[8] Ch. P. Ritz, E.J. Powers, T.L. Rhodes, R.D. Bengtson, K.W. Gentle, Hong Lin, P.E. Phillips, A. J. Wootton, D.L. Brower, N.C. Luhmann, Jr., W.A. Peebles, P.M. Schoch and R.L. Hickok. Advanced plasma fluctuation analysis techniques and their impact on fusion research. *Rev. Sci. Instrum.*, **59**: 1739, 1988.

[9] Y.C. Kim and E.J. Powers. Digital bispectral analysis of self-excited fluctuation spectra. *Phys. Fluids*, **21**: 1452, 1978.

[10] Y.C. Kim, J.M. Beall, E.J. Powers and R.W. Miksad. Bispectrum and nonlinear wave coupling. *Phys. Fluids*, **23**:258, 1980.

[11] B. Ph. van Milligen, C. Hidalgo and E. Sánchez. Nonlinear phenomena and intermittency in plasma turbulence. *Phys. Rev. Lett.*, **74**: 395, 1995.

[12] B. Ph. van Milligen, E. Sánchez, T. Estrada, C. Hidalgo, B. Brañas, B. Carreras and L. García. Wavelet bicoherence: a new turbulence analysis tool. *Phys. Plasmas*, **2**: 3017, 1995.

[13] I. Daubechies. *Ten Lectures on Wavelets*. (SIAM, Philadelphia, PA 1992)

[14] H.L. Pécseli and J. Trulsen. On the interpretation of experimental methods for investigating nonlinear wave phenomena. *Plasma Phys. Control. Fusion*, **35**: 1701, 1993.

[15] N. Mattor and P.W. Terry. Frequency spectrum in drift wave turbulence. *Phys. Fluids.*, B **4**: 1126, 1992.

[16] F.L. Hinton and R.D. Hazeltine. Theory of plasma transport in toroidal confinement systems. *Reviews of Modern Physics*, **48**: 239, 1976.

[17] W.M. Manheimer and C.N. Lashmore-Davies. *MHD and Microinstabilities in Confined Plasma*. (The Adam Hilger Series on Plasma Physics, IOP Publishing Ltd., Bristol, ISBN 0 85274 282 7, 1989).

[18] I. Pastor, V.M. Pérez-García, F. Encinas-Sanz and J.M. Guerra. Ordered and chaotic behaviour of two coupled van der Pol oscillators. *Phys. Rev. E*, **48**: 171, 1993.

[19] A. Thyagaraja. Is the Hartmann number relevant to tokamak physics? *Plasma Physics and Contr. Fusion*, **36**: 1037, 1994.

[20] A. Thyagaraja. Sources of nonadiabaticity in tokamak turbulence. *Physica Scripta*, **47**: 266, 1993.

[21] A. Thyagaraja. Direct numerical simulations of two-fluid plasma turbulence. *Journal de Physique IV, C6*, **5**: 105, 1995.

[22] A. Thyagaraja. *Global numerical simulations of turbulence and transport in a tokamak*. Invited paper, Proceedings International Joint Varenna-Lausanne Workshop on Fusion Plasmas, 1996.

[23] B.A. Carreras, K. Sidikman, P.H. Diamond, P.W. Terry and L. García. Theory of shear flow effects on long-wavelength drift wave turbulence. *Phys. Fluids*, B **4**: 3115, 1992.

[24] E. Ascasíbar, C. Alejaldre, J. Alonso, *et al.* Initial operation of the TJ-IU torsatron and theoretical studies for the flexible heliac TJ-II. *Plasma Phys. and Contr. Nucl. Fusion Research, IAEA-CN-60/A6-1*, **1**: 749, 1994.

[25] M.A. Pedrosa, C. Hidalgo, B.Ph. van Milligen, E. Sánchez, R. Balbín, I. García-Cortés, H. Niedermeyer and L. Giannone. Statistical properties of turbulent transport and fluctuations in tokamak and stellarator devices. *Proc. 23rd Eur. Conf. Kiev*, 1996.

[26] B.A. Carreras, C. Hidalgo, E. Sánchez, M.A. Pedrosa, R. Balbín, I. García-Cortés, B.Ph. van Milligen, D.E. Newman and V.E. Lynch. Fluctuation-induced flux at the plasma edge in toroidal devices. *Phys. Plasmas*, **3**: 2664, 1996.

[27] R.J. Taylor, M.L. Brown, B.D. Fried, H. Grote, J.R. Liberati, G.J. Morales, P. Pribyl, D. Darrow and M. Ono. H-Mode behavior induced by cross-field currents in a tokamak. *Phys. Rev. Lett.*, **63**: 2365, 1989.

[28] G.R. Tynan. Ph.D. Thesis. School of Engineering, University of California, Los Angeles, 1991.

[29] G.R. Tynan, L. Schmitz, R.W. Conn, R. Doerner and R. Lehmer. Steady-state convection and fluctuation-driven particle transport in the H-mode transition. *Phys. Rev. Lett.*, **68**: 3032, 1992.

7

Transfers and fluxes of wind kinetic energy between orthogonal wavelet components during atmospheric blocking

AIMÉ FOURNIER

Yale University Department of Physics,
New Haven CT 06520-8120, USA

Abstract

Atmospheric blocking is an irregularly recurring anomalous state of the atmospheric circulation which is large and spatially localized. Atmospheric blocking during three unusual winter months is studied by multiresolution analysis and a new periodic wavelet-based adaptation of traditional Fourier series-based energetics. New forms of the transfer functions of kinetic energy with the mean and eddy parts of the atmospheric circulation are introduced. These quantify the zonally localized conversion of energy between scales. A new accounting method for wavelet-indexed transfers permits the introduction of a physically meaningful zonally localized scale flux function. These techniques are applied to National Meteorological Center data. Blocking is found to be largely described by just the second-largest scale part of the multiresolution analysis. New support is found for the hypothesis that blocking is partially maintained by a particular kind of upscale cascade. Specifically, in both Atlantic and Pacific blocking cases there is a downscale (upscale) cascade west (east) of the block.

7.1 Introduction

Although wavelet analysis in the time domain has been applied to atmospheric boundary layer turbulence (e.g. [8]) and climatic time series (e.g. [3, 15, 17]), and in the space domain to numerically simulated turbulence [7, 9, 18], there has not been any application to observed global synoptic meteorological data. A broad review of wavelets applied to turbulence is presented by Farge *et al.*, this volume, Chapter 4. A collection of blocking studies is contained in [1]. During blocking, the normal progression of weather is locally inhibited. A definition of blocking is presented in Section 7.2.

Because of the compact organization of the block structure, wavelet-based analysis techniques are called for, rather then Fourier analysis. After a review of more conventional analyses, new forms of kinetic energy transfer and flux functions are introduced in Section 7.3. Section 7.4 presents the results of these analyses applied to blocking and nonblocking data.

7.2 Data and blocking description

The data for this study are the wind components u (eastward), v (northward) and geopotential height Z ($\equiv g^{-1}\Phi$, the height at which a given specific gravitational potential energy, relative to mean sea level, Φ would be attained if specific gravitational force were fixed at its global mean sea level value g) from National Meteorological Center (NMC) global analyses (e.g. [24]). Each of these variables depends on the independent coordinates longitude λ, latitude φ, pressure level p and time t. (The atmosphere is very close to hydrostatic equilibrium, in which case p may be taken as an independent coordinate instead of geometric height, which simplifies the mathematics and the data analysis.) The original coordinate grid is[†]

$$\lambda_l = \Delta\varphi(l-1) = 0, \cdots 360°, \quad l = 1, \cdots 145$$
$$\varphi_m = \Delta\varphi(m-1) = 0, \cdots 90°, \quad \Delta\varphi = 2.5°, \quad m = 1, \cdots 37$$
$$p_s = 10, 15, 20, 25, 30, 40, 50, 70, 85, 100 \text{ kPa}$$
$$t_i = \Delta t(i-1) = 0, \cdots 89.5\text{d}, \quad \Delta t = .5\text{d}, \quad i = 1, \cdots I = 180,$$
$$t_1 \doteq 1978 \text{ Dec. } 1, 0 \text{ UTC.}$$

Originally the λ and φ grid spacings were equal. Since the analysis requires grid size to be a power J of 2, the longitudes were cubically interpolated[‡] down to the periodic grid

$$\lambda_l = \Delta\lambda(l - 1 - 2^{J-1}), \quad \Delta\lambda = 2^{1-J}\pi, \quad l = 1, \cdots 2^J + 1 = 129, \qquad (7.1)$$

$J \equiv \lfloor \log_2 144 \rfloor = 7$ chosen to minimize interpolation artifacts. ($\lfloor x \rfloor \equiv$ greatest integer no greater than x.) The times $i = 57, 95, 110, 179$ were lost by NMC, and $i = 12, 20, 23$ contained a few physically unacceptable miscalculated u, v values at certain p. All fields at these times were replaced using $A_i \to 2^{-1}(A_{i-1} + A_{i+1})$. (Henceforth dependence on discrete λ, φ, p, t may be indicated by the respective positive indices l, m, s, i, and single indices of

---

[†]UTC stands for universal time coordinate.
[‡]Interpolation program rgrd1u.f documented at http://www.scd.ucar.edu/softlib/REGRIDPACK.html.

one coordinate may imply independence or simply notationally suppressed dependence on other coordinates, depending on the context.)

The three months of these data were marked by record or near-record cold weather in the USA, associated with quasi-persistent anomalous high pressure systems known as blocks [6, 23, 25]. There were five blocking events of 7 to 13 d duration [12]. In the present study the blocking events are taken directly from [12]. A blocking *event* was (somewhat arbitrarily) taken as a t-interval $\mathcal{T}$ longer than 7 d throughout which the longitudinal crest ${}_{55^\circ}^{80^\circ}\{Z(\lambda_m(t), p_7, t)\} \equiv \max_\lambda {}_{55^\circ}^{80^\circ}\{Z(\lambda, p_7, t)\}$ at $p_7 = 50$ kPa exceeds the zonal mean ${}_{55^\circ}^{80^\circ}\{Z_0(p_7, t)\}$ (7.2) by more than 250 m, and $\lambda_m(t)$ is continuous, where

$$
{}_{55^\circ}^{80^\circ}\{A\} \equiv \left(\sum_{\varphi_m=55^\circ\mathrm{N}}^{80^\circ\mathrm{N}} \cos\varphi_m \right)^{-1} \sum_{\varphi_m=55^\circ\mathrm{N}}^{80^\circ\mathrm{N}} A_m \cos\varphi_m.
$$

Let us define a blocking *indicator function*[†] $1_i^b \equiv 1 - 1_i^n$ which takes the value 1 if there exists $\mathcal{T} \ni t_i$, and 0 otherwise. Specific indicators 1^A and $1^P = 1^b - 1^A$ correspond to crests in the vicinities of the eastern Atlantic and Pacific oceans, respectively.[‡] The very definition of blocking is still a subject of much debate [16]. Part of the present study is aimed at characterizing blocking using multiresolution analysis techniques.

7.3 Analysis

This section presents several analysis techniques. The reader may wish to skip directly from Section 7.3.1 to Section 7.4.1 to acquire a rough picture of the blocking phenomenon motivating this research. Section 7.3.2 introduces the physical laws governing the system, and Sections 7.3.3 and 7.3.4 review traditional analyses, essentially an application of traditional turbulence statistical and spectral analyses applied to the special case of the earth's atmosphere. The relevent tools of basic wavelet analysis are reviewed in Section 7.3.5, and readers may skip over the previous three sections to here, and hence to Section 7.4.2. Section 7.3.6 generalizes the traditional analyses, and finally Section 7.3.7 introduces a useful manipulation of wavelet-indexed structures to provide a measure of localized flux across scales.

[†] Roman font is used to distinguish *labels* such as m,n,s from *indices* such as m, n, s.
[‡] For these data blocks appeared nowhere else, which is consistent with climatological studies.

7.3.1 Conventional statistics

The *zonal mean* and *zonal standard deviation* are

$$
A_0 \equiv 2^{-J} \sum_{l=1}^{2^J} A_l, \qquad \sigma_A \equiv \left(\left(1 - 2^{-J}\right)^{-1} \left(\left(A^2\right)_0 - A_0^2 \right) \right)^{1/2}. \tag{7.2}
$$

Define *blocking* and *nonblocking* averages[†]

$$
A^{\mathrm{b}} \equiv I_{\mathrm{b}}^{-1} \sum_{i=1}^{I} A_i 1_i^{\mathrm{b}}, \qquad A^{\mathrm{n}} \equiv I_{\mathrm{n}}^{-1} \sum_{i=1}^{I} A_i 1_i^{\mathrm{n}}, \tag{7.3}
$$

where the total number of blocking times is $I_{\mathrm{b}} \equiv \sum_{i=1}^{I} 1_i^{\mathrm{b}} = I_{\mathrm{A}} + I_{\mathrm{P}} = I - I_{\mathrm{n}}$. Blocking standard deviations would be

$$
\sigma_A^{\mathrm{b}} \equiv \left(\left(1 - I_{\mathrm{b}}^{-1}\right)^{-1} \left(\left(A^2\right)^{\mathrm{b}} - \left(A^{\mathrm{b}}\right)^2 \right) \right)^{1/2},
$$

and similarly for σ^{n}.

7.3.2 Fundamental equations

Assuming the data are samples of differentiable fields, the laws of physics may be applied to obtain equations governing those fields, and conserved quantities useful for diagnosing the atmospheric state. Generally speaking, conserved quantities are interesting because they constrain the available state space of a dynamical system. Dissipative systems such as the atmosphere may create locally organized structures in one quantity by sufficiently increasing the entropy of another [19]. While the governing partial differential equations admit several special kinds of invariants, this study shall be limited to the most familiar one, kinetic energy (henceforth KE).

For the rest of Section 7.3, to simplify equations let the units of length and time^{-1} be the earth's radius a and angular speed Ω, respectively. Thus the previously introduced quantities are symbolized without physical units by the redefinitions $(u, v) \leftarrow (a\Omega)^{-1}(u, v)$, $\Phi \leftarrow (a\Omega)^{-2}\Phi$, $t \leftarrow \Omega t$. To a good approximation for the present situation, the time evolution of u and v, assuming $\Phi \ll a^{-1}\Omega^{-2}g$ to avoid factors of $(1 + a^{-2}Z^2)^{-1/2}$, are given by the horizontal momentum conservation equations (derived from Newton's Second Law, neglecting the Coriolis and metric terms involving vertical motion),

---

[†]A stands for any field except 1.

$$u_t = - \sec \varphi u u_\lambda - v u_\varphi - \omega u_p + v(2 \sin \varphi + \tan \varphi u) - \sec \varphi \Phi_\lambda - X, \qquad (7.4)$$

$$v_t = - \sec \varphi u v_\lambda - v v_\varphi - \omega v_p - u(2 \sin \varphi + \tan \varphi u) - \Phi_\varphi - Y, \qquad (7.5)$$

where subscripts t, λ, φ and p stand for the corresponding partial derivatives. In isobaric coordinates, vertical motion is described by ω, denoting the rate of change (with respect to t) of a fluid element's pressure. The mass continuity equation, $\omega_p = - \sec \varphi \big(u_\lambda + (\cos \varphi v)_\varphi \big)$, diagnoses ω. The hydrostatic equilibrium condition for a perfect gas, $\Phi_p = -p^{-1} T$, diagnoses Φ in terms of air temperature, $(a\Omega)^2 R^{-1} T$. The thermodynamical energy conservation equation (derived from the First Law of Thermodynamics)

$$T_t = - \sec \varphi u T_\lambda - v T_\varphi - \omega T_p + \kappa p^{-1} \omega T + H \qquad (7.6)$$

predicts T. Other parameters are the frictional westward and southward accelerations $a\Omega^2 X$ and $a\Omega^2 Y$, the dry atmospheric isobaric specific heat $\kappa^{-1} R$, the specific diabatic heating rate $\kappa^{-1} a^2 \Omega^3 H$ and the dry atmospheric gas constant R. These equations may be found e.g. as equations (6.1–5) of [13], and are the starting point for countless analyses and predictions of synoptic- to global-scale atmospheric circulation.

7.3.3 Review of statistical equations

Introducing the operators

$$\overline{(\ )} \equiv (2\pi)^{-1} \int_{-\pi}^{\pi} (\ ) d\lambda, \qquad (\ )^* \equiv (\ ) - \overline{(\ )}, \qquad (7.7)$$

Saltzman re-derived from (7.4–7.5) an expression for $\overline{u u_t + v v_t} - \overline{u}\,\overline{u_t} - \overline{v}\,\overline{v_t}$ and hence the evolution equation for mean eddy kinetic energy [20] $K_e \equiv 2^{-1} \overline{u^{*2} + v^{*2}}$:

$$K_{et} = - \sec \varphi \left(\cos \varphi \overline{v 2^{-1} \big(u^{*2} + v^{*2} \big)} \right)_\varphi - \left(\overline{\omega 2^{-1} \big(u^{*2} + v^{*2} \big)} \right)_p \qquad (7.8)$$

$$- \overline{u^* v^*} \cos \varphi (\sec \varphi \overline{u})_\varphi - \overline{v^* v^*} \overline{v}_\varphi \qquad (7.9)$$

$$- \overline{u^* \omega^*} \overline{u}_p - \overline{v^* \omega^*} \overline{v}_p + \overline{u^* u^*} \tan \varphi \overline{v} \qquad (7.10)$$

$$- \sec \varphi \overline{u^* \Phi_\lambda^*} - \overline{v^* \Phi_\varphi^*} - \overline{u^* X^*} - \overline{v^* Y^*}. \qquad (7.11)$$

The terms (7.8) represent horizontal and vertical fluxes of K_e, which vanish under integration over a closed domain, and so include fluxes from outside any open domain considered. The terms (7.9–7.10) represent KE transfer between the mean flow $\overline{u}, \overline{v}$ and the all the eddies, appearing in the Reynolds stress components $\overline{u^* v^*}, \overline{v^* v^*}, \overline{u^* \omega^*}, \overline{v^* \omega^*}, \overline{u^* u^*}$. The first two terms of (7.11) give the conversion between potential and KE, while the last two terms of (7.11) measure the frictional energy dissipation. Physically, (7.8–7.11) describe the evolution of KE associated with the collection of zonally localized meteorological phenomena such as storms and quasi-persistent low pressure systems.

7.3.4 Review of Fourier based energetics

It is desirable to resolve the eddy processes described by (7.8–7.11) into contributions from atmospheric structures of distinct scales. Traditionally this is done with the Fourier series representation

$$u = \sum_{n=-\infty}^{\infty} \widehat{u}_n \mathrm{F}_n, \quad \widehat{u}_n \equiv \overline{u \mathrm{F}_{-n}}, \quad \mathrm{F}_n(\lambda - \pi) \equiv \mathrm{e}^{\iota n \lambda}, \quad \iota \equiv \sqrt{-1}. \tag{7.12}$$

In Section 7.3.6 the advantages of applying wavelet analysis to this problem will be shown, but first comes some review. Saltzman decomposed

$$K_e = \sum_{n=1}^{\infty} K_{en}, \quad K_{en} \equiv \widehat{u}_n \widehat{u}_{-n} + \widehat{v}_n \widehat{v}_{-n}, \tag{7.13}$$

and using (7.4–7.5) to express $2\mathrm{Re}(\widehat{u}_{-n}\widehat{u}_{nt} + \widehat{v}_{-n}\widehat{v}_{nt})$, introduced[†]

$$K_{ent} = L_n^{\mathrm{S}} + M_n + C_n - D_n, \tag{7.14}$$

where the present author writes the terms in the form[‡]

$$L_n^{\mathrm{S}} \equiv -2\mathrm{Re}\Bigg(\Big(\sec\varphi\big((uu^*)_\lambda + (u^*v\cos\varphi)_\varphi\big) + (u^*\omega)_p - \tan\varphi uv^* \Big)_n \widehat{u}_{-n}$$

$$+ \Big(\sec\varphi\big((uv^*)_\lambda + (v^*v\cos\varphi)_\varphi\big) + (v^*\omega)_p + \tan\varphi uu^* \Big)_n \widehat{v}_{-n} \Bigg) \tag{7.15}$$

$$M_n \equiv -2\mathrm{Re}\Big(\widehat{u}_{-n}\widehat{v}_n \cos\varphi(\sec\varphi\bar{u})_\varphi + \widehat{v}_{-n}\widehat{v}_n\bar{v}_\varphi \tag{7.16}$$

$$+ \widehat{u}_{-n}\widehat{\omega}_n\bar{u}_p + \widehat{v}_{-n}\widehat{\omega}_n\bar{v}_p - \widehat{u}_{-n}\widehat{u}_n \tan\varphi\bar{v}\Big) \tag{7.17}$$

$$C_n \equiv -2\mathrm{Re}\Big(\sec\varphi\iota n\widehat{u}_{-n}\widehat{\Phi}_n + \widehat{v}_{-n}\widehat{\Phi}_{\varphi n} \Big) \tag{7.18}$$

$$= -2\mathrm{Re}\Big(p^{-1}\widehat{\omega}_{-n}\widehat{T}_n + \sec\varphi\big(\cos\varphi\widehat{v}_{-n}\widehat{\Phi}_n\big)_\varphi + \big(\widehat{\omega}_{-n}\widehat{\Phi}_n\big)_p \Big), \tag{7.19}$$

$$D_n \equiv 2\mathrm{Re}\Big(\widehat{u}_{-n}\widehat{X}_n + \widehat{v}_{-n}\widehat{Y}_n \Big). \tag{7.20}$$

[†] L is denoted by T in the turbulence literature.
[‡] (7.4–7.5), (7.6), (7.8–7.11), (7.12), (7.13), (7.14–7.20) correspond to Saltzman's equations (1–2), (5), (23), (29,30), (45,46) and (47,48) respectively [20].

Note that each of the terms in (7.16), (7.17), (7.18, 7.20) corresponds to a term in (7.9), (7.10) and (7.11) respectively, but it is not clear what part of (7.15) corresponds to (7.8). Saltzman showed that

$$\int_{\varphi_S}^{\varphi_N} \int_0^{p_s} \sum_{n=1}^{\infty} L_n^S dp \, d \sin \varphi = 0, \tag{7.21}$$

where p_s is the surface pressure, for $\varphi_N = -\varphi_S = \pi/2$, and later suggested a reformulation of (7.15) for which the equivalent of (7.21) held for arbitrary $\varphi_{N,S}$, neglecting a term involving ω_p [14]. These null sums reflect the fact that nonlinear interactions act to transfer energy between wavenumbers n, but create or destroy no net energy. Hansen [11] has derived a formulation which this author writes as

$$L_n^S = L_n^H + B_{khn} + B_{kvn} + N_{khn} + N_{kvn}, \tag{7.22}$$

$$\cos \varphi L_n^H \equiv \mathrm{Re} \Bigg(\left(u^\star u^\star{}_\lambda + (u^\star u^\star)_\lambda + [v\partial_\varphi, u] + [\omega\partial_p, u] - 2 \sin \varphi u^\star v^\star \right)_n \widehat{u}_{-n}$$

$$+ \left(u^\star v^\star{}_\lambda + (u^\star v^\star)_\lambda + [v\partial_\varphi, v] + [\omega\partial_p, v] + 2 \sin \varphi u^\star u^\star \right)_n \widehat{v}_{-n} \Bigg), \tag{7.23}$$

$$B_{khn} \equiv -\sec \varphi \left(\cos \varphi \bar{v} K_{en} \right)_\varphi, \qquad B_{kvn} \equiv -\left(\bar{\omega} K_{en} \right)_p, \tag{7.24}$$

$$N_{khn} \equiv -\sec \varphi \mathrm{Re} \left(\cos \varphi \left(\widehat{u^\star v^\star}_n \widehat{u}_{-n} + \widehat{v^\star v^\star}_n \widehat{v}_{-n} \right) \right)_\varphi, \tag{7.25}$$

$$N_{kvn} \equiv -\mathrm{Re} \left(\widehat{u^\star \omega^\star}_n \widehat{u}_{-n} + \widehat{v^\star \omega^\star}_n \widehat{v}_{-n} \right)_p, \tag{7.26}$$

$$[A\partial_x, B] \equiv \cos \varphi A^\star (B^\star{}_x - B^\star \partial_x).$$

The advantages of this formulation are that $$\sum_{n=1}^{\infty} L_n^H = 0 \tag{7.27}$$

for every individual φ and p, and that the terms (7.24–7.26) correspond to (7.8), so that L_n^H (term (7.23)) better isolates the wave–wave interactions from the boundary effects (7.24–7.26) of an open domain.

7.3.5 Basic concepts from the theory of wavelet analysis

In this study the interesting dependence is on λ, a 2π-periodic coordinate, so in this subsection periodic orthogonal wavelet analysis is introduced. Any periodic, continuous or absolutely integrable, function $A(\lambda)$ may be expanded in a periodic orthonormal wavelet basis,[†]

$$A = \overline{A} + \sum_{j=0}^{\infty} \widetilde{A}_j, \qquad \widetilde{A}_j \equiv \sum_{k=1}^{2^j} \widetilde{A}_{jk} \psi_{jk}^{\mathrm{per}}, \tag{7.28}$$

$$\widetilde{A}_{jk} \equiv \overline{A \psi_{jk}^{\mathrm{per}}} = \sum_{n=-\infty}^{\infty} \widehat{A}_{-n} \widehat{\psi_{jk}^{\mathrm{per}}}_n, \tag{7.29}$$

$$\psi_{jk}^{\mathrm{per}}(2\pi x) \equiv 2^{j/2} \sum_{n=-\infty}^{\infty} \psi\bigl(2^j(x+n) - k + 1\bigr), \tag{7.30}$$

$$\overline{\psi_{jk}^{\mathrm{per}} \psi_{j'k'}^{\mathrm{per}}} = \delta_{j-j'}\delta_{k-k'}, \qquad \overline{\psi_{jk}^{\mathrm{per}}} = 0. \tag{7.31}$$

The Parseval Identity implies (recalling (7.7))

$$\overline{A^{\star}B^{\star}} = \sum_{j=0}^{\infty} \sum_{k=1}^{2^j} \widetilde{A}_{jk} \widetilde{B}_{jk}. \tag{7.32}$$

There are many possible 'mother wavelets' ψ on the real line which generate such a representation. The ψ used here cannot be written in terms of explicit functions, but is defined by the solution ϕ of the functional dilation equation

$$\phi(x) \equiv 2^{1/2} \sum_{k=-\infty}^{\infty} h_k \phi(2x - k), \tag{7.33}$$

$$\psi(x) \equiv 2^{1/2} \sum_{k=-\infty}^{\infty} (-1)^k h_{\mathrm{D}+1-k} \phi(2x - k). \tag{7.34}$$

Daubechies has shown that, with certain requirements on the sequence h, solutions exist which have compact support in x, smoothness (roughly, the highest order of existing derivative), and number of vanishing moments which all increase roughly linearly with the support length D of h [5]. This study uses the Daubechies-20 wavelet, with only D $= 20$ nonzero h_k. The ψ_{jk}^{per} are centred approximately at $\lambda_{jk} \equiv 2\pi(2^{-j}(k-1) - 2^{-1}) = \lambda_{l_{jk}}$, where

$$l_{jk} \equiv 2^{J-j}(k-1) + 1 \tag{7.35}$$

[†](7.28–7.32) all follow immediately from Section 9.3 of [5].

indicates the center on the grid (7.1). Figure 7.1 shows the ψ_{jk}^{per} in an arrangement which clarifies later figures.

Periodic wavelet analysis can also be visualized in the Fourier domain; the $\widehat{\psi_{jk}^{\mathrm{per}}}_{\,n}$ are shown in Figure 7.2. The magnitude $|\widehat{\psi_{jk}^{\mathrm{per}}}_{\,n}|$ is independent of k for a given j, and the bandpass bandcenter and bandwidth are both seen to increase with j, in accordance with the Heisenberg principle, as the spatial resolution increases (Figure 7.1). The phase arg $\widehat{\psi_{jk}^{\mathrm{per}}}_{\,n}$ is just a shift mod 2π of

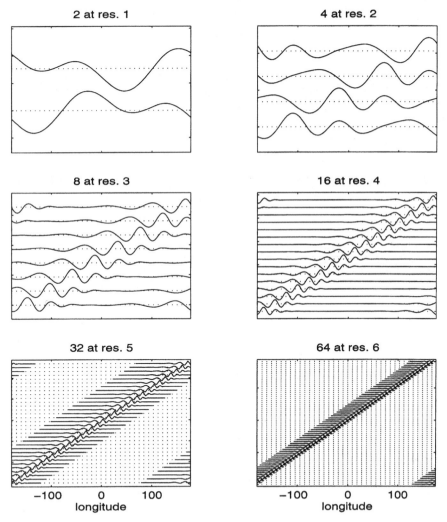

Fig. 7.1. Wavelet basis functions ψ_{jk}^{per} for $j = 1, \cdots J - 1 = 6$, as a function of λ. There are 2^{j} basis functions at resolution j. The curves are offset to avoid overlapping, and only the nonvanishing parts are plotted. Dotted lines show the offsets. Each function has mean zero.

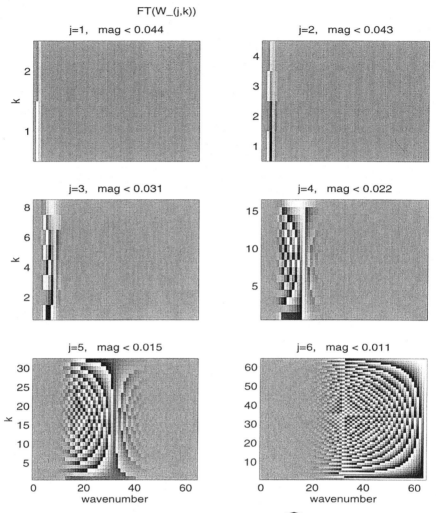

Fig. 7.2. Wavelet basis function Fourier coefficients $\widehat{\psi^{\mathrm{per}}_{jk}}{}_n$ for $j = 1, \ldots J - 1 = 6$, as a function of n (abscissa) and k (ordinate). Grayscale indicates $(\arg \widehat{\psi^{\mathrm{per}}_{jk}}{}_n)|\widehat{\psi^{\mathrm{per}}_{j.}}{}_n| / \max |\widehat{\psi^{\mathrm{per}}_{j.}}|$ from $-\pi$ (dark) to π (light).

$2^{1-j}\pi(1-k)n$ from the phase of the continuous Fourier transform of ψ at $2^{-j}n$, as can be seen from $\widehat{(7.30)}_n$.

In practice, given 2^J discrete samples $A(\lambda_{Jk})$ of A, then estimates of $\overline{A}$ and the $2^J - 1$ coefficients $\widetilde{A}_{jk}$, $j = 0, \cdots J - 1$, $k = 1, \cdots 2^j$ may be obtained in only $O(2^J)$ calculations involving just h, without ever having to evaluate ψ^{per}_{jk}. This *discrete wavelet transform* satisfies the discrete analogs of all the previous equations, just replacing the operator $\overline{()}$ by (7.2). Since both ψ^{per}_{jk} and $\widetilde{\psi^{\mathrm{per}}_{jk}}$ are localized, so $\widetilde{A}_{jk}$ selects the contribution to A from position $\sim 2^{-j}k$ and scale $\sim 2^{-j}$. The set $\{\overline{A}\} \bigcup_{j=0}^{J-1} \{\widetilde{A}_j\}$ defines a *multiresolution analysis* of A. It is essentially a sophisticated kind of bandpass filtering, a more familiar technique to climatologists [2, 21].

7.3.6 Energetics in the domain of wavelet indices (or any orthogonal basis)

The advantage of the wavelet over the Fourier formulation of energetics is the interpretation. As in the Fourier case, the nonlinear interactions between particular scales can be identified; but now the particular locations of the interacting scales are also represented, at least within a resolution corresponding to the scale. This information is not available in the Fourier representation.

To derive the wavelet form, first note that

$$K_{ejk} \equiv 2^{-1}\left(\widetilde{u}_{jk}{}^2 + \widetilde{v}_{jk}{}^2\right), \qquad K_e = \sum_{j=0}^{\infty} \sum_{k=1}^{2^j} K_{ejk} \qquad (7.36)$$

decomposes the eddy KE into contributions from distinct positions and scales. In his modification of the Fourier basis formulation mentioned above, Saltzman suggested collecting half the trilinear terms to form boundary fluxes, explicitly separating the nonlinear interactions which vanish under integration over a closed domain [14]. This same step was generally formulated in the Fourier basis energetics development of Hansen [11]. Following them, from (7.4–7.5) an expression for $\widetilde{u}_{jk}\widetilde{u}_{jk_t} + \widetilde{v}_{jk}\widetilde{v}_{jk_t}$ leads to

$$K_{ejkt} = L_{jk} + B_{khjk} + B_{kvjk} + N_{khjk} + N_{kvjk} + M_{jk} + C_{jk} + B_{ghjk} + B_{gvjk} - D_{jk},$$
$$(7.37)$$

$$\cos\varphi L_{jk} \equiv -2^{-1}\left(\left(u^{\star}u^{\star}{}_{\lambda} + (u^{\star}u^{\star})_{\lambda} + [v\partial_{\varphi}, u] + [\omega\partial_p, u] - 2\sin\varphi u^{\star}v^{\star}\right)_{jk}^{\widetilde{\;}} \widetilde{u}_{jk}\right.$$

$$\left. + \left(u^{\star}v^{\star}{}_{\lambda} + (u^{\star}v^{\star})_{\lambda} + [v\partial_{\varphi}, v] + [\omega\partial_p, v] + 2\sin\varphi u^{\star}u^{\star}\right)_{jk}^{\widetilde{\;}} \widetilde{v}_{jk}\right) \quad (7.38)$$

$$B_{khjk} \equiv -\sec\varphi\left(\cos\varphi\overline{v}K_{ejk}\right)_{\varphi}, \qquad B_{kvjk} \equiv -\left(\overline{\omega}K_{ejk}\right)_p, \qquad (7.39)$$

$$N_{khjk} \equiv -2^{-1}\sec\varphi\left(\cos\varphi\left(\widetilde{u^{\star}v^{\star}}_{jk}\widetilde{u}_{jk} + \widetilde{v^{\star}v^{\star}}_{jk}\widetilde{v}_{jk}\right)\right)_{\varphi}, \qquad (7.40)$$

$$N_{\mathrm{kv}jk} \equiv -2^{-1}\left(\widetilde{u^{\star}\omega^{\star}}_{jk}\tilde{u}_{jk} + \widetilde{v^{\star}\omega^{\star}}_{jk}\tilde{v}_{jk} \right)_{p},$$ (7.41)

$$M_{jk} \equiv - \cos\varphi\tilde{u}_{jk}\tilde{v}_{jk}(\sec\varphi\bar{u})_{\varphi} - \tilde{v}_{jk}\tilde{v}_{jk}\bar{v}_{\varphi}$$ (7.42)

$$- \tilde{u}_{jk}\tilde{\omega}_{jk}\bar{u}_{p} - \tilde{v}_{jk}\tilde{\omega}_{jk}\bar{v}_{p} + \tan\varphi\tilde{u}_{jk}\tilde{u}_{jk}\bar{v},$$ (7.43)

$$- \sec\varphi(\tilde{u}_{jk}\tilde{u}_{\lambda jk} + \tilde{v}_{jk}\tilde{v}_{\lambda jk})\bar{u},$$

$$C_{jk} \equiv -p^{-1}\tilde{\omega}_{jk}\widetilde{T}_{jk},$$ (7.44)

$$B_{\mathrm{gh}jk} \equiv - \sec\varphi\left(\cos\varphi\tilde{v}_{jk}\widetilde{\Phi}_{jk} \right)_{\varphi} - \sec\varphi\left(\tilde{u}_{jk}\widetilde{\Phi}_{\lambda jk} + \tilde{u}_{\lambda jk}\widetilde{\Phi}_{jk} \right),$$

$$B_{\mathrm{gv}jk} \equiv -\left(\tilde{\omega}_{jk}\widetilde{\Phi}_{jk} \right)_{p},$$ (7.45)

$$D_{jk} \equiv -\tilde{u}_{jk}\widetilde{X}_{jk} - \tilde{v}_{jk}\widetilde{Y}_{jk}.$$ (7.46)

Again, there is a one to one correspondance between the terms (7.39–7.41), (7.42–7.43), (7.44–7.46) and (7.8), (7.9–7.10), (7.11) respectively: *the physical processes have been resolved in both location and scale*. The fact (7.27) that wave–wave interactions create or destroy no net energy is now expressed by

$$\sum_{j=0}^{\infty}\sum_{k=1}^{2^{j}} L_{jk} = 0, \quad \text{and} \quad \sum_{j=0}^{\infty}\sum_{k=1}^{2^{j}} M_{jk} = (7.9, 7.10),$$ (7.47)

the total **KE** transfer from the mean flow to all the eddies.[†]

The above equations would be obtained for *any* orthogonal basis satisfying analogs of (7.28, 7.29, 7.32) the jk indices may be thought of as a single index.[‡]

7.3.7 Kinetic energy localized flux functions

Adapting the Fourier-based approach of [22], it is useful to construct a measure F_{jk} of the total flux of **KE** to scale j from larger scales $j' < j$, localized

---

[†]Although in the Fourier basis it is simple to expand the nonlinear terms into convolutions, and use $\widehat{A_{\lambda n}} = \imath n\widehat{A}_{n}$, in the wavelet basis it is simpler to perform the products and ∂_{λ} operations before the transform. In order to retain (7.46) to a good approximation, a 10th order scheme

$$A_{\lambda}(\lambda_{l}) = (\Delta\lambda)^{-1}\sum_{l'=-5}^{5} \gamma_{l'}A(\lambda_{l+l'}) + \mathrm{O}((\Delta\lambda)^{10})$$ (7.48)

was used, with $\gamma_{l} = -\gamma_{-l} = \frac{-1}{1260}, \frac{5}{504}, \frac{-5}{84}, \frac{5}{21}, \frac{-5}{6}, \cdots$ [4]. A first order scheme was used for A_{φ}.
[‡]Note that only the properties (7.28, 7.29, 7.32) were used to derive equations (7.37–7.45), that is not any uniquely 'wavelet' property such as (7.30, 7.33, 7.34). In the Fourier case, Saltzman used the *multiplication theorem* to advantage, a consequence of $\overline{F_{n}F_{n'}F_{n''}} = \delta_{n+n'+n''}$. There is no similar identity for wavelets, although $\psi_{jk}^{\mathrm{per}}\psi_{j'k'}^{\mathrm{per}}\psi_{j''k''}^{\mathrm{per}}$ is extremely sparse, and the Parseval corollary (7.32) still holds. For this reason all the trilinear terms are calculated by multiplication before wavelet transform.

at k. Local downscale (upscale) KE cascades correspond to $F_{jk} > 0 (< 0)$. The Fourier approach defines[†]

$$F_n \equiv - \sum_{n'=1}^{n} L_{n'}, \quad F_0 \equiv 0. \tag{7.49}$$

The wavelet construction proceeds as follows. The $2^J - 1$ elements L_{jk} form a pyramidal tableau with 2^j elements at level j. The author introduces a rectangular $J \times 2^{J-1}$ matrix equivalence (recalling (7.35))

$$L_j(\lambda_{2l}) \equiv 2^{j-J+1} L_{j K_{j2l}}, \qquad L_{jk} = 2^{J-j-1} L_j(\lambda_{l_{jk}+1}), \tag{7.50}$$

$$K_{jl} \equiv \lfloor 2^{j-J}(l-1) + 1 \rfloor = \lfloor 2^j((2\pi)^{-1}\lambda_l + 2^{-1}) + 1 \rfloor, \tag{7.51}$$

which is normalized to have the property $\sum_{l=1}^{2^{J-1}} L_j(\lambda_{2l}) = \sum_{k=1}^{2^j} L_{jk}$, which preserves the sum over the spatial index at each scale, but makes the left sum limits independent of scale.

To illustrate, for J temporarily taking the value 3, the L_{jk} tableau is

L_{01}			
L_{11}		L_{12}	
L_{21}	L_{22}	L_{23}	L_{24}

Then by (7.50) the elements $L_j(\lambda_{2l})$ form

$4^{-1}L_{01}$	$4^{-1}L_{01}$	$4^{-1}L_{01}$	$4^{-1}L_{01}$
$2^{-1}L_{11}$	$2^{-1}L_{11}$	$2^{-1}L_{12}$	$2^{-1}L_{12}$
L_{21}	L_{22}	L_{23}	L_{24}

The present author independently introduced a KE local scale flux function [10][‡]

[†]F is denoted by Π in the turbulence literature.
[‡]This may also be written $F_{jk} = -2^{-j} \sum_{j'=0}^{j} 2^{j'} L_{j K_{(j'-j+J)k}}$, but it is clearer, and computationally faster, to use the equation in the main text. However the latter form is proportional to an essentially equivalent formulation introduced earlier by Meneveau [18].

$$F_j(\lambda_{2l}) \equiv -\sum_{j'=0}^{j} L_{j'}(\lambda_{2l}),$$

(cf. (7.49).) By (7.47) and the choice of normalization,

$$\lim_{J\uparrow\infty} \sum_{l=1}^{2^{J-1}} F_{J-1}(\lambda_{2l}) = 0. \tag{7.52}$$

The rearrangement (7.50) (or an equivalent energetic bookkeeping) is necessary so that in the $\sum_{j'=0}^{j}$ no energy at larger scale (j', k') is 'double counted' for different k. Also, the K_{jl} is defined so that only those elements with k' accounting (at resolution j') for the same location as λ_{2l} will contribute to $F_j(\lambda_{2l})$. This makes $F_j(\lambda_{2l})$ a meaningfully local flux function.

Once calculated for each φ, p, t, the statistics described in (7.3) may be applied to $(M, L, F)_{jk}$. Other useful operations are[†]

$$\{A\} \equiv \left(\sum_{\varphi_m=30^{\circ}N}^{80^{\circ}N} \cos\varphi_m\right)^{-1} \sum_{\varphi_m=30^{\circ}N}^{80^{\circ}N} A_m \cos\varphi_m \tag{7.53}$$

and[‡] $\langle A\rangle \equiv g^{-1}\sum_{s=1}^{10} A_s\Delta p_s$. The quantities $\langle(M, L, F)_{jk}\rangle$ are then in W m^{-2}.

7.4 Results and interpretation

7.4.1 Time averaged statistics

Figure 7.3 shows the fields $Z^{n,b}$ at 70 kPa[§] superposed on the continental coastlines. The blocking conditions are clearly evidenced by the unusually strong ridges in the Atlantic and Pacific cases compared to the nonblocking cases. The blocks themselves are visible as localized high-Z structures (light grayscale) protruding[¶] northward from the tropics, resembling the nonlinear phenomenon of breaking waves on a fluid surface.

[†] A lower limit 30° N is used to reduce the meridional boundary fluxes B_{kh} and N_{kh}. Earlier the lower limit 55°N was used to focus on the block region.
[‡] $\Delta p_1 \equiv p_2 - p_1$, $\Delta p_s \equiv 2^{-1}(p_{s+1} - p_{s-1})$ $(s = 2, \cdots 9)$, $\Delta p_{10} \equiv p_{10} - p_9$, so that $\sum_{s=1}^{10} \Delta p_s = 100$ kPa, about 1 atmosphere.
[§] 70 kPa is convenient for comparing with [6, 23, 25], while the more representative 50 kPa was used for diagnosis in Section 7.2.
[¶] At lower p (higher altitude) these structures are cut off from the tropics.

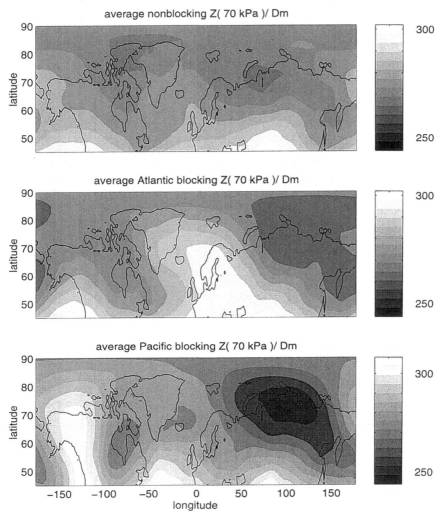

Fig. 7.3. Z^n (top), Z^A (middle), Z^P (bottom), value in Dm (10 m) indicted in gray-scale from dark (lows) to light (highs). All fields at $p = 70$ kPa as a function of λ (abscissa) and φ (ordinate).

Figure 7.4 shows $\sigma_Z^{n,b}$ superposed on the continental coastlines. The mainly diagonal structure of σ_Z^n is a result of the typical eastward-poleward progression of *synoptic weather patterns* such as the winter storm tracks seen here. Such diagonal structure is noticeably absent from σ_Z^b. Blocking is so-named for this reason: the progression of synoptic weather patterns is blocked. Instead of *progressive* (eastward translating) structures, what makes up the variance during blocking are small regions of (transient) 'eddy' activity in the vicinity of the block. There are two maxima of σ_Z^A, over the southern tip of

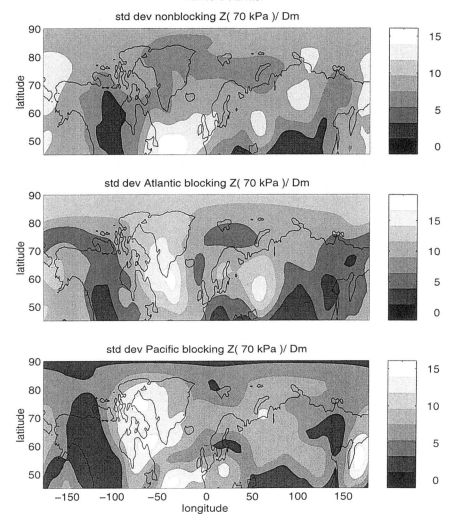

Fig. 7.4. σ_Z^n (top), σ_Z^A (middle), σ_Z^P (bottom), as in Fig. 7.3.

Greenland and western Russia, that is, one on each side of the ridge in Z^A over Norway, Fig. 7.3. Similarly σ_Z^P has several local maxima flanking the ridge of Z^P over Alaska, one just east of Japan and several along eastern Canada. These results are consistent with the idea that large blocking structures are supported energetically by smaller eddies around them. This will be investigated systematically in Section 7.4.3.

The zonal statistics $(u_0, v_0, \sigma_u, \sigma_v)^{n,b}$ are shown in Fig. 7.5. The jet stream is clearly seen in u_0 as a maximum near $(\varphi, p) = (30° \text{N}, 20\,\text{kPa})$. The pattern of $\sigma_{(u,v)}^b$ shows a bifurcation during blocking, consistent with the splitting of the jet stream around the block's anomalous high Z often observed then.

vertical profiles of u (gray) & v (contour)

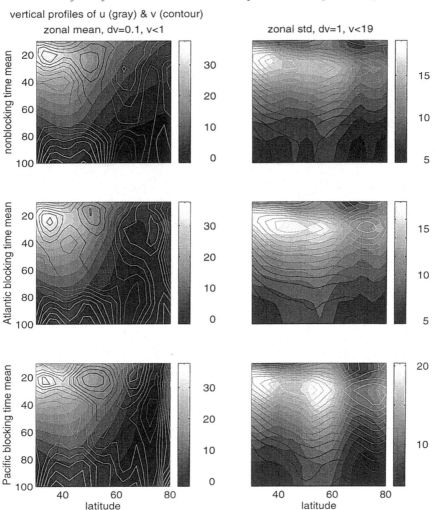

Fig. 7.5. $(u, v)_0$ (left), $\sigma_0(u, v)$ (right), for cases n (top), A (middle), P (bottom). Grayscale is value of u-statistics, m s^{-1}, contour intensity, the value of v-statistics, v_0 by 0.1 m s^{-1} up to 1 m s^{-1} (left), $\sigma_0(v)$ by 1 m s^{-1} up to 19 m s^{-1} (right). All fields depend on φ (abscissa) and p (ordinate).

7.4.2 *Time dependent multiresolution analysis at fixed* (φ, p)

From Fig. 7.3, the $\varphi = 65°\,\text{N}$ line was judged to be a reasonably representative latitude to discriminate blocking from nonblocking, since it cuts through the block structure. The Hovmöller diagram of $Z(\lambda, 65°\,\text{N}, 70\,\text{kPa}, t)$ is shown in Fig. 7.6. In such a figure, upper-left to lower-right diagonal structures are typical of progressive synoptic weather patterns. Vertical structures

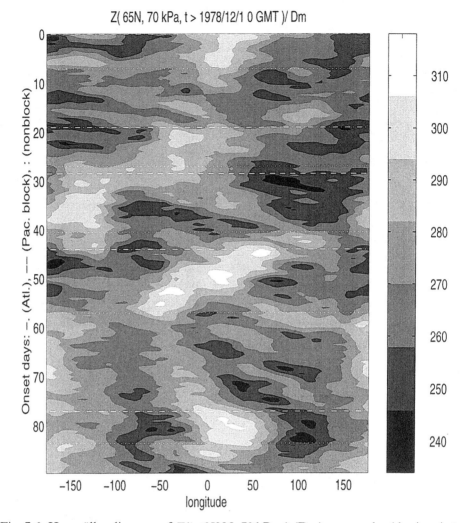

Fig. 7.6. Hovmöller diagram of $Z(\lambda, 65°\,\text{N}, 70\,\text{kPa}, t)$ (Dm), grayscale. Abscissa is λ, ordinate t (d), advancing downwards. Dotted, dash-dotted and dashed lines indicate the onset t of nonblocking, Atlantic and Pacific blocking t-intervals, respectively.

indicate stationary features, and conversely almost-horizontal structures indicate very rapidly moving features. A curve parallel to the main diagonal corresponds to a speed of $t_I^{-1}2\pi a\cos 65° \approx 8\,\text{km}\,\text{h}^{-1}$. The light regions of this figure very clearly depict four blocking events in the expected t-intervals $[0, 6.5]$, $[19, 26.5]$, $[44, 56.5]$, $[77, 83]$ d near the Atlantic (near $\lambda = 0$) and one in $[28.5, 40]$ d near the Pacific (near $\lambda = -150°$) [12]. The third Atlantic blocking event is very strong, and clearly moves in a retrograde manner (upper-right to lower-left). Apart from these features it is difficult to extract other insight from the complicated patterns. There is a range of scales displayed, since there are both regions with many close contour curves and broad, more uniform areas.

To be able to determine more from this figure, a multiresolution analysis (Section 7.3.5) was performed for each t: $Z(\lambda, t) = \sum_{j=0}^{J-1} \widetilde{Z}_j(\lambda, t)$, $\widetilde{Z}_j(\lambda, t) = \sum_{k=1}^{2^j} \widetilde{Z}_{jk}(t) \psi_{jk}^{\text{per}}(\lambda)$. Figure 7.7 shows the result. The sum of the six panels in Figure 7.7 reproduces Figure 7.6. Each panel labeled 'MRA j' shows a field $\widetilde{Z}_j(\lambda, t)$ generated by the linear combination of 2^j basis functions ψ_{jk}^{per} and coefficients $\widetilde{Z}_{jk}$, as in (7.28). The $j = 1$ (second-largest scale, $\frac{1}{2}$) level

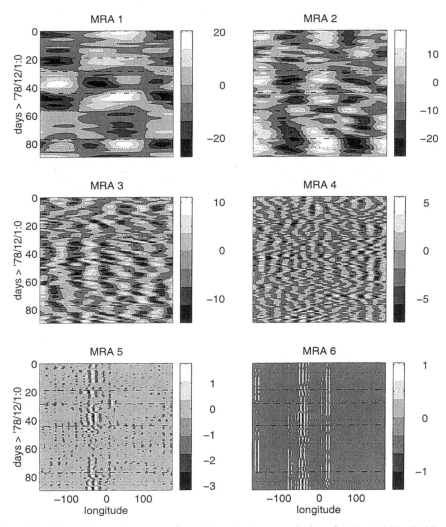

Fig. 7.7. Hovmöller diagram of multiresolution analysis of $Z(\lambda, 65° \text{N}, 70 \text{kPa}, t)$ (Dm), as in Fig. 7.6. Each panel labeled 'MRA j' corresponds to expansions $\widetilde{Z}_j(\lambda, t)$ (7.28) in ψ_{jk}^{per}.

by itself clearly shows the same blocking ridge pattern as Fig. 7.6. This suggests that information encoded in just the two numbers $\widetilde{Z}_{11}$ and $\widetilde{Z}_{12}$ may describe the presence of blocking, at least for these data. The intermediate scales $2^{-j} = \frac{1}{4}, \frac{1}{8}, \frac{1}{16}$ capture patterns associated with synoptic meteorology, from Atlantic-size ($j = 2$) down to Hudson Bay-size ($j = 4$). Generally speaking, there are more progressive structures evident during nonblocking. For example: $\widetilde{Z}_2$ for $\lambda < 0$ (western hemisphere) during the first nonblocking interval, $[7, 18.5]$ d, and for $0 < \lambda < 100°$ (over Europe and Russia) during the fourth nonblocking interval, $[57, 76.5]$ d; $\widetilde{Z}_3$ at these times and also globally during the fifth nonblocking interval, $[83.5, 89.5]$ d. Such progressive structures are absent from $\widetilde{Z}_1$, except for transitions from one mode to another, e.g. from $k = 1$ (western) to $k = 2$ (eastern hemisphere) close to the transition times (jumps in 1^b) 27 d and 57 d.[†]

To investigate whether the presence of blocking might be described by just the two coefficients $\widetilde{Z}_{11}$ and $\widetilde{Z}_{12}$, their time series are presented Fig. 7.8. For the most part, the curves tend to stay at large amplitudes, apart from each other during blocking, and to approach each other at small amplitudes during nonblocking. As might be expected, the eastern and western hemisphere wavelet modes exchange signs during the Pacific blocking relative to the Atlantic blocking.

The evolution of $\widetilde{Z}_{1k}(t)$, $k = 1, 2$, can also be described in a dynamical phase space, Fig. 7.9. Here it may be seen that Atlantic blocking is roughly characterized by counterclockwise, $\widetilde{Z}_{11} > 0$ leading $\widetilde{Z}_{12} < 0$ orbits in the fourth quadrant, while Pacific blocking is characterized by clockwise, $\widetilde{Z}_{11} < 0$ lagging $\widetilde{Z}_{12} > 0$ orbits in the second quadrant. Nonblocking points cluster nearer to the origin in all quadrants. The empty regions enclosed by the orbits suggest instabilities, although either longer time series or projection onto higher-dimensional phase space, or both, would be required to confirm this. If such diagnostic qualities were robust for more data, they could be useful in predicting blocking, an outstanding problem in extended-range weather prediction.

[†]It should be noted that there are exceptions to these observations. That is, occasionally $\widetilde{Z}_3$ or $\widetilde{Z}_4$ may appear stationary during nonblocking, or progressive during blocking.
Also, the apparent low-amplitude (a few m), small scale ($j > 4$) stationary structures over the eastern USA and Europe seem unphysical. These are probably artifacts, but not unique to this analysis; almost identical patterns are seen in Fourier band-pass filtering (i.e., $Z_j^{\mathrm{bp}}(\lambda, t) \equiv \sum_{|n|=2^j}^{2^{j+1}-1} \widetilde{Z}_n(t) F_n(\lambda)$ for high j) of this data (not shown). The stationary structures are aligned with mountain ranges, but the disturbance amplitudes are probably smaller than the original sensor sensitivity and so probably indicate mountain-generated artifacts from the NMC analysis procedure.

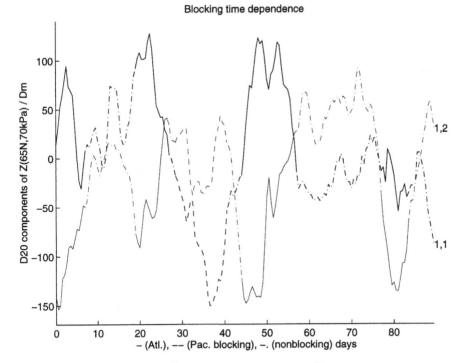

Fig. 7.8. Time series of $\tilde{Z}_{11}$ (dark curve) and $\tilde{Z}_{12}$ (light curve, Dm), at $(\varphi, p) = (65°\,\text{N}, 70\,\text{kPa})$. Abscissa is t (d). Dash-dotted, solid and dashed curves indicate nonblocking, Atlantic and Pacific blocking, respectively.

7.4.3 Kinetic energy transfer functions

Table 7.1 contains some numerical results of the analyses. The I are the observation counts of the various states. The $-\sum_{j=0}^{J-1}\sum_{k=1}^{2^j}\{\langle M_{jk}\rangle\}$ are the transfers of KE to the mean flow from all eddies. Although not below machine precision, the $\sum_{j=0}^{J-1}\sum_{k=1}^{2^j}\{\langle L_{jk}\rangle\}$ are negligible.

Table 7.1. *Numerical results*

		Blocking state	
	Non-	Atlantic	Pacific
I	87	69	24
$\sum_{j=0}^{J-1}\sum_{k=1}^{2^j}\{\langle L_{jk}\rangle\}(\text{mW m}^{-2})$	0.5	2.9	−0.4
$\sum_{j=0}^{J-1}\sum_{k=1}^{2^j}\{\langle M_{jk}\rangle\}(\text{mW m}^{-2})$	16.1	90.8	−317.0

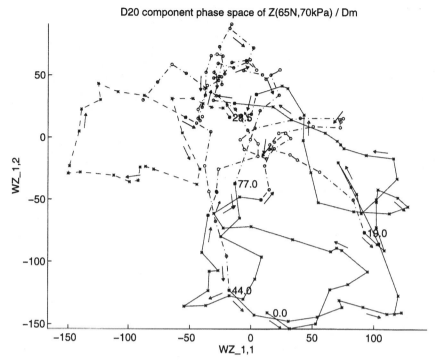

Fig. 7.9. $(\widetilde{Z}_{11}, \widetilde{Z}_{12})$ phase space (Dm). Dash-dotted, solid and dashed curves for n, A and P, respectively. Markers $\circ$ for nonblocking, $*$ for blocking. Initial blocking times (d) are labelled. Every sixth day an arrow to the right of the trajectory shows the direction of $\mathrm{d}(\widetilde{Z}_{11}, \widetilde{Z}_{12})/\mathrm{d}t$.

The global KE transfer function statistics $\{\langle (M, L)_{jk}^{\mathrm{n,b}} \rangle\}$ are presented in Figs. 7.10–7.13. Each panel of these four figures, as well as of Figs. 7.15–7.20, is labeled by resolution j, with 2^j abscissa values k for location. Comparing Atlantic blocking to nonblocking, Figs. 7.10–7.11, the following observations suggest eddies feeding energy to the block. The second-largest scale, eastern hemisphere part $(j, k) = (1, 2)$ gains KE mainly from the mean flow. At the scale $\frac{1}{4}(j = 2)$ there is a loss to both mean and eddies just to the west (upstream) of the block ($k = 2$), resulting in a gain $\{\langle L_{23}^{\mathrm{A}} \rangle\}$ to a large eddy at the block location ($k = 3$). (Throughout this chapter, block location in k, at a given resolution j, is indicated by a white bar.) Similarly the $j = 3$ scale shows enhanced KE losses to the mean ($\{\langle M_{34}^{\mathrm{A}} \rangle\}$) and eddy ($\{\langle L_{3k}^{\mathrm{A}} \rangle\}$, $k = 4, 5$) flows just upstream of the block, accompanied by smaller losses to the mean for $j = 4, 2 < k < 10$. The smaller scales $j = 5, 6$ contain negligible transfers (less than $50\,\mathrm{mW\,m^{-2}}$). There are sharply positive $\{\langle F_{jk}^{\mathrm{A}} \rangle\}$ (not shown) just upstream of the block, with broad negative $\{\langle F_{jk}^{\mathrm{A}} \rangle\}$ to the block's east (downstream).

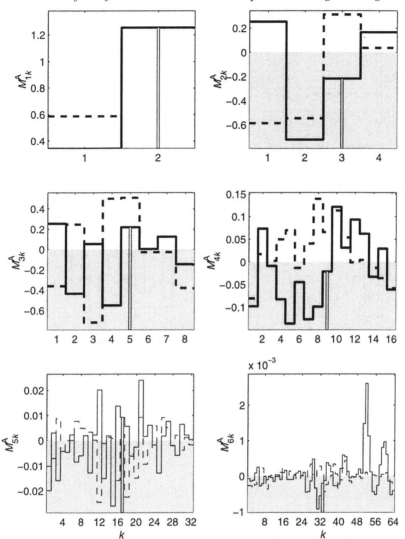

Fig. 7.10. $\{\langle M_{jk}^{\mathrm{n}}\rangle\}$ and $\{\langle M_{jk}^{\mathrm{A}}\rangle\}$, (dashed and solid curves respectively, W m^{-2}). Each panel shows the 2^j values $\{\langle M_{jk}\rangle\}$ corresponding to scale 2^{-j} of expansion (7.28). The abscissae are k, where each sequence $k = 1, \ldots 2^j$ corresponds to the *locations* λ_{jk} covering the entire circle, at resolution j. Negative ordinates are emphasized by a gray background. The block longitude $\lambda_{\mathrm{m}}^{\mathrm{b}}$ is indicated by a white bar extending up from the abscissa.

For the Pacific block, Figs. 7.12–7.13, at the second-largest scale there is a loss $(\{M_{12}^{\mathrm{P}}\})$ to the mean flow downstream ($k = 2$) accompanied by a larger gain $(\{\langle L_{11}^{\mathrm{P}}\rangle\})$ from the eddies upstream ($k = 1$). At $j = 2$ (scale $\frac{1}{4}$) there is a gain $\{\langle M_{21}^{\mathrm{P}}\rangle\}$ from the mean at the block ($k = 1$) and a loss $\{\langle M_{24}^{\mathrm{P}}\rangle\}$ to the mean upstream of the block ($k = 4$). For the eddy contribution, there are

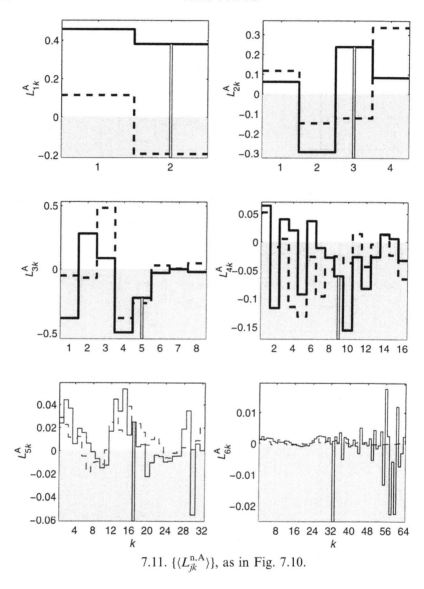

7.11. $\{\langle L_{jk}^{n,A}\rangle\}$, as in Fig. 7.10.

large gains again at the block $(k = 1, 4)$ accompanying losses elsewhere, especially from just downstream of the block $(k = 2)$. The mean flow feeds KE downstream of the block at the $\frac{1}{8}$ scale $(j = 3, k = 4)$, while the eddy flow removes KE at this scale everywhere but just at the block $(k = 1)$.[†] Most of the smaller scale mean transfers are negligible, but there are some significant eddy transfers located away from the block at $j = 4$, $k = 5, 8$. Although

---

[†]And curiously, also at the Atlantic block's location $(k = 5)$, inactive at this time in the data.

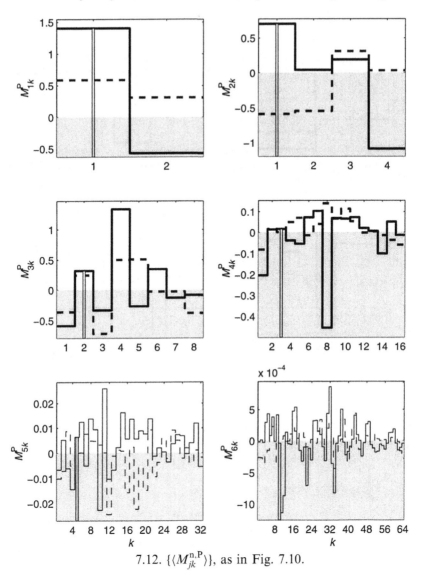

7.12. $\{\langle M_{jk}^{n,P}\rangle\}$, as in Fig. 7.10.

$\{\langle F_{jk}^{P}\rangle\}$ (not shown) is positive downstream of the block and negative at the block, not unlike $\{\langle F_{jk}^{A}\rangle\}$, $\{\langle F_{jk}^{P}\rangle\}$ is different in that the negative sector is only slightly broader than the positive sector, and lies more coincident to the block instead of distinctly upstream.

The content of Figs. 7.10–7.13 is summarized in the first two rows of Fig. 7.14, which depict $\{\langle (M, L)_j^{n,A,P}(\lambda_{2l})\rangle\}$. All the remarks made about Figs. 7.10–7.13 also pertain here, but this condensed representation less vividly shows the value of small transfers. The zonally (i.e. longitudinally) local

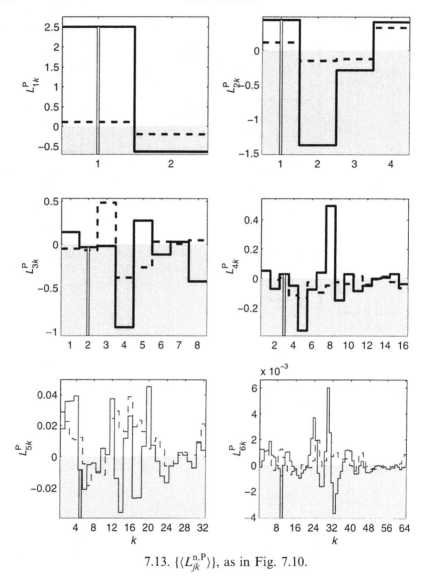

7.13. $\{\langle L_{jk}^{\mathrm{n,P}}\rangle\}$, as in Fig. 7.10.

flux functions $\{\langle F_j^{\mathrm{n,a,P}}(\lambda_{2l})\rangle\}$ (third row of Fig. 7.14) reveal a similarity between both Atlantic and Pacific blocking as viewed in space and scale simultaneously. Observe that (7.52) is obeyed. *For both locations of blocks,* there is a λ-interval with positive $\{\langle F_j^{\mathrm{b}}(\lambda_{2l})\rangle\}$ for larger j on the west (upstream), and another λ-interval with negative $\{\langle F_j^{\mathrm{b}}(\lambda_{2l})\rangle\}$ (very strong in the Pacific case), at or just downstream of the block. That is, there are a *localized downscale* KE *cascade upstream of the block, and a localized upscale*

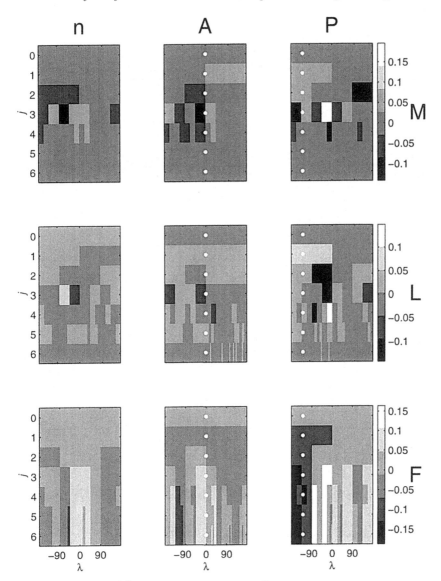

Fig. 7.14. $\{\langle (M, L, F)_j^{\mathrm{n,A,P}}(\lambda_{2l})\rangle\}$ (grayscale, W m^{-2}). The same grayscale value map is used for n,A,P. Rows are M, L, F, columns are nonblocking, Atlantic and Pacific blocking, respectively. Abscissae are longitude λ, ordinates resolution j, increasing downwards. The white dots indicate the block longitude $\lambda_{\mathrm{m}}^{\mathrm{b}}$.

KE *cascade downstream of the block.* In the Pacific case there are also localized downscale cascades further downstream of the block.

Still more detailed insight may be gained by inspecting the φ-dependence, but the reader also may skip to Section 7.5 now. Figures 7.15–7.20 show

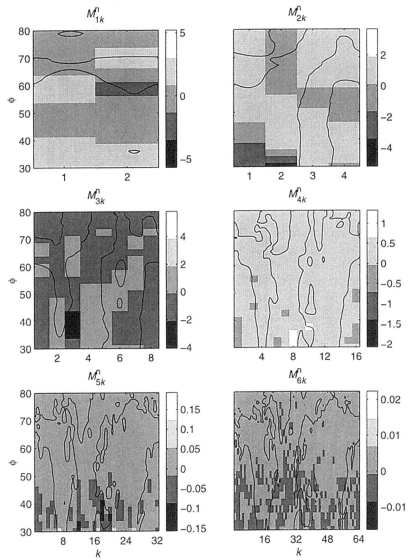

Fig. 7.15. Nonblocking $\langle M_{jk}^{\mathrm{n}} \rangle$ (grayscale, W m^{-2}), as a function of location k (abscissa) and φ (ordinate).

$\langle (M, L)_{jk}^{\mathrm{n,b}} \rangle$. The φ-averaging operation (7.53) applied to these figures yields the previous figures. In these figures the $\times$ indicates the block longitude $\lambda_{\mathrm{m}}^{\mathrm{b}}$, and block latitude, the latter estimated as the greatest φ_m for which

$$\left| Z^{\mathrm{b}}(\lambda_{\mathrm{m}}^{\mathrm{b}}, \varphi_m, 70\mathrm{kPa}) - Z^{\mathrm{b}}(\lambda_{\mathrm{m}}^{\mathrm{b}}, \varphi_{m-1}, 70\mathrm{kPa}) \right| < 1\,\mathrm{m} \tag{7.54}$$

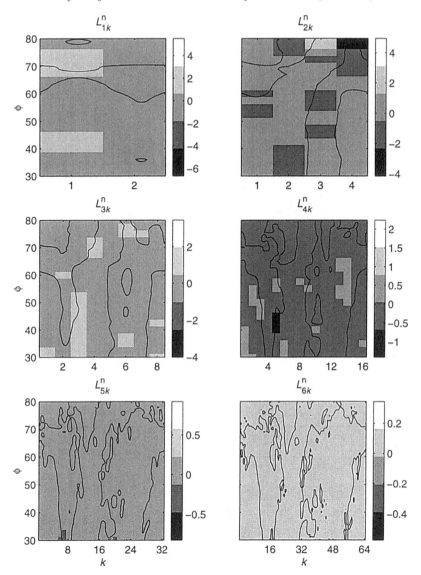

7.16. Nonblocking $\langle L_{jk}^{n} \rangle$, as in Fig. 7.15, only different grayscale value map.

which is a measure of the northernmost vanishing of the geostrophic wind zonal component.

To provide geographical reference and indicate the relevance of j, the continental boundary curves are drawn, following a resolution reduction $A_j \equiv A_0 + \sum_{j'=0}^{j-1} \tilde{A}_{j'}$ of the topography/bathymetry, which retains only the less resolved, $j' < j$ structure. The smoothed boundaries illustrate that location *and* shape are j-dependent in a multiresolution analysis. The

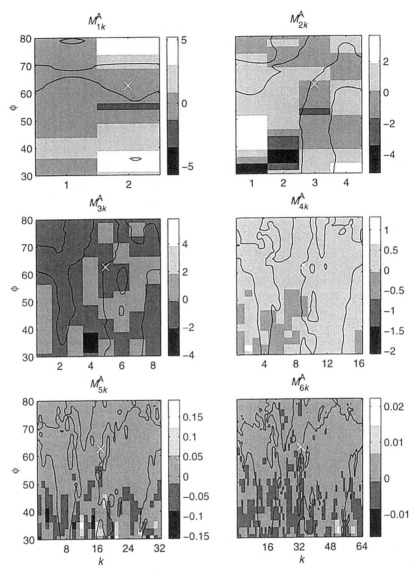

Fig. 7.17. Atlantic blocking $\langle M_{jk}^A \rangle$, as in Fig. 7.15, same grayscale value map. The $\times$ indicates the block longitude λ_m^b and latitude (7.54).

$(j, k) = (1, 2)$ gains seen in Figs. 7.10–7.11 from both mean and eddy flows are also similar in Figs. 7.17–7.18, which show their φ-dependence. The mean and eddy flows feed $(1, 2)$ north and south of the Atlantic block (Fig. 7.3 middle), around $\varphi = 75°$ N and $35°$ N (near the jet stream), while the mean also draws from $(1, 2)$ south of it, around $\varphi = 55°$ N. Although the losses from $(2, 2)$ just upstream of the block occur around the same φ in the south for both mean and eddy transfers, the eddy gain to $(2, 3)$ occurs in the north,

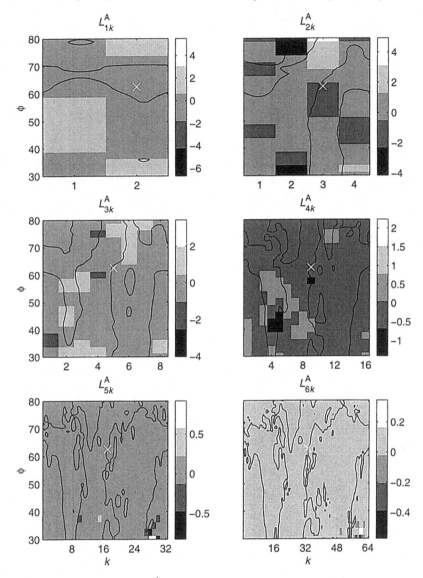

Fig. 7.18. Atlantic blocking $\langle L_{jk}^{\mathrm{A}} \rangle$, as in Fig. 7.16, same grayscale value map. The $\times$ indicates the block longitude $\lambda_{\mathrm{m}}^{\mathrm{b}}$ and latitude (7.54).

just downstream of an eddy loss. The $j = 3$, 4 scales show much spatial structure, with gains and losses side-by-side, especially for $\langle L_{jk}^{\mathrm{A}} \rangle$. *This zonal structure is not easily available to Fourier energetics analyses, since the latter discard phase information.* The φ-dependence of $\langle F_{jk}^{\mathrm{A}} \rangle$ (not shown) indicates that both the downscale and upscale cascades occur well north of the block.

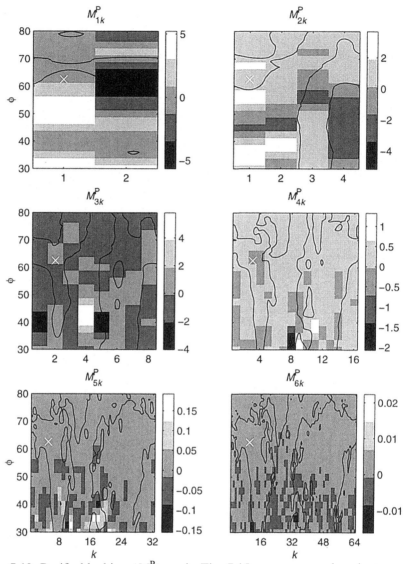

7.19. Pacific blocking $\langle M_{jk}^{\mathrm{P}} \rangle$, as in Fig. 7.15, same grayscale value map.

The φ-picture of $\langle (M, L)_{1k}^{\mathrm{P}} \rangle$ in Figs. 7.19–7.20 shows between $50°$ and $65°\,\mathrm{N}$ the same western gains as Figs. 7.12–7.13, but also reveals strong losses to eddies in the east, to the north of the block. Other sources and sinks, local in φ, are visible at $j = 3, 4$. Again in contradistinction to the Atlantic case, the cascades exhibited by $\langle F_{jk}^{\mathrm{P}} \rangle$ (not shown) are more to the south, further from the block.

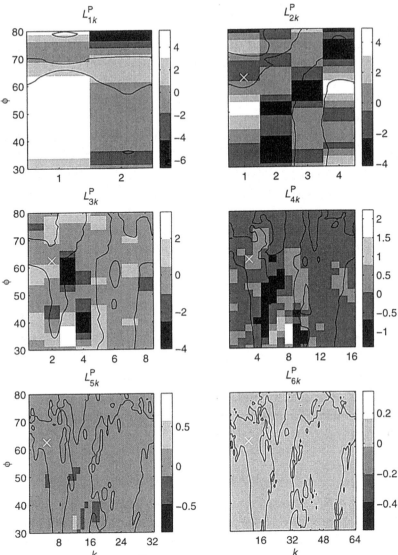

7.20. Pacific blocking $\langle L_{jk}^{P} \rangle$, as in Fig. 7.16, same grayscale value map.

7.5 Concluding remarks

Atmospheric blocking, like all nonlinear meteorological phenomena, involves the interaction of a range of scales. Blocks are localized structures, which are not well represented by truncated Fourier analysis, since the Fourier coefficients of λ-localized structures decay very slowly with increasing wavenumber n. This suggested to the author to translate the traditional Fourier analysis based atmospheric energetics [20] into a periodic orthogonal wavelet based energetics.

First of all, it was found that the second-largest scale (hemisphere) wavelet terms of a multiresolution analysis of geopotential height Z largely describe the state of blocking as defined here. The density and trajectory of these terms in phase space may provide clues to eventually understanding and predicting blocking.

The wavelet-indexed KE transfer and flux functions simultaneously represent activity localized in space and scale, an advantage over Fourier-based analogs, which represent global activity between scales. Applying this technique to the blocking and nonblocking data reveals new energetic characteristics of the phenomenon. In general, the eddy source term $\langle L_{jk}^{b} \rangle$ feeds KE to the block location at large scales, and extracts it from neighboring locations at intermediate scales. This supports the heuristic idea that blocks are maintained by an upscale KE energy cascade.

Depending on location, the transfer and flux to smaller scales can have either sign. Specifically, the analysis reveals downscale (upscale) cascades west (east) of the block.

The mean flow KE source ($\langle M_{jk}^{b} \rangle$) also generally feeds the block location at large scale. The Atlantic and Pacific cases were more dissimilar for $\langle M_{jk} \rangle$ than for $\langle L_{jk} \rangle$, so the former characteristic is more difficult to interpret.

This technique may be extended to include space-scale budgets for available potential energy, enstrophy, and other quantities. These calculations are in preparation.

Acknowledgements

Many thanks to B. Saltzman and A.R. Hansen for discussing extensions of their own works, to C.-F. Shih and G. Walters for data facilitation, to J. Buckheit and D. Donoho for WaveLab, to Hans van den Berg for editing. The author is supported by NSF-DMS-9420011 'Wavelet Modeling and Analysis of Atmospheric Dynamics and Energetics'.

References

[1] Benzi, R., Saltzman, B. & Wiin-Nielsen, A.C. (eds.) *Anomalous Atmospheric Flows and Blocking.* Adv. Geophys. 29. (Academic Press, San Diego, 1986).
[2] Blackmon, M. (1976) A climatological spectral study of the 500 mb geopotential height of the Northern Hemisphere, *J. Atmos. Sci.* **33**, 1607–23.
[3] Bolton, E.W., Maasch, K.A. & Lilly, J.M. (1995) A wavelet analysis of Plio-Pleistocene climate indicators: A new view of periodicity evolution, *Geophys. Res. Lett.* **22**, 2753–6.

[4] Celia, M.A. & Gray, W. G. (1992) *Numerical methods for differential equations*, chap. 2 (Prentice Hall, Englewood Cliffs).

[5] Daubechies, I. (1992) *Ten lectures on wavelets* (SIAM, Philadelphia).

[6] Dickson, R.R. (1979) Weather and circulation of February 1979— Near-record cold over the Northeast Quarter of the country, *Mon. Wea. Rev.* **107**, 624–30.

[7] Farge, M., Goirand, E., Meyer, Y., Pascal, F. & Wickerhauser, M.V. (1992) Improved predictability of two-dimensional turbulent flows using wavelet packet compression, *Fluid Dynamics Research* **10**, 229–50.

[8] Foufoula-Georgiou, E. & Kumar, P. (1994) *Wavelets in geophysics* (Academic Press, San Diego).

[9] Fournier, A. (1996) Wavelet analysis of observed geopotential and wind: Blocking and local energy coupling across scales, *Wavelet Applications in Signal and Image Processing IV*, ed. M.A. Unser, A. Aldroubi and A.F. Laine (SPIE, Bellingham).

[10] Fournier, A. and Stevens, D.E. (1996) Wavelet multiresolution analysis of numerically simulated 3d radiative convection, *Wavelet Applications III*, ed. H.H. Szu, (SPIE, Bellingham).

[11] Hansen, A.R. (1981). A diagnostic study of the spectral energetics of blocking, Iowa State University D. Phil. thesis.

[12] Hansen, A.R. & Sutera, A. (1984). A comparison of the spectral energy and enstrophy budgets of blocking versus nonblocking periods. *Tellus* **36A**, 52–63.

[13] Holton, J.R. (1992) *An Introduction to Dynamic Meteorology*, Third Edition (Academic Press, San Diego).

[14] Kanamitsu, M., Krishnamurti, T. & Depradine, C. (1972). On scale interactions in the tropics during northern summer, *J. Atmos. Sci.* **29**, 698–706.

[15] Lau, K.M. & Weng, H. (1995) Climate signal detection using wavelet transform: How to make a time series sing, *Bull. Amer. Meteor. Soc.* **76**, 2391–402.

[16] Lui, Q. (1994). On the definition and persistence of blocking, *Tellus* **46A**, 286–98.

[17] Mak, M. (1995) Orthogonal wavelet analysis: Interannual variability in the sea surface temperature. *Bull. Amer. Meteor. Soc.* **76**, 2179–86.

[18] Meneveau, C. (1991) Analysis of turbulence in the orthonormal wavelet representation, *J. Fluid Mech.* **232**, 469–520.

[19] Pandolfo, L. (1993) Observational aspects of the low-frequency intraseasonal variability of the atmosphere in middle latitudes, *Advances in Geophysics* **34**, 93–174.

[20] Saltzman, B. (1957) Equations governing the energetics of the larger scales of atmospheric turbulence in the domain of wave number, *J. Meteor.* **14**, 513–23.

[21] Saltzman, B. (1990) Three basic problems of paleoclimatic modeling: A personal perspective and review, *Climate Dyn.* **5**, 67–78.

[22] Steinberg, H.L., Wiin-Nielsen, A. & Yang, C.-H. (1971) On nonlinear cascades in large-scale atmospheric flow, *J. Geophys. Res.* **76**, 8629–40.

[23] Taubensee, R.E. (1979) Weather and circulation of December 1978–Record and near-record cold in the West, *Mon. Wea. Rev.* **107**, 354–60.

[24] Trenberth, K.E. & Olson, J.G., (1988) An evaluation and intercomparison of global analyses from NMC and ECMWF, *Bull. Amer. Meteor. Soc.* **69**, 1047–57.

[25] Wagner, A.J. (1979). Weather and circulation of January 1979–Widespread record cold with heavy snowfall in the Midwest, *Mon. Wea. Rev.* **107**, 499–506.

8

Wavelets in atomic physics and in solid state physics

J.-P. ANTOINE

Institut de Physique Théorique,
Université Catholique de Louvain, Belgium

Ph. ANTOINE and B. PIRAUX

Laboratoire de Physique Atomique et Moléculaire,
Université Catholique de Louvain, Belgium

Abstract

In the field of atomic and solid state physics, wavelet analysis has been applied so far in three different directions: (i) time–frequency analysis of harmonic generation in laser–atom interactions; (ii) *ab initio* electronic structure calculations in atoms and molecules; and (iii) construction of localized bases for the lowest Landau level of a 2-D electron gas submitted to a strong magnetic field. We survey these three types of applications, with more emphasis on methods than on precise results.

8.1 Introduction

There are two ways in which wavelets could play a role in atomic physics and possibly in solid state physics.

First one may envisage them as *physical objects*, namely quantum states or wave functions. It is commonplace to remark that coherent states (CS) have a privileged role in atomic physics. Laser–atom interactions, revival phenomena, Rydberg wave packets and various semi-classical situations are all instances in which a coherent state description is clearly well-adapted. Of course, what is implied here are *canonical* CS, associated to the harmonic oscillator or the electromagnetic field [36]. But wavelets are also coherent states, namely those associated to the affine groups in various space dimensions, as we have seen in Chapter 2 (see [1] for a review on coherent states). Thus wavelets could well be thought of as convenient substitutes for canonical CS. However, this suggestion is still speculative at the present moment, very little has been achieved in this direction.

Actually there have been so far only a few applications of wavelets in atomic physics and in solid state physics, and in all cases they were used as a *mathematical tool*. We will survey these applications in the present chapter, with more emphasis on methods than on actual results. More precisely, we will describe three main directions of research.

(1) When an atom is hit by a very intense, ultra-short laser pulse, it may emit light, in the form of harmonics of the incident electromagnetic field. The time profile of the emission spectrum reveals new information on the dynamics of this interaction process. Clearly, this time profile is out of reach of the traditional Fourier spectral methods, a time–frequency analysis is required. This phenomenon has received recently much attention, both experimentally and theoretically, although the numerical simulations are mostly limited to one-electron atoms. Notice that both the Continuous Wavelet Transform (CWT) and the Windowed Fourier Transform (just another name for canonical CS!) have been used.

(2) Both the extension of these phenomena to multi-electron atoms and the self-consistent electronic structure calculations (Hartree–Fock and generalizations) require the use of appropriate orthogonal bases for the description of the radial part of wave functions. Here (discrete) wavelet bases (or even frames) could adequately replace traditional plane waves or atomic orbital (LCAO) bases. The reason is that orthogonal wavelet bases with good localization properties will minimize the number of terms required for an accurate calculation of wave functions and related observable quantities. This program has been fulfilled in *ab initio* electronic structure calculations, in atoms, molecules and crystals. The crucial feature is the narrow support of the wavelets constituting an orthonormal basis. This is in fact the most active line of research with wavelets in solid state physics.

(3) Another application in solid state physics deals with a two-dimensional physical system, namely an electron gas submitted to a strong magnetic field, the set-up of the quantum Hall effect. Here wavelets yield various bases for the lowest Landau level, a necessary step for the description of the fractional quantum Hall effect.

Except for some marginal cases, these three topics are the most significant applications of wavelets in atomic and solid state physics. In our opinion, they are sufficiently promising to establish the credentials of wavelet methods in those fields of physics. Both the CWT, as a precise time-frequency analysis tool, and discrete (bi)orthogonal wavelet bases offer great potential for novel physical applications.

8.2 Harmonic generation in atom–laser interactions

8.2.1 The physical process

When an atom is exposed to a strong laser pulse, two competing processes may occur, the ionization of the atom and the emission of light. This emission process results from the oscillations of the atomic dipole at frequencies which are odd multiples (odd harmonics) of the driving field frequency (even harmonics are forbidden by parity conservation). Harmonic generation provides an efficient source of coherent soft X-rays [44], which explains the potential interest of the phenomenon for applications. Now for a full understanding of the emission mechanism, one would like to answer questions like: When are harmonics emitted during the optical cycle? What is the time evolution of the emission during the laser pulse? . . . . Clearly, this is beyond standard spectral methods, a time–frequency analysis is needed here.

Let us consider for simplicity the case of atomic hydrogen exposed to a strong laser pulse whose electric field is described classically as: $\mathbf{E} = \mathbf{E}_0(t) \cos \omega t$, where $\mathbf{E}_0$ is the pulse envelope. The atomic response to such a pulse is highly nonlinear, which leads to various unexpected phenomena. One of them is the emission by the atom of high order harmonics of the driving field. According to the semiclassical interpretation [18, 39], this harmonic generation results from the following two-step mechanism. The electron tunnels through the potential barrier formed by the combined Coulomb and electromagnetic (e.m.) fields. When it is outside, it is accelerated by the laser e.m. field and may be driven back to the residual ion. There, it may either be scattered, or recombine back into the ground state, emitting a harmonic photon. This interpretation is supported by quantum-mechanical models [41], in which the time-dependent dipole moment is expressed as the sum of the contributions from the electron trajectories in the continuum.

The resulting emission spectrum exhibits characteristic features which depend on the laser intensity. At low laser intensity, the spectrum of the emitted radiation decreases rapidly for increasing harmonic order, as expected. At high laser intensity, the spectrum changes drastically: after a rapid decrease for the first orders, it exhibits a long plateau, followed by a sharp cut-off. A spectacular example is given in Figure 8.1. Experimentally, harmonic orders as high as the 135th order have been observed with a Nd-Glass laser, which corresponds to a wavelength of 7.8 nm, i.e. in the soft X-ray (XUV) regime [43, 46]. The nonlinear character of the atomic response is manifested, for instance, by the fact that, in the spectrum of Figure 8.1, all the harmonics from the 69th to the 109th have almost equal intensity (hence the word 'plateau'), whereas the following ones drop dramatically.

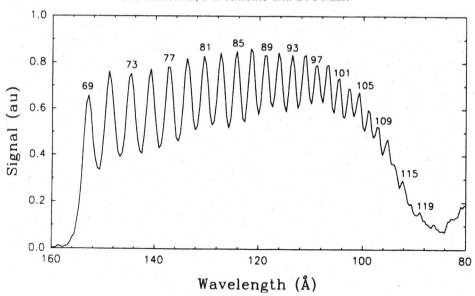

Fig. 8.1. Harmonic spectrum obtained in He for $\lambda = 1053$ nm and a laser intensity of 3×10^{14} Wcm^{-2}; in this case, the first harmonics up to about the order 67 are filtered out. Note the sharp cut-off around the order 113 (from [51]).

In the low intensity regime, the atomic response may be calculated by perturbation theory. But in the intense field regime, where the external field is of the same order or higher than the binding atomic field, nonperturbative methods are needed. A possible approach consists in solving (numerically) the time-dependent Schrödinger equation [38]. Atomic dipoles calculated by this method may be analysed with the standard Fourier spectral method, and the corresponding harmonic spectra exhibit the global features described above. However, a time–frequency analysis provides a deeper understanding of the mechanism: it allows to determine the time profile of each individual harmonic and from this one may deduce that harmonic emission takes place only when the electron is close to the nucleus.

8.2.2 Calculation of the atomic dipole for a one-electron atom

As mentioned above, the primary cause of harmonic emission is that the electron of the atomic hydrogen oscillates back and forth under the influence of the laser field, hence creates a dipole moment $d(t)$. Then, according to the Larmor formula, the energy radiated between frequencies ω and $\omega + d\omega$ is proportional to $|\widehat{a}(\omega)|^2$, where $\widehat{a}(\omega)$ is the Fourier transform of the dipole acceleration $a(t) = \ddot{d}(t)$. Therefore the problem consists in calculating (numerically) the acceleration $a(t)$ from the time-dependent Schrödinger equation:

$$i\frac{\partial}{\partial t}\psi(\mathbf{r}, t) = (H_{at} + \mathbf{A}(t) \cdot \mathbf{p})\,\psi(\mathbf{r}, t). \tag{8.1}$$

Here, H_{at} is the atomic Hamiltonian, $\mathbf{p}$ is the electron momentum, the vector potential $\mathbf{A}$ is written in the dipole approximation, i.e. $\mathbf{A}(\mathbf{r}, t) = \mathbf{A}(t)$ depends on time only, and is treated as a classical variable, and finally the quadratic term $\mathbf{A}^2$ has been gauged away. It is convenient to take the vector potential along the z-axis:

$$\mathbf{A}(t) = A_o(t)\,\sin\omega_L t\,\hat{\mathbf{e}}_z,$$

where $A_o(t)$ is the envelope of the pulse and ω_L the laser frequency. Notice that the shape of the pulse influences considerably the harmonic emission. In terms of the solution $\psi(\mathbf{r}, t)$ of the Schrödinger equation (8.1), the atomic dipole along the z-axis reads

$$d(t) = \langle \psi(\mathbf{r}, t) \mid z \mid \psi(\mathbf{r}, t)\rangle. \tag{8.2}$$

Using Ehrenfest's theorem, the corresponding acceleration may be written as:

$$a(t) = \ddot{d}(t) = -\langle\psi(\mathbf{r}, t) \mid \frac{z}{r^3} \mid \psi(\mathbf{r}, t)\rangle + \frac{\partial A_z}{\partial t}. \tag{8.3}$$

The next step is to expand $\psi(\mathbf{r}, t)$ in an appropriate basis. For obvious reasons, one takes spherical harmonics for the angular part. As for the radial part, a convenient choice is a Coulomb Sturmian basis $\{S_{nl}(r)\}$, because of its good convergence properties [50]. Notice that complex Sturmian functions are required, in order to reproduce the correct asymptotic behaviour (outgoing wave) of the wave function [34]. Thus we write:

$$\psi(\mathbf{r}, t) = \sum_{nlm} a_{nl}(t)\,\frac{S_{nl}(r)}{r}\,Y_l^m(\theta, \varphi). \tag{8.4}$$

Inserting the expansion (8.4) into the Schrödinger equation (8.1), one obtains a set of first order differential equations for the coefficients $a_{nl}(t)$, that can be readily solved numerically (but without approximation), to yield the dipole acceleration $a(t)$. A typical result is shown in Figure 8.2. The dipole acceleration presents a region of rapid oscillation, starting well before the maximum of the pulse (taken as the origin of time) is reached. This region corresponds to the generation of harmonics, as will be confirmed by the time–frequency analysis.

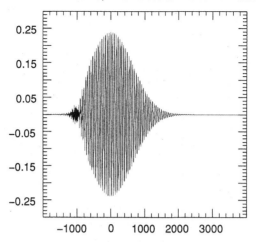

Fig. 8.2. Acceleration of the atomic dipole as a function of time (both in atomic units). This dipole is the result of the interaction of atomic hydrogen with a strong laser pulse of frequency $\omega_L = 0.118$ atomic units. The pulse envelope $A_o(t)$ is Gaussian and its full width at half maximum is 20 optical cycles. The peak intensity is 2×10^{15} Wcm^{-2}.

8.2.3 Time–frequency analysis of the dipole acceleration: H(1s)

Taking now the Fourier transform $\hat{a}(\omega)$ of the dipole acceleration $a(t)$, one obtains the power spectrum. Figure 8.3 shows two typical spectra. The left one corresponds to a rather low frequency e.m. field ($\omega_L = 0.047$ a.u.) and moderate intensity. The spectrum exhibits a large number of odd harmonics of the laser frequency, which form a long 'plateau', with a sharp cutoff beyond $\omega = 33\omega_L$. On the right, we show the spectrum corresponding to the high intensity pulse of Figure 8.2. Of course, no time localization is provided. When is each harmonic emitted? What is its time profile? Answering those questions requires a time–frequency analysis of the acceleration, as discussed in Chapter 1, section 1.1:

$$a(t) \mapsto \tilde{a}(\alpha, \tau) = \int_{-\infty}^{\infty} \overline{g_{\alpha\tau}(t)}\, a(t)\, dt. \tag{8.5}$$

In this expression $\alpha > 0$ is the scale parameter and τ the time parameter (usually denoted a and b, respectively, as in Chapter 1). Two types of time–frequency analysis have been used in the present problem, namely

- a Gabor transform, corresponding to $g_{\alpha\tau}(t) = e^{it/\alpha}\, e^{-\frac{1}{2}(t-\tau)^2}$
  ($g_{\alpha\tau}(t)$ is then a canonical coherent state)
- a wavelet transform, with a (truncated) Morlet wavelet $g(t) = e^{i\omega_o t}e^{-t^2/2}$ and
  $g_{\alpha\tau}(t) = \alpha^{-1/2}g(\alpha^{-1}(t - \tau))$.

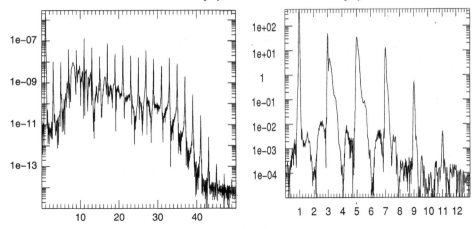

Fig. 8.3. (Left) Power spectrum (arbitrary units) as a function of harmonic order, in the case of the interaction of atomic hydrogen with a laser pulse of frequency $\omega_L = 0.047$ a.u. and (low) peak intensity 10^{14} Wcm^{-2}. The pulse has a 4 optical cycle sine-square turn-on and -off and a flat top of 16 optical cycles. (Right) Harmonic spectrum of the high intensity pulse of Fig. 8.2.

The information contained in the function $\tilde{a}(\alpha, \tau)$ may be exploited according to several different strategies. First, one may calculate the emission strength as a function of time, for a given frequency, that is, one evaluates $\tilde{a}(\alpha, \tau)$ for fixed α. One can also estimate the instantaneous frequency of emission of a given harmonic, as a function of time (τ). Alternatively, one may determine the full harmonic spectrum for a fixed time $\tau = t_o$, that is, consider $\tilde{a}(\alpha, t_o)$ as a function of α.

This technique has been applied successfully in the case of a hydrogen atom, both in its ground state and in the metastable state 2s, or a simplified model thereof (two-level atom), and considerable insight has been gained in the physical mechanism of harmonic generation. Let us give some details on the various aspects of the analysis.

8.2.3.1 Time dependence of harmonic emission in H(1s)

Let us begin with a hydrogen atom in its ground state. According to the semiclassical description, the so-called two-step model [18, 39], harmonic emission takes place only when the electron is close to the nucleus (see Section 8.2.1). Using a time–frequency analysis at fixed frequency, with a window whose bandwidth is smaller than $2\omega_L$, one is able to estimate the time profile of individual harmonics in the emitted radiation, as indicated by the following results.

(i) Time profile of harmonics: Choosing for α in (8.5) the inverse of a fixed odd multiple of the laser frequency, one obtains the time profile of the corresponding harmonic [3]. First, a Gabor analysis yields the global shape of each harmonic (Figure 8.4, left). Two interesting points are visible on this picture. First, the emission of each harmonic starts at a given characteristic time, which depends on the laser intensity. Then, as their order increases, the harmonics are emitted during shorter time intervals. This implies that the linewidth of higher harmonics should broaden with their order. The fact that harmonic emission stops before the field has reached its maximum amplitude (in $t = 0$) is due to the rapid excitation and ionization of the atom. This effect has been observed experimentally [52].

The picture may be refined by using a wavelet analysis (using a Morlet wavelet). Figure 8.4 (right) shows the time profile of the 9th harmonic of the same pulse. A fine structure appears, which is $2\omega_L$-periodic, in agreement with the semiclassical interpretation. Notice that, since the time resolution is better than half the optical period, the filter bandwidth is larger than $2\omega_L$. Therefore several harmonics are accepted simultaneously by the filter.

The same technique reveals also how the time profile depends on the position of the harmonic in the spectrum. Take for instance the case depicted in Figure 8.3 (left), which corresponds to a low frequency regime

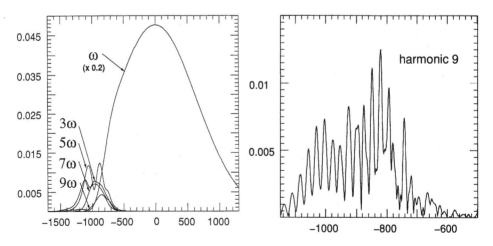

Fig. 8.4. (Left) Time profile (from Gabor analysis) of the odd harmonics $(1 - 9)$ produced as a result of the interaction of atomic hydrogen with the same laser pulse as in Fig. 8.2 (time is in atomic units). (Right) Time profile (from wavelet analysis) of the 9th harmonic emitted in the same pulse. Note that the frequency bandwidth of the pulse exceeds in this case $2\omega_L$.

($\omega_L = 0.047$ a.u.). The wavelet analysis shows that the time profile of harmonic 21, in the plateau (a), and that of harmonic 39, in the cutoff (b), are totally different (see Figure 8.5) [6]. The emission process is clearly more complex in the first case. In the case of a harmonic beyond the cutoff, one sees only one peak per half optical cycle. On the contrary, the time profile of harmonic 21 has two peaks for each half optical cycle. This behaviour is in agreement with the quantum-mechanical model of Lewenstein *et al.* [41], at least in the limit of high field and low frequency (the high frequency case is also interesting, but physically more complex). In the cutoff, there is only one electron trajectory contributing to the emission [42]. As a result, the decrease in the spectrum beyond the cutoff is very smooth and regular. By contrast, in the plateau where several electron trajectories lead to the same harmonic, the resulting interference leads to a highly structured spectrum as function of the harmonic order [42]. For instance, the presence of two peaks per half optical cycle in harmonic 21 reflects the existence of two return times contributing to the emission of the same harmonic, that is, two interfering electron trajectories. By analysing the harmonic emission in the time domain rather than in the frequency domain, the contributions of the different electron trajectories in the continuum are naturally separated [5]. In addition, the wavelet analysis also reveals a good agreement between the time-dependent Schrödinger equation model and the strong field approximation (SFA) model, for harmonics close to or beyond the cutoff [6].

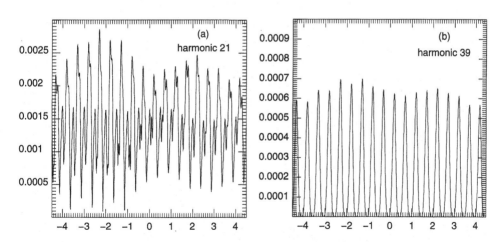

Fig. 8.5. (Left) Time profile (from wavelet analysis) of harmonic 21 (in the plateau) for the same conditions as in Fig. 8.3 (left); time is measured in optical periods. (Right) The same for harmonic 39.

(ii) Temporal control of harmonic emission: In the semiclassical description, the electron emits harmonics when it is close to the nucleus. Therefore, the harmonic emission may be controlled if one uses a laser beam with a time-varying polarization. When the polarization is linear, the electron oscillates back and forth, hence it comes periodically close to the nucleus, and harmonics are emitted. When the polarization is elliptic or circular, the electron stays far away and harmonic generation is suppressed. This polarization effect is demonstrated in Figure 8.6, which shows in parallel: (left) the time evolution of the harmonic 9, and (right) the projection of the full wave function on the bare 1s state of atomic hydrogen [4]. The latter measures the probability of the electron being close to the nucleus. Hence it oscillates when the field is linearly polarized, but remains constant in the circular or elliptic cases. As expected, the two curves are in perfect correspondence: the harmonic is totally suppressed when the polarization of the laser beam is circular and it reaches its maxima precisely when the polarization is linear.

In fact, this effect may be exploited further. It has been demonstrated on the basis of the strong field approximation (SFA), that the time profile of the harmonic emission consists of a regular attosecond pulse train [5] (1 attosecond $= 10^{-18}$s). By using a polarization which is linear during a very short period in place of a fixed polarization, it should be possible to select one of

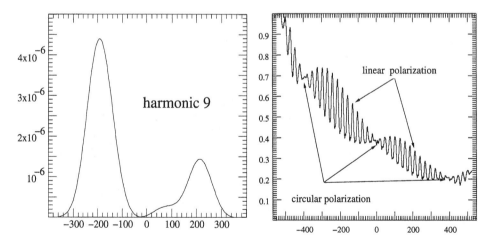

Fig. 8.6. The polarization effect. (Left) Intensity of the 9th harmonic as a function of time (in atomic units) in the case where the atom is exposed to two perpendicularly polarized laser pulses of 10^{14} Wcm^{-2}. The photon energies are 0.118 and 0.110 atomic units, respectively. Both pulses have a flat top and sine-square turn-on and -off. The total duration of the pulse is 20 optical cycles. (Right) Corresponding 1s population, measured by the projection of the full wave function on the bare 1s state of atomic hydrogen (from [4]).

these attosecond pulses of the train. This scheme opens the route to the production of a single intense attosecond pulse.

8.2.3.2 Harmonic emission in H(2s)

The same space localization effect of the harmonic generation takes place if the hydrogen atom is initially in the metastable state 2s [3]. When the latter interacts with a high intensity laser pulse, the system is excited into a linear superposition of many Rydberg states (mainly 8p, 9p and 10p). As a result, ionization is significantly suppressed and hence the atomic dipole does not vanish at the end of the interaction with the pulse. A similar situation occurs in H(1s), at low intensity, but the dynamics is much more complex now, because of the excitation of many atomic states. In order to get an insight into the time evolution of the process, we look again at the time profile of a typical harmonic (the third one), obtained by a Gabor analysis and shown in Figure 8.7(a). The curve shows two pronounced maxima. The left one corresponds to the emission of the harmonic. However the second one (around 1300 a.u.) is due to an atomic frequency which is almost degenerate with $3\dot{o}\omega_L$ and is present during the free evolution of the dipole after the interaction with the pulse. This interpretation is confirmed by an analysis of the population dynamics. We present in Figure 8.7(b) the time evolution of the 1s population (measured again by the projection of the total wave function on the 1s bare atomic state). This population is significant only at two moments: around 1000 a.u. before the maximum of the pulse and then again after

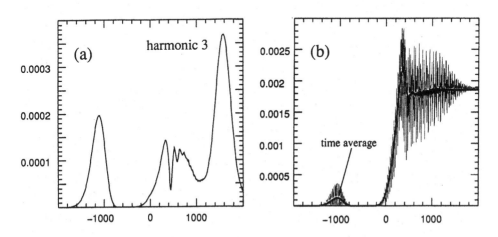

Fig. 8.7. (Left) Time profile (in atomic units) of the third harmonic emitted by atomic hydrogen initially in its 2s state and exposed to a Gaussian pulse of intensity 2×10^{14} Wcm^{-2} and width 20 optical cycles, and laser frequency $\omega_L = 0.118$ a.u. (Right) 1s population as a function of time for the same situation (from [3]).

interaction. In the first region, the 1s population oscillates with frequency $2\omega_L$ around its average, which corresponds to the back and forth oscillation of the electron through the central region of the atom, and the concomitant emission of the harmonic. As for the second region, it manifests a rapid exchange of population between the 1s state and excited states, with no harmonic emission.

8.2.3.3 *Instantaneous frequency of emission*

Actually, the frequencies of the harmonics are exact odd multiples of the frequency of the driving field only if the laser intensity remains constant. On the contrary, when the laser intensity increases, the frequencies are slightly above the exact multiples of ω_L (blueshift). To visualize that effect, we take a linear ramp as the pulse shape, which produces a long plateau, up to $41\omega_L$. On this spectrum, we perform a Gabor analysis rather than a wavelet one, in order to get a time resolution that is independent of the harmonic order. Then all the harmonics have the same behaviour: the Gabor coefficient grows rapidly in time ($\sim$ intensity) and then reaches a saturation intensity with some oscillations. For a given harmonic, the *instantaneous frequency* of emission is given by the time derivative $\frac{d}{d\tau}\phi(\alpha, \tau)$ of the phase $\phi(\alpha, \tau)$ of the corresponding Gabor coefficient. This notion is familiar in wavelet analysis, since it plays an essential role in the determination of spectral lines [20], but applies in the Gabor case as well. The computation fully confirms the effect: each harmonic is slightly blueshifted up to the moment (indicated by an arrow) when it reaches the saturation plateau (Figure 8.8) [54].

8.2.3.4 *Harmonic spectrum at fixed time*

Alternatively, one may choose a fixed time $\tau = t_o$ and determine the full harmonic spectrum. The computation has been performed on a two-state model for the atom, both with a Gabor analysis [24], and with a wavelet analysis [21]. The two methods yield very similar results, except that the Gabor spectrum contains more noise, which is due to the presence of the so-called hyper-Raman lines. For the wavelet case, two different wavelets have been used, the standard Morlet wavelet and another one, of compact support, which may have independent interest. This new wavelet is defined as:

$$F(t) = \begin{cases} \frac{1}{2}e^{it}[1 + \cos(t/N)], & t \in [-N\pi, N\pi]. \\ 0, & \text{otherwise} \end{cases} \tag{8.6}$$

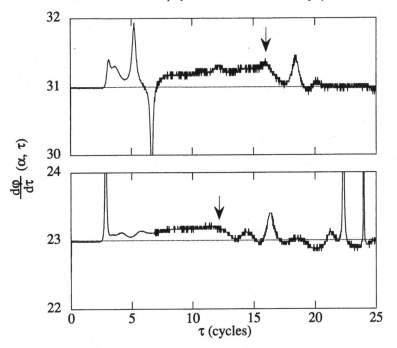

Fig. 8.8. Time derivative of the phase of the Gabor coefficients for $\alpha^{-1} = 23\omega_L$ (lower part) and $31\omega_L$ (upper part). The pulse is a linear ramp of 20 optical cycles and a maximum intensity of 10^{14} Wcm^{-2}. The arrows indicate the times when each harmonic reaches the saturation plateau (from [54]).

Here N is an integer, which gives the number of oscillations of $F(t)$ within its support (it is clear that one needs here a wavelet that oscillates rather fast, a Mexican hat, for instance, would be totally inadequate; typically one takes $N = 30$). Two examples of spectra are shown in Figure 8.9, taken respectively with an F wavelet (a) and a Morlet wavelet (b). As can be seen, there is no significant difference between the two spectra. This confirms that, here as in general, it is not the particular choice of the wavelet that matters, but rather the fact that its shape and the parameters chosen match closely the signal.

8.2.3.5 Which time–frequency method?

It is clear from this discussion that a time–frequency analysis is essential here. Harmonic generation is a highly nonstationary process, whose temporal evolution sheds much light on the actual physical phenomenon. This information is obviously inaccessible to the standard Fourier technique. In addition, as we pointed out already in Chapter 1, Fourier analysis is highly unstable with respect to perturbation, because of its global character. The remedy is to

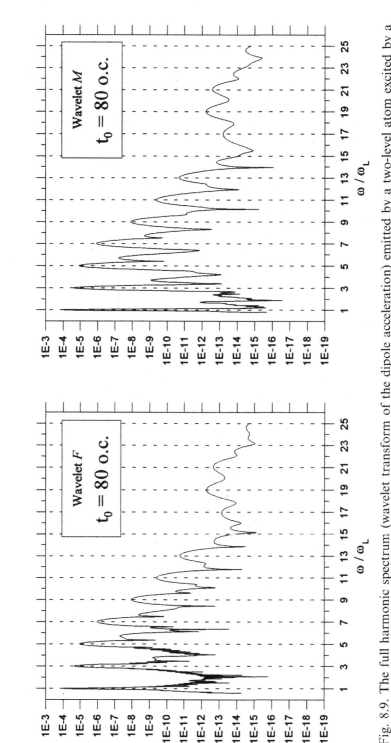

Fig. 8.9. The full harmonic spectrum (wavelet transform of the dipole acceleration) emitted by a laser pulse of intensity 8.8×10^{13} Wcm^{-2}, at time $t_0 \equiv \tau = 80$ optical cycles: (a) with the wavelet F of Eq. (8.6); (b) with a Morlet wavelet (private communication from S. De Luca).

represent the signal in terms of localized components, such as Gabor functions or wavelets.

Then a natural question is the choice between wavelets and a Gabor analysis. Globally, the two methods give similar results, provided the time resolution or the bandwidth of the analysing functions are identical. However, wavelet analysis is more appropriate if we want to study *simultaneously* the time profile of several frequencies, keeping the same analysing function. This results of course from the well-known property of wavelets, that $\Delta\omega/\omega$ is constant. As far as harmonic generation is concerned, a wavelet analysis is preferable in the following two cases: (i) for studying very high order harmonics with a time resolution better than the optical cycle, and (ii) when the atomic structure plays a more crucial role; in that case, hyper-Raman lines resulting from atomic transitions may occur in very short time intervals during the interaction with the pulse [3]. On the other hand, it may be inconvenient to make the time resolution depend on the frequency. For example, in order to demonstrate the slight blueshift of the harmonic frequency (Section 8.2.3.3), the time resolution has to be much larger than the optical period since the shift is much smaller than $2\omega_L$, but it should be the same for all harmonics.

So the answer to the question, wavelets or Gabor?, depends essentially on the physics of the problem at hand. The crucial choice is not that of a particular approach, Gabor or wavelets, or even that of a particular analysing function, what really matters is that the parameters of the analysing function be well-adapted to the signal. The case of the fixed time analysis of Section 8.2.3.4 is another confirmation. The conclusion is that *both* methods are needed in fact.

8.2.4 *Extension to multi-electron atoms*

The analysis of the interaction between a laser pulse and a one-electron atom, discussed in this section so far, rests on the possibility of solving numerically the time-dependent Schrödinger equation. The resulting wave function is then used to estimate various phenomena such as harmonic generation, as discussed above, but also multiphoton ionization, excitation of Rydberg wave packets, etc. In all these cases, the physical processes are complex and largely transient, so that a time–frequency analysis is necessary for a detailed description. Both the CWT and the Gabor analysis prove useful in this respect, as they provide otherwise inaccessible information that complements the traditional tools of atomic physics.

All that is even more true and interesting for two electron atoms, such as He or H$^-$. But now the situation is much more complicated, because of the correlation between the two electrons. It becomes very hard to solve directly the time-dependent Schrödinger equation and calculate a reasonable two-electron wave packet $\Psi(\mathbf{r}_1, \mathbf{r}_2, t)$, although a clever choice of coordinates may considerably alleviate the difficulty. Few results have been obtained so far. A case in point is a recent paper [40], which treats the case of harmonic generation and ionization from a one-dimensional two-electron He atom. Their conclusion is that, at least for a limited range of laser frequency and intensity, the harmonic generation spectrum in essentially the same as in the one-electron case (for higher intensity, double ionization becomes predominant). This result offers hope that the time–frequency method may be useful to a much wider range of physical situations than those analysed so far.

However, the full three-dimensional problem is still intractable. The difficulty is to find an appropriate basis for expanding the two-electron wave function, in such a way that numerical methods converge fast enough. In fact, only the radial part is subject to discussion, since one likes to keep the spherical harmonics for the angular part, so as to take advantage of the Racah algebra. One may think of several candidates for a good radial basis, such as generalized Sturmian functions, multiresolution-based orthogonal or bi-orthogonal wavelet bases, or even various kinds of frames, for instance those derived from Schrödinger coherent states. But in fact this a particular instance of a more general problem, namely the construction of multi-electronic wave functions, that we now discuss.

8.3 Calculation of multi-electronic wave functions

In any atomic process, the cross-section to be computed is proportional to $|\langle f \mid H_{int} \mid i\rangle|^2$, where H_{int} is the interaction Hamiltonian. The initial state $| i \rangle$ and the final state $| f \rangle$ are atomic states, possibly coupled to a continuum electronic state, in the case of an ionization process. As we have seen in Section 8.2, one may apply various approximations to the interaction Hamiltonian, such as the dipole approximation, the classical treatment of the e.m. field, etc. But the hard problem is to compute the initial and final wave functions. Many standard techniques are available for this purpose, such as variational methods, but the key ingredient is the expansion of the wave function into a suitable basis. Here again, the angular part will be described by spherical harmonics, only the radial part must be found. As mentioned above, a possibility is to use an orthogonal wavelet basis on $[0, \infty]$.

Of course, what we have here is a different use of wavelets, namely as convenient bases for expanding solutions of a partial differential equation. This is a familiar situation, for instance in the resolution of nonlinear PDEs [45]. The advantage of wavelets, as compared to usual bases, resides in their good localization properties. Let us go into some detail.

8.3.1 The self-consistent Hartree–Fock method (HF)

One of the best answers to such questions is given by the self-consistent methods, namely the well-known Hartree–Fock method and its descendants. We shall sketch it very briefly here in order to give some feeling for the technique and the possible use of wavelets for improving it. Further details may be found in standard textbooks, e.g. [15].

The basic idea is to obtain the wave function Ψ for an atom with N electrons by a variational computation

$$\delta \langle \Psi \mid H_{at} \mid \Psi \rangle = 0, \tag{8.7}$$

where H_{at} denotes the atomic Hamiltonian. The total antisymmetry of Ψ (Pauli principle) is enforced by taking it as a Slater determinant of one-electron wave functions:

$$\Psi = \frac{1}{N!} \begin{vmatrix} \phi_1(1) & \cdots & \phi_1(N) \\ \vdots & & \vdots \\ \phi_N(1) & \cdots & \phi_N(N) \end{vmatrix},$$

where $\phi_i(j)$ is a one-electron wave function ('orbital') of the jth electron. It has the form

$$\phi_i(\mathbf{r}, \text{spin}) = \frac{1}{r} P_{nl}(r) Y_l^{m_l}(\theta, \varphi) \chi_{m_s}, \tag{8.8}$$

where $Y_l^{m_l}$ is a spherical harmonic (thus allowing the Racah algebra for angular momentum), χ_{m_s} is a spin orbital, $P_{nl}(r)$ is an unknown radial function and $i \equiv (n, l, m_l, m_s)$ is a collective quantum number ($m_s = \pm \frac{1}{2}$; $m_l = -l, \ldots l$; $l = 0, 1, \ldots n-1$; $n = 1, 2, \ldots$). One usually imposes orthonormality conditions

$$\langle \phi_i \mid \phi_j \rangle = \delta_{ij}$$

$$\int_0^\infty P_{nl}(r) P_{n'l'}(r) \, dr = \delta_{nn'} \delta_{ll'}$$

and thus we rewrite the variational equation (8.7) with Lagrange multipliers $\{\lambda_{ij}\}$:

$$\delta \left\{ \langle \Psi \mid H_{at} \mid \Psi \rangle + \sum_{i<j} \lambda_{ij} \langle \phi_i \mid \phi_j \rangle \right\} = 0. \tag{8.9}$$

Inserting the ansatz (8.8) into (8.9), one obtains the so-called Hartree–Fock (HF) equations that determine the radial functions $\{P_{nl}(r)\}$:

$$\left\{ \frac{d^2}{dr^2} - \frac{l(l+1)}{r^2} + \frac{2}{r}[Z - Y(nl; r) - E_{nl}] \right\} P_{nl}(r) - \frac{2}{r}X(nl; r) =$$

$$= \sum_{n'} \lambda_{nln'l} P_{n'l}(r). \tag{8.10}$$

In this equation, Y is called the *direct* term and represents a local (spherical) potential, whereas the so-called *exchange* term X is nonlocal. These terms have the following form:

$$Y(nl; r) = \sum_{n'l'} \sum_k y_{nl,n'l'} \, Y^k(nl, n'l'; r)$$

$$X(nl; r) = \sum_{n'l'} \sum_k x_{nl,n'l'} \, Y^k(nl, n'l'; r) P_{n'l'}(r)$$

where $y_{nl,n'l'}$, $x_{nl,n'l'}$ are coefficients obtained from the angular momentum algebra and

$$Y^k(nl, n'l'; r) = \int_0^\infty \frac{r_<^k}{r_>^{k+1}} P_{nl}(r') P_{n'l'}(r') \, dr', \tag{8.11}$$

$$r_< = \min(r, r'), \quad r_> = \max(r, r').$$

Because of the exchange term X, (8.10) is a system of coupled, nonlinear equations in the unknown functions P_{nl}.

The usual technique consists in choosing for P_{nl} a simple trial function and solving the HF equation (8.10) by iteration. Typical choices are a Gaussian or a Slater function $r^{n-1}e^{-\alpha r}$. In general, many terms are needed in the expansion, which makes the method cumbersome or forces drastic truncations, thus leading to unrealistic results.

The alternative possibility that we consider here is to take instead for P_{nl} the elements of a suitable orthonormal wavelet basis. The idea is that the good localization properties of such bases may reduce considerably the number of terms needed. So far, however, this program is still purely speculative, even for two-electron systems. Yet essentially the same approach has been used successfully in the field of solid state physics, for calculating the electronic structure of various materials. We will discuss these results in the next section. Before that, we conclude our discussion of the self-consistent methods by indicating two possible extensions.

8.3.2 Beyond Hartree–Fock: inclusion of electron correlations

Clearly the HF approximation is too drastic and has to be improved by taking electron correlations into account. Atomic physicists have developed many methods to that effect, such as Configuration Interaction, the many-body perturbation theory, the R-matrix theory or the Multiconfiguration Hartree–Fock (MCHF) method [15]. Some of these methods are limited to correlations among discrete states, other ones include continuum states as well. In some cases at least (MCHF, for instance), the net result is to produce a wave function which is more concentrated around the origin, thus oscillates faster – and is therefore more likely to be well represented by wavelets. The computation usually requires a large number of terms. But again, considering the good localization properties of wavelets and their oscillatory behaviour, one may hope to reduce significantly the number of terms and accelerate the computation.

8.3.3 CWT realization of a 1-D HF equation

At the other extreme, a different wavelet method has been proposed recently [25] for studying the HF equation, albeit in a simplified one-dimensional version. The idea is to use the CWT in the same way as a Fourier or a Laplace transform, for obtaining a different realization of the differential equation to be solved.

One starts from the *radial* HF equation for the hydrogen atom, and extends it to $\mathbb{R}$ by antisymmetry, $x \mapsto -x$. Thus one obtains a 1-D differential equation:

$$-\frac{1}{2}f''(x) - \frac{f(x)}{|x|} = Ef(x), \quad f \in L^2(\mathbb{R}, dx). \tag{8.12}$$

Choosing as analysing wavelet the first derivative of a Gaussian, $w(x) \sim -x e^{-x^2/2}$, one takes the CWT of the two sides of (8.12), namely, with the usual notations ($a > 0, b \in \mathbb{R}$):

$$f(x) \mapsto \tilde{f}(a, b) \equiv \langle w_{ab} \mid f \rangle.$$

The result is a complicated integro-differential equation in the variables a and b, that one solves by iteration. The result is a marked improvement over the standard Slater or Gaussian inputs (the comparison is easy, since the exact solution is known!). One may of course object that the situation treated in this model is too simplified and bears little resemblance to the real physical problem, but the idea looks nevertheless interesting.

As a final remark, we may note that the same authors [26] have also studied the 1-D HF equation (8.12) by expressing it into an orthonormal wavelet basis and applying the fast wavelet transform [14]. Again their results compare favourably with those of a Slater basis. Actually this technique of exploiting the fast convergence of the wavelet algorithm in HF calculations has been used by several authors, as we will see in Section 8.5.3.

8.4 Other applications in atomic physics

8.4.1 Combination of wavelets with moment methods

For concluding this survey of possible applications of wavelets to atomic physics, we discuss briefly a new method for calculating energy levels in atoms, based on a clever combination of the WT with the well-known method of moments [32, 33].

In the simplest case, consider a one-dimensional Schrödinger eigenvalue equation, with a potential $V(x)$:

$$-\frac{d^2}{dx^2}\Psi(x) + V(x)\Psi(x) = E\Psi(x). \tag{8.13}$$

The wavelet transform of $\Psi(x)$ with respect to the Mexican hat wavelet $\psi(x) = (1 - x^2)\exp(-1/2x^2)$ reads:

$$W\Psi(a, b) = Na^{-1/2} \int_{-\infty}^{+\infty} \Psi(x + b)(1 - (x/a)^2)e^{-x^2/2a^2}\, dx. \tag{8.14}$$

Introducing the moments

$$\mu_{b,\gamma}(p) \equiv \int_{-\infty}^{+\infty} x^p\,\Psi(x + b)e^{-\gamma x^2}\, dx, \quad p \geq 0, \tag{8.15}$$

one may express the WT (8.14) as a linear combination of moments:

$$W\Psi(a, b) = N(2\gamma)^{1/4}[\mu_{b,\gamma}(0) - 2\gamma\mu_{b,\gamma}(2)], \quad \text{with } \gamma \equiv 1/2a^2. \tag{8.16}$$

On the other hand, these moments satisfy a first order differential equation in γ:

$$\frac{\partial}{\partial\gamma}\mu_{b,\gamma}(p) = -\mu_{b,\gamma}(p + 2). \tag{8.17}$$

The crucial point is that, if the potential $V(x)$ is a rational fraction (or may be transformed into one), then all the moments are linear combinations of the first $(1 + m_s)$ among them, the so-called initialization or missing moments. Inserting the corresponding expression into (8.17), one obtains a coupled set of linear differential equations in γ:

$$\frac{\partial}{\partial \gamma} \mu_{b,\gamma}(i) = \sum_{j=0}^{m_s} M_{ij}(\gamma, b, E) \, \mu_{b,\gamma}(j), \ 0 \le i \le m_s. \tag{8.18}$$

These equations may be integrated numerically, provided one has the value E of the ground state energy and the starting values $\mu_{b,0}(p)$, corresponding to the infinite scale limit $a = \infty$. Precisely, the Eigenvalue Moment Method (EMM), developed by Handy, Bessis and Morley [31], is able to give these initial data. Thus, the EMM yields the wavelet transform $W\Psi(a, b)$ of the unknown wave function $\Psi(x)$, which may then be obtained by a standard reconstruction formula (see Eq.(1.10) in Chapter 1, and more generally any textbook on wavelets, e.g. [19]).

This method, or variants thereof, has been applied in [32, 33] to the computation of energy levels and wave functions for a variety of one-dimensional potentials: the quartic anharmonic oscillator, the rational fraction $V(x) = gx^6(1 + \lambda x^2)^{-1}$, the Coulomb potential. In addition, the method probably extends to two or three dimensions. The results obtained in 1-D are reasonably good (the precision may be increased with more numerical effort), but, more important, this method introduces a totally new idea in the wavelet picture, which once again consists in combining wavelet techniques with existing methods. As such, this technique offers interesting perspectives for the future.

8.4.2 Wavelets in plasma physics

Since the physics of plasmas may be considered as a branch of atomic physics, it is appropriate to mention here an innovative application of the CWT to the analysis of intermittency in fusion plasmas [56]. Once again, the interesting methodological point is the combination of wavelet methods with a standard technique, in this case, bispectral methods. More precisely, the notion of bicoherence, which is a measure of the amount of phase coupling that occurs in the signal. Typically one studies the so-called wavelet bispectrum of a given signal $s(t)$, namely the function

$$B^W(a_1, a_2) = \int \overline{S(a, \tau)} \, S(a_1, \tau) \, S(a_2, \tau) \, d\tau, \tag{8.19}$$

the integral being taken over a finite time interval $\tau_1 \leq \tau \leq \tau_2$, and the following frequency sum rule holds

$$\frac{1}{a} = \frac{1}{a_1} + \frac{1}{a_2}. \tag{8.20}$$

The new point here consists in using the CWT of the signal in this definition, instead of its Fourier transform. Two remarks are in order here. First, this technique requires a complex wavelet, such as the Morlet wavelet, as always when phase information is essential. Second, the sum rule (8.20) can only be enforced if all frequencies are available, that is, if the continuous WT is used. Dyadic frequencies cannot in general satisfy the relation.

Using this tool in statistical analysis, one may detect the presence of intermittency and structure in the turbulent fusion plasmas [56]. A comprehensive description of this approach is contained in Chapter 6 of the present volume.

8.5 Electronic structure calculations

8.5.1 Principle

A basic problem in condensed-matter physics is the *ab initio* calculation of the electronic structure of a given material (ground state energy, wave function, etc.) Now, since the crystal is a 3-D periodic structure, it suffices to describe the electronic structure around a single lattice site and apply a Bloch transformation. Thus one comes back to the study of the electronic structure of a single atom or molecule, that is, to the problem discussed in Section 8.3. As explained there, the key is to find a good orthonormal basis, consisting of functions well adapted to the problem. Two standard methods are popular among chemists.

- *LCAO bases* (Linear Combination of Atomic Orbitals), based for instance on Slater or Gaussian orbitals. The method yields a good description of the electronic structure with relatively few terms, but it is difficult to improve it systematically (improving the LCAO method is sometimes described as an art!). In addition the expression for the forces is extremely complicated.
- *Plane wave bases.* These two difficulties disappear, but *a priori* plane waves are not suited for describing localized objects, since all information on space localization is lost. Yet the electronic structure of an atom is highly inhomogeneous in space, the wave function oscillates much more rapidly close to the nucleus. As a consequence a large number of terms is needed for describing the small inner region, but this increased precision is not necessary elsewhere.

In order to combine the advantages of the two methods, one should use a basis of localized functions, that allows to vary the precision according to the local electronic density. That suggests a wavelet basis, since wavelets are well localized, and adapt automatically to the scale of the object to be represented (the so-called automatic zoom effect). Several research groups have performed such an analysis, with different types of discrete wavelets, and we shall quickly review them.

8.5.2 A non-orthogonal wavelet basis

The first attempt [16] was based on a non-orthogonal multiresolution basis and its approximation by Gaussians and Mexican hats. The idea is to consider a multiresolution scale in three dimensions, $\{V_j; j \in \mathbb{Z}\} \subset L^2(\mathbb{R}^3)$, as described in Chapter 1, Section 1.3, but *without* the assumption of orthogonality:

$$V_{j+1} = V_j \dot{+} W_j, \tag{8.21}$$

where $\dot{+}$ denotes a direct sum, not necessarily orthogonal (hence the decomposition in (8.21) is still unique). The interpretation is the usual one: V_j describes the approximation at resolution 2^j, and W_j the additional details needed for passing from the resolution 2^j to the finer resolution 2^{j+1}. Thus one gets, as in Eq.(1.17):

$$L^2(\mathbb{R}^3) = V_{j_o} \dot{+} \left(\sum_{j \geq j_o} W_j \right), \tag{8.22}$$

where j_o corresponds to the lowest resolution considered and $\sum$ denotes again a direct sum. Then one chooses two orthogonal bases, localized around the nodes of a fixed 3-D lattice:

$$\{\phi_{j\mathbf{n}}(\mathbf{r})\} \in V_j \quad \text{and} \quad \{\psi_{j\mathbf{n}}(\mathbf{r})\} \in W_j, \tag{8.23}$$

where $\mathbf{n}$ denotes a lattice point and $j \in \mathbb{Z}$. From (8.22), the practical expansion of a general wave function into these bases reads:

$$f(\mathbf{r}) = \sum_{\mathbf{n}} a_{j_o\mathbf{n}}\phi_{j_o\mathbf{n}}(\mathbf{r}) + \sum_{j=j_o}^{j_{max}} \sum_{\mathbf{n}} d_{j\mathbf{n}}\psi_{j\mathbf{n}}(\mathbf{r}) \tag{8.24}$$

In this truncated expansion, $\phi_{j_o\mathbf{n}}$ is a scaling function, $\psi_{j\mathbf{n}}$ is a wavelet, both centred around the lattice point $\mathbf{n}$, and j_{max} is the finest resolution, corresponding to the desired precision. This expansion still has an infinite number of terms, since $\mathbf{n}$ runs over the whole lattice. However, both $\phi_{j\mathbf{n}}$ and $\psi_{j\mathbf{n}}$ are supposed to be well localized, so one has to keep only those functions which have significantly large coefficients $a_{j_o\mathbf{n}}$, resp. $d_{j\mathbf{n}}$, for the problem at hand. This means that one allows different resolutions for different localized regions. In particular, since the electronic wave function oscillates more rapidly in the atomic core region, one should add higher resolution scales j in the core region, where the precision must be higher, but only there. Globally the number of terms is thus considerably reduced.

The technique introduced in [16] for practical calculations runs as follows.

- One starts at resolution $j_o = 0$ with a cubic lattice L_0, of lattice spacing d_o small enough (the final basis should be a sufficiently tight frame). Then, for each resolution $j = 1, 2 \ldots, j_{max}$, one considers the refined lattice L_j of spacing $2^{-j}d_o$. The successive basis functions will be localized on the nodes of the finest lattice $L_{j_{max}}$, with no overlap between different resolutions. Notice that this lattice is fixed in space, independently of the position of the atomic nuclei.
- In order to adapt the resolution to the electronic density, one draws around each nucleus a sequence of concentric spheres S_j of radii $r_j = 2^{-j}r_0$, $j = 0, \ldots, j_{max}$. Then the finite (approximate) basis consists of the following functions: (i) scaling functions $\phi_{0\mathbf{n}}$ centred on the nodes of L_0 localized inside S_0; (ii) wavelets $\psi_{0\mathbf{n}}$ on the nodes of L_0 localized inside S_1, (iii) wavelets $\psi_{1\mathbf{n}}$ on the nodes of $L_1 \setminus L_0$ localized inside S_2; and so on until $j = j_{max}$. This arrangement is schematized in Figure 8.10 in a 2-D version.

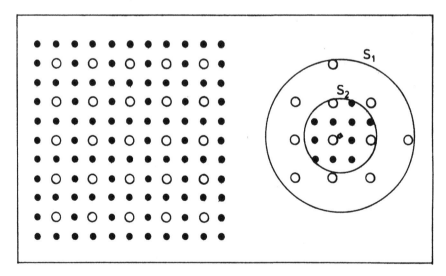

Fig. 8.10. The geometrical arrangement of [16] in a 2-D version. (left) The lattice L_0, corresponding to the resolution $j = 0$ (open circles) and the nodes of $L_1 \setminus L_0$, corresponding to $j = 1$ (black circles). (right) The spheres S_1 and S_2, both centred on an atomic nucleus (lozenge), and containing centres of wavelets with $j = 0$ and $j = 0, 1$, respectively.

The net result is a highly flexible finite basis, with higher resolutions introduced only where they are needed, namely close to the nuclear cores.

The next problem is to choose basis functions which are both efficient and simple to work with. Cho *et al.* [16] use Gaussians for $\phi_{0,\mathbf{n}}$ and Mexican hat wavelets for $\psi_{j,\mathbf{n}}$. Strictly speaking, this choice is not allowed, since these functions do not generate an orthogonal basis, but only a tight frame [19]. However they give an extremely good approximation to such a basis, as results from the following argument [9].

The crucial input in the construction of a wavelet basis is the pair m_0, m_1 of 2π-periodic functions defining the two-scale relations, for the scaling function and the wavelet, respectively (see [19] and Chapter 1, section 1):

$$\widehat{\phi}(2\omega) = m_0(\omega)\widehat{\phi}(\omega), \quad \widehat{\psi}(2\omega) = m_1(\omega)\widehat{\phi}(\omega) \tag{8.25}$$

(in the orthogonal case, one usually takes $m_1(\omega) = \overline{m_0(\omega + \pi)}e^{i\omega}$). According to [17, Theorem 5.16], the decomposition (8.21) is unique (but without orthogonality in general) iff

$$m_0(\omega)m_1(\omega + \pi) - m_1(\omega)m_0(\omega + \pi) \neq 0, \quad \text{for all } \omega \in [0, 2\pi). \tag{8.26}$$

Cho *et al.* [16] choose the following functions :

$$m_0(\omega) = \sum_{j\in\mathbb{Z}} \exp\left(-\frac{3}{2}\sigma^2(\omega - 2\pi j)^2\right),$$

$$m_1(\omega) = -e^{i\omega}\left(\omega - 2\pi\left[\frac{\omega + \pi}{2\pi}\right]\right)^2 m_0(\omega),$$

where [.] denotes the integer part function. It is readily seen by direct computation that these functions satisfy the criterion (8.26), thus guaranteeing a unique decomposition (8.21). Furthermore, and this is the interesting point, for $\sigma = 1.35$, m_0 and m_1 generate a scaling function $\phi(x)$ and a 'wavelet' $\psi(x)$ that match, respectively, a Gaussian $G(x)$ and a Mexican hat $-G''(x)$ within an absolute error of 10^{-10}. This is indeed a good approximation, which tends to confirm the practical efficiency of the tight frame based on the Mexican hat wavelet.

As a test, the method is applied to the hydrogen atom [16]. The best result is that the (known) ground state radial wave function is reproduced within 0.3% with an 85 function basis ($j_{max} = 1, r_o = 2$ a.u.). What is more convincing, the ground state energies of all elements, from hydrogen to uranium, may be estimated within 3% using a single 67 function basis, with $j_{max} = 10$. Next the total energy of a hydrogen molecular ion H_2^+ is calculated as a function of the separation R between the two protons. With a basis of 141 to 167 functions (depending on R), the exact values are reproduced within

1%. Finally, the analysis is extended [8] to the full carbon atom, in the local density approximation [37], and it yields again rather accurate energies and wave functions.

The conclusion is that a (quasi-)wavelet basis offers a very good alternative to standard methods for calculating the electronic structure of atoms or molecules, because it allows to vary the spatial resolution with space, contrary to the LCAO or plane wave methods. The analysis extends immediately to periodic systems by introducing a Bloch transformation. But, of course, the precision obtained in this approach is not entirely satisfactory.

One reason is the conflict between the spherical geometry of the atom and the Cartesian geometry of the lattice used in the wavelet expansion. Some progress in this direction has been made in the recent work of the MIT group [10], with a more isotropic scheme based on the so-called *interpolets* and combining wavelets with finite element methods (see also [7]). Another improvement, due to the same group, consists of combining wavelets with traditional multigrid methods, which also allow to vary the spatial resolution. Thus once again, the lesson is that wavelets yield optimal results when they are combined with standard methods, well adapted to the problem at hand (and usually the result from a long practice).

8.5.3 Orthogonal wavelet bases

Instead of the non-orthogonal scheme of [16], several groups have considered genuine orthonormal wavelet bases, e.g. Daubechies or Meyer bases. Once again, the key to efficiency lies in a clever selection of the most significant expansion coefficients.

8.5.3.1 Diagonalizing the LDA Hamiltonian in a Daubechies basis

Wei and Chou [57] essentially diagonalize the Hamiltonian in the local density approximation (LDA [37]), based on the Kohn–Sham equation (instead of the Schrödinger equation) :

$$\left(-\frac{\hbar^2}{2m}\Delta + V[\mathbf{r}; \rho(\mathbf{r})]\right)\psi_n(\mathbf{r}) = \epsilon_n\psi_n(\mathbf{r}), \tag{8.27}$$

where $\rho(\mathbf{r}) = \sum_n f_n|\psi_n(\mathbf{r})|^2$ is the local electronic density and $V[\mathbf{r}; \rho(\mathbf{r})]$ an effective potential. The idea is to expand the solution of the effective equation (8.27) in a suitable wavelet basis, so as to get a small number of significant coefficients (the same numerical exigence underlies the solution of the Schrödinger equation described in Section 8.2). The authors of [57] choose a Daubechies D6 wavelet basis of compact support, which generates a 3-D orthonormal basis by a threefold tensor product. Physical quantities are then calculated in this basis, using fast wavelet transform algorithms. For instance, they evaluate matrix elements of the Hamiltonian, $\int \psi_l H \psi_m d^3 x$, where ψ_l stands for the wavelet basis vectors, and in particular potential terms $\int \psi_l U \psi_m d^3 x$, where U may be a local potential (ionic, Hartree or exchange-correlation) or a nonlocal separable pseudopotential. The grid is the standard multiresolution grid, and its position relative to the centres of the atoms is essentially irrelevant. As compared to [16], no artificial cutoff is put by hand, the rapid convergence of the calculation follows from the smallness of the compact support of the D6 wavelet, which yields sparse matrices.

With this technique, one computes [57] the total energies for a (fictitious) hydrogen atom and for a dimer molecule, H_2 or O_2 (as a function of the bond length). In all cases, the method requires relatively few basis functions (1000 or 2000) as compared to the plane wave method ($\sim$ 8700), with similar results. Thus it offers hope for a serious improving of standard calculations. One may remark, however, that the plane wave calculation that serves for comparison has not been optimized (for instance, the energy cutoff of 100 Ry on the pseudopotential is abnormally high, and thus leads to a larger number of terms than necessary), so that the improvement reported may be smaller than claimed. Also the case of small diatomic molecules is specially unfavourable for a plane wave basis (too much void in the supercell). It is plausible that the two methods would have similar performances for larger molecules [28].

8.5.3.2 Molecular dynamics algorithm in a Daubechies basis

Tymczak and Wang [55] also use a 3-D Daubechies wavelet basis for performing electronic structure calculations in a local density approximation, starting again from the Kohn–Sham equation (8.27). However, instead of trying to squarely diagonalize the Kohn–Sham Hamiltonian in the wavelet basis, as the previous authors, they resort to a standard algorithm of molecular dynamics (Car–Parrinello) to obtain the eigenfunctions iteratively (this algorithm is similar to the dynamic simulated annealing method familiar in statistical mechanics). Of course, as in the other approaches, the key point for combining speed and precision of the calculation is to select adequately the most significant wavelet coefficients to be kept. In order to do so, one may exploit the self-similar behaviour of wavelet coefficients: from each scale to the next finer one, all coefficients are multiplied by a common small factor (which gets smaller for an increasing number of vanishing moments of the wavelet). Hence those coefficients that are negligible at a given scale lead to negligible ones at finer scales, no significant coefficients reemerge. Then, combining the selection of significant wavelet coefficients and the Car–Parrinello algorithm at each successive scale, from coarse to fine, one obtains fast convergence to the approximate eigenvalues and reasonably good values for the latter. Both the convergence and the compression rates (percentage of coefficients being kept) increase with the grid size, and so does the advantage of the method over the conventional plane wave approach. Only simple systems are treated in [55], namely the 3-D harmonic oscillator, the hydrogen atom, and the LDA to the helium atom and the hydrogen dimer H_2. However, the method seems powerful enough for attacking real multi-electron systems.

8.5.3.3 Galerkin method in a Meyer basis

Yamaguchi and Mukoyama [58] solve the *radial* Schrödinger equation by a variational (Galerkin) method, using an effective one-electron local potential and extending the equation to $\mathbb{R}$ by antisymmetry $x \mapsto -x$, as in Section 8.3.3. In order to formulate the variational equations (see Section 8.3), they expand the wave function into an antisymmetrized Meyer wavelet basis (C^∞, symmetric, all moments vanishing, but noncompact support), keeping only the most significant terms, as usual. By this technique they compute energy eigenvalues and wave functions for hydrogen, neon and argon atoms, and continuum wave functions (corresponding to pseudostates) for the argon atom. As a test of the numerical quality of their wave functions, they calculate, respectively, radiative transition rates in neon and argon atoms, and partial photoionization rates for an argon atom in its 1s, 2s and 3s state. The results are in excellent agreement with those obtained by the traditional Hartree–Fock–Slater method. However, one may notice that a large number of basis vectors is necessary for a good precision (475 for Ar), especially for the excited states (3s, 3p), which are much less localized.

An interesting remark is that each type of atomic state (lower bound state, Rydberg state, continuum state) has a characteristic distribution in the wavelet parameter space. This permits one to choose for each kind of physical process an adequate trial function before performing the variational procedure.

8.5.4 Second generation wavelets

The Daubechies wavelets used in [55] and [57] have compact support, which is numerically convenient, but they are not very smooth, and this becomes a drawback when it comes to solving differential equations, for instance diagonalizing a Laplacian. An elegant way of avoiding this problem is to use *biorthogonal* wavelets, which offer the most flexible version of wavelet techniques, and are also the most widely used in the wavelet community.

In the case of quantum physics applications, this step was made very recently in a paper by Goedecker and Ivanov [27], in which they treat the full Coulomb problem by pure wavelet methods. Namely they solve the Poisson equation $\Delta V = 4\pi\rho$, by expanding both the potential V and the electronic density ρ in a biorthogonal basis, in the present case a second generation wavelet basis (8th order lifted Lazy wavelet) [53]. Varying the resolution according to the electronic density, as in [16], and using BCR fast wavelet algorithms [14], they obtain rather spectacular results, definitely improving upon [57]. For instance, they are able to treat the potential arising from a fully 3-D all-electron uranium dimer. To give an idea of the power of the method, this problem involves length scales that differ by more than 3 orders of magnitude, and so does the potential. The resolutions involved differ by 7 orders of magnitude, and the potential is obtained with 6 significant digits throughout the whole region. As far as we know, this is the most successful application so far of wavelet methods in an atomic structure calculation.

8.6 Wavelet-like orthonormal bases for the lowest Landau level

As mentioned already, the electronic structure calculations described in the previous section give information on the bulk properties of solids, via a Bloch transformation [10]. Besides these calculations, wavelets have found applications in two other problems of condensed-matter physics.

The first one is a striking similarity between wavelets and Wannier functions of a 1-D crystal [35]. The context is the study of *inflation*, which means the following. Any one-dimensional periodic system of period a may be viewed as a $2a$-periodic system. The question is, how does the dynamics change? In particular, how do Bloch and Wannier functions transform under inflation? It turns out that both types of functions obey two-scale relations, characteristic of multiresolution wavelets [17, 19]. In particular, the Wannier functions of a free electron in the 1-D periodic system coincide with the Littlewood–Paley wavelets (see Section 8.6.3.2). It remains to be seen whether this is a mere curiosity or physically useful information.

The other application pertains to a 2-D electron gas submitted to a strong magnetic field, that is, the system in which the Quantum Hall Effects (integer or fractional) take place. We will devote the rest of this section to this problem and the promising role of wavelets in that context.

8.6.1 The Fractional Quantum Hall Effect setup

The system considered in the Fractional Quantum Hall Effect (FQHE) is a (quasi)-planar gas of electrons in a strong magnetic field perpendicular to the plane (see [48] for a review and the original references). The first problem to tackle is to find the ground state of the system. As in the electronic structure calculations described in Section 8.5, the key physical parameter is the electron density, which is measured by the so-called *filling factor* v. As shown in [11], good energy values are obtained for small ($0 < v < \frac{1}{5}$) or high ($\frac{4}{5} < v < 1$) electron densities with a Hartree–Fock description of a system of N two-dimensional electrons.

The first step in the HF procedure is to select an adequate wave function for a single electron in the magnetic field. As it is well known [48], the energy levels, the so-called Landau levels, are infinitely degenerate, and there arises the problem of finding a good basis in the corresponding Hilbert subspace. In particular, the ground state belongs to the lowest Landau level (LLL). A general method has been proposed for constructing an orthogonal basis for the LLL, starting from standard 1-D orthogonal wavelet bases [2, 12, 13]. We will describe this construction below, but we shall first recall the physical background of the problem.

Consider a single electron confined in the xy-plane and subjected to a strong magnetic field in the z-direction. In the symmetric gauge, the Hamiltonian reads (we use units such that $\hbar = M = e|\vec{H}|/c = 1$):

$$H_o = \frac{1}{2}(p_x - y/2)^2 + \frac{1}{2}(p_y + x/2)^2. \qquad (8.28)$$

Introducing the canonical variables

$$P' - p_x - y/2, \quad Q' - p_y + x/2, \qquad (8.29)$$

this can be written in the form of the Hamiltonian of a harmonic oscillator:

$$H_o = \tfrac{1}{2}(Q'^2 + P'^2). \qquad (8.30)$$

Therefore the eigenstates of the Hamiltonian (8.28) can be found explicitly, and they have the following form:

$$\Phi_{mn}(x, y) \sim e^{(x^2+y^2)/4}(\partial_x + i\partial_y)^m (\partial_x - i\partial_y)^n e^{-(x^2+y^2)/2}, \quad m, n = 0, 1, 2, \ldots,$$
$$(8.31)$$

corresponding to the eigenvalues $E_{mn} \equiv E_n = n + 1/2$. Thus the energy levels are all degenerate in m, so that the ground level (LLL) is spanned by the set $\{\Phi_{m0}(x, y)\}$, which forms an orthonormal basis. However these wave functions are not very well localized, since the mean value of the distance from the origin, $r \equiv \sqrt{x^2 + y^2}$, increases with m. Yet the physics of the problem requires that the wave functions be fairly well localized, since the system tends, as $T \to 0$, to the configuration of the Wigner crystal, that is, a triangular lattice [11]. Thus the basis $\{\Phi_{m0}(x, y)\}$ of the LLL is inadequate for the present purposes.

8.6.2 The LLL basis problem

In order to find another basis of eigenfunctions, orthogonal or not, spanning the same energy level, one may use the method introduced in [11], which is based on a technique introduced by Moshinsky and Quesne [49]. The transformation (8.29) can be seen as a part of a canonical transformation from the variables x, y, p_x, p_y into the new ones Q, P, Q', P', where

$$P = p_y - x/2, \quad Q = p_x + y/2. \tag{8.32}$$

These operators satisfy the following commutation relations:

$$[Q, P] = [Q', P'] = i, \tag{8.33}$$

$$[Q, P'] = [Q', P] = [Q, Q'] = [P, P'] = 0. \tag{8.34}$$

Then a wave function in the (x, y)-space is related to its PP'-expression by the formula

$$\Phi(x, y) = \frac{e^{ixy/2}}{2\pi} \iint_{\mathbb{R}^2} e^{i(xP' + yP + PP')} \Psi(P, P') \, dP dP'. \tag{8.35}$$

In virtue of the expression (8.30) of H_o, the Schrödinger equation $H_o \Psi = \frac{1}{2}(Q'^2 + P'^2) \Psi = E\Psi$ admits factorized solutions $\Psi(P, P') = f(P') h(P)$. Thus the ground state wave function of (8.30) must have the form

$$\Psi_0(P, P') = f_0(P') h(P), \tag{8.36}$$

where $f_0(P') = \pi^{-1/4} e^{-P'^2/2}$, and the function $h(P)$ is arbitrary, which manifests the infinite degeneracy of the LLL.

Depending on the choice of $h(P)$, several types of bases for the LLL may be obtained, according to the following general scheme. Inserting (8.36) into the integral (8.35), the Gaussian integration on P' can be performed exactly. Next, taking a wave function $\Psi_n(P, P') = f_0(P') h_n(P)$, where $\{h_n(P)\}$ is an arbitrary basis in $L^2(\mathbb{R})$, we define:

$$h_n^{(2)}(x, y) = \frac{e^{ixy/2}}{\sqrt{2\pi^{3/4}}} \int_{-\infty}^{\infty} e^{iyP} e^{-(x+P)^2/2} h_n(P) \, dP. \tag{8.37}$$

Then the set $\{h_n^{(2)}(x, y)\}$ is a basis for the LLL, and it is orthonormal iff $\{h_n(P)\}$ is orthonormal in $L^2(\mathbb{R})$. This follows from the canonicity of the change of variables given in Eqs. (8.29), (8.32) or simply by an explicit calculation of the matrix element $\langle h_n | h_m \rangle$, using the integral (8.37). Several examples of this construction have been presented in the literature.

(1) Bagarello *et al.* [11] choose for the ground state a Gaussian $h_0(P) = f_0(P)$, which yields, by (8.37), $h_0^{(2)}(x, y) = \Phi_{00}(x, y) \sim \exp -(x^2 + y^2)/4$. Then they construct a complete set of basis functions of the LLL by acting on $\Phi_{00}(x, y)$ with the so-called magnetic translation operators. The resulting basis vectors have Gaussian localization around the sites of a regular two-dimensional lattice, and thus the basis is lattice-translation invariant. However, each vector has a well-defined, fixed (essential) support, so that there is no possibility of modifying the mutual overlap for fixed electron density. In addition, this basis is not orthogonal, since coherent states are in general not mutually orthogonal. Enforcing orthogonality (by the Gram–Schmidt method, for instance) spoils much of the simplicity of the basis functions, and in particular the localization properties for intermediate fillings and the lattice-translation invariance property.

(2) In order to keep the good localization properties and some sort of translation invariance, Ferrari [23] has constructed an orthonormal basis for the LLL by taking infinite superpositions of the above (coherent) states. The resulting basis vectors are Bloch functions, which may be made translation invariant over the nodes of a given lattice, typically triangular or hexagonal (remember that the Wigner crystal is a triangular lattice). Clearly this basis describes very well the two-dimensional low-density system of electrons of the FQHE, but its construction is rather involved and ad hoc.

(3) Since one wants basis wave functions which are both well localized *and* orthogonal, obvious candidates are orthogonal wavelets. They do enjoy good localization properties, and the latter are easily controlled by varying the scale parameter, in contrast to the Gaussian-like functions of [11]. In addition, the physical problem has an intrinsic hierarchical structure [29, 30]. In particular, the filling factor may take arbitrary rational values, and this suggests a fractal behaviour, as remarked recently [47]. All this points again to wavelets. In the next section, we will review several examples of this construction, as proposed in [2, 12, 13].

8.6.3 Wavelet-like bases

8.6.3.1 The Haar basis

Let us look first at the LLL basis generated by the Haar wavelet basis [19]. Since the mother wavelet $h(x)$ is a discontinuous function, its localization in frequency space is poor, it decays as ω^{-1}. However, the transformation (8.37) is not a Fourier transform, hence it is not clear *a priori* that the corresponding functions $\{h_{jk}^{(2)}(x, y)\}$ will also have a poor localization in both variables. In fact it is *not* the case, as can be seen by investigating the asymptotic behaviour of the basis functions.

Using standard results on Gaussian integrals, one finds for the ground state wave function:

$$H_{00}(x, y) = \frac{e^{-ixy/2}e^{-y^2/2}}{2\pi^{1/4}} \{2\,\Xi(x - iy + 1/2) - \Xi(x - iy) - \Xi(x - iy + 1)\},$$

(8.38)

where

$$\Xi(z) = \mathrm{erf}(z/\sqrt{2}) = \frac{2}{\sqrt{\pi}} \int_0^{z/\sqrt{2}} e^{-t^2}\, dt, \quad z \in \mathbb{C}.$$

Using the asymptotic expansion of the error function, we get, for $|x|, |y| \gg 1$:

$$H_{00}(x, y) \simeq \frac{e^{ixy/2}e^{-x^2/2}}{2\pi^{1/4}} \sqrt{\frac{2}{\pi}} \left(\frac{1}{x - iy} + \frac{e^{-1/2-x+iy}}{x - iy + 1} - 2\frac{e^{-1/8-(x-iy)/2}}{x - iy + 1/2} \right), \quad (8.39)$$

which displays the Gaussian localization of the wave function in the variable x and the rather poor one in y (decay as y^{-1}). This behaviour is confirmed in Figure 8.11 (left), which shows the modulus of $H_{00}(x, y)$ in a 3-D perspective. Clearly the function $H_{00}(x, y)$ is much better localized in the x variable than in y.

An analogous behaviour can be obtained for the generic function $H_{jk}(x, y)$, which may also be calculated exactly. Using (8.37), it is easily seen that the asymptotic behavior of $h_n^{(2)}(x, y)$ in x is governed by the asymptotic behaviour of $h_n(P)$, and the one in y by that of the Fourier transform of $h_n(P)$. Since in the present case, $h_{jk}(P)$ has compact support, we expect $H_{jk}(x, y)$ to be strongly localized in x and delocalized in y, and that its decay in x gets faster for smaller j. This is indeed the case, as may be seen on the figures presented in [12].

8.6.3.2 *The Littlewood–Paley and other wavelet bases*

Another simple example of an orthonormal wavelet basis in $L^2(\mathbb{R})$, also coming from MRA, is the Littlewood–Paley basis [19], generated from the mother wavelet

$$\Psi(x) = (\pi x)^{-1}(\sin 2\pi x - \sin \pi x). \tag{8.40}$$

The behaviour of this function is complementary to that of the Haar wavelet: it has a compact support in frequency space but it decays like $1/x$ in configuration space.

An analogous complementary behaviour is found also for the corresponding LLL wave functions. They are exponentially localized in the y-variable, while in the other variable they behave like $1/x$. This is manifest on the asymptotic behaviour of $\Psi_{00}(x, y)$ for $|x|, |y| \gg 1$. By the same method as before, one finds:

$$\Psi_{00}(x, y) \simeq \frac{e^{-ixy/2}e^{-y^2/2}}{2\pi^{5/4}} \left\{ -\frac{e^{2\pi(y+ix)}e^{-2\pi^2}}{|2\pi - y - ix|} + \frac{e^{\pi(y+ix)}e^{-\pi^2/2}}{|\pi - y - ix|} \right.$$
$$\left. -\frac{e^{-2\pi(y+ix)}e^{-2\pi^2}}{|2\pi + y + ix|} + \frac{e^{-\pi(y+ix)}e^{-\pi^2/2}}{|\pi + y + ix|} \right\}, \tag{8.41}$$

which displays the exponential decay of $|\Psi_{00}(x, y)|$ in y and the slow decay in x, as observed in Figure 8.11 (right).

Similar considerations can be made for the LLL bases obtained by (8.37) from other 1-D wavelet orthonormal bases; for instance [12], the following.

- The *Journé basis*, which does not come from MRA (see [19], p. 136). This case is very similar to the Littlewood–Paley basis, since here too the mother wavelet has compact support in frequency space.
- *Spline bases*, for instance the order 1 splines, coming from the triangle or 'tent' function as scaling function.

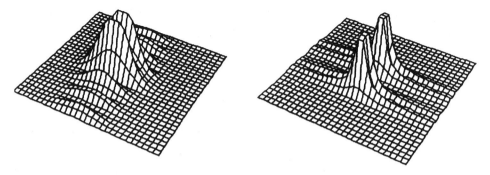

Fig. 8.11. 3-D perspective view of the modulus of the lowest LLL basis functions; the x-axis runs from left to right, the y-axis from front to back. (Left) The Haar function $H_{00}(x, y)$. (Right) The Littlewood–Paley function $\Psi_{00}(x, y)$.

8.6.3.3 Outcome

In conclusion we may say that the orthonormal bases obtained so far are not yet sufficient for a good solution of the LLL problem, except at very low electron density. However, the construction sketched above suggests a general method for designing good bases, with localized functions and respecting the symmetry of a given lattice, for instance, a triangular lattice. This goes in the right direction, since the whole QHE may be characterized as the transport of electrons in a (local) Wigner crystal [47]. The most promising point of such wavelet bases is the possibility of controlling the width of the (essential) supports, hence the overlap between basis functions at neighbouring points, with help of the scaling parameter.

8.6.4 Further variations on the same theme

In a further paper, Bagarello [13] has investigated another aspect of the trial wave function introduced in [11] for a two-dimensional system of electrons in Coulomb interaction. Namely he compares the ground state energy given by harmonic oscillator wave functions with that obtained with particular wavelets. It turns out that the latter always give results that can easily be interpreted as localization properties of the wave function.

Consider an N-electron system in $\mathbb{R}^2$, with Hamiltonian

$$H = \sum_{i=1}^{N} H_o(i) + \frac{1}{2} \sum_{i \neq j} \frac{1}{|\mathbf{r}_i - \mathbf{r}_j|} \tag{8.42}$$

$$H_o = \frac{1}{2}(p_x^2 + x^2) + \frac{1}{2}p_y^2 + p_x p_y. \tag{8.43}$$

Then, under the following canonical transformation:

$$\begin{array}{llll} Q &=& p_x + p_y, & \qquad P &=& -x, \\ Q' &=& p_y, & \qquad P' &=& x - y, \end{array} \tag{8.44}$$

the free Hamiltonian turns into that of a harmonic oscillator, as before:

$$H_o = \tfrac{1}{2}(Q^2 + P^2), \tag{8.45}$$

and again one has an integral transform relating the QQ' wave function to the original one:

$$\Psi(x, y) = \frac{1}{2} \iint_{\mathbb{R}^2} e^{i[Q'(x-y)+Qx]} \, \Phi(Q, Q') \, dQ \, dQ'. \tag{8.46}$$

Here also the free Schrödinger equation admits factorized solutions

$$\Phi(Q, Q') = \pi^{-1/4} e^{-Q^2/2} \phi(Q'), \tag{8.47}$$

where $\phi(Q')$ is an arbitrary function. Choosing for the latter various basis functions, such as elements of the Littlewood–Paley, Haar or harmonic oscillator basis, one may compute the ground state energy E_c of the N-electron system, with the wave function taken again as a Slater determinant, in the familiar Hartree–Fock manner. Actually E_c is the sum of two terms, the direct one and the exchange term (see Section 8.3.1), but the latter is much smaller and may be neglected. As a preparation for the FQHE, one should also consider the electrons localized on the nodes of a lattice generated by magnetic translations, as explained in the preceding section.

Calculations of this type are presented in [13] for $N = 2$ (this case already displays the general features) and several configurations. For instance, two electrons localized at the origin (with different wave functions, otherwise the Slater determinant would vanish identically); or two electrons on the y-axis, separated by a magnetic translation. In both cases, the value obtained for the energy of the trial ground state shows that the wavelet wave function is better localized than the harmonic oscillator wave function.

All these results strongly suggest that wavelet bases, localized around the nodes of the triangular Wigner crystal, may be extremely useful for finding the ground state wave function of the FQHE.

8.7 Outcome: what have wavelets brought to us?

Looking in retrospect at the discussion above, we may conclude that wavelets can be used profitably in various problems in atomic physics and in solid state physics, under different aspects and for different purposes. For instance, we may state the following.

- The detailed description and physical understanding of harmonic generation (and similar transient phenomena) is inaccessible to standard spectral methods, it requires a *time–frequency representation*, by wavelets or Gabor analysis.
- The computation of N-electron atomic wave functions (HF and relatives) demands a good *orthogonal basis* on the half line $\mathbb{R}^+$, and a wavelet basis seems well-adapted.
- Finally, there is the possibility of using 2-D orthogonal wavelet bases for the description of 2-D phenomena in solid state physics, such as the FQHE.

In all these applications, the key property of wavelets is their good localization, both in space and in frequency, and the possibility of controlling it by scaling (the automatic zooming property). This property permits, for instance, to vary the precision of electronic structure calculations in space, depending on the value of the local electronic density. The net result is a reduction of the number of terms required for the expansion of wave functions in the chosen bases. This is analogous to the higher compression rates achieved with wavelet bases in the synthesis and transmission of signals.

As a final point, we may remark that all the applications described in this chapter consider the WT as a mathematical tool, whereas wavelets might also be used as genuine physical entities, exactly as coherent states – which they are after all! Many interesting phenomena could be described in that language (Rydberg atoms, semiclassical limit, revivals, . . .). Finally there is also the possibility to use different wavelets, such as the CS associated to the Schrödinger group [22].

Our conclusion will be that the applications of wavelets in atomic physics and in solid state physics are a new field (it is characteristic that most of the papers quoted in this chapter have appeared in the past two years). Many promising results have been obtained already, much work remains to be done, and hopes for progress are reasonably well-founded. In addition, it is likely that new applications in various domains of quantum (or classical) physics will be found in the near future.

Acknowledgements

It is a pleasure to thank T.A. Arias for communicating additional information about his work, and Z. Felfli, X. Gonze, C. Handy, R. Murenzi, B. Torrésani, C.J. Tymczak and X.Q. Wang for interesting discussions and comments about several aspects of this chapter.

References

[1] S.T. Ali, J.-P. Antoine, J.-P. Gazeau and U.A. Mueller, Coherent states and their generalizations: A mathematical overview, *Reviews Math. Phys.*, **7**: 1013–1104, (1995)

[2] J.-P. Antoine and F. Bagarello, Wavelet-like orthonormal bases for the lowest Landau level, *J. Phys. A: Math. Gen.*, **27**: 2471–2481, (1994)

[3] Ph. Antoine, B. Piraux and A. Maquet, Time profile of harmonics generated by a single atom in a strong electromagnetic field, *Phys. Rev. A*, **51**: R1750–R1753, (1995)

[4] Ph. Antoine, B. Piraux, D.B. Milošević and M. Gajda, Generation of ultrashort pulses of harmonics, *Phys. Rev. A*, **54**: R1761–R1764, (1996)

[5] Ph. Antoine, A. L'Huillier and M. Lewenstein, Attosecond pulse trains using high-order harmonics, *Phys. Rev. Lett.*, **77**: 1234–1237, (1996)

[6] Ph. Antoine, B. Piraux, D.B. Milošević and M. Gajda, Temporal profile and time control of harmonic generation, *Laser Physics*, **7**: 594–601, (1997)

[7] Proc. March 1996 Meeting of the Amer. Phys. Soc., Session R19, *Bull. Amer. Phys. Soc.*, **41**: 746–748, (1996)

[8] T.A. Arias, K.J. Cho, P.K. Lam, J.D. Joannopoulos and M.P. Teter, Wavelet-transform representation of the electronic structure of materials, in *Toward Teraflop Computing and New Grand Challenge Applications*, p. 23; ed. P.K. Kalia and P. Vashishta (Nova Scient. Publ., 1995)

[9] T.A. Arias, private communication

[10] T.A. Arias, http://web.mit.edu/muchomas/www/home.html

[11] F. Bagarello, G. Morchio and F. Strocchi, Quantum corrections to the Wigner crystal: A Hartree–Fock expansion, *Phys. Rev. B*, **48**: 5306–5314, (1993)

[12] F. Bagarello, More wavelet-like orthonormal bases for the lowest Landau level: some considerations, *J. Phys. A: Math. Gen.*, **27**: 5583–5597, (1994)

[13] F. Bagarello, Applications of wavelets in quantum mechanics: a pedagogical example, *J. Phys. A: Math. Gen.*, **29**: 565–576, (1996)

[14] G. Beylkin, R. Coifman and V. Rokhlin, Fast wavelet transforms and numerical algorithms, *Comm. Pure Appl. Math.*, **44**: 141–183, (1991)

[15] B.H. Bransden and C.J. Joachain, *Physics of Atoms and Molecules* (Longman, London and New York, 1983)

[16] K. Cho, T.A. Arias, J.D. Joannopoulos and P.K. Lam, Wavelets in electronic structure calculations, *Phys. Rev. Lett.*, **71**: 1808–1811, (1993)

[17] C.K. Chui, *An Introduction to Wavelets* (Academic Press, San Diego, 1992)

[18] P.B. Corkum, Plasma perspective on strong-field multiphoton ionization, *Phys. Rev. Lett.*, **71**: 1994–1997, (1993)

[19] I. Daubechies, *Ten Lectures on Wavelets* (SIAM, Philadelphia, PA, 1992)

[20] N. Delprat, B. Escudié, Ph. Guillemain, R. Kronland-Martinet, Ph. Tchamitchian and B. Torrésani, Asymptotic wavelet and Gabor analysis: Extraction of instantaneous frequencies, *IEEE Trans. Inform. Theory*, **38**: 644–664, (1992)

[21] S. De Luca and E. Fiordilino, Wavelet temporal profile of high order harmonics emitted by a two-level atom in the presence of a laser pulse, *J. Phys. B: At. Mol. Opt. Phys.*, **29**: 3277–3292, (1996)

[22] D.H. Feng, J.R. Klauder and M. Strayer (eds.), *Coherent States: Past, Present and Future (Proc. Oak Ridge 1993)* (World Scientific, Singapore, 1994)

[23] R. Ferrari, Two-dimensional electrons in a strong magnetic field: A basis for single-particle states, *Phys. Rev. B*, **42**: 4598–4609, (1990)

[24] E. Fiordilino and V. Miceli, Temporal evolution of the spectrum emitted by a two-level atom in the presence of a laser field, *J. Mod. Optics*, **43**: 735–751, (1996)

[25] P. Fischer and M. Defranceschi, Iterative process for solving Hartree–Fock equations by means of a wavelet transform, *Appl. Comput. Harm. Anal.*, **1**: 232–241, (1994)

[26] P. Fischer and M. Defranceschi, Representation of the atomic Hartree–Fock equations in a wavelet basis by means of the BCR algorithm, in *Wavelets: Theory, Algorithms, and Applications*, pp. 495–506; ed. C.K. Chui, L. Montefusco and L. Puccio (Academic Press, San Diego, 1994)

[27] S. Goedecker and O.V. Ivanov, Linear scaling solution of the Coulomb problem using wavelets, *Solid State Commun.*, **105**: 665–669, (1998)

[28] X. Gonze, private communication

[29] M. Greiter and I.A. McDonald, Hierarchy of quantized Hall states in double layer electron systems, *Nucl. Phys. B*, **410**: 521–534, (1993)

[30] F.D.M. Haldane, Fractional quantization of the Hall effect: A hierarchy of incompressible quantum fluid states, *Phys. Rev. Lett.*, **51**: 605–608, (1983)

[31] C.R. Handy, D. Bessis and T.D. Morley, Generating quantum energy bounds by the moment method: A linear-programming approach, *Phys. Rev. A*, **37**: 4557–4569, (1988), and related papers

[32] C.R. Handy and R. Murenzi, Continuous wavelet transform analysis of one dimensional quantum bound states from first principles, *Phys. Rev. A*, **54**: 3754–3763, (1996)

[33] C.R. Handy and R. Murenzi, Continuous wavelet transform analysis of quantum systems with rational potentials, *J. Phys. A*, **30**: 4709–4729, (1997)

[34] E. Huens, B. Piraux, A. Bugacov and M. Gajda, Numerical studies of the dynamics of multiphoton processes with arbitrary field polarization: methodological considerations, *Phys. Rev. A*, **55**: 2132–2143, (1997)

[35] K. Kaneda and T. Odagaki, Two-scale relations in one-dimensional crystals and wavelets, *J. Phys. A: Math. Gen.*, **28**: 4389–4406, (1995)

[36] J.R. Klauder and B.S. Skagerstam, *Coherent States – Applications in Physics and Mathematical Physics* (World Scientific, Singapore, 1985)

[37] W. Kohn and L.J. Sham, Self-consistent equations including exchange and correlation effects, *Phys. Rev. A*, **140**: 1133–1138, (1965)

[38] K.C. Kulander, Multiphoton ionization of hydrogen: A time-dependent theory, *Phys. Rev. A*, **35**: 445–447, (1987)

[39] K.C. Kulander, K.J. Schafer and J.L. Krause, Dynamics of short-pulse excitation, ionization and harmonic conversion, in *Super-Intense Laser-Atom Physics*, pp. 95–110; ed. B. Piraux, A. L'Huillier and K. Rzażewski, NATO ASI Series B, **316** (Plenum Press, New York, 1993)

[40] D.G. Lappas, A. Sanpera, J.B. Watson, K. Burnett, P.L. Knight, R. Grobe and J.H. Eberly, Two-electron effects in harmonic generation and ionization from a model He atom, *J. Phys. B: At. Mol. Opt. Phys.*, **29**: L619–L627, (1996)

[41] M. Lewenstein, Ph. Balcou, M.Yu. Ivanov, A. L'Huillier and P.B. Corkum, Theory of high-harmonic generation by low-frequency laser fields, *Phys. Rev. A*, **49**: 2117–2132, (1994)

[42] M. Lewenstein, P. Salières and A. L'Huillier, Phase of the atomic polarization in high-order harmonic generation, *Phys. Rev. A*, **52**: 4747–4754, (1995)

[43] A. L'Huillier and Ph. Balcou, High-order harmonic generation in rare-gases with a 1-ps 1053-nm laser, *Phys. Rev. Lett.*, **70**: 774–777, (1993)

[44] A. L'Huillier, T. Auguste, Ph. Balcou, B. Carré, P. Monot, P. Salières, C. Altucci, M.B. Gaarde, J. Larsson, E. Mevel, T. Starczewski, S. Svanberg, C.-G. Wahlström, R. Zerne, K.S. Budil, T. Ditmire and M.D. Perry, High-order harmonics: a coherent source in the XUV range. *J. Nonlinear Optical Physics and Materials*, **4**: 647–664, (1995)

[45] J. Liandrat, V. Perrier and Ph. Tchamitchian, Numerical resolution of the regularized Burgers equation using the wavelet transform, in *Wavelets and Applications*, pp.420–433; ed. Y. Meyer (Springer, Berlin, and Masson, Paris, 1992)

[46] J.J. Macklin, J.D. Kmetec and C.L. Gordon III, High-order harmonic generation using intense femtosecond pulses, *Phys. Rev. Lett.*, **70**: 766–769, (1993)

[47] R.G. Mani and K. von Klitzing, Fractional quantum Hall effect as an example of fractal geometry in nature, *Z. Phys. B*, **100**: 635–642, (1996)

[48] G. Morandi, *The Role of Topology in Classical and Quantum Physics*, Lecture Notes Phys. **m7** (Springer, Berlin, Heidelberg, 1992)

[49] M. Moshinsky and C. Quesne, Linear canonical transformations and their unitary representations, *J. Math. Phys.*, **12**: 1772–1780, (1971)

[50] M. Rotenberg, Theory and application of Sturmian functions, *Adv. At. Mol. Phys.*, **6**: 233–268, (1970)

[51] R.A. Smith, J.W.G. Tisch, M. Ciarrocca, S. Augst and M.H.R. Hutchinson, Angularly resolved ultra high harmonic generation experiments with picosecond laser pulses, in *Super-Intense Laser-Atom Physics*, pp. 31–41; ed. B. Piraux, A. L'Huillier and K. Rzcazewski, NATO ASI Series B, **316** (Plenum Press, New York, 1993)

[52] T. Starczewski, J. Larsson, C.-G. Wahlström, J.W.G. Tisch, R.A. Smith, J.E. Muffet and M.H.R. Hutchinson, Time-resolved harmonic generation in an ionizing gas, *J. Phys. B: At. Mol. Opt. Phys.*, **27**: 3291–3301, (1994)

[53] W. Sweldens, The lifting scheme: a custom-design construction of biorthogonal wavelets, *Applied Comput. Harm. Anal.*, **3**: 1186–1200, (1996)

[54] R. Taieb, A. Maquet, Ph. Antoine and B. Piraux, Time dependence of harmonic generation by a single atom, in *Super-Intense Laser-Atom Physics IV*; pp. 445–454; ed. H.G. Muller and M.V. Fedorov, NATO ASI Series 3, **13**: (Kluwer, Dordrecht, 1996)

[55] C.J. Tymczak and X.Q. Wang, Orthonormal wavelet bases for quantum molecular dynamics, *Phys. Rev. Lett.*, **78**: 3654–3657, (1997)

[56] B.Ph. van Milligen, C. Hidalgo and E. Sánchez, Nonlinear phenomena and intermittency in plasma turbulence, *Phys. Rev. Lett.*, **74**: 395–398, (1995)

[57] S. Wei and M.Y. Chou, Wavelets in self-consistent electronic structure calculations, *Phys. Rev. Lett.*, **76**: 2650–2653, (1996)

[58] K. Yamaguchi and T. Mukoyama, Wavelet representation for the solution of radial Schrödinger equation, *J. Phys. B: At. Mol. Opt. Phys.*, **29**: 4059–4071, (1996)

9

The thermodynamics of fractals revisited with wavelets

A. ARNEODO[1], E. BACRY[2] and J.F. MUZY[1]

[1]*Centre de Recherche Paul Pascal,*
Avenue Schweitzer, 33600 Pessac, France

[2]*C.M.A.P., Ecole Polytechnique,*
91128 Palaiseau Cedex, France

Abstract

The multifractal formalism originally introduced to describe statistically the scaling properties of singular measures is revisited using the wavelet transform. This new approach is based on the definition of partition functions from the wavelet transform modulus maxima. We demonstrate that very much like thermodynamic functions, the generalized fractal dimensions D_q and the $f(\alpha)$ singularity spectrum can be readily determined from the scaling behaviour of these partition functions. We show that this method provides a natural generalization of the classical box-counting techniques to fractal signals, the wavelets playing the role of 'generalized boxes'. We illustrate our theoretical considerations on pedagogical examples, e.g., devil's staircases and fractional Brownian motions. We also report the results of some recent applications of the wavelet transform modulus maxima method to fully developed turbulence data. Then we emphasize the wavelet transform as a mathematical microscope that can be further used to extract microscopic information about the scaling properties of fractal objects. In particular, we show that a dynamical system which leaves invariant such an object can be uncovered from the space-scale arrangement of its wavelet transform modulus maxima. We elaborate on a wavelet based tree matching algorithm that provides a very promising tool for solving the inverse fractal problem. This step towards a statistical mechanics of fractals is illustrated on discrete period-doubling dynamical systems where the wavelet transform is shown to reveal the renormalization operation which is essential to the understanding of the universal properties of this transition to chaos. Finally, we apply our technique to analyse the fractal hierarchy of DLA azimuthal Cantor sets

This paper appeared originally in *Physica A* **213** (1995), 232–75. It is reprinted here slightly revised, with kind permission of Elsevier Science–NL.

defined by intersecting the inner frozen region of large mass off-lattice diffusion-limited aggregates (DLA) with a circle. This study clearly lets out the existence of an underlying multiplicative process that is likely to account for the Fibonacci structural ordering recently discovered in the apparently disordered arborescent DLA morphology.

9.1 Introduction

In the real world, it is often the case that a wide range of scales is needed to characterize physical properties. Actually, multi-scale phenomena seem to be ubiquitous in nature. A paradigmatic illustration of such a situation are *fractals* which are complex mathematical objects that have no minimal natural length scale. The relevance of fractals to physics and many other fields was pointed out by Mandelbrot [1, 2] who demonstrated the richness of fractal geometry and stimulated many theoretical, numerical and experimental studies. There are many phenomena in physics that are characrerized by complicated singular measures or singular functions exhibiting self-similar scaling properties [3–12]. In particular, scale invariance is commonly encountered in the context of critical phenomena [13, 14] where the divergence of the correlation length leads to universality. Systems maintained far from equilibrium [15, 16] also display scale invariance in the way they organize spatially as well as in their dynamical evolution [3–12, 17–21].

The aim of a quantitative theory of fractal objects is to provide mathematical concepts and numerical tools for the description of the scaling properties of these objects based on some limited amount of information. For fractal objects which display a recursive hierarchical structure, the knowledge of a few steps of refinement of the object is sufficient for carrying on the refinement ad infinitum [1, 2]. Unfortunately, fractals that appear in nature do not generally exhibit, at least at the first glance, such a well ordered architecture. There are two levels of description that one can hope to achieve which are formally and computationally equivalent to *thermodynamics* and *statistical mechanics* in the theory of many-body systems [22]. On the one hand, one can seek for a global thermodynamic characterization of a fractal object as seen as a *macroscopic* system in terms of intensive variables (temperature, pressure, . . .) and thermodynamic functions (free energy, entropy, . . .) [23–25]. On the other hand, one can look for *microscopic* information about the local scaling properties of fractals in order to define the equivalent of the Hamiltonian from which statistical mechanics tells us how to calculate these thermodynamic functions [26–28]. This amounts to solve what is called the *inverse fractal problem*, i.e., to extract from the data a dynamical system (or its main characteristics) which accounts for the construction rule in the sense that it leaves the object invariant.

Recently, a phenomenological approach to the characterization of fractal objects has been proposed and advanced: the *multifractal formalism* [29, 30]. In its original form, this aproach is essentially adapted to describe statistically the scaling properties of *singular measures* [19, 26–28, 30–37]. Notable examples of such measures include the invariant probability distribution on a strange attractor [30, 34, 36], the distribution of voltage drops across a random resistor network [3, 6, 7, 12], the distribution of growth probabilities along the boundary of diffusion-limited aggregates [7, 8, 38] and the spatial distribution of the dissipation field of fully developed turbulence [19, 29–41]. The multifractal formalism consists in considering a fractal measure as a 'multi-singularity' system. More specifically, a fractal measure can be decomposed into interwoven sets which are characterized by their singularity strength α and their Hausdorff dimension $f(\alpha)$ [30]. The so-called $f(\alpha)$ singularity spectrum has been shown to be intimately related to the generalized fractal dimensions D_q [42–46]. The link between the multifractal formalism and thermodynamics can be understood as follows: the variables q and $\tau(q) = (q - 1)D_q$ play the same role as the inverse of temperature and the free energy, while the Legendre transform $f(\alpha) = \min_q[q\alpha - \tau(q)]$ indicates that instead of the energy and the entropy, we have α and $f(\alpha)$ as the thermodynamic variables conjugated to q and $\tau(q)$ [33–35, 47]. Most of the rigorous mathematical results concerning the multifractal formalism have been obtained in the context of dynamical system theory [34, 36]. It has recently been developed into a powerful technique accessible also to experimentalists. Successful applications have been reported in various fields and the pertinence of the multifractal approach seems, nowadays, to be well admitted in the scientific community at large.

However, in physics as well as in other applied sciences, fractals appear not only as singular measures, but also as *singular functions*. The examples range from graphs of various kinds of random walks, e.g., Brownian signals [48, 49], to financial time series [50–52], to geologic shapes [1, 53], to rough interfaces developing in far from equilibrium growth processes [11], to turbulent velocity signals [54, 55] and to DNA 'walk' coding of nucleotide sequences [56]. There have been several attempts to extend the concept of multifractality to singular functions [29, 57]. In the context of fully developed turbulence, the multiscaling properties of the recorded turbulent velocity signals have been investigated by calculating the moments $S_p(\ell) = \langle \delta v_p \ell \rangle \sim \ell^{\zeta p}$ of the probability density function of longitudinal velocity increments $\delta v_\ell(x) = v(x + \ell) - v(x)$ over inertial separation [54, 55]. By

Lengendre transforming the scaling exponents ζ_p of the structure functions S_p of order p, one expects to get the Hausdorff dimension $D(h)$ of the subset of $\mathbb{R}$ for which the velocity increments behave as $\delta v_\ell \sim \ell^h$ [29]. In a more general context, $D(h)$ will be defined as the spectrum of Hölder exponents of the singular signal under study and thus will have a similar status to the $f(\alpha)$ singularity spectrum for singular measures [58]. But there are some fundamental limitations to the structure function approach which intrinsically fails to fully characterize the $D(h)$ singularity spectrum [58–61]. Actually, only the singularities of Hölder exponents $0 < h < 1$ are potentially amenable to this method; singularities in the derivatives of the signal are not identified. Moreover it has fundamental drawbacks which may introduce drastic bias in the estimate of the $D(h)$ singularity spectrum, e.g., divergencies in $S_p(\ell)$ for $p < 0$.

Our purpose here, is to elaborate on a novel strategy that we have recently proposed and which is likely to provide a thermodynamics of multifractal distributions including measures and functions [59–63]. This approach relies on the use of a mathematical tool introduced in signal analysis in the early 1980s: the *wavelet transform* [64–70]. The wavelet transform has been proved to be very efficient to detect singularities [71–74]. In that respect, it is, a priori, a rather promising technique to study fractal objects [75–81]. Since a wavelet can be seen as an oscillating variant of the characteristic function of a box (i.e., a 'square' function), we will show, as a first step, that one can generalize in a rather natural way the multifractal formalism by defining some partition functions in terms of the wavelet coefficients [59–63, 82]. In particular, by choosing a wavelet which is orthogonal to polynomial behaviour up to some order N, one can make the wavelet transform blind to regular behaviour, remedying in this way one of the main failures of the classical approaches (e.g., the box-counting method in the case of measures and the structure function method in the case of functions). The other fundamental advantage of using wavelets is that the skeleton defined by the *wavelet transform modulus maxima* [74] provides an adaptative space-scale partition of the fractal distribution under study from which one can extract the $D(h)$ singularity spectrum via the scaling behaviour of some partition functions defined on this skeleton [59–63, 82].

As a second step, we will demonstrate that the wavelet transform can be further considered to collect additional information concerning the hierarchy that governs the spatial distribution of the singularities of a fractal object. The wavelet transform can be used as a mathematical microscope [71, 75–81]; increasing the magnification one gains insight into the intricate internal structure of these objects. In this context, we will elaborate on a wavelet based tree matching algorithm which provides a very attractive method for solving the inverse fractal problem [82–84]. This method amounts to extracting from the wavelet transform modulus maxima skeleton, a one-dimensional map which accounts for the construction process of the considered fractal. In that prospect, it constitutes a very promising alternative methodology to the approaches developed in the theory of Iterated Function Systems (IFS) [85–87]. Along the line of the analogy with the physics of multi-body systems, this microscopic description of multifractals is the counter-part of classical statistical mechanics based on the knowledge of the Hamiltonian of the system.

9.2 The multifractal formalism

The multifractal formalism has been introduced to provide a statistical description of singular measures in terms of thermodynamic functions such as the generalized fractal dimensions D_q and the $f(\alpha)$ singularity spectrum [30–37]. In this section, we review both the microcanonical method of computing the $f(\alpha)$ singularity spectrum directly from the data and the canonical method which consists in determining $f(\alpha)$ from the estimation of the D_qs.

9.2.1 Microcanonical description

9.2.1.1 The f(α) singularity spectrum

Usually, when dealing with fractal objects on which a measure μ is defined, the dimension D is introduced to describe the increase of the mass $\mu(B_x(\epsilon))$ with size ϵ:

$$\mu(B_x(\epsilon)) = \int_{B_x(\epsilon)} d\mu(y) \sim \epsilon^D, \qquad (9.1)$$

where $B_x(\epsilon)$ is the ball centred at x and of size ϵ (in $\mathbb{R}$, $B_x(\epsilon)$ is an ϵ-inteval). In general, however, fractal measures display multifractal properties in the sense that they scale differently from point to point. Then one is led to consider local scaling behaviour [30, 46, 88]:

$$\mu(B_x(\epsilon)) \sim \epsilon^{\alpha(x)}, \qquad (9.2)$$

where the exponent $\alpha(x)$ represents the *singularity strength* of the measure μ at the point x. The smaller the exponent $\alpha(x)$, the more singular the measure around x and the 'stronger' the singularity. Let us note that the prefactor in the right-hand side of Eq. (9.2) can be a function of ϵ which varies slower than any power of ϵ. The $f(\alpha)$ *singularity spectrum* describes the statistical distribution of the singularity exponent $\alpha(x)$. If we cover the support of the measure μ with balls of size ϵ, the number of such balls that scale like ϵ^α, for a given α, behaves like [30, 37]:

$$N_\alpha(\epsilon) \sim \epsilon^{-f(\alpha)}. \tag{9.3}$$

Thus $f(\alpha)$ describes how the 'histogram' $N_\alpha(\epsilon)$ varies when ϵ goes to zero. In the limit $\epsilon \to 0^+$, $f(\alpha)$ is defined as the Hausdorff dimension of the set of all points x such that $\alpha(x) = \alpha$ [34, 36]:

$$f(\alpha) = d_H\{x \in \text{supp } \mu, \quad \alpha(x) = \alpha\}. \tag{9.4}$$

At this point one can distingish two main classes of singular measures. *Homogeneous measures* [1, 2, 30, 46, 89] are characterized by a singularity spectrum supported by a single point $((\alpha)_0, f(\alpha_0))$: only one 'sort' of singularity is present in the measure. *Multifractal measures* [30–37, 46, 47] involve singularities of different strengths; in this case the $f(\alpha)$ spectrum has generally a single humped shape which extends over a finite interval $[\alpha_{min}, \alpha_{max}]$, where α_{min} (resp. α_{max}) correspond to the strongest (resp. weakest) singularities.

For singular measures which possess a recursive multiplicative structure, the $f(\alpha)$ singularity spectrum can be calculated analytically [30]. But this is not the case in general and one must have recourse to numerical algorithms for computing the $f(\alpha)$ spectrum. The most natural way would consist in scanning the support of μ, measuring $\alpha(x)$ at each point x by estimating the slope of the curve $\ln \mu(B_x(\epsilon))$ as a function of $\ln \epsilon$; then by using the so-called *box-counting method* [46, 88, 90], one could try to compute the fractal dimension $f(\alpha)$ of the subset of points where the measure scales with the exponent α. However, such a method would lead to dramatic errors since, for any ϵ, $\mu(B_x(\epsilon))$ takes into account a lot of points with very different singularity exponents. Moreover the presence of oscillations in the log-log plot procedure makes extremely unstable the estimate of $\alpha(x)$ on a finite range of scales [90, 91]. One can use a slightly different method called the *histogram method* [46, 90–92]. It consists in covering the support of the measure μ with balls $\{B_i(\epsilon)\}_i$ of size ϵ. For each ball $B_i(\epsilon)$, we define the exponent $\alpha_i(\epsilon) = \ln \mu(B_i(\epsilon))/\ln \epsilon$. This exponent is like a singularity exponent 'seen' at the scale ϵ. Then, if $N_\alpha(\epsilon)$ is the histogram of the values $\{\alpha_i(\epsilon)\}_i$, $f(\alpha)$ can be computed using Eq. (9.3). Even though the histogram method is stable under certain conditions, the convergence when ϵ goes to 0^+ is very slow [90]. In most cases, the range of scales available in the numerical data is too small and this method leads to rather approximate results because of scale dependent prefactors. This is due to the fact that this method is based on the computation of scaling exponents which represent 'local' quantities that can vary a lot from one point to another. Basically, this is a characteristic deficiency of this microcanonical description which intrinsically suffers from finite-size effects [46, 90–96].

9.2.1.2 The generalized fractal dimensions

The *generalized fractal dimensions* D_q [42–46], which correspond to scaling exponents for the qth moments of the measure μ, provide an alternative description of singular measures. Once again, if we cover the support of μ with boxes $B_i(\epsilon)$ of size ϵ, one can define a series of exponents $\tau(q)$ from the scaling behaviour of the *partition function*:

$$\mathcal{Z}(q, \epsilon) = \sum_{i=1}^{N(\epsilon)} \mu_i^q(\epsilon), \qquad (9.5)$$

where $\mu_i = \mu(B_i)\epsilon))$. In the limit $\epsilon \to 0^+$, $\mathcal{Z}(q, \epsilon)$ behaves as a power law:

$$\mathcal{Z}(q, \epsilon) \sim \epsilon^{\tau(q)}. \qquad (9.6)$$

The spectrum of generalized fractal dimensions is obtained from the knowledge of the exponents $\tau(q)$ by the following relation [43–45]:

$$D_q = \tau(q)/(q - 1). \qquad (9.7)$$

For certain values of q one can recognize well known quantities. D_0 corresponds to the capacity (box dimension) [97] of the support of μ. D_1 characterizes the scaling behaviour of the information $I(\epsilon) = \sum_i \mu_i(\epsilon) \ln \mu_i(\epsilon)$: it is called the information dimension [42]. Moreover, for q integer ≥ 2, the D_qs can be related to the scaling behaviour of the q-point correlation integrals [43–45]. In fact, it is easy to see that varying q in Eq. (9.5) amounts to characterize selectively the nonhomogeneity of the measure, positive $q's$ accentuate the 'densest' regions while negative $q's$ accentuate the 'smoothest' ones.

Let us see how one can relate the $f(\alpha)$ singualrity spectrum to the $\tau(q) = (q - 1)D_q$ spectrum. At the scale ϵ, if we consider that the distribution of the $\alpha's$ is of the form $\rho(\alpha)\epsilon^{-f(\alpha)}$ and if we use this expression in Eq. (9.5), it follows [29, 30]:

$$\mathcal{Z}(q, \epsilon) \sim \int \rho(\alpha)\epsilon^{q\alpha - f(\alpha)} d\alpha. \qquad (9.8)$$

In the limit $\epsilon \to 0^+$, this sum is dominated by the term $\epsilon^{\min_\alpha(q\alpha - f(\alpha))}$. Then, from the definition of $\tau(q)$, one obtains

$$\tau(q) = \min_\alpha(q\alpha - f(\alpha)). \qquad (9.9)$$

Thus the $\tau(q)$ spectrum, and in turns the D_qs, are obtained by Legendre transforming the $f(\alpha)$ singularity spectrum. When $f(\alpha)$ and D_q are smooth functions, the relation (9.9) can be rewritten in the following way:

$$\begin{cases} q &= df/d\alpha, \\ \tau(q) = q\alpha - f(\alpha). \end{cases} \tag{9.10}$$

This relationship is a natural consequence of a deep analogy with thermodynamics [26–28, 33–36, 47, 89]. As just pointed out, q can be identified with a Boltzmann temperature ($\beta = 1/kT$) which allows us to examine the different *self-similarity phases* of our multi-singularity measure system. The limit $\epsilon \to 0^+$ can be seen as the *thermodynamic limit* of infinite volume ($V = \ln 1/\epsilon \to +\infty$). Then by identifying $\alpha_i = -\ln \mu_i/\ln(1/\epsilon)$ to the energy E_i per unit of volume of a microstate i, one can rewrite the partition function (9.5) under the familiar form:

$$\mathcal{Z}(\beta) = \sum_i \exp(-\beta E_i). \tag{9.11}$$

From the definition (9.3), $f(\alpha) = \ln N_\alpha(\epsilon)/\ln(1/\epsilon)$ plays the role of the entropy (per unit of volume). Similarly, since the partition function can be reexpressed as an exponential times a free energy (by convention we absorb the temperature dependence in the free-energy function itself), $\tau(q)$ can be identified to the free energy $F(\beta)$ (per unit of volume).

The computation of the $f(\alpha)$ curve can thus be understood as the computation of the entropy versus internal energy curve of a multi-body system. When using a single dominant term approximation in evaluating the integral in Eq. (9.8) via steepest descent, one explicitly computes thermodynamic averages via *microcanonical ensembles* [22]. This assumes that the most probable value is also the average value and is correct only in the thermodynamic limit. The severe *finite-size effects* encountered when computing the $f(\alpha)$ singularity spectrum with the histogram method arise precisely due to this assumption [46, 90–96] and can be taken care of using a canonical method as explained below.

9.2.2 Canonical description

The canonical counterpart of the microcanonical method described in Section 9.2.1, consists in computing the $f(\alpha)$ spectrum as the Legendre transform of the $\tau(q)$ exponents extracted from the scaling behaviour of the partition function defined in Eqs. (9.5) and (9.6) [30–37]. Explicitly, this amounts to considering the quantities α and $f(\alpha)$ as mean quantities defined in a *canonical ensemble*, i.e. with respect to their Boltzmann weights [26–28, 33, 95, 96]:

$$\mu_i(q, \epsilon) = \frac{\mu_i^q(\epsilon)}{\sum_j \mu_j^q(\epsilon)} = \frac{e^{-\beta E_i(V)}}{\mathcal{Z}(\beta, V)}. \tag{9.12}$$

Then one computes expectation values:

$$\langle \alpha \rangle(q) = \sum_i \frac{\ln \mu_i(\epsilon)}{\ln \epsilon} \mu_i(q, \epsilon). \tag{9.13}$$

It is easy to see that $\langle \alpha \rangle(q)$ is related to the scaling exponents $\tau(q)$ of the partition function $\mathcal{Z}(q, \epsilon)$ (Eq. (9.6)) in the following way:

$$\langle \alpha \rangle(q) = \partial \tau(q)/\partial q. \tag{9.14}$$

In addition, if $f(q)$ is defined by

$$f(q) = \sum_i \mu_i(q, \epsilon) \frac{\ln \mu_i(q, \epsilon)}{\ln \epsilon}, \tag{9.15}$$

then one has

$$f = q\langle \alpha \rangle(q) - \tau(q). \tag{9.16}$$

Eqs. (9.14) and (9.16) provide a relationship between a mean entropy f and an average singularity strength $\langle \alpha \rangle$ as implicit functions of the temperature parameter q. These thermodynamic relations clearly demonstrate that the $f(\alpha)$ singularity spectrum can be determined by first computing $\tau(q)$ and then Legendre transforming it in order to get a canonical average of the entropy. In the thermodynamic limit $\epsilon \rightarrow 0^+$, one recovers the 'principle of ensemble equivalence', i.e. the canonical $f(\alpha)$ equals the microcanonical $f(\alpha)$ singularity spectrum.

From a practical point of view, there are however some difficulties in the actual computation of $\tau(q) = (q - 1)D_q$. These difficulties mainly arise from intrinsic properties of fractals, namely, lacunarity [1] and nonhomogeneity [30, 46, 89]. Lacunarity manifests itself as intrinsic oscillations in the usual linear regressions of the log-log procedure used to extract $\tau(q)$ from the scaling behaviour of the partition function $\mathcal{Z}(q, \epsilon)$ [90, 91, 98–100]. Multifractality requires the simultaneous characterization of the most concentrated ($D_{+\infty}$) and the most rarified ($D_{-\infty}$) regions of the support of μ which is a rather difficult task because of poor sampling statistics. Moreover, the Legendre transform (Eqs. (9.14) and (9.16)) requires first a smoothing of the $\tau(q)$ curve. This procedure has a main disadvantage. The smoothing operation prevents the observation of any non-analycity in the curves $\tau(q)$

and $f(\alpha)$ and the interesting physics of *phase transitions* in the scaling properties of a fractal measure [33, 46, 101] can be completely missed. In that respect, Eqs. (9.13) and (9.15) provide an alternative definition of the singularity spectrum which can be used to compute the $f(\alpha)$ curve directly from the experimental data without the intermediate explicit Legendre transform of the (free energy) $\tau(q)$ curve.

9.3 Wavelets and multifractal formalism for fractal functions

There have been some attempts to generalize the multifractal formalism to self-affine functions [29, 57]. The structure function method [54, 55] is a very interesting first step in this direction despite some intrinsic fundamental limitations [58–61] which explain why a thermodynamic description of fractal signals is still missing. Our goal, in this section, is to demonstrate that the wavelet transform [64–70] is the appropriate technical tool needed to process a 'multi-singularity' function system.

9.3.1 The wavelet transform

The wavelet transform is a mathematical technique introduced for analysing seismic data and accoustic signals [102, 103]. Since then, it has been the subject of considerable theoretical developments and practical applications in a wide variety of fields [64–70]. The wavelet transform (WT) of a function s consists in decomposing it into elementary space-scale contributions, associated to the so-called *wavelets* which are constructed from one single function, the analysing wavelet ψ, by mean of translations and dilations. The WT of s is defined as [102, 103]:

$$W_\psi[s](b, a) = \frac{1}{a} \int\limits_{-\infty}^{+\infty} \overline{\psi}\left(\frac{x - b}{a}\right) s(x)dx, \qquad (9.17)$$

where $a \in \mathbb{R}^{+*}$ is a scale parameter, $b \in \mathbb{R}$ is a space parameter and $\overline{\psi}$ is the complex conjugate of ψ. The analysing wavelet ψ is generally chosen to be well localized in both space and frequency. Usually, ψ is only required to be of zero mean but, for the particular purpose of singularity tracking that is of interest here, we will further require ψ to be orthogonal to some low-order polynomials [71–74]:

$$\int_{-\infty}^{+\infty} x^m \psi(x) dx = 0, \quad \forall m, \quad 0 \le m < n_\psi. \tag{9.18}$$

There are almost as many analysing wavelets as applications of the WT. A class of commonly used real-valued analysing wavelets which satisfies the above condition is given by the successive derivatives of the Gaussian function [90]:

$$\psi^{(N)}(x) = \frac{d^N}{dx^N} e^{-x^2/2}, \tag{9.19}$$

for which $n_\psi = N$.

9.3.2 Singularity detection and processing with wavelets

The strength of a singularity of a function is usually defined by an exponent called *Hölder exponent*. The Hölder exponent $h(x_0)$ of a function s at the point x_0 is defined as the largest exponent such that there exists a polynomial $P_n(x)$ of order n satisfying [62, 72–74]:

$$|s(x) - P_n(x - x_0)| \le C|x - x_0|^h, \tag{9.20}$$

for x in a neighbourhood of x_0. If $h(x_0) \in]n, n+1[$, one can easily prove that s is n times but not $n+1$ times differentiable at the point x_0. The polynomial $P_n(x)$ corresponds to the Taylor series of s around $x = x_0$, up to the order n. Thus $h(x_0)$ measures how irregular the function s is at the point x_0. The higher the exponent $h(x_0)$, the more regular the function s.

This definition of the singularity strength naturally leads to a generalization of the $f(\alpha)$ singularity spectrum introduced for fractal measures (Eq. (9.4)). Henceforth we will denote $D(h)$ the Hausdorff dimension of the set where the Hölder exponent is equal to h [58–63, 82]:

$$D(h) = d_H\{x, \quad h(x) = h\}, \tag{9.21}$$

where h can take, a priori, positive as well as negative real values (e.g., the Dirac distribution $\delta(x)$ corresponds to a Hölder exponent $h(0) = -1$).

Remark. The results reported in this work apply to fractal distributions, including measures and functions. However, we will consider only distributions whose singularities are not oscillating, i.e., satisfying $\forall x, f' = df/dx$ is Hölder $h - 1$ iff f is Hölder h [74].

If one uses an analysing wavelet ψ that satisfies the condition (9.18), the local behaviour of s in Eq. (9.20) is mirrored by the wavelet transform which locally behaves like [62, 71–74]:

$$W_\psi[s](x_0, a) \sim a^{h(x_0)}, \tag{9.22}$$

in the limit $a \to 0^+$, provided n_ψ satisfy $n_\psi > h(x_0)$. Therefore, one can extract the exponent $h(x_0)$ from a log-log plot of the WT amplitude versus the scale a. Moreover, if $n_\psi < h(x_0)$ (e.g., $s \in C^\infty$ at x_0), one could prove that we would still get a power law behaviour but with a scaling exponent n_ψ:

$$W_\psi[s](x_0, a) \sim a^{n_\psi}. \tag{9.23}$$

Thus, around a given point x_0, the faster the wavelet transform decreases when the scale a goes to zero, the more regular s is around that point.

9.3.3 The wavelet transform modulus maxima method

The situation is somewhat more intricate when investigating fractal functions. The characteristic feature of these singular signals is the existence of a hierarchical distribution of singularities [59–63, 71, 82]. Locally, the Hölder exponent $h(x_0)$ is then governed by the singularities which accumulate at x_0. This results in unavoidable oscillations around the expected power-law behaviour of the WT amplitude [58–60, 104, 105]. The exact determination of h from log-log plots on a finite range of scales is therefore somewhat uncertain. Of course, there have been many attempts to circumvent these difficulties [104–106]; nevertheless, there exist fundamental limitations (which are not intrinsic to the WT technique) to the local measurement of the Hölder exponents of a fractal function [58–60]. Therefore the determination of statistical quantities like the $D(h)$ singularity spectrum requires a method which is more feasible and more appropriate than a systematic investigation of the WT local scaling behaviour as experienced in Refs. [104, 105].

9.3.3.1 Determination of the singularity spectrum of fractal functions from wavelet analysis

A natural way of performing a multifractal analysis of fractal functions consists in noticing that a wavelet ψ can actually be seen as an oscillatory variant of the box function $\chi_{[0,1]}$ (i.e., the characteristic function of the interval [0, 1]). Indeed, $\mu(B_x(\epsilon))$ in Eq. (9.1), is nothing but the 'wavelet transform' of μ using χ as the analysing wavelet. In this way, Eq. (9.22) can be seen as a generalization of Eq. (9.2), in the sense that when using our freedom in the choice of the 'generalized box' analysing wavelet, one can hope to get rid of possible smooth behaviour that could mask the singularities or perturb the estimate of their strength h (let us note that $n_x = 0$, which means that only negative Hölder exponents $h < 0$ are amenable to box-counting techniques). Our aim in this section is to revisit the multifractal formalism described in Section 9.2, substituting the box functions by wavelets.

A simple method would thus rely on the following definition of the partition functions in terms of wavelet coefficients [59–63, 71]:

$$\mathcal{Z}(q, a) = \int |W_\psi[s](x, a)|^q dx, \tag{9.24}$$

where $q \in \mathbb{R}$. This would be a rather naive generalization of Eq. (9.5) since nothing prevents $W_\psi[s](b, a)$ from vanishing at some points (b, a) of the space-scale half-plane. The partition function would then diverge for $q < 0$. One thus needs to define the equivalent of a covering of the signal in terms of wavelets.

The wavelet transform modulus maxima (WTMM) method [59–63, 82] consists in changing the continuous sum over space in Eq. (9.24), into a discrete sum over the local maxima of $|W_\psi[s](x, a)|$ considered as a function of x. In Figure 9.1, we show the space-scale arrangement of the WTMM of the devil's staircase (i.e. the distribution function $s(x) = \mu([0, x])$ of the uniform measure lying on the triadic Cantor set). These WTMM are disposed on connected curves called *maxima lines* [74]. let us define $\mathcal{L}(a_0)$ as the set of all the maxima lines that exist at the scale a_0 and which contain maxima at any scale $a \le a_0$. An important feature of these maxima lines is that, each time the analysed signal has a local Hölder exponent $h(x_0) < n_\psi$, there is at least one maxima line pointing towards x_0 along which Eq. (9.22) holds [59–63, 74]. In the case of fractal signals, we thus expect that the number of maxima lines will diverge in the limit $a \to 0^+$. In fact, as emphasized in Refs. [59, 60], the branching structure of the WTMM skeleton in the (x, a) half-plane, enlightens the hierarchical organization of the singularities. This is clearly illustrated in Figure 9.1d where the WTMM skeleton of the devil's staircase shown in Figure 9.1a, is a tree whose branching structure reveals the construction rule of the triadic Cantor set: at the scale $a = a_0 3^{-n}$, each one of the $k_0 2^n$ WT modulus maxima simultaneously bifurcates into two new maxima (k_0 is a constant which depends upon the analysing wavelet ψ).

The WTMM method consists in taking advantage of the space-scale partitioning given by this skeleton to define the following partition functions [59–63, 82]:

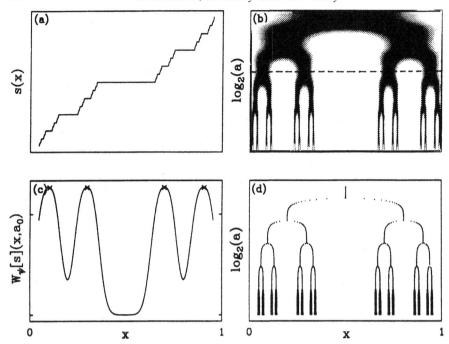

Fig. 9.1. Continuous wavelet transform of the devil's staircase corresponding to the uniform triadic Cantor set. (a) Graph of the function. (b) Wavelet transform computed with the analysing wavelet $\psi^{(1)}$; the amplitude is coded, independently at each scale a, using 32 grey levels from white ($W_\psi[s](x, a) < 0$) to black ($\max_x W_\psi[s](x, a)$). (c) Definition of the modulus maxima at a given scale a_0 corresponding to the dashed line in (b). (d) The skeleton of the wavelet transform, i.e., the set of all the maxima lines. In (b) and (d) the large scales are at the top.

$$\mathcal{Z}(q, a) = \sum_{\ell \in \mathcal{L}(a)} \left(\sup_{(x,a') \in \ell} |W_\psi[s](x, a')|^q \right), \qquad (9.25)$$

where $q \in \mathbb{R}$. As compared to Eq. (9.5), the analysing wavelet ψ plays the role of a generalized box, the scale a defines its size (ϵ in Eq. (9.5)), while the WTMM skeleton indicates how to position our 'oscillating boxes' to obtain a partition at the considered scale. Without the 'sup' in Eq. (9.25), one would have implicitly considered a uniform partition with wavelets of the same size a (Figure 9.2a). As illustrated in Figure 9.2b, the 'sup' can be regarded as a way of defining a scale-adaptative partition which will prevent divergences from showing up in the calculation of $\mathcal{Z}(q, a)$ for $q < 0$.

One can again define the (free energy) exponents $\tau(q)$ from the scaling behaviour of the partition functions:

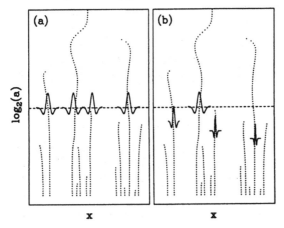

Fig. 9.2. Representation of the uniform and scale-adapted partitions. (a) Uniform partition: $\mathcal{Z}(q, a)$ involves wavelets of the same size a. (b) Scale-adapted partition: $\mathcal{Z}(q, a)$ involves wavelets of different sizes $a' \leq a$. The large scales are at the top.

$$\mathcal{Z}(q, a) \sim a^{\tau(a)}. \tag{9.26}$$

Then, by using both the behaviour of the WT along the maxima lines (Eq. (9.22)) and the definition (9.21) of the $D(h)$ singularity spectrum, one can show that in the thermodynamic limit $a \to 0^+$, $D(h)$, like the entropy, can be computed by Legendre transforming $\tau(q)$:

$$D(h) = \min_{q}(qh - \tau(q)). \tag{9.27}$$

As pointed out in Section 9.2.2, in the framework of this canonical description, one can avoid some practical difficulties that occur when directly performing the Legendre transform of $\tau(q)$, by first computing the following Boltzmann weights from the WTMM [81, 82]:

$$\widehat{W}_\psi[s](q, \ell, a) = \frac{|\sup_{(x,a')\in\ell} W_\psi[s](x, a')|^q}{\mathcal{Z}(q, a)}, \tag{9.28}$$

where $\mathcal{Z}(q, a)$ is the partition function defined in Eq. (9.25). Then one computes expectation values (e.g. Eqs. (9.13) and (9.15)):

$$h(q, a) = \sum_{\ell \in \mathcal{L}(a)} \ln \left| \sup_{(x,a')\in\ell} W_\psi[s](x, a') \right| \widehat{W}_\psi[s](q, \ell, a), \tag{9.29}$$

and

$$D(q, a) = \sum_{\ell \in \mathcal{L}(a)} \widehat{W}_\psi[s](q, \ell, a) \ln \widehat{W}_\psi[s](q, \ell, a), \qquad (9.30)$$

from which one extracts $h(q) = h(q, a)/\ln a$ and $D(q) = D(q, a)/\ln a$, and therefore the $D(h)$ singularity spectrum.

Remark. It is worth pointing out the meaning of $\tau(q)$ for some specific values of q [58]. In full analogy with standard box-counting arguments, $-\tau(0)$ can be identified to the capacity of the set of singularities of s: $-\tau(0) = d_c(\{x, h(x) < +\infty\})$. $\tau(1)$ is related to the capacity of the graph $\mathcal{G}$ of the considered function: $d_c(\mathcal{G}) = \max(1, 1 - \tau(1))$. Finally $\tau(2)$ is related to the scaling exponent β of the spectral density, $\widehat{S}(k) = |\widehat{s}(k)|^2 \sim k^{-\beta}$, with $\beta = 2 + \tau(2)$.

9.3.3.2 *Application of the WTMM method to recursive fractal functions*

The class of fractal and multifractal signals that possess an exact recursive structure provides analytically tractable cases. It is thus well adapted to test the efficiency of the WTMM method. The devil's staircases and more generally the characteristic functions of singular measures can be used as a guinea-pig for our approach [58–62, 82], since one can easily show that the partition function scaling exponents $\tau(q)$ (Eq. (9.26)) are identical to the spectrum $\tau_\mu(q)$ of the underlying measure μ (Eq. (9.6)). In Figure 9.3 we report the results of the analysis of the triadic devil's staircase (Figure 9.1a). The partition functions $\mathcal{Z}(q, a)$ are computed from the WTMM skeleton (Figure 9.1d) of this continuous and almost everywhere constant signal. Figure 9.3a displays some plots of $\log_2 \mathcal{Z}(q, a)$ versus $\log_2 a$ for different values of q. Besides the presence of periodic oscillations of period $\log_2 3$ which reflects the invariance of the Cantor set under discrete dilations by a factor 3, these plots clearly display a linear behaviour on the whole range of scales and this for any q. Using a linear regression fit, we then obtain the slopes $\tau(q)$ of these graphs. As shown in Figure 9.3b, $\tau(q)$ follows a linear curve, the slope of which provides an accurate estimate of the unique Hölder exponent $h = \partial\tau/\partial q = \ln 2/\ln 3$, which characterizes the uniform triadic Cantor set. Actually, the data in Figure 9.3b are in remarkable agreement with the theoretical result $\tau_\mu = (q - 1)\ln 2/\ln 3$. This result is corroborated in Figure 9.3c where $h(q)$ is determined, for different values of q, by plotting $h(q, a)$ versus $\log_2 a$ (Eq. (9.29)). The slope of these graphs is $h = \ln 2/\ln 3$, independently of q. Then, by Legendre transforming $\tau(q)$ (Eq. (9.27)), one gets, up to the experimental uncertainty, that the singularity spectrum reduces to a single point $D(h = \ln 2/\ln 3) = \ln 2/\ln 3$, i.e., the Hausdorff dimension of the triadic Cantor set [59–62, 82]. Let us note that although

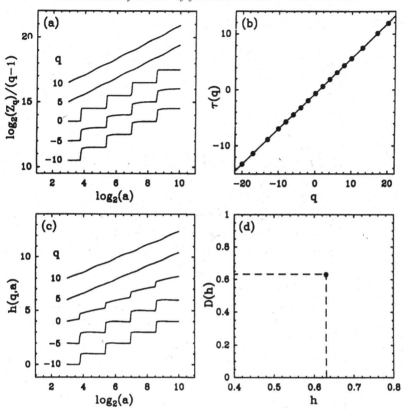

Fig. 9.3. Determination of the multifractal spectra of the devil's staircase associated to the uniform triadic Cantor set using the WTMM method. (a) $\log_2 \mathcal{Z}(q, a)/(q - 1)$ versus $\log_2 a$. (b) $\tau(q)$ versus q; the solid line corresponds to the theoretical curve $\tau(q) = (q - 1) \ln 2 / \ln 3$. (c) Determination of the exponents $h(q)$; $h(q, a)$ is plotted versus $\log_2 a$ according to Eq. (9.29). (d) $D(h)$ versus h. The analysing wavelet is $\psi^{(1)}$.

this example could seem too 'simple', it is a basic example for which the use of the WT maxima lines to partition the signal is crucial. Indeed, as the singularities of s are lying on a set of Lebesgue measure 0, a continuous sum (Eq. (9.24)) over the whole domain [0, 1] would lead to drastic errors [61].

We have reproduced this multifractal analysis for generalized devil's staircases associated to self-similar signed measures [59–62, 82]. For example, let us consider the measure μ constructed recursively as follows: at each step of the construction, each interval is divided into 4 sub-intervals of same length on which we distribute respectively the weights $p_1 = 0.69$, $p_2 = -p_3 = 0.46$ and $p_4 = 0.31 (\sum_{i=1}^{4} p_i = 1)$. Let us note that in the case of a distribution function of a signed measure, the relation $\tau(1) = 0$ does not hold a priori since the 'norm' $\sum_{\mathcal{L}(a)} |W_\psi[s](b_\ell(a), a|$ is no longer conserved through the scales. Indeed $1 - \tau(1)$ is the fractal dimension of the graph of the function. Actually, one can prove that in this particular case, $\tau(q) = -\ln_4(\sum_{i=1}^{4} |p_i|^q)$. Figure 9.4 displays the distribution function $s(x) = \mu([0, x])$ and its wavelet transform. Figure 9.5 shows the distribution function $s_r(x) = \mu_r([0, x])$ which is constructed exactly in the same way as s except that, at each step of the construction, the order of the weights is chosen randomly. Its WT in Figure 9.5b can be compared to the WT of the deterministic function in Figure 9.4b. In the case of the random function s_r, the partition function is averaged over the realizations of the random process, i.e.,

$$\mathcal{Z}_r(q, a) = \langle \mathcal{Z}(q, a) \rangle_{real} \sim a^{\tau_r(q)}. \tag{9.31}$$

Clearly, as the analytical expression of $\tau(q)$ does not depend on the specific order of p_1, p_2, p_3 and p_4, one deduces easily that $\tau_r(q) = \tau(q)$. The results of the multifractal analysis of s and s_r using the WTMM method are reported in Figure 9.6 [59, 60]. As shown in Figure 9.6b, $\tau(q)$ and $\tau_r(q)$ are nonlinear convex increasing functions. The numerical data for both the deterministic (•) and the random (▲) signal match perfectly the theoretical prediction. The corresponding $D(h)$ spectra are displayed in Figure 9.6d; their single humped shapes are characteristic of multifractal signals. The support of $D(h)$ extends

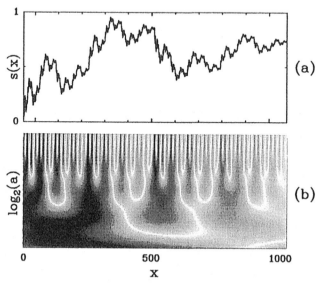

Fig. 9.4. (a) Graph of the deterministic generalized devil's staircase $s(x) = \mu([0, x])$. (b) Continuous wavelet transform of $s(x)$ computed with the first derivative $\psi^{(1)}$ of the Gaussian function. Same coding as in Fig. 9.1b. Small scales are at the top.

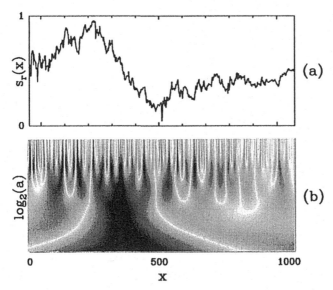

Fig. 9.5. (a) Graph of the random generalized devil's staircase $s_r(x) = \mu_r([0, x])$. (b) Continuous wavelet transform of $s_r(x)$ computed with the first derivative $\psi^{(1)}$ of the Gaussian function. Same coding as in Fig. 9.1b. Small scales are at the top.

over a finite interval $h_{\min} \leq h \leq h_{\max}$. This non-uniqueness of the Hölder exponent is confirmed in Figure 9.6c, where the exponent $h(q)$, computed directly from Eq. (9.29), clearly evolves from the value $h_{\min} \simeq 0.28$ to $h_{\max} \simeq 0.82$ when q varies from $q = 10$ to $q = -10$. The maximum of the $D(h)$ curve is obtained for $q = 0$: $D(h(q = 0)) = -\tau(0) = D_F = 1$. The generalized devil's staircases in Figs. 9.4a and 9.5a are thus singular signals that display multifractal properties; the fractal dimension of the support of the set of singularities of these distribution functions is $D_F = 1$.

9.3.4 Phase transition in the multifractal spectra

In the context of thermodynamics, a phase transition corresponds to some nonanalyticity in the thermodynamic functions, e.g., the free energy and the entropy [22, 107]. Phase transitions in the multifractal scaling properties of singular measures are now well documented in the literature [33, 36, 101, 107–111]. In this section we will illustrate this phenomenon on the singularity spectrum of multifractal functions as induced by the presence of smooth behaviour [61–63, 82].

In the previous sections, we have pointed out that the WTMM method is very efficient as far as we use an analysing wavelet with a number n_ψ of

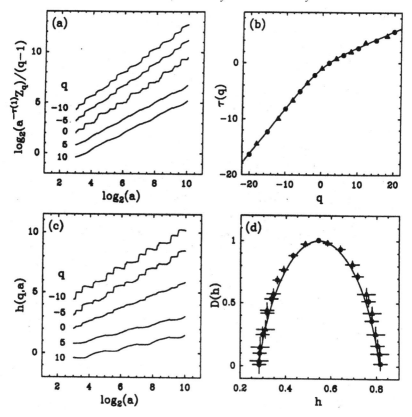

Fig. 9.6. Determination of the multifractal spectra of the devil's staircases displayed in Figs. 9.4 and 9.5 using the WTMM method. (a) $\log_2(a^{-\tau(1)}\mathcal{Z}(q, a))/(q-1)$ versus $\log_2 a$. (b) $\tau(q)$ versus q; the solid line corresponds to the theoretical prediction. (c) Determination of the exponents $h(q)$; $h(q, a)$ is plotted versus $\log_2 a$ according to Eq. (9.29). (d) $D(h)$ versus h; the solid line corresponds to the theoretical spectrum. The analysing wavelet is $\psi^{(2)}$. In (b) and (d) the symbols correspond to the data obtained for the deterministic ($\bullet$) and the random ($\blacktriangle$) signal.

vanishing moments which is greater than $h_{\max} = \max_h\{h, D(h) \neq -\infty\}$. Let us see what happens when this is not possible, e.g., if $h_{\max} = +\infty$. This would mean that the analysed function is C^∞ at some points. For the sake of simplicity, we will assume that the signal $f(x) = s(x) + r(x)$ is a superposition of a multifractal singular part $s(x)$ with $h_{\max} < +\infty$ living on a Cantor set ($s(x)$ is assumed to be constant on each interval on which it is not singular), and a C^∞ regular part $r(x)$. Let $\tau_s(q)$ and $D_s(h)$ be the multifractal spectra which characterize the function $f(x)$. At each scale a_0, the set of maxima lines $\mathcal{L}_f(a_0)$ of the WT of f can be basically decomposed into two disjoint sets of maxima lines, $\mathcal{L}_s(a_0)$ and $\mathcal{L}_r(a_0)$ corresponding to the lines created respectively by $s(x)$ (and which are slightly perturbated by the presence of $r(x)$) and by the C^∞ function $r(x)$. It can be established [74] that along each line created by $r(x) \in \mathcal{L}_r(a_0)$, the WT behaves like $W_\psi[r] \sim a^{n_\psi}$ in the limit $a \to 0^+$ (Eq. (9.23)), while along the other maxima lines ($\in \mathcal{L}_s(a_0)$) $W_\psi[s] \sim a^h$ provided n_ψ is larger than the upper bound of the singularity range of $s(x)$. Then, the partition functions defined in Eq. (9.25) split into two parts [61, 62]:

$$\mathcal{Z}_f(q, a) = \mathcal{Z}_s(q, a) + \mathcal{Z}_r(q, a) \sim a^{\tau_s(q)} + a^{qn_\psi}, \tag{9.32}$$

where $\mathcal{Z}_s$ and $\mathcal{Z}_r$ are the partition functions corresponding respectively to summing over the maxima lines in $\mathcal{L}_s$ and $\mathcal{L}_r$. Thus one deduces easily that $\tau(q) = \min\{\tau_s(q), qn_\psi\}$. In other words, there exists a critical value $q_{crit} < 0$ so that

$$\tau(q) = \begin{cases} \tau_s(q) & \text{for } q > q_{crit}, \\ qn_\psi & \text{for } q < q_{crit}. \end{cases} \tag{9.33}$$

One thus predicts the existence of a singularity in the $\tau(q)$ spectrum. This nonanalyticity in the function $\tau(q)$ expresses the breaking of the self-similarity of the singular signal $s(x)$ by the C^∞ perturbation $r(x)$. Below the critical value q_{crit} (which is the analogue of the inverse of the transition temperature) one observes a regular phase, whereas for $q > q_{crit}$ one switches to the multifractal phase. Let us note that the $\tau(q)$ spectrum in the 'C^∞ *phase*' is governed by the number n_ψ of vanishing moments of the analysing wavelet. Therefore, checking whether $\tau(q)$ is sensitive to some change in the order n_ψ of ψ, constitutes a very good test for the presence of highly regular part in the signal [61, 62].

This phase transition phenomenon is illustrated in Figure 9.7. The analysed function $f(x)$ is the sum of $r(x) = R\sin(8\pi x)$ and a generalized devil's staircase $s(x)$ which is the distribution function of a measure nonuniformly distributed on the triadic Cantor set with the weights $p_1 = 0.6$ and $p_2 = 0.4$. The function $f(x)$ is represented in Figure 9.7a. The $\tau(q)$ and $D(h)$ spectra computed with the WTMM method are displayed in Figures 9.7b and 9.7c respectively. The data obtained for $\tau(q)$ when using two different analysing wavelets $\psi^{(1)}$ ($n_\psi = 1$) and $\psi^{(2)}$ ($n_\psi = 2$) are in remarkable agreement with the theoretical spectrum $D_s(h)$ for $q > Q_{crit}(n_\psi)$. For $q \le q_{crit}(n_\psi)$, however, $D(h)$ displays a linear fall off towards the limiting value $h = 1$ for $\psi^{(1)}$ and $h = 2$ for $\psi^{(2)}$ (indeed $h = N$ for $\psi^{(N)}$) where D vanishes. This linear part is tangent to the theoretical $D_s(h)$ spectrum (dashed line) and has a slope equal to q_{crit} (n_ψ). This is the signature of the phase transition phenomenon described above [61–63].

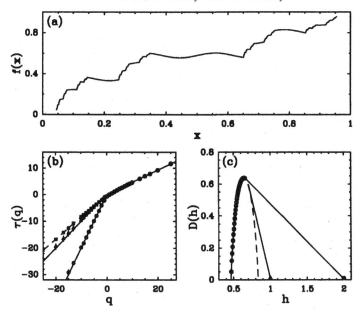

Fig. 9.7. WTMM analysis of a signal which is not singular on some intervals. (a) Graph of the signal $f(x) = s(x) + r(x)$, with $r(x) = R\sin(8\pi x)$ and $s(x)$ is a multifractal devil's staircase (see text). (b) $\tau(q)$ vs q as obtained with $\psi^{(1)}$ ((o) and (▲)), $\psi^{(2)}$ ((o) and (●)) and $\psi^{(4)}$ ((o) and (■)); the solid lines correspond to the theoretical predictions (Eq. (9.33)); the dashed line is the part $q < q_{crit}$ of $\tau_s(q)$. (c) $D(h)$ vs h from the Legendre transform of $\tau(q)$; the symbols are the same as in (b).

Remark 1. Since the wavelet coefficients behave like a^{n_ψ} along the maxima lines created by the C^∞ function, by choosing n_ψ large enough and/or choosing a numerical threshold below which any local maximum is not considered, one can remove all the C^∞ maxima lines in $\mathcal{L}_r(a)$ and thus numerically 'restore' the self-similarity of $s(x)$. The whole $\tau_s(q)$ and $D_s(h)$ spectra can then be estimated.

Remark 2. If one considers the analysing wavelet $\Delta^{(1)} = \delta(x - 1) - \delta(x)$, the wavelet transform is nothing else than the increments used in the structure function method: $\delta s_\ell(x) = s(x + \ell) - s(x) = W_{\Delta^{(1)}}[s](x, \ell)$. Since $n_{\Delta^{(1)}} = 1$, this explains why this method fails to capture singularities with Hölder exponents $h \notin [0, 1]$.

9.4 Multifractal analysis of fully developed turbulence data

The central problem of three-dimensional fully developed turbulence is the energy cascading process. It has resisted all attempts at a full understanding or mathematical formulation. The main reasons for this failure are related to the large hierarchy of scales involved, the highly nonlinear character inherent in the Navier-Stokes equations and the spatial intermittency of the dynamical active regions [40, 41, 112]. In this context, statistical and scaling properties have been the basic concepts used to characterize turbulent flows [113]. One of the striking signatures of the so-called *intermittency phenomenon*, is the non-Gaussian statistics at small scales. The energy transfer towards small scales is related to the non-zero skewness of the probability distribution function (PDF) of the velocity increments and the large flatness of the PDF (kurtosis) corresponds to the presence of strong bursts in the energy dissipation. This fine-scale intermittency is responsible for some departure to the classical $k^{-5/3}$ theory of Kolmogorov [114] which neglects the presence of fluctuations in the energy transfer. Mandelbrot [39] was the first one to advocate the use of fractals in turbulence. Some of his early multiplicative cascade models contain all the ingredients of the multifractal formalism described in Section 9.2. During the past few years, considerable effort has been devoted to the multifractal analysis of high Reynolds number turbulence [40, 41]. But the problem of comparing the predictions of multifractal cascade models [31, 39, 115–118] with experimental data comes from the fact that three-dimensional processing of turbulent flows is at the moment feasible only for numerical simulations which are unfortunately limited in Reynolds numbers to regimes where the scaling just begins to manifest itself. Present experimental techniques have access to the two-dimensional structure of passive scalars [119, 120] and only to the one-dimensional cuts of the velocity field [54, 55, 121, 122]. Here, we are only interested in the statistical analysis of single-point data based on hot-wire techniques in the presence of a mean flow (wind tunnels, jets, etc. . . . .).

Very recently, there has been increasing interest in applying the wavelet analysis to turbulence data [123]. In this section, we report on the first such analysis performed on single point velocity data from high Reynolds number 3D turbulence [77, 104]. The data were obtained by Gagne and collaborators [54, 55, 121, 122] in the large wind tunnel S1 of ONERA at Modane. The Taylor scale based Reynolds number is $R_\lambda = 2720$ and the extent of the inertial range following approximately the Kolmogorov $k^{-5/3}$ law is almost three decades (integral scale $\ell_0 = 15\,\text{m}$, dissipation scale $\ell_d = 0.3\,\text{mm}$). The results reported here concern the analysis in the inertial range of about 100 integral length scales of the recorded experimental signal.

9.4.1 Wavelet analysis of local scaling properties of a turbulent velocity signal

The application of the continuous WT to investigate the local scaling exponent fluctuations that characterize the multifractal nature of a turbulent
velocity field at inertial range scales has been initiated in Ref. [104]. Figure
9.8 illustrates the wavelet transform of a sample of the velocity signal of
length of about two integral scales. The WTMM sekeleton in Figure 9.8c
is actually hardly distinguishable from the WTMM arrangement obtained in
Figure 9.9c for a fractional Brownian signal $B_{1/3}(x)$ which has a $k^{-5/3}$ power
spectrum like the turbulent signal. However, when using the additional information given by the WT amplitude in Figures 9.8b and 9.9b respectively, this
discrimination becomes easier. By analysing the behaviour of $W_\psi[s](x, a)$
versus a along the WTMM lines, one can estimate the value of the local

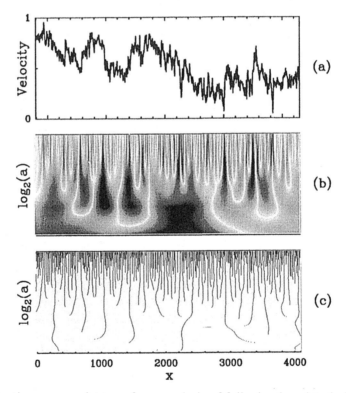

Fig. 9.8. Continuous wavelet transform analysis of fully developed turbulence from
wind tunnel data. (a) The turbulent velocity signal over about two integral length
scales. (b) WT of the turbulent signal; the amplitude is coded like in Fig. 9.1b. (c)
WTMM skeleton. The analysing wavelet is $\psi^{(2)}$. In (b) and (c) the small scales are at
the top.

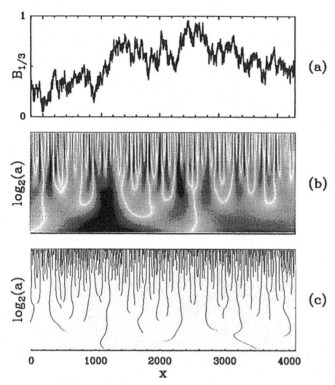

Fig. 9.9. Continuous wavelet transform of a Brownian signal. (a) A realization of the fractional Brownian motion $B_{1/3}$. (b) WT of the Brownian signal; same coding as in Fig. 9.1b. (c) WTMM skeleton. The analysing wavelet is $\psi^{(2)}$. In (b) and (c) the small scales are at the top.

Hölder exponent $h(x)$ according to Eq. (9.22). Regardless of some fluctuations due to finite size effects [106], the Hölder exponent of the Brownian signal $B_{1/3}(x)$ does not depend on x: $h = H = 1/3$. In contrast, for the turbulent velocity signal, h is actually found to fluctuate in a wide range between -0.3 and 0.7 [60, 104], thereby suggesting that the multifractal picture proposed by Parisi and Frisch [29] is appropriate. Statistically, the most frequent exponents are close to the Kolmogorov value $h = 1/3$. Let us stress the observation of negative exponents down to -0.1 and beyond, which correspond to rare but very active events. Negative exponents do not seem to have been previously reported in the literature. One interpretation tossed in Ref. [104] is the occasional passage nearby the probe of slender vortex filaments of the sort observed in recent experiments [124, 125] and 3D numerical simulations [126–130].

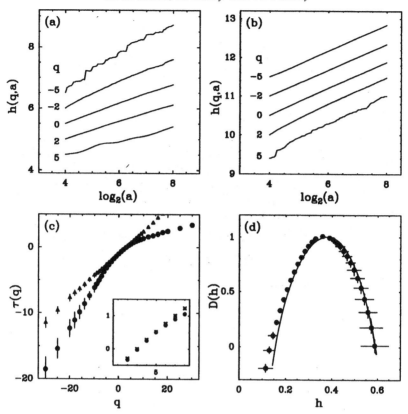

Fig. 9.10. WTMM measurement of the $\tau(q)$ and $D(h)$ spectra of both the Modane turbulent velocity signal and the Brownian signal $B_{1/3}$. Determination of the exponent $h(q)$ from Eq. (9.29) for (a) the turbulent velocity signal and (b) the Brownian signal $B_{1/3}$. (c) $\tau(q)$ vs q; the symbols ($\bullet$) and ($\blacktriangle$) correspond to the turbulent and Brownian signals respectively; the symbols (x) correspond to $\tau(q) = \zeta_q - 1$ obtained when computing the scaling exponents ζ_q with the structure function method. (d) $D(h)$ vs h; the solid line corresponds to the average singularity spectrum obtained from dissipation field data via the Kolmogorov scaling relation (9.34). The results reported in this figure concern the analysis in the inertial range of about 100 integral length scales of the turbulent velocity signal. The analysing wavelet is $\psi^{(2)}$.

9.4.2 Determination of the singularity spectrum of a turbulent velocity signal with the WTMM method

In Figure 9.10 are shown the results of the multifractal analysis of the Modane turbulent velocity signal performed with the WTMM method [59, 60, 63]. The analysis of the Brownian signal $B_{1/3}(x)$ is shown for comparison. As reported in Figure 9.10c, when plotted versus q, the scaling exponent $\tau(q)$ of the partition function $\mathcal{Z}(q, a)$ (Eq. (9.25)) obtained for the Gaussian process, remarkably falls on a straight line $\tau(q) = q/3 - 1$ of slope $h = 1/3$. From the Legendre transform (Eq. (9.27)) of the data for $\tau(q)$, one gets $D(h) = 1$. Thus, as expected theoretically [48, 49] we find that the Brownian signal is everywhere singular with a unique Hölder exponent $h = 1/3$.

In contrast to the fractional Brownian motion ((▲)) in Figure 9.10c), the $\tau(q)$ spectrum obtained for the experimental turbulent signal ((•) in Figure 9.10c) unambiguously deviates from a straight line. Let us note that the results previously derived with the structure function method ($\tau(q) = \zeta_q - 1$ for $q > 0$ exclusively) [55, 121] are in good agreement ((x) in Figure 9.10c) with the nonlinear behaviour of $\tau(q)$ found with the WTMM method. The values of $h = \partial\tau(q)/\partial q$ when varying q from $+30$ to -30 range in the interval $[0.10, 0.62]$. This result is corroborated by the scaling behaviour of $h(q, a)$ (Eq. (9.29)) which clearly depends on q in Figure 9.10a, on the opposite to what is observed in Figure 9.10b for the fractional Brownian motion $B_{1/3}(x)$. The corresponding $D(h)$ singularity spectrum obtained by Legendre transforming $\tau(q)$ is shown in Figure 9.10d. Its characteristic single humped shape over a finite range of Hölder exponents ($h \in [0.11, 0.60]$) is a clear signature of the multifractal nature of the turbulent velocity signal. For $q = 0$, the largest dimension is attained for singularities of exponent $h(q = 0) = 0.335 \pm 0.005$, i.e., a value which is very close to the Kolmogorov prediction $h = 1/3$. Moreover, the correspnding maximum of the $D(h)$ curve, $D(h(q = 0)) = -\tau(0) = 1.000 \pm 0.001$ does not deviate substantially from $D_F = 1$. This suggests that the turbulent signal could be everywhere singular. This possibility seems to be confirmed by the robustness of the $D(h)$ data with respect to changes in the shape of the analysing wavelet: similar quantitative estimates of the $\tau(q)$ and $D(h)$ spectra are obtained when using the first ($\psi^{(1)}$), the second ($\psi^{(2)}$) and the fourth ($\psi^{(4)}$) derivative of the Gaussian function and no wavelet dependent phase transition of the type described in Section 9.3.4 is observed.

In Figure 9.10d, the $D(h)$ singularity spectrum of the wind tunnel velocity signal is compared to a solid curve which actually corresponds to a common fit of dissipation field data at lower Reynolds number [41]. This curve has been deduced from the experimental average $f(\alpha)$ spectrum of the energy dissipation $\epsilon(x) = (dv/dx)^2$ (considered as a measure) of laboratory and atmospheric turbulent flows by using the local Kolmogorov scaling relation [131]:

$$\frac{1}{\ell} \int_{x-\ell}^{x+\ell} \epsilon(x)dx \simeq \frac{\partial v_\ell^3}{\ell}, \tag{9.34}$$

where $\simeq$ means that the two quantities have the same scaling laws. The fact that, for similar statistical samples, one cannot discriminate between these two singularity spectra within the experimental uncertainty, can be interpreted a posteriori as an experimental verification of the above Kolmogorov hypothesis. This observation can be understood also as an experimental confirmation of the universality of the multifractal singularity spectrum of fully developed turbulence with respect to Reynolds number. However, it is clear that considerable further work is needed to get definitive conclusions. In particular, long term statistical analysis must be carried out in order to capture more accurately the latent part ($D(h) < 0$) [132] of the singularity spectrum, including possible violent rare events (rare as compared to the integral scale l_0) corresponding to singularities of negative Hölder exponents. This WT analysis is likely to provide crucial information about the conjectured interpretation of these very energetic localized events in terms of the slender vortex filaments recently observed in hydrodynamic laboratory experiments [133].

9.5 Beyond multifractal analysis using wavelets

The issue of carrying out a statistical mechanics of fractal objects has been mainly addressed in the context of dynamical system theory [26–28, 33–36, 47]. In particular Feigenbaum has shown that the microscopic information about a deterministic multiplicative dynamical system and its scaling properties is contained in the so-called scaling function [134, 135], which describes the scaling or contractions of the various elements of the attractor in time. This scaling function can be seen as the analogue of the Hamiltonian. From the knowledge of this function one can use the transfer matrix technique [26–28] to compute the thermodynamic functions of interest, i.e., the partition function exponents $\tau(q)$ (Eq. (9.26)) and the $D(h)$ singularity spectrum (Eq. (9.27)).

On a more general ground, for any fractal object that can be observed in nature, there is a need to go beyond simple statistical averages and eventually to extract some 'microscopic' information about their underlying hierarchical structure. In many cases, the self-similarity properties of fractal objects can be expressed in terms of a dynamical system which leaves the object invariant. The inverse problem consists in recovering this dynamical system (or its main characteristics) from the data representing the fractal object. This problem has been previously approached within the theory of Iterated Function Systems (IFS) [85–87]. But the methods developed in this context are based on the search of a 'best fit' within a prescribed class of IFS attractors (mainly linear homogeneous attractors). In that sense, they approximate the self-similarity properties more than they reveal them. In this section, we show that, in many situations, the space-scale representation of the wavelet transform of a fractal object can be used to extract some dynamical system which accounts for its construction process [83, 84].

9.5.1 Solving the inverse fractal problem from wavelet analysis

The class of fractal objects we will use to carry out our demonstration are the invariant measures of 'cookie-cutters'. A cookie-cutter [36] is a map on $A = [0, 1]$ which is hyperbolic ($|T'|) > 1$) and so that $T^{-1}(A)$ is a finite union of s disjoint subintervals $(A_k)_{1 \leq k \leq s}$ of A. For each k, $T_k = T|_{A_k}$ is a one to one map on A. An invariant measure μ associated to T is a measure which satisfies $\mu \circ T^{-1} = \mu$. We will suppose that μ is multiplicatively distributed on A:

$$\mu \circ T_k^{-1} = p_k \mu, \quad \forall k \in \{1, \ldots, s\}, \tag{9.35}$$

where $\sum p_k = 1$. These self-similar measures are also referred to as Bernoulli invariant measures of expanding Markov maps [34]. These measures have been the subject of considerable mathematical interest [34, 36, 62]. Practically, they have been widely used for modelizing a large variety of highly irregular physical distributions; notable examples include strange repllers which characterize transient behaviour of nonlinear dynamical systems [36] and the spatial distribution of the dissipation field in fully developed turbulent flows [19, 41].

The 1D continuous wavelet transform of a measure μ according to the analysing wavelet ψ is defined as [75, 76, 78–82]:

$$W_\psi[\mu](b, a) = \int_A \psi\left(\frac{x - b}{a}\right) d\mu, \tag{9.36}$$

where $a \in \mathbb{R}^{+*}$ is the scale parameter and $b \in \mathbb{R}$ is the space parameter. As we have seen ψ is usually chosen to have some vanishing moments, up to a certain order, so that it is orthogonal to possible regular (i.e. polynomial) behaviour of μ. In the particular case of invariant measures of cookie-cutters, there is no such behaviour so we will use a simple 'smoothing wavelet'

$\psi^{(0)} = \exp(-x^2)$. By combining Eqs. (9.35) and (9.36), a straightforward calculation at the first order in a ($a \ll 1$) leads to the following 'self-similarity' relation [62, 82]:

$$W_\psi[\mu](b, a) = \frac{1}{p_k} W_\psi[\mu](T_k^{-1}(b), T_k^{-1'}(b)a), \quad \forall k \in \{1, \ldots, s\}, \qquad (9.37)$$

where $T_k^{-1'}$ is the first derivative of T_k^{-1}. This relation can be interpreted as describing the self-similarity properties of the wavelet transform itself in the (b, a) half-plane [75, 76, 78–82]. Our goal is to study the self-similarity properties of μ through those of its wavelet transform $W_\psi[\mu]$. For that purpose, we are not going to deal with the whole wavelet transform but only with its restriction to the local maxima of its modulus (see Figure 9.1d). In fact, one can easily prove that the self-similarity relation (9.37) still holds when restricted to the set of modulus maxima of the WT. For more details, we refer the reader to our previous work in Refs. [62, 83, 84] and to a recent preprint by W.L. Hwang and S. Mallat [136] where an alternative approach to recover the self-similarity parameters through a voting procedure based on Eq. (9.37) is reported.

9.5.1.1 Linear cookie-cutters

For the sake of simplicity we will first consider the case of linear cookie-cutters such that the T_k^{-1}'s, are linear, i.e., $T_k^{-1}(x) = r_k x + t_k$, where $r_k < 1$. Then the self-similarity relation (9.37) becomes [83, 84]:

$$W_\psi[\mu](b, a) = \frac{1}{p_k} W_\psi[\mu](r_k b + s_k, r_k a), \quad \forall k \in \{1, \ldots, s\}. \qquad (9.38)$$

The meaning of this relation is illustrated in Figure 9.11, for the particular model parameters: $s = 2$, $p_1 = 0.7$, $p_2 = 0.3$, $T_1(x) = 5x/3$ and $T_2(x) = 5x - 4$. The corresponding invariant measure is shown in Figure 9.11a. As previously noticed, one can see that the part of the space-scale plane displayed in Figure 9.11b (the entire rectangle $[0, 1] \times]0, a_0]$) is 'similar' to the two rectangles delimited by the dashed lines ($[0, 3/5] \times]0, 3a_0/5]$ and $[4/5, 1] \times]0, a_0/5]$), up to a global rescaling of the modulus of the wavelet transform. Let us describe on this particular example our technique for recovering from the wavelet transform modulus maxima, the discrete (cookie-cutter) dynamical system T. We call *bifurcation point* any point in the space-scale plane located at a scale where a maxima line appears and which is equidistant to this line and to the closest longer line. The bifurcation points at coarse scales are displayed in Figure 9.11b using the symbols ($\bullet$). They lie on a binary tree whose root is the bifurcation point at the coarsest scale. Each

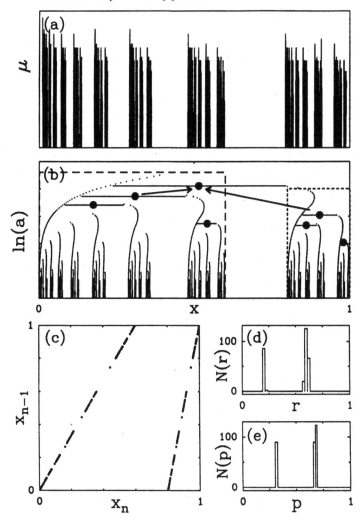

Fig. 9.11. (a) Invariant measure of the two branch cookie-cutter $T_1(x) = 5x/3$, $T_2(x) = 5x - 4$, distributed with the weights $p_1 = 0.7, p_2 = 0.3$ on the interval $[0, 1]$. (b) Position in the (x, a) half-plane of the local maxima of the modulus of the wavelet transform of the measure shown in (a), using a Gaussian analysing wavelet; the large scales are at the top. According to the self-similarity relation (Eq. (9.38)), the maxima line arrangement in the two dashed rectangles is the same as in the original rectangle. The bifurcation points associated to each rectangle are represented by the symbols ($\bullet$). Arrows indicate the matching of these bifurcation points according to the self-similarity relation (9.38). (c) 1D map that represents the position x_{n-1} of an order $n - 1$ bifurcation point versus the position x_n of the associated order n bifurcation point following the tree matching defined in (b). The graph of this map corresponds exactly to the original cookie-cutter. (d) Histogram of scale ratios $r = a_n/a_{n-1}$ between the scales of two associated bifurcation points. (e) Histogram of amplitude ratios $p = |W_\psi[\mu](x_n, a_n)|/ |W_\psi[\mu](x_{n-1}, a_{n-1})|$ computed from two associated bifurcation points.

bifurcation point defines naturally a subtree which can be associated to a rectangle in the space-scale half-plane. This root corresponds to the original rectangle $[0, 1] \times]0, a_0]$, whereas its two sons correspond to reduced copies delimited by the dashed lines. As illustrated in Figure 9.11b, the self-similarity relation (Eq. (9.38)) amounts to matching the 'root rectangle' with one of the 'son rectangles', i.e., the whole tree with one of the subtrees. More generally, this relation associates any bifurcation point (x_n, a_n) of an order n subtree to its hierarchical homologous (x_{n-1}, a_{n-1}) of an order $n-1$ subtree. It follows from Eq. (9.38) that $x_n = r_k x_{n-1} + s_k$ and $a_n = r_k a_{n-1}$. Thus by plotting x_{n-1} versus x_n, one can expect to recover the original cookie-cutter T. This reconstructed 1D map is displayed in Figure 9.11c. As one can see, the two branches T_1 and T_2 of the cookie-cutter T provide a remarkable fit of the numerical data. Let us point out that the nonuniform repartition of the data points on the theoretical curve results from the lacunarity of the measure induced by the 'hole' between the two branches T_1 and T_2. In Figure 9.11d, we show the histogram of the (contracting) scale ratio values between the scales of two bifurcation points of successive generations, $r = a_n/a_{n-1}$, as computed when investigating systematically the WTMM skeleton. As expected, it displays two peaks corresponding to the slopes $r_1 = 3/5$ and $r_2 = 1/5$ of T_1^{-1} and T_2^{-1} respectively. Note that the peak corresponding to the smallest value of r is lower than the other one; this is a direct consequence of the finite cut-off we use in our wavelet transform calculation at small scales. On a finite range of scales, the construction process involves less steps with the smallest scale ratio r_2 than steps with the largest one r_1. (The so-computed histogram can be artificially corrected in order to account for these finite size effects, by plotting $N(r) \ln(1/r)$ instead of $N(r)$.) Figure 9.11e displays the histogram of amplitude ratio values $p = |W_\psi[\mu](x_n, a_n)|/|W_\psi[\mu](x_{n-1}, a_{n-1})|$; one clearly distinguishes two peaks in good agreement with the weights $p_1 = 0.7$ and $p_2 = 0.3$.

At this point, let us mention that the distribution $N(r)$ of scale ratios is in a way redundant with the 1D map, since it is basically made of two Diracs located at the inverse of the slopes of the two branches of this piece-wise linear map. On the contrary, the distribution $N(p)$ of amplitude ratios brings a very important piece of information which is not present in the 1D map: the repartition of the weights at each construction step. When this repartition is uniform, we get a histogram $N(p)$ which reduces to a single point $p = 1/2$. When the repartition is not uniform, as in Figure 9.11, one can furthermore study the joint law of p with r in order to find out the specific 'rules' for associating a p with a r.

Now, having extracted the analogue of the Hamiltonian, i.e. the dynamical system which accounts for the exact recursive structure of the considered measure, one can compute analytically the generalized fractal dimensions D_q (Eq. (9.7)) and the $f(\alpha)$ singularity spectrum (Eq. (9.16)) using the following partition function:

$$\Gamma(q, \tau, r) = \sum_{i=1}^{s} p_i^q / r_i^\tau, \qquad (9.39)$$

where $r = \max r_i$. As pointed out in Ref. [30], for such measures, this partition function at the first level of refinement will generate all the others at finer levels:

$$\Gamma(q, \tau, r^n) = \Gamma(q, \tau, r)^n. \qquad (9.40)$$

In the spirit of the original definition of Hausdorff [137], the dimensions D_q are obtained on requiring the partition function to be of the order of unity:

$$\Gamma(q, \tau, r) = 1. \qquad (9.41)$$

In the case illustrated in Figure 9.11, $D_q = \tau(q)/(q-1)$ are obtained by solving the implicit equation:

$$(0.7)^q (\tfrac{2}{3})^\tau + (0.3)^q 5^\tau = 1. \qquad (9.42)$$

Then by Legendre transforming $\tau(q)$ one gets the $f(\alpha)$ singularity spectrum.

Remark. In the case where s is no longer equal to 2, one can easily adapt the wavelet based technique by trying to match not only the root bifurcation point on its sons but also on its grandsons and so on [83, 84] . . . For instance, in the case $s = 3$, we will match the root with one of its sons and with each of the two sons of its other son. The general algorithm uses a 'best matching' procedure so that it automatically performs the matching which is the most consistent (e.g., so that the different derivatives of $W_\psi[\mu]$ follow the same self-similarity rules as $W_\psi[\mu]$). Thus the algorithm is not looking for a given number s of branches that the user would have guessed a priori, it automatically comes up with the 'best' value of s. In Figure 9.12 are shown the 1D map and the histograms of scale and amplitude ratios obtained in the linear case where $s = 3$, $p_1 = p_2 = p_3 = 1/3$ and $r_1 = 0.2$, $r_2 = 0.3$, $r_3 = 0.5$. All these values are very accurately recovered by our algorithm. Let us notice that we have considered in this work only measures which do not involve any 'memory' effect in their hierarchical structure. i.e., the successive (backward) iterations always consist in applying the same dynamical system T, independently of the previous iterations. However, in a certain

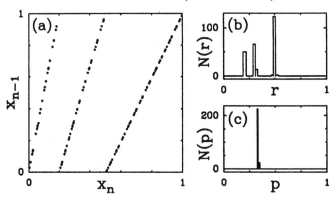

Fig. 9.12. (a) Inverse problem for the invariant measure of the three branch cookie-cutter $T_1(x) = 5x$, $T_2(x) = 10x/3 - 2/3$, $T_3(x) = 2x - 1$, distributed with equal weights $p_1 = p_2 = p_3 = 1/3$ on the interval $[0, 1]$. (b) Histogram of scale ratios $r = a_n/a_{n-1}$. (c) Histogram of amplitude ratios $p = |W_\psi[\mu](x_n, a_n)|/|W_\psi[\mu] (x_{n-1}, a_{n-1})|$.

way, a memory component can be accounted by increasing the number s of branches of a 'no-memory' map T. As illustrated in Figure 9.12, this class of dynamical systems is directly amenable to our WT algorithmic procedure. Nevertheless, it is important to emphasize that it is meaningless to look for dynamical systems with a rather high number of branches; generally, there would not be enough scales in the data in order to ensure the theoretical validity of the outcoming discrete map.

9.5.1.2 Nonlinear cookie-cutters

In the former examples, we have described our wavelet based technique to solve the inverse fractal problem for piece-wise linear cookie-cutters. Since locally in the space-scale plane, the self-similarity relation (9.37) looks like Eq. (9.38), we can apply exactly the same technique for nonlinear expanding maps [83, 84]. Let us point out that the hyperbolicity condition is a priori required for the first derivative of T_k^{-1} involved in the right-hand side of Eq. (9.37) to be finite. Figure 9.13 displays the 1D map extracted from the WTMM skeleton of the uniform Bernoulli measure associated to a nonlinear cookie-cutter made of two inverse hyperbolic tangent branches. Once again, the numerical results match perfectly the theoretical curve. In this case, the histogram of amplitude ratios is still concentrated at a single point $p = 1/2$. But the histogram of scale ratios $N(r)$ involves more than simply two scale ratios as before, since the non-linearity of the map implies that new scale ratios are actually operating at each construction step. This

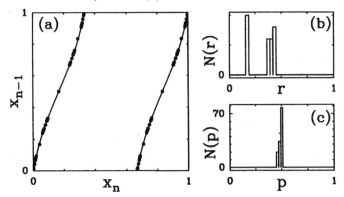

Fig. 9.13. Inverse problem for a non-linear cookie-cutter made of two inverse hyperbolic tangent branches. (a) 1D map obtained with the same wavelet transform tree matching analysis as in Fig. 9.11; the original nonlinear dynamical system (solid lines) is recovered accurately. (b) Histogram of scale ratios $r = a_n/a_{n-1}$. (c) Histogram of amplitude ratios $p = |W + \psi[\mu](x_n, a_n)|/|W_\psi[\mu](x_{n-1}, a_{n-1})|$.

explains the thickness of the two peaks observed in Figure 9.13b. A careful analysis of the fine structure of this histogram would require the investigation of a large number of construction steps; but this out of the scope of the present study.

9.5.2 Wavelet transform and renormalization of the transition to chaos

As a first step towards fully developed turbulence, the transition to chaos in dissipative systems [17, 18] presents a strong analogy with second-order phase transitions [138–141]. Among the different scenarios from ordered to disordered temporal patterns, the most popular are undoubtedly the cascade of period-doubling bifurcations [134, 135, 142–144] and the transition to chaos from quasiperiodicity with irrational winding numbers [145–147]. In this section, we will focus on the period-doubling scenario and we refer the reader to our original work in Refs. [75, 76, 78] for a preliminary analysis of the scenario from quasiperiodicity with golden mean winding number.

Dissipative dynamical systems that exhibit the cascade of period-doubling bifurcations are in practice well modelled by one-dimensional maps with a single quadratic extremum [142–144, 148] such as the map (Figure 9.14):

$$x_{n+1} = \Phi_R(x_n) = 1 - Rx_n^2, \qquad (9.43)$$

or quadratic maps of the form $\Phi_R(x) = Rx(1 - x)$, $R \sin \pi x \cdots$ As one increases the parameter R which determines the height of the maximum of

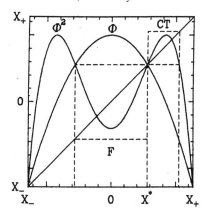

Fig. 9.14. Sketch of the quadratic map Φ_R defined in Eq. (9.43). In the squares F and CT, the second iterate $\Phi_R^{(2)}$ of this map has a similar shape to Φ_R in the initial square. This observation is at the origin of the definition of the renormalization operations $\mathcal{R}_I$ (Eq. (9.44)) and $\mathcal{R}_{II}$ (Eq. (9.45)).

Φ_R at $x = x_c = 0$, one observes an infinite sequence of subharmonic bifurcations at each stage of which the period of the limit cycle is doubled. This period-doubling cascade accumulates at $R_c = 1.40115\cdots$ where the system possesses a 2^∞-orbit that displays scale invariance. Beyond this critical value, the attractor becomes chaotic, even though there still exist parameter windows of periodic behaviour. As originally emphasized by Feigenbaum [134, 135, 142] and Coullet and Tresser [143, 144], this scenario presents strong analogy with second-order phase transition in critical phenomena. Above criticality ($R > R_c$), the envelope of the Lyapunov characteristic exponent (which provides a quantitative estimate of chaos) displays a universal 'order parameter' like behaviour $\overline{L}(R) \sim (R - R)c)^\nu$, where ν is a universal exponent in the sense that it does not depend on the explicit form of the map but only on the quadratic nature of its maximum. Below criticality ($R < R_c$), the period of the bifurcating cycles is a 'characteristic time' which diverges at the transition according to the scaling law $P(R) \sim (R_c - R)^{-\nu}$, with the same critical exponent $\nu = \ln 2/\ln \lambda$ as for the Lyapunov exponent. This universal behaviour results from the observation that the bifurcation parameter values R_n from an orbit of period 2^n to an orbit of period 2^{n+1}, converge to $R_c = R_\infty$ according to the geometric law $(R_c - R_n) \sim \lambda^{-n}$, where $\lambda = 4.669\cdots$ for quadratic maps. Very much like in critical phenomena, these universal properties can be understood using renormalization group techniques [134, 135, 142–144].

Indeed, at criticality $R = R_c$, the attractor of the quadratic map (43) exhibits scale invariance: the adherence of the asymptotic orbit of almost all initial conditions in the invariant interval is a Cantor set. The iterates of the critical point $x_c = 0$ form this Cantor set, with half of the iterates falling between $\Phi_{R_c}^{(3)}(0)$ and $\Phi_{R_c}(0)$, the other half between $\Phi_{R_c}^{(2)}(0)$ and $\Phi_{R_c}^{(4)}(0)$. At the next stage of the construction process, each subinterval is again divided into two subintervals with equal probability and so on. Consequently, the visiting probability measure is symmetrically distributed with the weights $p_1 = p_2 = 1/2$. The multifractal scaling properties of the corresponding

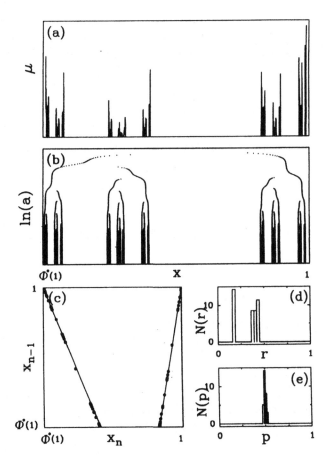

Fig. 9.15. (a) Invariant measure associated to the critical period-doubling dynamical system Φ^* (see text). (b) WTMM skeleton computed with a Gaussian analysing function. (c) 1D map obtained with our tree matching algorithm; the solid lines represent the theoretical prediction (Eq. (9.46)). (d) Histogram of scale ratios $r = a_n/a_{n-1}$. (e) Histogram of amplitude ratios $p = |W + \psi[\mu] (x_n, a_n)|/|W_\psi[\mu](x_{n-1}, a_{n-1})|$.

invariant measure (Figure 9.15a), can be understood from the two renormalization operations that have been proposed. The renormalization operation $\mathcal{R}_I$, originally discovered by Feigenbaum [134, 135, 142]:

$$\mathcal{R}_I(\Phi_R(x)) = \alpha \Phi_R o \Phi_R(x/a), \qquad (9.44)$$

indicates that up to some dilation by a scale factor $\alpha = 1/\Phi_R(1)$, the second iterate of Φ_R in the interval noted F in Figure 9.14, looks like Φ_R itself. The renormalization operation $\mathcal{R}_{II}$ introduced by Coullet and Tresser [143, 144]:

$$\mathcal{R}_{II}(\Phi_R(x)) = \alpha^2 \left[\Phi_R \left(\Phi_R \left(\frac{x}{\alpha^2} + \eta \right) \right) - \eta \right], \qquad (9.45)$$

results from the observation that, up to some translation by η and some dilation by α^2, the second iterate $\Phi_R o \Phi_R$ in the interval noted CT in Figure 9.14, looks like Φ_R in the original interval. The universality of the multifractal properties of the invariant measure of Φ_{R_c} [30, 34] is the consequence of the fact that Φ_{R_c} belongs to the stable manifolds of the two fixed points of $\mathcal{R}_I$ and $\mathcal{R}_{II}$ respectively. The complexity of the ocnstruction rule of the period-doubling Cantor set is actually contained in this subtle interplay between $\mathcal{R}_I$ and $\mathcal{R}_{II}$ [134, 135, 142–144]. Ledrappier and Misiurewicz [149] have succeeded to prove that this measure can be considered as the invaraint measure obtained by iterating backward the following dynamical system defined on the interval $A = [\Phi_*(1), 1]$:

$$T(x) = \begin{cases} T_1(x) = x/\Phi_*(1) & \text{if } x \in [\Phi_*(1), x^*] \\ T_2(x) = \Phi_*(x)/\Phi_*(1) & \text{if } x \in [x^*, 1] \end{cases} \qquad (9.46)$$

where

$$\Phi_*(x) = 1 - 1.5276 \cdots x^2 + 0.1048 \cdots x^4 + \cdots \qquad (9.47)$$

is the fixed point of the renormalization $\mathcal{R}_I$ ($\Phi_* = \mathcal{R}_I \Phi_*$) and x^* is the point in A such that $\Phi_*(x^*) = x^*$ [134, 135, 142, 148, 150]. Let us point out that, as compared to the Bernoulli measures distributed on generalized Cantor sets, the self-similarity properties of the invariant measure of critical period-doubling dynamical systems depend dramatically on the fact that, while $T_1(x)$ is linear (a simple dilation like $\mathcal{R}_I$), the second branch $T_2(x)$ is nonlinear (as the consequence of the fact that $\mathcal{R}_{II}$ is not a simple dilation since it involves also a translation).

As a first application of the *wavelet based tree matching algorithm* described in Section 9.5.1, to a physical problem, we report in Figure 9.15 the results obtained when analysing the natural measure associated to the iteration of the quadratic unimodal map $\Phi_R(x)$ defined in Eq. (9.43) at the accumulation of period-doublings [82–84]. A well defined 1D map with two distinct hyperbolic branches is numerically reconstructed in Figure 9.15c. A computation at a finer resolution would reveal that the left-hand branch is linear with a slope $1/r = 1/\Phi_*(1) \simeq -2.5$, whereas the right-hand one is nonlinear. A close inspection of the scale ratio histogram in Figure 9.15d confirms this observation. The amplitude ratio histogram computed in Figure 9.15e displays a unique peak at $p = 1/2$ which suggests that the weights associated to the two branches of the 1D map are equal: $p_1 = p_2 = 1/2$. The critical period-doubling natural measure can thus be seen as the invariant measure of the cookie-cutter shown in Figure 9.15c with uniform probability distribution. The solid lines shown in this figure correspond to the dynamical system defined in Eq. (9.46). Our numerical data are in remarkable agreement with the theoretical prediction.

9.6 Uncovering a Fibonacci multiplicative process in the arborescent fractal geometry of diffusion-limited aggregates

The diffusion limited aggregation (DLA) model introduced by Witten and Sander [151] about a decade ago, has become the basic paradigm for fractal pattern forming phenomena [8, 38, 152]. This prototype model mimics two-dimensional Laplacian growth processes according to the following algorithm: particles originating from far away are added, one at a time, to a growing cluster via random walk trajectories in the plane. Extensive on-lattice and off-lattice computer simulations have produced complex branched fractals that bear a striking resemblance to the tenuous tree-like structures observed in viscous fingering, electrodeposition, bacterial and neuronal growths [3–12, 38]. The appealing simplicity of the DLA model and its relevance to various experimental situations have stimulated considerable experimental, numerical and theoretical interest [3–12]. But having regard to the efforts spent, the progress in capturing the screening mechanisms that govern DLA growth has been very limited. Actually, only a little is known about the ramified DLA morphology which is still very mysterious to many extents. In particular, we do not know whether some structural order is hidden in the apparently disordered geometry of DLA clusters. More generally, we still appear to be quite far from a physical understanding of Laplacian growth phenomena. This explains why, after more than ten years of extensive inquiry, the DLA model remains one of the most exciting theoretical challenges in the physics of structure formation.

One of the main obstacles to theoretical progress lies in the lack of structural characterization of the growing clusters. Most of the previous studies have mainly focused on the multifractal analysis of either the DLA geometry or the growth probability distribution along its boundary [8, 38, 151–155]. But, as pointed out in the previous sections, the estimate of the generalized fractal dimensions D_q and the $f(\alpha)$ singularity spectrum provides only 'macroscopic' statistical information about the self-similarity properties of fractal objects. The incompleteness of the multifractal description lies in the fact that, to some extent, the 'microscopic' information concerning the hierarchical architecture of these arborescent morphologies has been filtered (averaged) out.

To achieve a more elaborated structural analysis of the DLA clusters, we have recently advocated the use of the continuous 2D wavelet transform [79–81]. With this mathematical microscope, we have discovered the existence of Fibonacci sequences in the internal fractal branching of large mass off-lattice DLA clusters [156–158]. This analysis also reveals that this fascinating hierarchy is likely to be related to a predominant structural five-fold symmetry [157]. Our aim here, is to establish the statistical relevance of the golden mean arithmetic to the structural fractal ordering of DLA clusters. For that purpose, we will use the wavelet based tree matching method presented in Section 9.5.1 for solving the inverse fractal problem. This method turns out to be a very efficient tool to extract, directly from one-dimensional cuts of large mass aggregates, a discrete dynamical system (1D map) which accounts for their multiplicative construction rule [84]. In this section, we apply this method to 50 azimuthal Cantor sets obtained by intersecting off-lattice DLA clusters containing 10^6 particles (Figure 9.16a) with a circle of

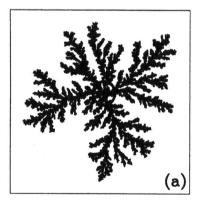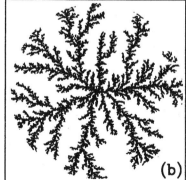

Fig. 9.16. (a) A 10^6 particle DLA cluster computed with an off-lattice random walker model. (b) The inner frozen region delimited by the circle sketched in (a); about $8 \, 10^4$ particles are contained in a disk of radius $R = 480$ particles sizes.

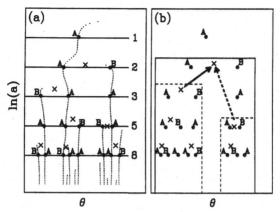

Fig. 9.17. WTMM skeleton of a part of the azimuthal Cantor set corresponding to one main branch of an off-lattice DLA cluster. The analysing wavelet is the Gaussian function; the large scales are at the top. (a) Symbolic coding of the WTMM skeleton according to the Fibonacci recursive process; the horizontal lines mark the scales $a_n = r^{*n}$ with $r^* = 0.44$. (b) Illustration of the tree-matching algorithm after transforming the bifurcation points ($\times$) in such a way that the symbols A emerge systematically on the left. According to the self-similarity relation (9.37), the 2 dashed rectangles are mapped into the original rectangle. Arrows indicate the matching of the bifurcation points with the maps T_A (——) and T_B (- - - -) respectively.

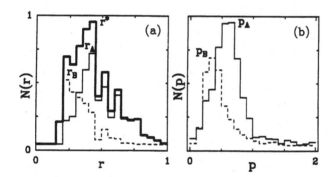

Fig. 9.18 (a) Histogram $N(r)$ (——) of the values of the scale ratio between two successive bifurcation points in the WTMM skeletons of 50 DLA azimuthal Cantor sets. A maximum is observed for $r* = 1/\lambda* \simeq 0.44$ ($\lambda* \simeq 2.2$). $N(r_A)$ (——) and $N(r_B)$ (- - - -) are the histograms obtained when considering bifurcation points mapped by T_A and T_B respectively. (b) The corresponding histograms of amplitude ratio values $N(p_A)$ (——) and $N(p_B)$ (- - - -).

radius $R \sim R_g/3$ (where R_g is the gyration radius) centred at the origin and that somehow delimits their inner frozen region (Figure 9.16b) [159].

The wavelet transform modulus maxima of an azimuthal DLA Cantor set are shown in Figure 9.17a. Let us first proceed to a systematic investigation of the contracting scale ratio $r = a_n/a_{n-1}$ (< 1) between the scales of two bifurcation points of successive generations in the WTMM skeleton. The results of the statistical analysis of 50 off-lattice DLA clusters are shown in Figure 9.18a [156, 159]. The scale ratio histogram $N(r)$ displays a maximum at the value $r^* = 1/\lambda^* = 0.44 \pm 0.03$ ($\lambda^* = 2.2 \pm 0.2$). Let us note that similar histograms have been obtained by H.L. Hwang and S. Mallat in Ref. [136]. The 'generations' of branching are thus expected to occur preferentially at the scales $a_n = a_0 r^{*n} = a_0(0.44)^n$, where a_0 is a macroscopic scale that is determined by the size of the DLA branch under study. The horizontal lines in the (θ, a) half-plane in Figure 9.17a are drawn as guide marks for those successive generations.

As seen in Figure 9.17a, by assigning a symbol A or B to each maxima line, one obtains a coding of the WTMM skeleton that complies with the Fibonacci recursive process [160]:

$$A \to AB, \quad B \to A \tag{9.48}$$

Thus if one starts with the symbol B at the generation $n = 0$, one gets A at the generation $n = 1$, and successively $AB, ABA, ABAAB, ABAABABA\ldots$ The population F_n at the generation n can be deduced from the populations F_{n-1} and F_{n-2} at the two preceding generations, according to the iterative law:

$$F_n = F_{n-1} + F_{n-2}, \tag{9.49}$$

with $F_0 = F_1 = 1$. A remarkable property of the Fibonacci series $\{F_n\} = \{1, 1, 2, 3, 5, 8, 13, 21, \ldots\}$ is that the ratio of two consecutive Fibonacci numbers converges to the golden mean ϕ:

$$\lim_{n \to +\infty} F_{n+1}/F_n = \phi = (1 + \sqrt{5})/2 = 1.618\ldots \tag{9.50}$$

Now, if one uses the general formula established for the WTMM skeletons of one-scale Cantor sets [82], one gets the following estimate for the fractal dimension D_F^A of the DLA azimuthal Cantor sets [156]:

$$D_F^A = \frac{\ln \phi}{\ln 1/r^*} \simeq \frac{\ln 1.62}{\ln 2.2} \simeq 0.61, \tag{9.51}$$

where we have identified the branching ratio and the scale ratio to the values that have been recorded the most frequently in our statistical study, namely ϕ and $1/r^*$ respectively. This numerical value for D_F^A is in good agreement with our previous measurements based on classical box-counting technique: $D_F^A = 0.62 \pm 0.03$ [156–158].

The spreading of the histogram in Figure 9.18a around $r^* \sim 0.44$ indicates the existence of important fluctuations in the scale ratio value. These fluctuations can be related to some local departure from the Fibonacci structural

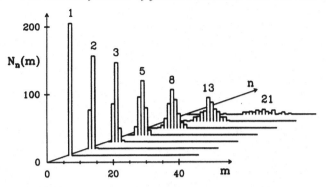

Fig. 9.19. Statistical distributions of the number of WTMM lines that exist at scales $a_n = (0.44)^n$ for successive generations from $n = 1$ to 7.

ordering. The presence of these structural defects raises the question of the statistical pertinence of this Fibonaccian architecture. In Figure 9.19 are reported the results of a systematic analysis of the WTMM skeletons associated to the 240 main DLA branches identified in our statistical sample. The angular width of each of these branches has been normalized to 1 before computing the WTMM skeleton. The different histograms represented in Figure 9.19 correspond to the statistical distribution of the number of WTMM lines that exist at scales $a_n = r*^n = (0.44)^n$ for successive generations. Each of these histograms displays a well defined maximum at the value F_n given by the Fibonacci series. This is a quantitative confirmation that the Fibonacci structural ordering is a generic statistical characteristic of the azimuthal DLA Cantor sets and not some feature recognized on particular realizations [156–158].

The technique introduced in Section 9.5.1 provides a very attractive method to push further this analysis and to extract some 'mean 1D map' which could explain and quantify the presence of a predominant statistical Fibonaccian structural hierarchy in the DLA Cantor sets. In order to carry out this analysis [84, 159], we have first proceeded to a systematic investigation of the symbolic coding of the WTMM skeletons of the azimuthal Cantor sets. A close inspection of this coding reveals some randomness in the relative position of the symbols A and B at each bifurcation $A \rightarrow AB$. Out of 1586 bifurcation points for which the coding has been achieved, 747 (47%) correspond to the A branch being on the left and 839 (53%) to the A branch being on the right. Moreover, the analysis of the correlations between two successive bifurcation points does not indicate any memory effect. Actually, within the statistical uncertainty, one cannot distinguish the random occurrence of the symbols A and B at each bifurcation point from a fair tossing coin.

In order to adapt the tree matching WTMM technique to the presence of this statistical left-right symmetry, we 'flip' the relative position of A and B whenever the A branch is found on the right of B, so that the skeleton actually processed is made only of A branches emerging on the left (Figure 9.17b) [84, 154]. Then our tree-matching algorithm consists in extracting the map T which is made of two branches T_A and T_B and which leaves μ invariant (i.e. $\mu \circ T_A^{-1} = p_A \mu$ and $\mu \circ T_B^{-1} = p_B \mu$) from the 'self-similarity' relation (9.37). The 1D map $T(x)$ reconstructed from scanning the 50 azimuthal Cantor set WTMM skeletons is shown in Figure 9.20a. The data points obviously do not fall on a well defined 1D map. However, the set of data points clearly separates into two distinct 'noisy' branches. The solid lines in this figure correspond to the piece-wise linear 1D map:

$$T(x) = \begin{cases} T_A(x) = \lambda_A^* x & x \in [0, r_A^*] \\ & \text{for} \\ T_B(x) = \lambda_B^*(x-1) + 1 & x \in [1 - r_B^*, 1] \end{cases} \qquad (9.52)$$

where $\lambda_A^* \sim r_A^{*-1} \sim 2.2$ and $\lambda_B^* \sim r_B^{*-1} \sim \lambda_A^{*2} \sim 4.8$. This 1D map is made of two linear branches whose slopes correspond to the inverse of the preferential scale ratios found when splitting the histogram $N(r)$ in Figure 9.18a into two histograms $N(r_A)$ and $N(r_B)$. These two histograms account for the scale ratio fluctuations observed in the WTMM skeletons when one com-

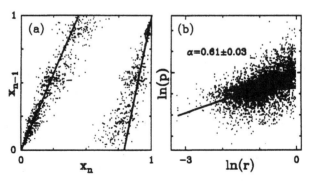

Fig. 9.20. (a) 1D map extracted from the WTMM skeletons of 50 DLA azimuthal Cantor sets using the tree matching algorithm described in Fig. 9.17. The solid lines correspond to the two branches of the linear cookie-cutter (9.52) with the respective slopes $\lambda_A^* = 2.2$ and $\lambda_B^* = \lambda_A^{*2} = 4.8$. (b) $\ln p$ versus $\ln r$, where $r = a_n/a_{n-1}$ ($p = |W_\psi[\mu](x_n, a_n)|/|W_\psi[\mu](x_{n-1}, a_{n-1})|$) is the ratio between the scales (amplitudes) of two bifurcation points that are associated by our tree matching algorithm. A linear regression fit of the data provides a slope $\alpha = 0.61 \pm 0.03$.

putes the scale ratio for A branches and B branches separately. They both display a maximum for $r_A^* = 0.44 \pm 0.03$ and $r_B^* = 0.21 \pm 0.03$ respectively. The fact that $r_B^* \simeq r_A^{*2}$ has a remarkable consequence. A straightforward computation shows that if this equality holds, then the number of cylinders (subintervals) of a given size r_A^k generated by iterating T^{-1} is exactly the Fibonacci number F_k [58, 84, 159]. A 1D map model as simple as the piecewise linear map (9.52) therefore provides a rather natural understanding of the origin of the Fibonacci structural hierarchy discovered on individual realizations (Figure 9.17a). The accumulation of data points around the solid lines in Figure 9.20a can thus be regarded as a quantitative indication of the existence of a statistically predominant multiplicative process, whereas the 'noise' around these lines is the signature of the importance of the structural defects to the Fibonacci ordering. There is moreover some randomness in this multiplicative process since at each bifurcation point in the WTMM skeletons, there are as many chances for $T(x)$ (A on the left) as for its 'flipped' version $\widehat{T}(x) = 1 - T(1 - x)$ (A on the right) to be iterated backward.

The WTMM tree matching algorithm gives also access to the histogram of amplitude ratio values $p = |W_\psi[\mu](x_n, a_n)|/|W_\psi[\mu](x_{n-1}, a_{n-1})|$. The histograms $N(p_A)$ and $N(p_B)$, corresponding respectively to the histograms of scale ratios $N(r_A)$ and $N(r_B)$, are shown in Figure 9.18b. Let us note that for the Bernoulli invariant measures of the piece-wise linear cookie-cutter model (9.52) to be homogeneous, the respective weights p_A and p_B, distributed multiplicatively at each iteration, have to satisfy the requirement $p_A^* = p_B^{*2}$. Since $p_A^* + p_B^* = 1$, one gets exactly $p_A^* = \phi^{-1} \simeq 0.618$ and $p_B^* = \phi^{-2} \sim 0.382$. Both histograms in Figure 9.18b display a maximum in very good agreement with those expected values for p_A^* and p_B^*. This is an indication that the DLA azimuthal Cantor sets are likely to be homogeneous fractals. Furthermore we show in Figure 9.20b that the random variables $\ln r$ and $\ln p$ are strongly correlated according to the law $p = Cr^{0.61}$. This result is in remarkable agreement with previous WT measurements of the local scaling exponent $\alpha = D_F^A = 0.61 \pm 0.03$ of the DLA azimuthal Cantor sets [157]. The scatter of points around the solid line in Figure 9.20b might explain some weak multifractal departure from statistical homogeneity as noticed in previous box-counting calculations [155].

To summarize, we believe that the set of results reported in this work is a very attractive breakthrough on the main challenge raised by the puzzling DLA morphology. To our knowledge this is the first time that some statistical evidence for the existence of a multiplicative construction process hidden in the DLA geometry is reported in the literature. The cookie-cutter T

defined in Eq. (9.52) accounts for the presence of a statistically predominant Fibonacci structural ordering. Moreoever, we have shown that there exist mainly two sources of randomness superimposed to this structural ordering. The first one results from some (left-right) symmetry in the Fibonacci multiplicative process itself. The second one appears as intrinsic noise in the reconstructed 1D map and can be understood as structural defects in the Fibonacci fractal hierarchy.

9.7 Conclusion

To summarize, we have presented in this paper a first theoretical step towards a unified theory of singular distributions, including multifractal measures and multifractal functions based on wavelet analysis. Indeed we believe that the WTMM method for determining the singularity spectrum of a fractal signal is likely to become as useful as the well-known phase portrait reconstruction, Poincaré section and first return map techniques for the analysis of chaotic time series [17, 18]. The reported results of a preliminary analysis of a fully developed turbulent velocity signal show that this method is readily applicable to experimental situations. We have also shown that one can further use the wavelet transform to go beyond this thermodynamic description of fractal objects and eventually to extract from the data some dynamical system which accounts for its multiplicative hierarchical structure. The reported application of a wavelet based tree matching algorithm to characterize the fractal properties of DLA azimuthal Cantor sets has revealed the existence of a Fibonacci multiplicative process in the apparently disordered arborescent morphology of DLA clusters. This discovery is a very spectacular manifestation of the statistical relevance of the golden mean arithmetic to Laplacian fractal growth phenomena. We are convinced that further applications of this wavelet based thermodynamics (the WTMM method) and statistical mechanics (the wavelet based tree matching algorithm for solving the inverse fractal problem) will lead to similar major breakthroughs in various fields where multi-scale phenomena are ubiquitous. Applications to hydrodynamic turbulent dynamics (2D and 3D), critical fluctuations in colloidal systems, surface roughening in noise driven growth processes and DNA 'walks' nucleotide sequences are currently in progress.

In the present study we have mainly focused on fractal objects which are made of a more or less complicated hierarchy of non-oscillating singularities. However there are many examples in nature where oscillating singularities play an important role, e.g., spiral vortices in turbulent flows. In our canonical thermodynamic description of fractal signals involving non-oscillating singularities, we have mainly considered the fluctuations in the value of the Holder exponent h from one maxima line to the next in the WTMM skeleton since the distance between two adjoining maxima lines behaves systematically as the scale a of the analysing wavelet. A straightforward calculation shows that for a spiral-type signal of the form: $s(x) = x^\alpha \sin(1/x^\beta)$, the distance between two adjoining maxima lines in the WTMM skeleton scales like a^φ with $\varphi = 1/(\beta + 1)$. In a forthcoming publication, we hope to elaborate on a grand canonical description which will take also into account the fluctuations of the exponent φ in the branching process of the WTMM skeleton. This extended multifractal formalism is likely to provide a general framework for a unified thermodynamic theory of a large class of fractal distributions involving non-oscillating as well as oscillating singularities.

Acknowledgement

We are very grateful to F. Argoul, J. Elezgaray and M. Tabard for stimulating discussions. This work was supported by the Direction des Recherches, Etudes et Techniques (DRET) under contrat no. 92/097 and by the Centre National des Etudes Spatials (CNES) under contrat no. 92/0225.

References

[1] B.B. Mandelbrot, *Fractals: Form, Chance and Dimension* (Freeman, San Francisco, 1977)

[2] B.B. Mandelbrot, *The Fractal Geometry of Nature* (Freeman, San Francisco, 1982)

[3] H.E. Stanley and N. Ostrowsky, eds., *On Growth and Form: Fractal and Nonfractal Patterns in Physics* (Martinus Nijhof, Dordrecht, 1986)

[4] L. Pietronero and E. Tosatti, eds., *Fractals in Physics* (North-Holland, Amsterdam, 1986)

[5] W. Guttinger and D. Dangelmayr, eds., *The Physics of Structure Formation* (Springer-Verlag, Berlin, 1987)

[6] H.E. Stanley and N. Ostrowsky, eds., *Random Fluctuations and Patterns Growth* (Kluwer, Dordrecht, 1988)

[7] J. Feder, *Fractals* (Pergamon, New York, 1988)

[8] T. Vicsek, *Fractal Growth Phenomena* (World Scientific, Singapore, 1989)

[9] A. Aharony and J. Feder, eds., Fractals in Physics, Essays in Honour of B.B. Mandelbrot, *Physica D* **38** (1989)

[10] L. Pietronero, ed., *Fractals' Physical Origin and Properties* (Plenum, New York, 1989)

[11] F. Family and T. Vicsek, *Dynamics of Fractal Surfaces* (World Scientific, Singapore, 1991)

[12] A. Bunde and S. Havlin, *Fractals and Disordered Systems* (Springer-Verlag, Berlin, 1991)

[13] S.K. Ma, *Modern Theory of Critical Phenomena* (Benjamin Reading, Mass., 1976)

[14] D. Amit, *Field Theory, the Renormalization Group and Critical Phenomena* (McGraw-Hill, New York, 1978)

[15] G. Nicolis and I. Prigogine, *Self-Organization in Non-Equilibrium Systems* (Wiley, New York, 1977)

[16] H. Haken, *Synergetics: An Introduction* (Springer-Verlag, New York, 1976); *Advanced Synergetics* (Springer-Verlag, New York, 1983)

[17] P. Cvitanovic, ed., *Universality in Chaos* (Hilger, Bristol, 1984)

[18] B.L. Hao, ed., *Chaos* (World Scientific, Singapore, 1984)

[19] G. Paladin and A. Vulpiani, *Phys. Rep.* **156** (1987) 148

[20] P. Manneville, *Dissipative Structures and Weak Turbulence* (Academic Press, New York, 1990)

[21] M.C. Cross and P.C. Hohenberg, *Rev. Mod. Phys.* **65** (1993) 851

[22] R. Balian, *Du Microscopique au Macroscopique* (Ecole Polytechnique, Ellipses, Palaiseau, 1982)

[23] Y.G. Sinai, *J. Russ. Math. Surveys* **166** (1972) 21

[24] R. Bowen, *Lect. Notes in Maths.* **470** (Springer-Verlag, New York, 1975) 1

[25] R. Ruelle, *Thermodynamic Formalism* (Addison Wesley, Reading, 1978)

[26] M.J. Feigenbaum, M.H. Jensen and I. Procaccia, *Phys. Rev. Lett.* **57** (1986) 1503

[27] M.H. Jensen, L.P. Kadanoff and I. Procaccia, *Phys. Rev.* A **36** (1987) 1409

[28] A.B. Chhabra, R.V. Jensen and K.R. Sreenivasan, *Phys. Rev.* A **40** (1989) 4593

[29] U. Frisch and G. Parisi, in: *Turbulence and Predictability in Geophysical Fluid Dynamic and Climate Dynamics*, eds., M. Ghil, R. Benzi and G. Parisi (North-Holland, Amsterdam, 1985) p. 84

[30] T.C. Halsey, M.H. Jensen, L.P. Kadanoff, I. Procaccia and B.I. Shraiman, *Phys. Rev.* A **33** (1986) 1141

[31] R. Benzi, G. Paladin, G. Parisi and A. Vulpiani, *J. Phys.* A **17** (1984) 3521

[32] E.B. Vul, Ya.G. Sinai and K.M. Khanin, *J. Russ. Math. Surv.* **39** (1984) 1

[33] R. Badii, Thesis, University of Zurich (1987)

[34] P. Collet, J. Lebowitz and A. Porzio, *J. Stat. Phys.* **47** (1987) 609

[35] M.J. Feigenbaum, *J. Stat. Phys.* **46** (1987) 919, 925

[36] R. Rand, *Ergod. Th. and Dyn. Sys.* **9** (1989) 527

[37] B.B. Mandelbrot, Fractals and Multifractals: Noise, Turbulence and Galaxies, Selecta Vol. 1 (Springer-Verlag, New York, 1989); in: *Random Fluctuations and Patterns Growth*, H.E. Stanley and N. Ostrowsky, eds. (Kluwer, Dordrecht, 1988), p. 279

[38] P. Meakin, in: *Phase Transitions and Critical Phenomena*, eds., C. Domb and J.L. Lebowitz (Academic Press, Orlando, 1988), Vol. 12, p. 355

[39] B.B. Mandelbrot, *J. Fluid Mech.* **62** (1974) 331

[40] U. Frisch and S.A. Orzag, Turbulence: challenges for theory and experiments, *Physics Today* (1990) 24

[41] C. Meneveau and K.R. Sreenivasan, *J. Fluid Mech.* **224** (1991) 429

[42] A. Renyi, *Probability Theory* (North-Holland, Amsterdam, 1970)

[43] P. Grassberger, *Phys. Lett.* A **97** (1983) 227

[44] H.G.E. Hentschel and I. Procaccia, *Physica D* **8** (1983) 435

[45] P. Grassberger and I. Procaccia, *Physica D* **13** (1984) 34

[46] P. Grassberger, R. Badii and A. Politi, *J. Stat. Phys.* **51** (1988) 135

[47] T. Bohr and T. Tèl, in: *Direction in Chaos*, Vol. 2, ed., B.L. Hao (World Scientific, Singapore, 1988)

[48] B.B. Mandelbrot and J.W. Van Ness, *S.I.A.M. Rev.* **10** (1968) 422

[49] H.O. Peitgen and D. Saupe, eds., *The Science of Fractal Image* (Springer-Verlag, New York, 1987)

[50] B.B. Mandelbrot, *J. Bus. Univ. Chicago* **40** (1967) 393

[51] B.B. Mandelbrot and H.M. Taylor, *Oper. Res.* **15** (1967) 1057

[52] W. Li, *Int. J. of Bifurcation and Chaos* **1** (1991) 583

[53] M. Goodchild, *Math. Geo.* **12** (1980) 85

[54] F. Anselmet, Y. Gagne, E. Hopfinger and R. Antonia, *J. Fluid Mech.* **140** (1984) 63

[55] Y. Gagne, E. Hopfinger and U. Frisch, in *New Trends in Nonlinear Dynamics and Pattern Forming Phenomena: The Geometry of Nonequilibrium*, eds., P. Huerre and P. Coullet (Plenum, New York, 1988)

[56] C.K. Peng, S.V. Buldyrev, A.N. Golberger, S. Havlin, F. Sciortino, M. Simons and H.E. Stanley, *Nature* **356** (1992) 158

[57] A.L. Barabási and T. Vicsek, *Phys. Rev. A* **44** (1991) 2730

[58] J.F. Muzy, Thesis, University of Nice (1993)

[59] J.F. Muzy, E. Bacry and A. Arneodo, *Phys. Rev. Lett.* **67** (1991) 3515

[60] A. Arneodo, E. Bacry and J.F. Muzy, *Wavelet analysis of fractal signals: direct determination of the singularity spectrum of fully developed turbulence data* (Springer-Verlag, Berlin, 1991) to appear

[61] J.F. Muzy, E. Bacry and A. Arneodo, *Phys. Rev. E* **47** (1993) 875

[62] E. Bacry, J.F. Muzy and A. Arneodo, *J. Stat. Phys.* **70** (1993) 635

[63] J.F. Muzy, E. Bacry and A. Arneodo, in: *Progress in Wavelet Analysis and Applications*, Y. Meyer and S. Roques, eds. (Frontières, Gif sur Yvette, 1993), p. 323

[64] J.M. Combes, A. Grossmann and P. Tchamitchian, eds., *Wavelets* (Springer-Verlag, Berlin, 1988)

[65] Y. Meyer, *Ondelettes* (Herman, Paris, 1990)

[66] P.G. Lemarié, *Les Ondelettes en 1989* (Springer-Verlag, Berlin, 1990)

[67] Y. Meyer, ed., *Wavelets and Applications* (Springer-Verlag, Berlin, 1992)

[68] I. Daubechies, *Ten Lectures on Wavelets* (S.I.A.M., Philadelphia, 1992)

[69] M.B. Ruskai, G. Beylkin, R. Coifman, I. Daubechies, S. Mallat, Y. Meyer and L. Raphael, eds., *Wavelets and Their Applications* (Jones and Bartlett, Boston, 1992)

[70] Y. Meyer and S. Roques, eds., *Progress in Wavelet Analysis and Applications* (Frontières, Gif sur Yvette, 1993)

[71] M. Holschneider, *J. Stat. Phys.* **50** (1988) 963; Thesis, University of Aix-Marseille II (1988)

[72] S. Jaffard, *C.R. Acad. Sci. Paris*, **308**, série 1 (1989) 79; **315**, série 1 (1992) 19

[73] M. Holschneider and P. Tchamitchian, in: *Les Ondelettes en 1989*, P.G. Lemarié, ed. (Springer-Verlag, Berlin, 1990), p. 102

[74] S. Mallat and W.L. Hwang, *IEEE Trans. on Information Theory* **38** (1992) 617

[75] A. Arneodo, G. Grasseau and M. Holschneider, *Phys. Rev. Lett.* **61** (1988) 2281; in: *Wavelets*, J.M. Combes, A. Grossmann and P. Tchamitchian, eds. (Springer-Verlag, Berlin, 1988), p. 182

[76] A. Arneodo, F. Argoul, J. Elezgaray and G. Grasseau, in: *Nonlinear Dynamics*, G. Turchetti, ed. (World Scientific, Singapore, 1989) p. 130

[77] F. Argoul, A. Arneodo, G. Grasseau, Y. Gagne, E. Hopfinger and U. Frisch, *Nature* **338** (1989) 52

[78] A. Arneodo, F. Argoul and G. Grasseau, in: *Les Ondelettes en 1989*, P.G. Lemarié, ed. (Springer-Verlag, Berlin, 1990), p. 125

[79] F. Argoul, A. Arneodo, J. Elezgaray, G. Grasseau and R. Murenzi, *Phys. Lett. A* **135** (1989) 327; *Phys. Rev. A* 41 (1990) 5537

[80] A. Arneodo, F. Argoul, E. Bacry, J. Elezgaray, E. Freysz, G. Grasseau, J.F. Muzy and B. Pouligny, in: *Wavelets and Applications*, Y. Meyer, ed. (Springer-Verlag, Berlin, 1992), p. 286

[81] A. Arneodo, F. Argoul, E. Freysz, J.F. Muzy and B. Pouligny, in: *Wavelets and Their Applications*, M.B. Ruskai, G. Beylkin, R. Coifman, I. Daubechies, S. Mallat, Y. Meyer and L. Raphael, eds. (Jones and Bartlett, Boston, 1992), p. 241

[82] J.F. Muzy, E. Bacry and A. Arneodo, *Int. J. of Bifurcation and Chaos* **4** (1994) 245

[83] A. Arneodo, E. Bacry and J.F. Muzy, *Europhys. Lett.* **25** (1994) 479

[84] A. Arneodo, F. Argoul, E. Bacry, J.F. Muzy and M. Tabard, *Fractals* **1** (1993) 629

[85] M.F. Barnsley and S.G. Demko, *Proc. R. Soc. London A* **399** (1985) 243

[86] M.F. Barnsley, *Fractals Everywhere* (Academic Press, New York, 1988)

[87] C.R. Handy and G. Mantica, *Physica D* **43** (1990) 17

[88] J.D. Farmer, E. Ott and J.A. Yorke, *Physica D* **7** (1983) 153

[89] T. Tèl, *Z. Naturforsch.* **43a** (1988) 1154

[90] G. Grasseau, Thesis, University of Bordeaux I (1989)

[91] A. Arneodo, G. Grasseau and E. Kostelich, *Phys. Lett. A* **124** (1987) 426

[92] R. Badii and G. Broggi, *Phys. Lett. A* **131** (1988) 339

[93] C. Meneveau and K. R. Sreenivasan, *Phys. Lett..A* **1** (1989) 103

[94] W. Vander der Water and P. Schram, *Phys. Rev. A* **37** (1989) 3118

[95] A.B. Chhabra and R.V. Jensen, *Phys. Rev. Lett.* **62** (1989) 1327

[96] A.B. Chhabra, C. Meneveau, R.V. Jensen and K.R. Sreenivasan, *Phys. Rev. A* **40** (1989) 5284

[97] A.N. Kolmogorov, *Dokl. Akad. Nauka. USSR* **119** (1958) 861

[98] R. Badii and A. Politi, *Phys. Lett. A* **104** (1984) 303

[99] L.A. Smith, J.D. Fournier and E.A. Spiegel, *Phys. Lett. A* **114** (1986) 465

[100] D. Bessis, J.D. Fournier, G. Servizi, G. Turchetti and S. Vaienti, *Phys. Rev. A* **36** (1987) 920

[101] P. Cvitanovic, in: *Proc. Group Theoretical Methods in Physics*, ed., R. Gilmore (World Scientific, Singapore, 1987)

[102] P. Goupillaud, A. Grossmann and J. Morlet, *Geoexploration* **23** (1984) 85

[103] A. Grossmann and J. Morlet, *S.I.A.M. J. Math. Anal.* **15** (1984) 723; in: Mathematics and Physics, Lectures on Recent Results, ed. L. Streit (World Scientific, Singapore, 1985)

[104] E. Bacry, A. Arneodo, U. Frisch, Y. Gagne and E.J. Hopfinger, in *Turbulence and Coherent Structures*, eds. M. Lesieur and O. Metais (Kluwer, Dordrecht, 1991), p. 203

[105] M. Vergassola, R. Benzi, L. Biferale and D. Pissarenko, *J. Phys. A* **26** (1993) 6093

[106] M. Vergassola and U. Frisch, *Physica D* **54** (1991) 58

[107] F.T. Dyson, *Comm. Math. Phys.* **21** (1971) 269

[108] D. Katzen and I. Procaccia, *Phys. Rev. Lett.* **58** (1987) 1169

[109] P. Szépfaluzy, T. Tèl, A. Csordás and Z. Kovács, *Phys. Rev.* A **36** (1987) 523
[110] T. Bohr and M.H. Jensen, *Phys. Rev.* A **36** (1987) 4904
[111] T. Bohr, P. Cvitanovic and M.H. Jensen, *Europhys. Lett.* **6** (1988) 445
[112] A.S. Monin and A.M. Yaglom, *Statistical Fluid Mechanics*, Vol. II (MIT Press, 1971)
[113] U. Frisch, *Physica Scripta* **T9** (1985) 137; From global (Kolmogorov 1941) scaling to local (multifractal) scaling in fully developed turbulence, *Proc. R. Soc.* A, special issue on A.N. Kolmogorov, in press
[114] A.N. Kolmogorov, *C.R. Acad. Sci. USSR* **30** (1941) 301
[115] E.A. Novikov and R.W. Stewart, *Izv. Akad. Nauk. USSR, Goefiz.* **3** (1964) 408
[116] U. Frisch, P.L. Sulem and M. Nelkin, *J. Fluid Mech.* **87** (1978) 719
[117] B. Mandelbrot, *J. Stat. Phys.* **34** (1984) 895
[118] M. Nelkin, *J. Stat. Phys.* **54** (1989) 1
[119] R.R. Prasad, C. Meneveau and K.R. Sreenivasan, *Phys. Rev. Lett.* **61** (1989) 74
[120] P. Miller and P. Dimotakis, *Phys. Fluids* A **3** (1991) 168
[121] Y. Gagne, Thesis, University of Grenoble (1987)
[122] B. Castaing, Y. Gagne and E.J. Hopfinger, *Physica D* **46** (1990) 177
[123] M. Farge, *Annu. Rev. Fluid Mech.* **24** (1992) 395
[124] S. Douady, Y. Couder and M.E. Brachet, *Phys. Rev. Lett.* **67** (1991) 983
[125] S. Fauve, C. Laroche and B. Castaing, *J. Phys. II France* **3** (1993) 271
[126] E. Siggia, *J. Fluid Mech.* **107** (1981) 375
[127] M.E. Brachet, *C.R. Acad. Sci. Paris* **311** (1990) 775; *Fluid Dyn.* **8** (1991) 1
[128] Z.S. She, E. Jackson and S.A. Orzag, *Nature* **344** (1990) 6263, 226
[129] A. Vincent and M. Meneguzi, *J. Fluid Mech.* **255** (1991) 1
[130] J. Jiménez, A.A. Wray, P.G. Saffman and R.S. Rogallo, *J. Fluid Mech.* **255** (1993) 65
[131] A.N. Kolmogorov, *J. Fluid Mech.* **13** (1962) 82
[132] B.B. Mandelbrot, *Pure Appl. Geophys.* **131** (1989) 5
[133] A. Arneodo, J.F. Muzy, S. Roux, O. Cadot, Y. Couder and S. Douady, in preparation
[134] M.J. Feigenbaum, *J. Stat. Phys.* **21** (1979) 669
[135] M.J. Feigenbaum, *Los Alamos Sci.* **1** (1980) 4; *Comm. Math. Phys.* **77** (1980) 65
[136] W.L. Hwang and S. Mallat, *Characterization of self-similar multifractals with wavelet maxima*, New York University Technical Report no. 641 (Computer Science Department, 1993)
[137] F. Hausdorff, *Mathematische Annalen* **79** (1919) 157
[138] B. Hu, *Phys. Rep.* **91** (1982) 232
[139] J.P. Crutchfield, J.D. Farmer and B.A. Huberman, *Phys. Rep.* **92** (1982) 47
[140] P. Coullet, in *Chaos and Statistical Methods*, ed. E. Kuramoto (Springer-Verlag, Berlin, 1984)
[141] F. Argoul and A. Arneodo, in Lyapunov Exponents, eds. L. Arnold and V. Wichstutz, *Lect. Notes* in *Maths.* **1186** (1986) 338
[142] M.J. Feigenbaum, *J. Stat. Phys.* **19** (1978) 25
[143] P. Coullet and C. Tresser, *J. Physique*, Coll. **39** (1978) C5; in: *Field Theory, Quantization and Statistical Physics*, ed. E. Tirapegui (Reidel, Dordrecht, 1981) p. 249
[144] C. Tresser and P. Coullet, *C.R. Acad. Sci.* **287** (1978) 577
[145] S.J. Shenker, *Physica D* **5** (1982) 405
[146] M.J. Feigenbaum, L.P. Kadanoff and S.J. Shenker, *Physica D* **5** (1982) 370

[147] S. Ostlund, D. Rand, J.P. Sethna and E.D. Siggia, *Phys. Rev. Lett.* **49** (1982) 132; *Physica D* **8** (1983) 303

[148] P. Collet and J.P. Eckmann, *Iterated Maps on the Interval as Dynamical Systems* (Birkhauser, Boston, 1980)

[149] F. Ledrappier and M. Misiurewicz, *Ergod. Theor. Dyn. Syst.* **5** (1985) 595

[150] B. Derrida, A. Gervois and Y. Pomeau, *J. Phys.* A **12** (1979) 269

[151] T.A. Witten and L.M. Sander, *Phys. Rev. Lett.* **47** (1981) 1400; *Phys. Rev.* B **27** (1983) 5686

[152] H.E. Stanley, A. Bunde, S. Havlin, J. Lee, E. Roman and S. Schwarzer, *Physica A* **168** (1990) 23

[153] F. Argoul, A. Arneodo, G. Grasseau and H.L. Swinney, *Phys. Rev. Lett.* **61** (1988) 2558; **63** (1989) 1323

[154] G. Li, L.M. Sander and P. Meakin, *Phys. Rev. Lett.* **63** (1989) 1322

[155] T. Vicsek, F. Family and P. Meakin, *Europhys. Lett.* **12** (1990) 217

[156] A. Arneodo, F. Argoul, E. Bacry, J.F. Muzy and M. Tabard, *Phys. Rev. Lett.* **68** (1992) 3456; in: *Growth Patterns in Physical Sciences and Biology*, eds. J.M. Garcia-Ruiz, E. Louis and L.M. Sander (Plenum, New York, 1993) p. 191

[157] A. Arneodo, F. Argoul, J.F. Muzy and M. Tabard, *Phys. Lett.* A **171** (1992) 31; *Physica A* **188** (1992) 217

[158] A. Arneodo, F. Argoul, E. Bacry, J. Elezgaray, J.F. Muzy and M. Tabard, in: *Progress in Wavelet Analysis and Applications*, Y. Meyer and S. Roques, eds. (Frontières, Gif sur Yvette, 1993), p. 21

[159] A. Arneodo, F. Argoul, E. Bacry, J.F. Muzy and M. Tabard, *Uncovering a multiplicative process in one-dimensional cuts of diffusion-limited aggregates*, preprint (1993)

[160] T.H. Garland, *Fascinating Fibonacci: Mystery and Magic in Numbers* (Dayle Seymour, Palo Alto, 1987)

10

Wavelets in medicine and physiology

P.Ch. IVANOV[1], A.L. GOLDBERGER[2],
S. HAVLIN[1,3], C.-K. PENG[1,2], M.G. ROSENBLUM[1]
and H.E. STANLEY[1]

[1]*Center for Polymer Studies and Department of Physics,
Boston University, Boston, MA 02215, USA*

[2]*Cardiovascular Division, Harvard Medical School, Beth Israel
Hospital, Boston, MA 02215, USA*

[3]*Gonda-Goldschmid Center and Department of Physics
Bar-Ilan University, Ramat-Gan 52900, Israel*

Abstract

We present a combined wavelet and analytic signal approach to study biological and physiological nonstationary time series. The method enables one to reduce the effects of nonstationarity and to identify dynamical features on different time scales. Such an approach can test for the existence of universal scaling properties in the underlying complex dynamics. We applied the technique to human cardiac dynamics and find a universal scaling form for the heartbeat variability in healthy subjects. A breakdown of this scaling is associated with pathological conditions.

10.1 Introduction

The central task of statistical physics is to study macroscopic phenomena that result from microscopic interactions among many individual components. This problem is akin to many investigations undertaken in biology. In particular, physiological systems under neuroautonomic regulation, such as heart rate regulation, are good candidates for such an approach, since: (i) the systems often include multiple components, thus leading to very large numbers of degrees of freedom, and (ii) the systems usually are driven by competing forces. Therefore, it seems reasonable to consider the possibility that dynamical systems under neural regulation may exhibit temporal structures which are similar, under certain conditions, to those found in physical systems. Indeed, concepts and techniques originating in statistical physics are showing promise as useful tools for quantitative analysis of complicated physiological systems.

An unsolved problem in biology is the quantitative analysis of a nonstationary time series generated under free-running conditions [1–3]. The signals obtained under these constantly varying conditions raise serious challenges to both technical and theoretical aspects of time series analyses. A central question is whether such noisy fluctuating signals contain dynamical patterns essential for understanding underlying physiological mechanisms.

Representative examples of complex dynamical behaviour under physiologic and pathologic conditions are shown in Figure 10.1. Figure 10.1(a) shows a physiologic cardiac interbeat time series—the output of a spatially and temporally integrated neuroautonomic control system. The time series shows erratic fluctuations and 'patchiness'. These fluctuations are usually ignored in conventional studies which focus on averaged quantities. In fact, these fluctuations are still often labelled as 'noise' to distinguish them

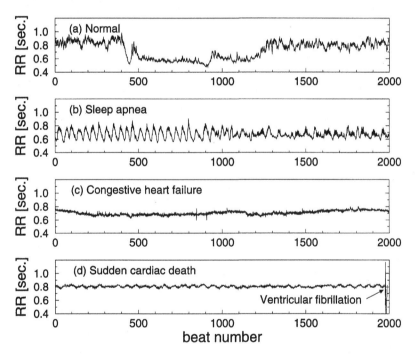

Fig. 10.1. Representative complex physiological fluctuations. Cardiac interbeat interval (normal sinus rhythm) time series of 2000 beats from (a) a healthy subject, (b) a subject with obstructive sleep apnea, (c) a subject with congestive heart failure and (d) a sudden cardiac death subject with ventricular fibrillation. Note the nonstationarity (patchiness) of these time series [most apparent in (a) and (b)]. Although these patches clearly differ in their amplitude and frequency of variations, their quantitative characterization remains an open problem and limits the applicability not only of traditional methods of analysis and modelling, but also newer techniques based on 'chaos' theory.

from the true 'signal' of interest. Furthermore, these patterns change with pathological perturbations (shown in Figures 10.1(b)–10.1(d)). However, with the recent adaption and extension of methods developed in statistical physics and nonlinear mathematics, it has been found that the physiological fluctuations shown in Figure 10.1(a) exhibit an unexpected hidden scaling structure [4–8]. These findings raise the possibility that understanding the origin of such temporal structures and their alterations may (i) elucidate certain basic features of heart rate control mechanisms, and (ii) have practical value in clinical monitoring.

When analysing complex cardiac fluctuations of the type shown in Figure 10.1(a), we must carefully exclude two obvious explanations for these observed structures: (i) they are simply an epiphenomenon of random (uncorrelated) trends, or (ii) they are a trivial consequence of the fact that cardiac function under neuroautonomic control is actually modulated by independent mechanisms with many time scales. To address the first possibility, researchers have recently developed and implemented methods to deal with the technical issue of nonstationarity in cardiac time series. To test the second possibility, numerically simulated systems with multiple time scales were studied, leading to the conclusion that robust scaling structures *cannot* be generated trivially from systems modulated by multiple time scales [8]. Instead, certain unique conditions are required to yield the structures observed. Furthermore, these two 'mechanisms' will not account for the observation of consistent changes in scaling patterns under pathological conditions, where complex nonstationarity and multiple time scale modulation are also present, but in altered form.

Among the difficulties associated with research on biomedical systems is not only the extreme variability of the signals but also the necessity of operating on a case-by-case basis. Often one does not know *a priori* which information is pertinent and on what scale it is located. Another important aspect of biomedical signals is that the information of interest is often a combination of features that are well-localized (temporally or spatially) and others that are more diffuse. As a result, the problems require the use of methods sufficiently robust to handle events that can be at opposite extremes in terms of their time–frequency localization. In the past few years, researchers have developed powerful wavelet methods for multiscale representation and analysis of signals [9–17]. These new tools differ from the traditional Fourier techniques in that they localize information in the time–frequency plane and are especially suitable for the analysis of nonstationary data signals.

Due to the wide variety of signals and problems encountered in medicine and biology, the spectrum of applications of the wavelet transform has been extremely large. It ranges from signal processing analysis of physiological signals in bioacoustics (e.g., turbulent heart sounds) [18–27], electrocardiography [28–42], and electroencephalography [43–53] to applications for compression [54–57] and enhancement [58–60] in biomedical imaging, noise reduction [61–63], detecting microcalcifications in mammograms [64–69], detection and reconstruction techniques for X-ray tomography [70, 71], magnetic resonance imaging [72–75], positron emission tomography [76], human vision [77–80], and human DNA [81, 82]. Extensive reviews of these applications have been recently published [83–86].

In this chapter, we present a method to analyse the properties of human cardiac activity by means of a wavelet transform and analytic signal approach designed to address nonstationary behaviour [7]. We find a universal scaling function for the distribution of the variations in the beat-to-beat intervals for healthy subjects. However, such a scaling function does not exist for a group with a cardiopulmonary instability due to sleep apnea (a condition in which breathing abnormalities during sleep affect cardiac activity). This scaling form allows us to express the global characteristics of a highly heterogeneous time series of interbeat intervals of each healthy individual with a single parameter. We find also that the observed scaling represents the Fourier phase correlations attributable to the underlying nonlinear dynamics. This approach has the potential to quantify the output of other nonlinear biological signals.

10.2 Nonstationary physiological signals

A time series is *stationary* if its statistical characteristics such as the mean and the variance are invariant under time shifts, i.e., if they remain the same when t is replaced by $t + \Delta$, where Δ is arbitrary. Then the probability densities, together with the moment and correlation functions, do not depend on the absolute position of the points on the time axis, but only on their relative configuration [87]. Non-stationarity, an important feature of biological variability, can be associated with regimes of different drifts in the mean value of a given signal, or with changes in its variance which may be gradual or abrupt.

Time series of beat-to-beat (RR) heart rate intervals (Figure 10.2(a)) obtained from digitized electrocardiograms are known to be nonstationary and exhibit extremely complex behaviour [88]. A typical feature of such nonstationary signals is the presence of 'patchy' patterns which change over time (Figure 10.2(b)). The mechanism underlying this complex heart rate variability is related to competing neuroautonomic inputs [89, 90]. Parasympathetic stimulation decreases the firing rate of pacemaker cells in

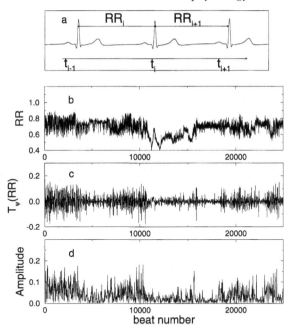

Fig. 10.2. (a) Segment of electrocardiogram showing beat-to-beat (RR_i) intervals. (b) Plot of RR-time series vs. consecutive beat number for a period of 6 h ($\approx 2.5 \times 10^4$ beats). Nonstationarity (patchiness) is evident over both long and short time scales. (c) Wavelet transform $T_\psi(RR)$ of the RR-signal in (b) using the second derivative of the Gaussian function $\psi^{(2)}$ as analysing wavelet with scale $a = 8$ beats. Nonstationarities related to constants and linear trends have been filtered. (d) Instantaneous amplitudes $A(t)$ of the wavelet-transform signal in (c); $A(t)$ calculated using the Hilbert transform measures the cumulative variations in the interbeat intervals over an interval proportional to the wavelet scale a.

the heart's sinus node. Sympathetic stimulation has the opposite effect. The nonlinear interaction (coupling) of the two branches of the nervous system is the postulated mechanism for the type of erratic heart rate variability recorded in healthy subjects [91–93]. We focus our studies on interbeat interval variability as an important tool for elucidating possibly non-homeostatic cardiac variability because (i) the heart rate is under direct neuroautonomic control, (ii) interbeat interval variability is readily measured by non-invasive means, and (iii) analysis of these heart rate dynamics may provide important diagnostic and prognostic information.

Even under healthy, basal conditions, the cardiovascular system shows erratic fluctuations resembling those found in dynamical systems driven away from a single equilibrium state [94]. Do such 'nonequilibrium' fluctuations [95] simply reflect the fact that physiological systems are being constantly perturbed by external and intrinsic noise? Or, do these fluctuations actually contain useful information about the underlying nonequilibrium control mechanisms?

Traditional approaches – such as the power spectrum and correlation analysis [96, 97] – are not suited for such nonstationary (patchy) sequences. In particular, they do not carry information stored in the Fourier phases which is crucial for determining nonlinear characteristics [98–100].

To address these problems, we develop a method – 'cumulative variation amplitude analysis' (CVAA) – to study the subtle structure of physiological time series. This method comprises sequential application of a set of algorithms based on wavelet and Hilbert transform analysis.

10.3 Wavelet transform

We first apply the wavelet transform (Figure 10.2(c)), because it does not require stationarity and it preserves important Fourier phase information. The wavelet transform [9, 101, 102] of a time series $s(t)$ is defined as

$$T_\psi(t_0, a) \equiv \frac{1}{a} \int_{-\infty}^{+\infty} s(t)\psi\left(\frac{t - t_0}{a}\right) dt, \qquad (10.1)$$

where the analysing wavelet ψ has a width of the order of the scale a and is centred at t_0. As pointed out in previous chapters, the wavelet transform is sometimes called a 'mathematical microscope' because it allows one to study properties of the signal on any chosen scale a. For high frequencies (small a), the ψ functions have good localization (being effectively non-zero only on small sub-intervals), so short-time regimes or high-frequency components can be detected by the wavelet analysis. However, a wavelet with too large a value of scale a (low frequency) will filter out almost the entire frequency content of the time series, thus losing information about the intrinsic dynamics of the system. We focus our 'microscope' on a scale $a = 8$ beats which smooths locally very high-frequency variations and best probes patterns of duration 30 s to 1 min. The wavelet transform is attractive because it can eliminate local polynomial behaviour (trends) in the nonstationary signal by an appropriate choice of the analysing wavelet ψ [103].

In our study we use derivatives of the Gaussian function,

$$\psi^{(n)} \equiv \frac{d^n}{dt^n} e^{-\frac{1}{2}t^2}. \qquad (10.2)$$

The first derivative is orthogonal to segments of the time series with an approximately constant local average. This results in fluctuations of the wavelet transform values around zero with highest spikes at the positions where a sharp transition occurs (Figure 10.3(b)). Thus, the larger spikes indicate the boundaries *between* regimes with different local average in the signal, and the smaller fluctuations represent variations of the signal within a given regime. With increasing wavelet scale a, the fluctuations become broader and reflect the dominant structures (variations) in the signal (Figure 10.3(c)) Since $\psi^{(1)}$ is not orthogonal to linear (non-constant) trends, the presence of consecutive linear trends (Figure 10.3(d)) in the RR-intervals will give rise to fluctuations of the wavelet transform values around different nonzero levels corresponding to the slopes of the linear trends (Figure 10.3(e)). The second derivative $\psi^{(2)}$ of the Gaussian function and higher order derivatives can eliminate the influence of linear as well as nonlinear trends in the fluctuations of the wavelet transform values (Figure 10.3(f)).

The wavelet transform allows one to 'extract' from the data particular features. The object is to probe the *variations* in the heart rate signal at different time scales. The particular choice of the derivatives of the Gaussian function as analysing wavelets allows us to extract these variations. One can argue that the same can be done by simply subtracting consecutive interbeat intervals by analysing the increments only, but such standard analysis does *not* distinguish healthy from unhealthy cardiac dynamics [5]. The reason is that the wavelet transform in addition to extracting the variations over given time-scale in the heart rate signal reduces masking effects of the nonstationarities since the analysing wavelet is orthogonal to local polynomial trends. The wavelet also filters out the very high-frequency noise in the original signal, preserving at the same time the sharpness of the edges separating different patterns in the signal, thus minimizing possibly artificial errors in the statistical analysis. Moreover, we find that the scale of the wavelet is crucial for extracting the *hidden* patterns in the cardiac dynamics. Thus, the ability of the wavelet transform to probe the signal on different scales is important for detecting essential features of cardiac dynamics under healthy as well as pathologic conditions.

The wavelet transform is thus a cumulative measure of the variations in the heart rate signal over a region proportional to the wavelet scale a, so the study of the behaviour of the wavelet values can reveal intrinsic properties of the dynamics masked by nonstationarity.

10.4 Hilbert transform

The wavelet transform signal at a fixed scale (Figure 10.2(c)) shows segments of different duration and amplitudes. So the next step of the CVAA is to

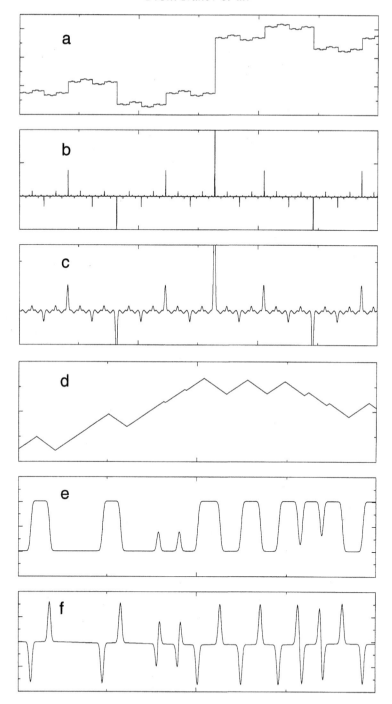

Fig. 10.3. Derivatives of the Gaussian function as analysing wavelet extract the singularities (variations) from a signal with (a) constant and (d) linear trends. Wavelet transform of the signal in (a) using $\psi^{(1)}$ as analysing wavelet with (b) smaller and (c) larger time scale. (e) $\psi^{(1)}$ and (f) $\psi^{(2)}$ are used on the signal in (d) at the same time scale.

extract the amplitudes of the variations in the beat-to-beat signal by means of an analytic signal approach [96, 104] which also does *not* require stationarity. This general approach, based on the Hilbert transform and originally introduced by Gabor [105], unambiguously gives the instantaneous phase and amplitude for a given signal $s(t)$ (in our case the wavelet transform of the interbeat interval time series) via construction of the analytic signal $S(t)$, which is a complex function of time defined as

$$S(t) \equiv s(t) + i\tilde{s}(t) = A(t)e^{i\phi(t)}. \tag{10.3}$$

Here $\tilde{s}(t)$ is the Hilbert transform of $s(t)$,

$$\tilde{s}(t) = \pi^{-1}\text{P.V.} \int_{-\infty}^{+\infty} \frac{s(\tau)}{t-\tau}d\tau \tag{10.4}$$

where P.V. means that the integral is taken in the sense of the Cauchy principal value. The amplitude is defined as

$$A(t) \equiv \sqrt{s^2(t) + \tilde{s}^2(t)} \tag{10.5}$$

and the phase as

$$\phi(t) \equiv \tan^{-1}(\tilde{s}(t)/s(t)). \tag{10.6}$$

The Hilbert transform $\tilde{s}(t)$ of $s(t)$ can be considered as the convolution of the functions $s(t)$ and $1/\pi t$. This means that the Hilbert transform can be realized by an ideal filter whose amplitude response is unity, and phase response is a constant $\pi/2$ lag at all frequencies [96]. A harmonic oscillation $s(t) = A\cos\omega t$ is often represented in the complex notation as $A\cos\omega t + jA\sin\omega t$. This means that the real oscillation is complemented by the imaginary part which is delayed in phase by $\pi/2$, and which is related to $s(t)$ by the Hilbert transform. The analytic signal is the direct and natural extension of this technique, as the Hilbert transform performs the $-\pi/2$ phase shift for every frequency component of an arbitrary signal.

Why do we need the instantaneous amplitude (envelope) of the signal? Suppose that our wavelet transform signal for a given scale consists of two segments (patches), both being sine waves with the amplitudes A and A'. Then the values of the signal for the first patch are distributed from $-A$ to A, and for the second patch from $-A'$ to A' ($A' > A$).

So the distributions of the data points values along the two patches of the signal overlap between $-A$ and A. However, if we consider the distributions of the instantaneous amplitudes of the data points from these two segments, then they *do not* overlap; they are, actually, two points, $P(A)$ and $P(A')$ with values reflecting the number of data points in each segment (Figure 10.4). By changing the wavelet scale we can learn about the distribution of patches with different duration.

10.5 Universal distribution of variations

Quantifying the probability distribution of variation amplitudes in the interbeat intervals can provide insights into the underlying dynamical processes because the distribution of interbeat intervals is directly related to the mechanisms which control heart rate variability. Therefore, by finding consistent features of the distribution which are robust with respect to different healthy subjects, we can quantify physiologic dynamics. However there are important technical difficulties which must first be overcome before such robust features can be found.

Among the possible reasons why an interbeat interval histogram can differ from case to case are the following. (i) Histograms can differ because they have different means and standard deviations but follow the same functional form. (ii) Histograms are described by different functional forms, i.e. they belong to different classes of processes. The first type of difference is commonly observed (especially in physiological data where significant variation

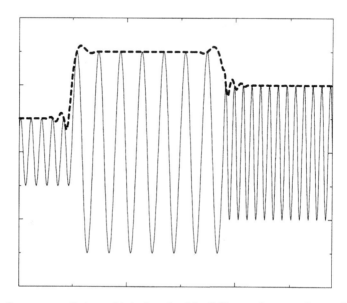

Fig. 10.4. Segments of sinusoidal signal with different frequencies and amplitudes (solid line) and their envelope obtained from Hilbert transform (dashed line).

between individuals is expected) and should be taken care of by properly 'renormalizing' (with respect to the mean and standard deviation) the histogram. If we assume that heart rate control mechanisms in healthy subjects follow the same general set of dynamical rules, then we expect that some variables of the system's output will be described by a single, well-defined distribution function. Functional differences between distributions, on the other hand, can be a result of altered mechanisms, and could be indicative of pathological behaviour.

We analysed the distribution of the amplitudes of the beat-to-beat variations (Figure 10.2(d)) for a group of healthy subjects ($N = 18$: 5 males and 13 females; age 20–50, mean 34) and a group of subjects [106] with obstructive sleep apnea [107, 108] ($N = 16$ males; age 32–56, mean 43). To minimize nonstationarity due to changes in the level of activity, we begin by considering night phase (12 p.m.–6 a.m.) records of interbeat intervals ($\approx 10^4$ beats) for both groups.

Inspection of the distribution functions of the amplitudes of the cumulative variations reveals marked differences between individuals (Figure 10.5(a)). These differences are not surprising given the underlying physiological differences among healthy subjects.

For the healthy group, we find that these distributions are well fit by the *generalized homogeneous* form [109] (the Gamma distribution):

$$P(x, b) = \frac{b^{\nu+1}}{\Gamma(\nu + 1)} x^\nu e^{-bx}, \tag{10.7}$$

where $b \equiv \nu/x_0$, $\Gamma(\nu + 1)$ is the Gamma function, x_0 is the position of the peak $P = P_{\max}$, and ν is a fitting parameter (Figure 10.6(a)). A function $P(x, b)$ is a generalized homogeneous function if there exist two numbers α and β – called scaling powers – such that for all positive values of the parameter λ

$$P(\lambda^\alpha x, \lambda^\beta b) = \lambda P(x, b). \tag{10.8}$$

Generalized homogeneous functions are defined as solutions of this functional equation. One can see that in our case, $P(x, b)$ satisfies (10.8) with $\alpha = -1$ and $\beta = 1$.

Functions describing physical systems near their critical points are known to be generalized homogeneous functions [110]. Data collapse is among the key properties of generalized homogeneous functions. Instead of data points falling on a family of curves, one for each value of b, data points can be made to *collapse* onto a single curve given by the scaling function

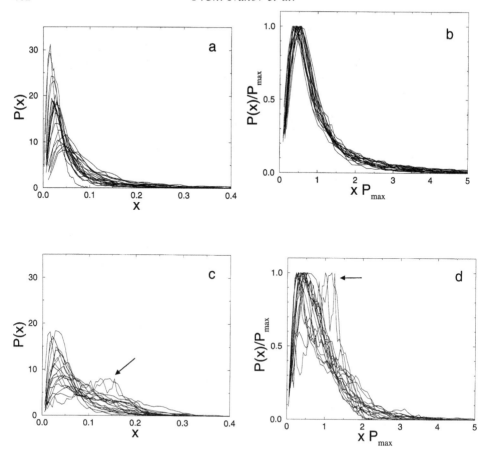

Fig. 10.5. (a) Probability distributions $P(x)$ of the amplitudes of heart rate variations $x \equiv A(t)$ for a group of 18 healthy adults (after wavelet transform with $\psi^{(2)}$ and scale $a = 8$ beats). Individual differences are reflected in the different average value and widths (standard deviations) of these distributions. All distributions are normalized to unit area. (b) Same probability distributions as in (a) after rescaling: $P(x)$ by P_{max}, and x by $1/P_{max}$ to preserve the normalization to unit area. This rescaling is equivalent to the scaling procedure discussed in the text (Eq. 10.9), since $P(x) \equiv P(x, b)$ and $P_{max} \propto b$. We are able to describe the distributions using a single curve, indicating a robust, consistent scaling mechanism for the nonequilibrium dynamics. (c) Probability distributions for a group of 16 subjects with obstructive sleep apnea. We note that the second (rightward) peak (arrow) in the distributions for the sleep apnea subjects corresponds to the transient emergence of characteristic pathologic oscillations in the heart rate associated with periodic breathing (Fig. 10.1b). (d) Distributions for the apnea group after the same rescaling as in (b). These distributions *cannot* be well described by a single curve, indicating that the nonequilibrium dynamics are altered.

$$\tilde{P}(u) \equiv \frac{P(x, b)}{b}, \tag{10.9}$$

where the number of independent variables is reduced by defining the scaled variable $u \equiv bx$. Our results show that a *common scaling* function $\tilde{P}(u)$ defines the probability density of the magnitudes of the variations in the beat-to-beat intervals for each healthy subject. Note that it is sufficient to specify *only one* parameter b in order to characterize the heterogeneous heartbeat variations for *each subject* in this group.

To test the hypothesis that there is a hidden, possibly universal, structure to these heterogeneous time series, we rescale the distributions and find for all healthy subjects that the data conform to a single scaled plot ('data collapse') (Figure 10.5(b)). Such behaviour is reminiscent of a wide class of well-studied physical systems with universal scaling properties [110, 111]. In contrast, the subjects with *sleep apnea* show individual probability distributions that *fail* to collapse (Figure 10.5(d)). The collapse of the individual distributions for all healthy subjects after rescaling their 'individual' parameter is indicative of a 'universal' structure. The term 'universal' is used in the sense that a closed mathematical scaling form is established to describe in a unified quantitative way the cardiac dynamics of all studied healthy subjects.

An analysis of the heart rate dynamics for healthy subjects during the daytime (noon–6 p.m.) indicates that the observed, apparently universal, behaviour holds not only for the night phase but for the day phase as well (Figure 10.6(b)). Semilog plots of the averaged distributions show a systematic deviation from the exponential form (slower decay) in the tails of the night-phase distributions, whereas the day-phase distributions follow the exponential form over practically the entire range. Note that the tail of the observed distribution for the night phase indicates higher probability of larger variations in the healthy heart dynamics during sleep hours in comparison with the daytime dynamics.

We observe for the healthy group good data collapse with a *stable* scaling form for wavelet scales $a = 2$ up to $a = 64$ (Figure 10.6(c)). However, for very small scales ($a = 1, 2$), the group average of the rescaled distributions of the apnea subjects is indistinguishable from the average of the rescaled distributions of the healthy group. Thus, very high frequency variations are equally present in the signals from both groups. Our analysis yields the most robust results when a is tuned to probe the collective properties of patterns with duration of $\approx \frac{1}{2} - 1$ min in the time series ($a = 8, 10$). The subtle difference in the tail of the distributions between day and night phases is also best seen for this scale range.

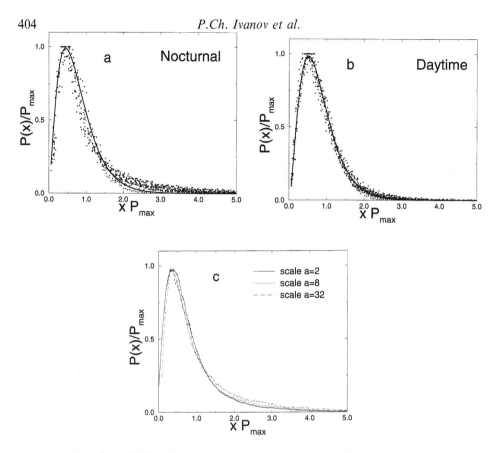

Fig. 10.6. (a) The solid line is an analytic fit of the rescaled distributions of the beat-to-beat variation amplitudes of the 18 healthy subjects during sleep hours to a stable Gamma distribution with $\nu = 1.4 \pm 0.1$. (b) Data for 6 h records of RR intervals for the day phase of the same control group of 18 healthy subjects demonstrate similar scaling behaviour with a Gamma distribution and $\nu = 1.8 \pm 0.1$, thereby showing that the observed common structure for the healthy heart dynamics is not confined to the nocturnal phase. (c) Group average of the rescaled distributions of the cumulative variation amplitudes for the healthy individuals during nocturnal hours. Note that the observed Gamma scaling is *stable* for a wide range of the wavelet transform scales a.

We note that direct analysis of interbeat interval histograms does *not* lead to data collapse or separation between the healthy and apnea group. Such histograms measured directly for each subject do not converge to a single representative curve describing healthy dynamics, because the interbeat interval time series is highly nonstationary. Even rescaling the time series to give all histograms identical means and variances does not lead to a common curve for the histograms. Moreover, the direct application only of the Hilbert transform yielding the probability distribution of the instantaneous amplitudes of the original signal does *not* distinguish clearly healthy from abnormal cardiac dynamics. Hence, the wavelet transform, with its ability to be orthogonal to polynomial trends and to probe the signal on different time scales, proves crucial to extract dynamical properties hidden in the cumulative variations, since different patterns can be observed on different time scales.

10.6 Wavelets and scale invariance

Differences between healthy and abnormal cardiac dynamics are known to be reflected in different correlations and power spectra [4–6, 97]. However, it is currently widely assumed in the literature that the difference in time series of interbeat intervals in sick and healthy adults *lies not in the distribution of the interbeat variations but rather in their time ordering*. This assumption is based on more conventional studies of interbeat increments [112]. These studies essentially amount to taking derivatives of the heart rate signal and thus extracting pointwise characteristics. Also, it has been hypothesized that even if the interbeat variations are different (e.g. smaller) during illness, the pattern of heart rate variability might be otherwise very similar to that during health, so that the interbeat variations for normal and abnormal cardiac dynamics, once normalized, would have the same distribution. Our study clearly rejects this hypothesis, showing the presence of scaling in the distributions of the variation amplitudes for the healthy (Figure 10.5(b)) and a breakdown of this scaling for abnormal dynamics (Figure 10.5(d)). Moreover, the stability of this scaling form (Figure 10.6(c)) indicates that the underlying dynamical mechanisms regulating the healthy heart beat have similar statistical properties on different time scales. Such statistical self-similarity is an important characteristic of fractal objects [98, 113]. The wavelet decomposition of beat-to-beat heart rate signals can be used to provide a visual representation of this fractal structure (Figure 10.7). The wavelet transform, with its ability to remove local trends and to extract interbeat variations on different time scales, enables us to identify self-similar patterns (arches) in these variations even when the signals change as a result of background interference. Data from sick hearts lack these patterns. Fractal characteristics of the cardiac dynamics and other biological signals can be successfully studied with the generalized multifractal formalism based on the wavelet transform modulus maxima method (WTMM) presented in Chapter 9.

Similar time scale invariance was observed in the experiments of Rodieck on the interspike intervals of a single neuron cell whose distribution was analysed by Gerstein and Mandelbrot [114]. For several types of single

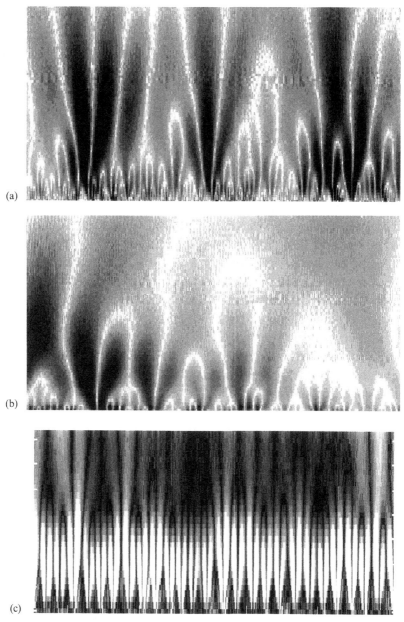

(a)

(b)

(c)

Fig. 10.7. Colour coded wavelet analysis of *RR* signals. (Colours referred to in this caption are shown at www.cambridge.org/resources/0521533538.) The *x*-axis represents time (≈ 2000 beats) and the *y*-axis indicates the scale of the wavelet used ($a = 1, 2, \ldots, 60$) with large scales at the top. The brighter colours indicate larger values of the wavelet amplitudes. (a) The wavelet analysis performed with $\psi^{(2)}$ (the Mexican hat) as an analysing wavelet uncovers a hierarchical scale invariance quantitatively expressed by the stability of the scaling form on Fig. 10.6(c). This wavelet decomposition reveals a self-similar fractal structure in the healthy cardiac dynamics – a magnification of the central portion of the top panel (a) with 200 beats on the *x*-axis and wavelet scale $a = 1, 2, \ldots, 25$ on the *y*-axis presented in (b) shows identical branching patterns. (c) Loss of this fractal structure in cases with sleep apnea (lower panel).

neuron cells Gerstein and Mandelbrot find that the interspike intervals distributions remain invariant with the time scale. However the heartbeat variations, unlike the single neuron dynamics, represent the integrated output of spatially and temporally distributed feedback system.

Analysis of the variance of the distributions for healthy cardiac dynamics at different time scales shows a power law behaviour with an exponent close to zero. This relates to previous studies reporting long-range anticorrelations in the heartbeat variations [5]. The findings that correlation functions and distributions describing physiological systems are not characterized by a single time scale become more plausible if we consider the survival advantage conferred upon organisms that evolved with an infinite hierarchy of time scales compared to organisms that evolved with a single characteristic time scale. Organisms with a physiologic control system generated by a single time scale are analogous, formally, to the famous Tacoma Narrows bridge, which survived many years until by chance a wind storm occurred that happened to correspond to the characteristic frequency (inverse of the characteristic time scale). Organisms that have survived millions of years have plausibly evolved some feature to render them immune from the analogue of the Tacoma bridge disaster, and this feature would seem to be the absence of any characteristic time scales (compare Figure 10.1(a) with 10.1(b) and 10.1(d), which show pathologic mode-locking).

10.7 A diagnostic for health vs. disease

We employ the Kolmogorov–Smirnov test to measure how similar two probability distributions are. A mathematical relation exists which links the Kolmogorov–Smirnov parameter $D(KS)$ to the corresponding statistical significance level [115]. The larger the value of $D(KS)$, the more unlikely it is that the two data sets were obtained from the same probability distribution (the null hypothesis).

The Kolmogorov–Smirnov test provides a simple measure that is defined as the *maximum value* of the absolute difference between two cumulative distribution functions.

The K–S test is defined as follows.

(i) Once the probability distribution $P(x)$ is found for a subject which we want to compare to a fit $P_0(x)$, the cumulative probability distribution $W(x)$ for the subject is found using the relation

$$W(x) \equiv \int_0^x P(x')dx',$$

and similarly for the the cumulative probability distribution $W_0(x)$ of the fit $P_0(x)$.

(ii) The absolute difference $\Delta W(x) \equiv |W(x) - W_0(x)|$ is found.

(iii) The maximum value of this absolute difference is defined as the K–S parameter (Figure 10.8(a)): $D(KS) \equiv \max[\Delta W(x)]$.

Once the distributions for the subjects and a fit for healthy subjects are found, we apply the K–S test to see how different each subject's distribution is from the fit. Comparing the individual distributions of the healthy and sleep apnea subjects with the reported scaling form (Eq. (10.9)) for the healthy dynamics, we find that the Kolmogorov–Smirnov test can serve as a potentially useful tool to separate healthy from abnormal cardiac dynamics (Figure 10.8(a)). The question of diagnostics motivates us to look more closely at the first and second moments of the distributions of the variation amplitudes for both groups. We find that a simple presentation of the values for these moments can be also effectively used to separate quantitatively the two groups. We present these results in Figure 10.8(b) – the first and second moments of the healthy distributions exhibit lower values with good linear fit, whereas for the apnea group these values are higher and dispersed with almost no overlap with the healthy data.

10.8 Information in the Fourier phases

Correlation functions measure how the value of some function depends on its value at an earlier time. Many simple systems in nature have correlation functions that decay with time in an exponential way. For systems comprised of many interacting subsystems, physicists discovered that such exponential decays do not occur. Rather, correlation functions were found to decay with a power-law form. The implication of this discovery is that in complex systems, there is no single characteristic time [119, 120]. If correlations decay with a power-law form, we say the system is 'scale free' since there is no characteristic scale associated with a power law. Since at large time scales a power law is always larger than an exponential function, correlations described by power laws are termed 'long-range' correlations – they are of longer range than exponentially-decaying correlations.

In physiological systems, recent work has suggested that such 'long-range' power-law correlations occur in a range of physiological systems [118, 121, 122] including, most remarkably, the intervals between successive heartbeats [5, 6]. The discovery of long-range correlations in these intervals is all the

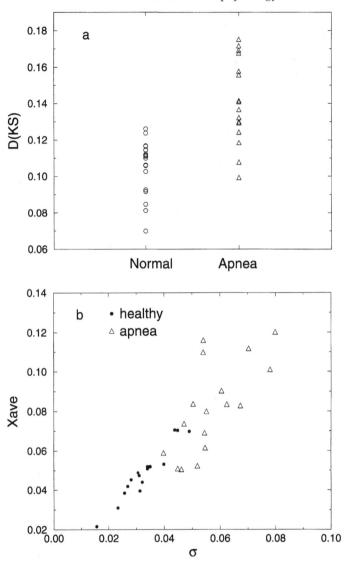

Fig. 10.8. (a) The Kolmogorov–Smirnov parameter $D(KS)$ and (b) the values of the first moments (mean and standard deviation σ) of the cumulative variation amplitude distributions can be used as a diagnostic of the healthy vs. apnea subjects with more then 80% true-positive recognition.

more interesting because it appears that these correlations are not present in certain disease states.

What are the possible adaptive advantages of the apparently far-from-equilibrium behaviour that appears to characterize the free-running dynamics of certain neural control systems? First, we note that complex erratic fluctuations shown in Figure 10.1(a) are *not* inconsistent with the general concept that physiological systems must operate with certain bounds. However, an intriguing possibility is that these complex nonequilibrium dynamics, rather than classical homeostatic *constancy*, may be a mechanism for maintaining physiologic stability. Such complex multiscale variability keeps the system from becoming locked to a dominant frequency (mode locking), a common manifestation of pathologic dynamics (Figure 10.1(b)). At the same time, long-range fractal correlations underlying these complex fluctuations may provide an important organizational mechanism for systems that lack a characteristic spatial or temporal scale. Finally, the intrinsic 'noisiness' of far-from-equilibrium dynamics may facilitate coping with unpredictable environmental stimuli.

However, these fractal correlations detected by Fourier and fluctuation analysis techniques, ignore information related to the phase interactions of component modes. The nonlinear interaction of these modes accounts for the visually 'patchy' appearance of the normal heartbeat time series.

To ascertain whether the observed scaling of the distributions for healthy subjects is an intrinsic property of normal heart beat dynamics, we test the cumulative variation amplitude analysis on artificially generated signals with known properties. Our analysis of uniformly distributed random numbers in the interval [0, 1] and of Gaussian-distributed noise with and without long-range power law correlations shows that after the wavelet transform the amplitude distributions follow the Rayleigh probability distribution

$$R(x) = \left(\frac{x}{\sigma^2}\right) e^{-x^2/\sigma^2}.$$

This finding agrees with the central limit theorem, which can be expressed as a property of convolutions (in our case wavelet transforms): the convolution of a large number of positive functions is approximately a Gaussian function, and the instantaneous amplitudes of a Gaussian process follow the Rayleigh probability distribution [87].

We perform parallel analysis on surrogate data obtained from a healthy subject by Fourier transforming the original time series, preserving the amplitudes of the Fourier transform but *randomizing the phases*, and performing an inverse Fourier transform (Figure 10.9(c)). Thus, both the original and surrogate signals have *identical* power spectra. Application of the CVAA method on this surrogate signal results again in a Rayleigh distribution, whereas the original time series has a distribution with an exponential tail. This test clearly indicates the important role of *phase correlations* in the RR time series. The presence of these correlations is most likely related to the

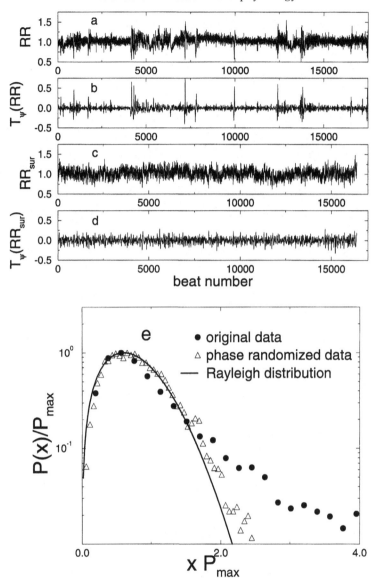

Fig. 10.9. (a) Original time series *RR* as a function of beat number. (b) Wavelet transform $T_\psi(RR)$ of this series. (c) Surrogate signal (RR_{sur}) after phase randomization. (d) Wavelet transform of the surrogate signal which is more homogeneous (less patchy) in comparison with (b). (e) Probability distributions of the amplitudes of variations after wavelet transform of the original and surrogate signals, as well as the theoretical Rayleigh distribution. The theoretical Rayleigh agrees with the distribution of the wavelet transform of the surrogate signal with randomized phases.

underlying nonlinear dynamics [117, 123]. The observed breakdown of this scaling pattern in the sleep apnea cases – a common and important instability of cardiopulmonary regulation – is possibly related to pathological mode locking associated with periodic breathing dynamics [116].

These tests show that the observed scaling in the variations of interbeat intervals for healthy dynamics actually represents the Fourier phase correlations. This result is non-trivial since it adds to an ongoing discussion about whether nonlinear phase interactions are present in healthy cardiac dynamics [91]. Furthermore, this finding suggests that, for healthy individuals, there may be a common structure to this nonlinear phase interaction. Also, the tests demonstrate that the scaling is not an artificial result of our approach in that it gives the expected results for known processes, i.e., a Rayleigh distribution for the amplitudes of uniformly distributed random numbers and for Gaussian noise as well. The basis of this robust temporal structure remains unknown and presents a new challenge to the understanding of nonlinear mechanisms of heartbeat control.

10.9 Concluding remarks

(i) Heart rate dynamics under normal conditions display nonequilibrium fluctuations that reveal a remarkable physiological structure when analysed using wavelets and methods adapted from statistical physics.

(ii) There is a hitherto unknown scaling pattern to interbeat interval variations in healthy subjects. This finding allows us to express the global characteristics of the highly heterogeneous heart rate time series of each healthy individual with only a *single parameter*. This scaling property cannot be explained by activity, since we analysed data from subjects during nocturnal hours. Moreover, it cannot be accounted for by sleep stage transitions, since we found a similar pattern during day-time hours.

(iii) This scaling is related to the intrinsic nonlinear dynamics of the control mechanism because it is due to information in the phase relationships. This information is *not* in the $1/f$ power spectrum on which all previous heart rate scaling is based, and any realistic attempt to model heart rate control will need to account for this scaling behaviour.

(iv) The reported results are also the first that clearly show a *difference* in the distributions of the interbeat variations for normal and abnormal heart dynamics. However, to observe it, one must:

(a) properly reduce masking effects of nonstationarity;
(b) account for the importance of time scales to reveal *hidden* scaling.

In both aspects the wavelet analysis proves superior to other more conventional techniques.

(v) The observation of nonlinear dynamics is *not* accounted for by traditional physiological mechanisms and motivates new modelling strategies to understand nonequilibrium control systems under healthy and pathologic conditions.

(vi) The wavelet-based method we present can be applied to other complex, non-stationary time series.

Acknowledgements

We thank Alain Arneodo, Hernán A. Makse, Martin Meyer, Luis A. N. Amaral and Alison Hill for discussions. This work was supported by NIH, NIMH, NASA, Binational Science Foundation USA-Israel and The G. Harold and Leila Y. Mathers Charitable Foundation.

References

[1] G. E. P. Box, G. M. Jenkins and G. C. Reinsel, *Time series analysis: forecasting and control* (Prentice-Hall, Englewood Cliffs, 1994)

[2] M. F. Shlesinger, 'Fractal time and $1/f$ noise in complex systems,' *Ann. NY Acad. Sci.* **504**, 214–228 (1987)

[3] L. S. Liebovitch, 'Testing fractal and Markov models of ion channel kinetics,' *Biophys. J.* **55**, 373–377 (1989)

[4] R. G. Turcott and M. C. Teich, 'Fractal character of the electrocardiogram: distinguishing heart-failure and normal patients,' *Ann. Biomed. Eng.* **24**, 269–293 (1996)

[5] C.-K. Peng, J. Mietus, J. M. Hausdorff, S. Havlin, H. E. Stanley and A. L. Goldberger, 'Long-range anti-correlations and non-Gaussian behaviour of the heartbeat,' *Phys. Rev. Lett.* **70**, 1343–1346 (1993)

[6] C.-K. Peng, S. Havlin, H. E. Stanley and A. L. Goldberger, 'Quantification of scaling exponents and crossover phenomena in nonstationary heartbeat time series,' in *Proc. NATO dynamical disease conference*, edited by L. Glass *Chaos* **5**, 82–87 (1995)

[7] P. Ch. Ivanov, M. G. Rosenblum, C.-K. Peng, J. Mietus, S. Havlin, H. E. Stanley and A. L. Goldberger, 'Scaling behaviour of heartbeat intervals obtained by wavelet-based time-series analysis,' *Nature* **383**, 323–327 (1996)

[8] J. M. Hausdorff and C.-K. Peng, 'Multiscaled randomness: a possible source of $1/f$ noise in biology,' *Phys. Rev. E* **54**, 2154–2157 (1996)

[9] I. Daubechies, 'Orthonormal bases of compactly supported wavelets,' *Comm. Pure and Appl. Math.* **41**, 909–996 (1988)

[10] M. Vetterli and J. Kovacevic, *Wavelets and Subbane Coding* (Prentice-Hall, Englewood Cliffs, NJ, 1995)

[11] H. Bray, K. McCormick, R. O. Wells, Jr. and X. D. Zhou, 'Wavelet variations on the Shannon sampling theorem,' *Biosystems* **34**, 249–257 (1995)

[12] H. L. Resnikoff, 'Analytic representation of compactly supported wavelets,' *Biosystems* **34**, 259–272 (1995)

[13] M. Fridman and J. M. Steele, 'Three statistical technologies with high potential in biological imaging and modeling,' *Basic Life Sci.* **63**, 199–224 (1994)

[14] G. Mayer-Kress, 'Localized measures for nonstationary time-series of physiological data,' *Integr. Physiol. Behav. Sci.* **29**, 205–210 (1994)

[15] L. Keselbrener and S. Akselrod, 'Time-frequency analysis of transient signals — application to cardiovascular control,' *Physica A* (1997) (in press)

[16] T. A. Gyaw and S. R. Ray, 'The wavelet transform as a tool for recognition of biosignals,' *Biomed. Sci. Instrum.* **30**, 63–68 (1994)

[17] S. Mallat and W. L. Hwang,' 'Singularity detection and processing with wavelets,' *IEEE Trans. Inform. Theory* **38**, 617–643 (1992)

[18] M. Akay, Y. M. Akay, W. Welkowitz and S. Lewkowicz, 'Investigating the effects of vasodilator drugs on the turbulent sound caused by femoral artery stenosis using short-term Fourier and wavelet transform methods,' *IEEE Trans. Biomed. Eng.* **41**, 921–928 (1994)

[19] L. Khadra, M. Matalgah, B. el-Asir and S. Mawagdeh, 'The wavelet transform and its applications to phonocardiogram signal analysis,' *Med. Inf. (London)* **16**, 271–277 (1991)

[20] M. S. Obaidad, 'Phonocardiogram signal analysis: techniques and performance,' *J. Med. Eng. and Techol.* **17**, 221–227 (1993)

[21] C. Heneghan *et al.*, 'Investigating the nonlinear dynamics of cellular motion in the inner ear using the short-time Fourier and continuous wavelet transforms,' *IEEE Trans. Signal Process.* **42**, 3335–3352 (1994)

[22] W. J. Lammers, A. el-Kays, K. Arafat and T. Y. el-Sharkawy, 'Wave mapping: detection of co-existing multiple wavefronts in high-resolution electrical mapping,' *Med. Biol. Eng. Comput.* **33**, 476–481 (1995)

[23] E. G. Pasanen, J. D. Travis and R. J. Thornhill, 'Wavelet-type analysis of transient-evoked otoacoustic emissions,' *Biomed. Sci. Instrum.* **30**, 75–80 (1994)

[24] H. P. Wit, P. van Dijk and P. Avan, 'Wavelet analysis of real ear and synthesized click evoked otoacoustic emissions,' *Hear. Res.* **73**, 141–147 (1994)

[25] J. R. Bulgrin, B. J. Rubal, C. R. Thompson and J. M. Moody, 'Comparison of short-time Fourier, wavelet and time-domain analyses of intracardiac sounds,' *Biomed. Sci. Instrum.* **29**, 465–478 (1993)

[26] X. L. Xu, A. H. Tewfik and J. F. Greenleaf, 'Time delay estimation using wavelet transform for pulsed-wave ultrasound,' *Ann. Biomed. Eng.* **23**, 612–621 (1995)

[27] M. C. Teich, C. Heneghan, S. M. Khanna, A. Flock, M. Ulfendahl and L. Brundin, 'Investigating routes to chaos in the guinea-pig cochlea using the continuous wavelet transform and the short-time Fourier transform,' *Ann. Biomed. Eng.* **23**, 583–607 (1995)

[28] C. Li and C. Zheng, 'QRS Detection by wavelet transform,' in *Proc. Annu. Conf. on Eng. in Med. and Biol.* **15**, 330–331 (1993)

[29] L. Khadr, H. Dickhaus and A. Lipp, 'Representations of ECG late potentials in the time frequency plane,' *J. Med. Eng. & Technol.* **17**, 228–231 (1993)

[30] H. Dickhaus, L. Khadra and J. Brachmann, 'Time-frequency analysis of ventricular late potentials,' *Methods Inform. Med.* **33**, 187–195 (1994)

[31] O. Meste, H. Rix, P. Caminal and N. V. Thakor, 'Ventricular late potentials characterization in time-frequency domain by means of a wavelet transform,' *IEEE Trans. Biomed. Eng.* **41**, 625–634 (1994)

[32] L. Senhadji, G. Carrault, J. J. Bellanger and G. Passariello, 'Comparing wavelet transforms for recognizing cardiac patterns,' *IEEE Eng. in Med. and Biol. Mag.* **14**, 167–173 (1995)

[33] M. Karrakchou, C. V. Lambrecht and M. Kunt, 'Analysing pulmonary capillary-pressure: more accurate using mutual wavelet packets for adaptive filtering,' *IEEE Eng. in Med. and Biol. Mag.* **14**, 179–185 (1995)

[34] Z. Li, B. J. Grant and B. B. Lieber, 'Time-varying pulmonary arterial input impedance via wavelet decomposition,' *J. Appl. Physiol.* **78**, 2309–2319 (1995)

[35] N. V. Thakor, X. R. Guo, Y. C. Sun and D. F. Hanley, 'Multiresolution wavelet analysis of evoked potentials,' *IEEE Trans. Biomed. Eng.* **40**, 1085–1094 (1993)

[36] D. Morlet, J. P. Couderc, P. Touboul and P. Rubel, 'Wavelet analysis of high-resolution ECGs in post-infarction patients: role of the basic wavelet and of the analysed lead,' *Int. J. Biomed. Comput.* **39**, 311–325 (1995)

[37] A. B. Geva, H. Pratt and Y. Y. Zeevi, 'Spatio-temporal multiple source localization by wavelet-type decomposition of evoked potentials,' *Electroencephalogr. Clin. Neurophysiol.* **96**, 278–286 (1995)

[38] H. Dickhaus, L. Khadra and J. Brachmann, 'Quantification of ECG late potentials by wavelet transformation,' *Comput. Methods Programs Biomed.* **43**, 185–192 (1994)

[39] D. Morlet, F. Peyrin, P. Desseigne, P. Touboul and P. Rubel, 'Wavelet analysis of high-resolution signal-averaged ECGs in postinfarction patients,' *J. Electrocardiol.* **26**, 311–320 (1993)

[40] D. L. Jones, J. S. Touvannas, P. Lander and D. E. Albert, 'Advanced time-frequency methods for signal-averaged ECG analysis,' *J. Electrocardiol.* **25 Suppl.**, 188–194 (1992)

[41] L. Reinhardt, M. Mäkijärvi, T. Fetsch, J. Montonen, G. Sierra, A. Martínez-Rubio, T. Katila, M. Borggrefe and G. Breithardt, 'Predictive value of wavelet correlation functions of signal-averaged electrocardiogram in patients after anterior versus inferior myocardial infarction,' *J. Am. Coll. Cardiol.* **27**, 53–59 (1996)

[42] M. Karrakchou and M. Kunt, 'Multiscale analysis for singularity detection in pulmonary microvascular pressure transients,' *Ann. Biomed. Eng.* **23**, 562–573 (1995)

[43] M. Akay, Y. M. Akay, P. Cheng and H. H. Szeto, 'Investigating the effects of opioid drugs on electrocortical activity using wavelet transform,' *Biol. Cybern.* **72**, 431–437 (1995)

[44] M. Akay, Y. M. Akay, P. Cheng and H. H. Szeto, 'Time-frequency analysis of the electrocortical activity during maturation using wavelet transform,' *Biol. Cybern.* **71**, 169–176 (1994)

[45] S. J. Schiff, J. Milton, J. Heller and A. L. Weinstein, 'Wavelet transforms and surrogate data for electroencephalographic and seizure localization,' *Opt. Eng.* **33**, 2162–2169 (1994)

[46] S. J. Schiff, A. Aldroubi, M. Unser and S. Sato, 'Fast wavelet transformation of EEG,' *Electroencephalogr. Clin. Neurophysiol.* **91**, 442–455 (1994)

[47] I. Clark, R. Biscay, M. Echeverría and T. Virués, 'Multiresolution decomposition of non-stationary EEG signals: a preliminary study,' *Comput. Biol. Med.* **25**, 373–382 (1995)

[48] V. J. Samar, K. P. Swartz and M. R. Raghuveer, 'Multiresolution analysis of event-related potentials by wavelet decomposition,' *Brain Cong.* **27**, 398–438 (1995)

[49] A. W. Przybyszewski, 'An analysis of the oscillatory patterns in the central nervous system with the wavelet method,' *J. Neurosci. Methods* **38**, 245–257 (1991)

[50] E. A. Bartnik and K. J. Blinowska, 'Wavelets: a new method of evoked potential analysis [letter],' *Med. Biol. Eng. Comput.* **30**, 125–126 (1992)

[51] R. Sartenc *et al.*, 'Using wavelet transform to analyse cardiorespiratory and electroencephalographic signals during sleep,' in *Proc. IEEE EMBS Workshop on Wavelets in Med. and Biol.* (Baltimore, 1994), pp. 18a–19a

[52] M. Akay and H. H. Szeto, 'Investigating the relationship between fetus EEG, respiratory, and blood pressure signals during maturation using wavelet transform,' *Ann. Biomed. Eng.* **23**, 574–582 (1995)

[53] L. Senhadji, J. L. Dillenseger, F. Wendling, C. Rocha and A. Kinie, 'Wavelet analysis of EEG for three-dimensional mapping of epileptic events,' *Ann. Biomed. Eng.* **23**, 543–552 (1995)

[54] A. S. Lewis and G. Knowles, 'Image compression using the 2-D wavelet transform,' *IEEE Trans. Image Process* **1**, 244–250 (1992)

[55] M. Antonini, M. Barland, P. Mathieu and I Danbechies, 'Image coding using wavelet transform,' *IEEE Trans. Image Process* **1**, 205–220 (1992)

[56] J. A. Crowe, N. M. Gibson, M. S. Woolfson and M. G. Somekh, 'Wavelet transform as a potential tool for ECG analysis and compression,' *J. Biomed. Eng.* **14**, 268–272 (1992)

[57] J. G. Daugman, 'Complete discrete 2-D Gabor transforms in neural networks for image analysis and compression,' *IEEE Trans. Acoust., Speech and Signal Process.* **36**, 1169–1179 (1988)

[58] A. F. Laine and S. Song, 'Multiscale wavelet representations for mammographic feature analysis,' in *Proc. SPIE Conf. Mathemat. Methods in Med. Imag.* **1768**, 306–316 (1992)

[59] R. A. Kiltie, J. Fan and A. F. Laine, 'A wavelet-based metric for visual texture discrimination with applications in evolutionary ecology,' *Math. Biosci.* **126**, 21–39 (1995)

[60] D. M. Healy, J. Lu and J. B. Weaver, 'Two applications of wavelets and related techniques in medical imaging,' *Ann. Biomed. Eng.* **23**, 637–665 (1995)

[61] L. M. Lim, M. Akay and J. A. Daubenspeck, 'Identifying respiratory-related evoked-potentials,' *IEEE Eng. in Med. and Biol. Mag.* **13**, 174–178 (1995)

[62] O. Bertrand, J. Bohorquez and J. Pernier, 'Time-frequency digital filtering based on an invertible wavelet transform: an application to evoked potentials,' *IEEE Trans. Biomed. Eng.* **41**, 77–88 (1994)

[63] R. Carmona and L. Hudgins, 'Wavelet de-noising of EEG signals and identification of evoked response potentials,' in *Proc. SPIE Conf. Wavelet Applicat. in Signal and Image Process. II, Vol. 2303* (San Diego, July 1994), pp. 91–104

[64] R. N. Strickland and H. I. Hahn, 'Detection of microcalcifications in mammograms using wavelets,' in *Proc. SPIE Conf. Wavelet Applicat. in Signal and Image Process. II, Vol. 2303* (San Diego, July 1994), pp. 430–441

[65] B. J. Lucier, M. Kallergi, W. Qian, R. A. De Vore, R. A. Clark, E. B. Saff and L. P. Clarke, 'Wavelet compression and segmentation of digital mammograms,' *J. Digit. Imaging* **7**, 27–38 (1994)

[66] W. Qian, M. Kallergi, L. P. Clarke, H. D. Li, P. Venugopal, D. Song and R. A. Clark, 'Tree structured wavelet transform segmentation of microcalcifications in digital mammography,' *Med. Phys.* **22**, 1247–1254 (1995)

[67] L. P. Clarke, M. Kallergi, W. Qian, H. D. Li, R. A. Clark and M. L. Silbiger, 'Tree-structured non-liner and wavelet transform for microcalcification segmentation in digital mammography,' *Cancer Lett.* **77**, 173–181 (1994)

[68] W. Qian *et al.*, 'Digital mammography: *m*-channel quadrature mirror filters (QMFs) for microcalcification extraction,' *Computerized Med. Imaging and Graphics* **18**, 301–314 (1994)

[69] D. Wei, H. P. Chan, M. A. Helvie, B. Sahiner, N. Petrick, D. D. Adler and M. M. Goodsitt, 'Classification of mass and normal breast tissue on digital mammograms: multiresolution texture analysis,' *Med. Phys.* **22**, 1501–1513 (1995)

[70] M. A. Goldberg, M. Pivovarov, W. W. Mayo-Smith, M. P. Bhalla, J. G. Blickman, R. T. Bramson, G. W. Boland, H. J. Llewellyn and E. Halpern, 'Application of wavelet compression to digitized radiographs,' *AJR Am. J. Roentgenol.* **163**, 463–468 (1994)

[71] A. H. Delaney and Y. Bresler, 'Multiresolution tomographic reconstruction using wavelets,' *IEEE Trans. Image Process.* **6**, 799–813 (1995)

[72] L. P. Panych and F. A. Jolesz, 'A dynamically adaptive imaging algorithm for wavelet-encoded MRI,' *Magn. Reson. Med.* **32**, 738–748 (1994)

[73] J. B. Weaver, X. Yansun, D. M. Healy, and J. R. Driscoll, 'Wavelet-encoded MR imaging,' *Magn. Reson. Med.* **24**, 275–287 (1992)

[74] J. B. Weaver, X. Yansun, D. M. Healy Jr. and L. D. Cromwell, 'Filtering noise from images with wavelet transforms,' *Magn. Reson. Med.* **21**, 288–295 (1991)

[75] D. M. Healy and J. B. Weaver, 'Two applications of wavelet transforms in magnetic resonance,' *IEEE Trans. Inform. Theory* **38**, 840–860 (1992)

[76] U. E. Ruttimann, M. Unser, D. Rio and R. R. Rawlings, 'Use of the wavelet transform to investigate differences in brain PET images between patients,' in *Proc. SPIE Conf. Mathemat. Methods in Med. Imag. II, Vol. 2035* (San Diego, July 1993), pp. 192–203

[77] J. G. Daugman, 'Entropy reduction and decorrelation in visual coding by oriented neural receptive fields,' *IEEE Trans. Acoust., Biomed. Eng.* **36**, 107–114 (1989)

[78] L. Gaudart, J. Crebassa and J. P. Petrakian, 'Wavelet transform in human visual channels,' *Applied Optics* **32**, 4119–4127 (1993)

[79] M. Porst and Y. Y. Zeevi, 'Localized texture processing in vision: analysis and synthesis in Batorian space,' *IEEE Trans. Biomed. Eng.* **36**, 115–129 (1989)

[80] C. Tallon, O. Bertrand, P. Bouchet and J. Pernier, 'Gamma-range activity evoked by coherent visual stimuli in humans,' *Eur. J. Neurosci.* **7**, 1285–1291 (1995)

[81] A. Arneodo, Y. d'Aubenton-Carafa, E. Bacry, P. V. Graves, J. F. Muzy and C. Thermes, 'Wavelet based fractal analysis of DNA sequences,' *Physica D* **96**, 291–320 (1996)

[82] A. A. Tsonis, P. Kumar, J. B. Elsner and P. A. Tsonis, 'Wavelet analysis of DNA sequences,' *Phys. Rev. E* **53**, 1828–1834 (1996)

[83] M. Unser and A. Aldroubi, 'A review of wavelets in biomedical applications,' in *Proceedings of the IEEE, Vol. 84, No. 4* (1996)

[84] A. Aldroubi and M. Unser, eds., *Wavelets in Medicine and Biology* (CRC Press, Boca Raton, 1996)

[85] M. Akay, 'Introduction: wavelet transforms in biomedical engineering,' *Ann. Biomed. Eng.* **23**, 529–530 (1995)

[86] M. Akay, 'Wavelets in biomedical engineering,' *Ann. Biomed. Eng.* **23**, 531–542 (1995)

[87] R. L. Stratonovich, *Topics in the theory of random noise, vol. I* (Gordon and Breach, New York, 1981)

[88] R. I. Kitney, D. Linkens, A. C. Selman and A. H. McDonald, *Automedica* **4**, 141–153 (1982)

[89] M. N. Levy, 'Sympathetic-parasympathetic interactions in the heart,' *Circ. Res.* **29**, 437–445 (1971)

[90] M. Malik and A. J. Camm, eds., *Heart rate variability* (Futura, Armonk NY, 1995)

[91] G. Sugihara, W. Allan, D. Sobel and K. D. Allan, 'Nonlinear control of heart rate variability in human infants,' *Proc. Natl. Acad. Sci. USA* **93**, 2608–2613 (1996)

[92] J. T. Bigger, Jr., C. A. Hoover, R. C. Steinman, L. M. Rolnitzky and J. L. Fleiss, 'Autonomic nervous system activity during myocardial ischemia in man estimated by power spectral analysis of heart period variability,' *Am. J. Cardiol.* **21**, 729–736 (1993)

[93] D. C. Michaels, E. P. Matyas and J. Jalife, 'A mathematical model of the effects of acetylcholine pulses on sino-atrial pacemaker activity,' *Circ. Res.* **55**, 89–101 (1984)

[94] C.-K. Peng, S. V. Buldyrev, J. M. Hausdorff, S. Havlin, J. E. Mietus, M. Simons, H. E. Stanley, and A. L. Goldberger, 'Nonequilibrium dynamics as an indispensable characteristic of a healthy biological system,' *Integr. Physiol. Behavioral Sci.* **29**, 283–298 (1994)

[95] E. W. Montroll and M. F. Shlesinger, 'The wonderful world of random walks,' in *Nonequilibrium phenomena II: from stochastics to hydrodynamics*, edited by L. J. Lebowitz and E. W. Montroll (North-Holland, Amsterdam, 1984), pp. 1–121

[96] D. Panter, *Modulation, noise and spectral analysis* (McGraw-Hill, New York, 1965)

[97] S. Akselrod, D. Gordon, F. A. Ubel, D. C. Shannon, A. C. Barger, and R. J. Cohen, 'Power spectrum analysis of heart rate fluctuation: a quantitative probe of beat-to-beat cardiovascular control,' *Science* **213**, 220–222 (1981)

[98] J. B. Bassingthwaighte, L. S. Liebovitch and B. J. West, *Fractal Physiology* (Oxford University Press, New York, 1994)

[99] A. Bezerianos, T. Bountis, G. Papaioannou, and P. Polydoropoulos, 'Nonlinear time series analysis of electrocardiograms,' *Chaos* **5**, 95–101 (1995)

[100] D. Hoyer, K. Schmidt, R. Bauer, U. Zwiener, M. Köhler, B. Lüthke, and M. Eiselt, 'Nonlinear analysis of heart rate and respiratory dynamics,' *IEEE Eng. Med. Bio.* **xx**, 31–39 (January/February 1997)

[101] A. Grossmann and J. Morlet, *Mathematics and physics: lectures on recent results* (World Scientific, Singapore, 1985)

[102] J. F. Muzy, E. Bacry and A. Arneodo, 'The multifractal formalism revisited with wavelets,' *Int. J. Bifurc. Chaos* **4**, 245–302 (1994)

[103] A. Arneodo, E. Bacry, P. V. Graves and J. F. Muzy, 'Characterizing long-range correlations in DNA sequences from wavelet analysis,' *Phys. Rev. Lett.* **74**, 3293–3296 (1995)

[104] L. A. Vainshtein and D. E. Vakman, *Separation of frequencies in the theory of oscillations and waves* (Nauka, Moscow, 1983)

[105] D. Gabor, 'Theory of communication,' *J. Inst. Elect. Engrs.* **93**, 429–457 (1946)

[106] *MIT-BIH polysomnographic database CD-ROM, second edition* (MIT-BIH Database Distribution, Cambridge, 1992), see Appendix B

[107] C. Guilleminault, S. Connolly, R. Winkle, K. Melvin and A. Tilkian, 'Cyclical variation of the heart rate in sleep apnea syndrome,' *Lancet* **1**, 126–131 (1984)

[108] P. J. Strollo, Jr. and R. M. Rogers, 'Obstructive sleep apnea,' *N. Engl. J. Med.* **334**, 99–104 (1996)

[109] D. Stauffer and H. E. Stanley, *From Newton to Mandelbrot: a primer in theoretical physics, second edition* (Springer-Verlag, Heidelberg & New York, 1996)

[110] H. E. Stanley, *Introduction to phase transitions and critical phenomena* (Oxford University Press, London, 1971)

[111] T. Vicsek, *Fractal growth phenomena, second edition* (World Scientific, Singapore, 1992)

[112] A. A. Aghili, X. X. Rizwan-uddin, M. P. Griggin and J. R. Moorman, 'Scaling and ordering of neonatal rate variability,' *Phys. Rev. Lett.* **74**, 1254–1257 (1995)

[113] A. Bunde and S. Havlin, *Fractals in science* (Springer-Verlag, Berlin, 1994)

[114] G. L. Gerstein and B. B. Mandelbrot, 'Random walk models for the spike activity of a single neuron,' *Biophys. J.* **4**, 41–68 (1964)

[115] W. H. Press, *Numerical recipes in C: the art of scientific computing* (Cambridge University Press, Cambridge, 1988)

[116] L. A. Lipsitz, F. Hashimoto, L. Pl. Lubowsky, J. E. Mietus, G. B. Moody, A. Appenzeller, and A. L. Goldberger, 'Heart rate and respiratory rhythm dynamics on ascent to high altitude,' *Br. Heart J.* **74**, 340–396 (1995)

[117] J. Theiler, S. Eubank, A. Longtin, B. Galdrikian and J. D. Farmer, 'Testing for nonlinearity in time series: the method of surrogate data,' *Physica D* **58**, 77–94 (1992)

[118] C.-K. Peng, S. V. Buldyrev, S. Havlin, M. Simons, H. E. Stanley and A. L. Goldberger, 'On the mosaic organization of DNA sequences,' *Phys. Rev. E* **49**, 1691–1695 (1994)

[119] P. Bak, C. Tang and K. Wiesenfeld, 'Self-organized criticality: an explanation of 1/f noise,' *Phys. Rev. Lett.* **59**, 381–384 (1987)

[120] L. P. Kadanoff, *From order to chaos* (World Scientific, Singapore, 1993)

[121] S. M. Ossadnik, S. V. Buldyrev, A. L. Goldberger, S. Havlin, R. N. Mantegna, C.-K. Peng, M. Simons and H. E. Stanley, 'Correlation approach to identify coding regions in DNA sequences,' *Biophys. J.* **67**, 64–70 (1994)

[122] J. M. Hausdorff, C.-K. Peng, Z. Ladin, J. Y. Wei and A. L. Goldberger, 'Is walking a random walk? Evidence for long-range correlations in the stride interval of human gait,' *J. Appl. Physiol.* **78**, 349–358 (1995)

[123] A. L. Goldberger, D. R. Rigney, J. Mietus, E. M. Antman and M. Greenwald, 'Nonlinear dynamics in sudden cardiac death syndrome: heart rate oscillations and bifurcations,' *Experientia* **44**, 983–987 (1988).

11

Wavelet dimensions and time evolution

CHARLES-ANTOINE GUÉRIN
and
MATTHIAS HOLSCHNEIDER

Centre de Physique Théorique, C.N.R.S - Luminy - Case 907,
F-13288 Marseille Cedex 9, France

Abstract

In this chapter, we study some aspects of the chaotic behaviour of the time evolution generated by Hamiltonian systems, or more generally, dynamical systems. We introduce a characteristic quantity, namely the lacunarity dimension, to quantify the intermittency phenomena that can arise in the time evolution. We then focus on the time evolution of wave packets according to the Schrödinger equation with time independent Hamiltonian. We introduce a set of fractal dimensions constructed by means of the wavelet transform, the (generalized) wavelet dimensions. We show that the lacunarity dimension of the wave packets can be obtained via the wavelet dimensions of the spectral measure of the Schrödinger operator. This establishes a precise link between the long time chaotic behaviour of the wave packets and the small scale spectral properties of the Hamiltonian.

11.1 Introduction

In this chapter, we are interested in the characterization of some intermittency phenomena that can arise in chaotic dynamical systems. Our aim is to introduce parameters to quantify the strength of intermittency in a turbulent signal. To motivate the discussion, let us begin with a simple example. Consider a particle whose motion in $X \subset \mathbb{R}^n$ is governed by some Hamiltonian system

$$\frac{\partial q}{\partial t} = \frac{\partial H}{\partial p},$$

$$\frac{\partial p}{\partial t} = -\frac{\partial H}{\partial q},$$

421

where $q(t) \in \mathbb{R}^n$ and $p(t) \in \mathbb{R}^n$ are the conjugate generalized coordinates at time t. Denote by $T^*(X) \subset \mathbb{R}^{2n}$ the phase space associated to the motion and $x(t) = (q(t), p(t))$ the position of the particle in phase space. If the Hamiltonian H is time independent, the evolution of $x(t)$ is given by a flow Φ_t, that is a one parameter semi-group of transformations

$$x(t + s) = \Phi_t(x(s)), \quad t, s \geq 0.$$

By Liouville's theorem, the area in phase space is conserved under the Hamiltonian flow. Precisely, we have for any bounded region A in $T^*(X)$:

$$\int_A dp dq = \int_{\Phi_t A} dp dq.$$

Thus the 'surface' measure (this is actually a surface for $n = 1$) on the phase space

$$\mu(A) = \int_A dp dq$$

is invariant under Φ_t. Furthermore, if the phase space $T^*(X)$ is compact, then μ is finite.

Now suppose we can evaluate the location of the particle in phase space periodically in time (with some period say τ) by means of some stroboscopic system, that is we are given a discrete set of values $x_n = x(n\tau)$. The passage from x_n to x_{n+1} reads

$$x_{n+1} = F(x_n),$$

where $F = \Phi_\tau$ is the evolution operator over one period. Thus the system $(T^*(X), \mu, F)$ is a discrete dynamical system associated to the finite invariant measure μ. It follows from the Poincaré recurrence theorem that μ-almost every point of a region in phase space is recurrent. Precisely, for all $A \subset T^*(X)$, there is a set $B \subset A$ with $\mu(B) = \mu(A)$ such that for all $x_0 \in B$, the sequence (x_{n+1}) returns infinitely many times in A.

Now a natural question arises. How frequently does the particle return to the same region A of phase space? This can be visualized by forming the function

$$h(t) = \chi_A(x(t)),$$

where χ_A is the characteristic function of A

$$\chi_A(x) = \begin{cases} 1 \text{ if } x \in A \\ 0 \text{ elsewhere.} \end{cases}$$

The recurrent motion of the particle is mirrored in the intermittent behaviour of $h(t)$ (Figure 11.1). The more lacunary this function is, the sparser is the come back in region A. Thus, the strength of intermittency is characterized by the degree of lacunarity of $h(t)$.

Now let us state the problem in a more abstract and general framework. Consider a particle whose motion $x(t)$ in some phase space, possibly unbounded, is given by an arbitrary dynamical system and as before test if the particle is present or not in some fixed region A by looking at the function $h(t) = \chi_A(x(t))$. The physical windowing system which corresponds to the characteristic function may not be perfect, so it is more natural to take $h(t) = \varphi(x(t))$, where φ is some smooth positive function well localized in region A (Figure 11.2).

At instant T, the fraction of time $\langle h \rangle_T$ spent by the particle in region A is

$$\langle h \rangle_T = \frac{1}{T} \int_0^T dt \, h(t).$$

If $\langle h \rangle_T$ converges toward some finite constant as $T \to \infty$, the limit can be interpreted as a rate of presence in region A. If the particle never returns in A, then $\langle h \rangle_T \sim T^{-1}$, $T \to \infty$. In the general case where the particle returns intermittently in A, we may expect some overall decrease of the form

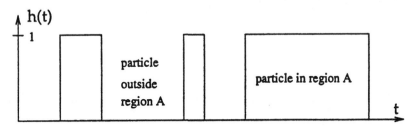

Fig. 11.1. Theoretical characteristic function of the motion.

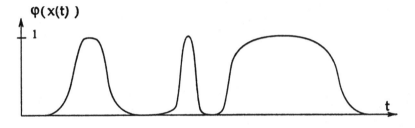

Fig. 11.2. Observed characteristic function of the motion.

$\langle h \rangle_T \sim T^{-\alpha}$, $T \to \infty$. The scaling may or may not exist. However, we can always define the following exponents

$$d_0^+[h] = \limsup_{T \to \infty} \frac{\log\left(\int_0^T dt\, h(t)\right)}{\log T}, \quad d_0^-[h] = \liminf_{T \to \infty} \frac{\log\left(\int_0^T dt\, h(t)\right)}{\log T}.$$

The problem is that $\langle h \rangle_T$ is an average quantity and therefore only gives a rough idea of the real time evolution. Indeed, for given exponents $d_0^\pm[h]$ several scenarios are possible. For instance, think of a particle going further and further away from its initial localization in phase space so that $\langle h \rangle_T \sim T^{-1}$ as $T \to \infty$ and therefore $d_0^+[h] = d_0^-[h] = 0$. Another situation is a particle wandering somewhere in phase space but returning infinitely many times in the same region A with more and more time needed for each come-back in such a way that the fraction of time spent in A still scales like T^{-1} whence again $d_0^+[h] = d_0^-[h] = 0$.

Thus, it appears that the exponents d_0^+ and d_0^- are not capable to detect the intermittent nature of the motion. To get a sharper description, we propose to consider not only the mean value $\langle h \rangle_T$ but also the higher momenta

$$\langle t^m h \rangle_T = \frac{1}{T^{m+1}} \int_0^T dt\, t^m\, h(t), \quad m = 1, 2, \ldots$$

and the associated upper and lower exponents

$$d_m^+[h] = \limsup_{T \to \infty} \frac{\log\left(\int_0^T dt\, t^m\, h(t)\right)}{\log T}, \quad d_m^-[h] = \liminf_{T \to \infty} \frac{\log\left(\int_0^T dt\, t^m\, h(t)\right)}{\log T}.$$

Note that the above exponents are invariant under a time translation $h(t) \to h(t + t_0)$, that is the time origin that we have taken to be 0 can actually be any arbitrary constant. In the next section, we will prove that the limit

$$d_{lac}[h] = \lim_{m \to \infty} \frac{d_m^-[h]}{m}$$

exists. We will call it *lacunarity dimension* because it measures, in some sense, the degree of lacunarity of a positive function. Then, we will focus on a case of quantum chaos and show that the lacunary character of the time evolution can be related to fractal spectral properties of the corresponding Hamiltonian via the *fractal wavelet dimensions*.

11.2 The lacunarity dimension

Since the above definition of the lacunarity dimension is not at all intuitive, let us motivate it by looking at the following simple example.

Example 1 Consider the function:

$$h(t) = \sum_{n=0}^{\infty} \delta(t - b_n),$$

where $\delta(t)$ is the Dirac function

$$\delta(t) = \begin{cases} 1 & \text{if } t = 0, \\ 0 & \text{else.} \end{cases}$$

This can be seen as the characteristic function of a motion with infinitely short times of sojourn in some region of recurrence, the b_n corresponding to the successive instants of return. Here, we choose a sequence (b_n) which becomes more and more lacunary as n increases, precisely

$$b_{n+1} \sim b_n^{\gamma}, \quad n \to \infty,$$

with $\gamma > 1$ and $b_0 > 1$. In this case, $d_m^+[h]$ and $d_m^-[h]$ can be computed explicitly. Indeed we have, for all $T \geq b_0$,

$$\int_0^T dt \; t^m h(t) = \sum_{b_n \leq T} b_n^m \sim b_N^m, \quad T \to \infty,$$

where N is the unique integer such that $b_N \leq T < b_{N+1}$. The log-log diagram of the function $\int_0^T dt \; t^m h(t)$ is plotted on Figure 11.3. Clearly, it appears that:

$$d_m^+[h] = \lim_{N \to \infty} \frac{\log\left(\int_0^{b_N} dt \; t^m h(t)\right)}{\log b_N} = m,$$

and

$$d_m^-[h] = \lim_{N \to \infty} \frac{\log\left(\int_0^{b_N} dt \; t^m h(t)\right)}{\log b_{N+1}} = \frac{m}{\gamma},$$

that is the upper and lower exponents $d_m^+[h]$ and $d_m^-[h]$ have different rates of growth in m. Now this example supplies motivation for the following.

Theorem 11.2.1 *Let h be a positive measurable function such that $d_0^+[h] < \infty$. Then the limit*

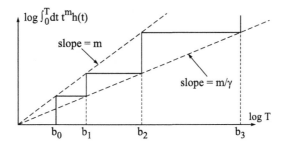

Fig. 11.3. Computation of $d_m^{\pm}[h]$ in a simple case.

$$d_{lac}[h] = \lim_{m \to +\infty} \frac{d_m^-[h]}{m}$$

exists and satisfies $0 \le d_{lac}[h] \le 1$. *Moreover, the limit*

$$\lim_{m \to +\infty} \frac{d_m^+[h]}{m}$$

also exists and is trivial in the sense that it is either 0 or 1. We call $d_{lac}[h]$ *the lacunarity dimension of h and we say the function h is* lacunary *if* $d_{lac} < 1$.

Since the proof is quite heavy, although not difficult, we have deferred it to the appendix.

The example given above to introduce the lacunarity dimension is instructive but not realistic because the true characteristic function of a motion cannot be expressed in terms of Dirac functions (the speed of the particle is finite!). Therefore, the example needs to be refined by taking account of the time of sojourn in the region of recurrence. We now consider the following.

Example 2 Let $h(t)$ be a positive function which can be written as a superposition of polynomially localized bumps centred at instants b_n

$$h(t) = \sum_{n=0}^{\infty} \varphi(t - b_n),$$

where

$$\varphi(t) = (1 + |t|)^{-K}.$$

We take $K > 1$ and again we assume the b_n to scale asymptotically like $b_{n+1} \sim b_n^{\gamma}$, $n \to \infty$, with $\gamma > 1$ and $b_0 > 1$. Such a function is illustrated in Figure 11.4. Straightforward computations lead to the following expressions for $d_m^+[h]$ and $d_m^-[h]$. If $m - K + 1 \le m/\gamma$, as can occur for small m, then

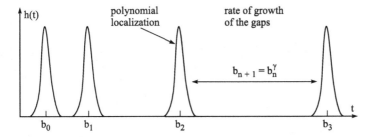

Fig. 11.4. Lacunary function.

$$d_m^+[h] = m \quad \text{and} \quad d_m^-[h] = \frac{m}{\gamma}, \tag{11.2.1}$$

else if $m/\gamma \le m - K + 1$ (for large m), then

$$d_m^+[h] = m, \quad \text{and} \quad d_m^-[h] = m - K + 1. \tag{11.2.2}$$

If the function φ is exponentially localized, $\varphi(t) = e^{-|\sigma t|}$, we obtain $d_m^+[h] = m$ and $d_m^-[h] = \frac{m}{\gamma}$ for all m. The proof is given in the appendix.

Again, we see that the introduction of a weight t^m in the averages tends to separate the upper and lower exponents $d_m^+[h]$ and $d_m^-[h]$, at least for the lowest momenta, and thus makes the lacunarity more visible. Note that here $d_0^+[h] = d_0^-[h] = 0$. Therefore the classical averages $\int_0^T dt\, h(t)$ do not reveal the chaotic behaviour of the function h, whereas the higher momenta do. Indeed, the rate of growth of $d_m^-[h]$ as m increases in the first regime (small m) gives access to γ. This parameter tells how fast the gaps enlarge with the time, that is it quantifies the strength of intermittency in the time evolution. The value of m for which the regime transition occurs gives access to the parameter K, which measures the accuracy of the bumps, that is the form of the window φ. In this example, we have a competition between the lacunarity of the sequence (b_n) and the localization of the function $\varphi(t)$. When m increases, the bumps $t^m\varphi(t)$ become less and less well separated and so the lacunarity becomes less and less apparent. This explains why for large m the exponent $d_m^-[h]$ does not depend anymore on the parameter γ if φ is only polymomially localized. In that case, we have actually $d_{lac}[h] = 1$ and thus the lacunary behaviour of $h(t)$ is not shown up with our definition. However, we can observe $d_m^-[h] \propto m/\gamma$ on some range (see Figure 11.5), from which we deduce that h is lacunary but with a bad localization. Note that the same kind of problem often arises with fractal dimensions in physics. Some natural objects can be assimilated to fractals up to a certain scale, but the fractality

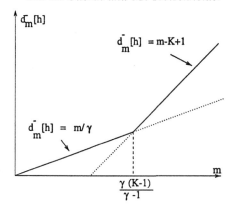

Fig. 11.5. Different regimes for $d_m^-[h]$.

breaks down when one looks at too small a scale. For these objects, the fractal dimension with a theorical definition is trivial although a certain scaling law exists in some range of scales.

Now let us make some comments on the choice of the sequence b_n. In the above example, we took the instants of return b_n to grow like $b_{n+1} \sim b_n^\gamma$, $\gamma > 1$, and with this assumption we obtained $d_{lac}[h] = 1/\gamma$ (at least for exponentially localized window function φ). This example can appear somewhat artificial and restrictive. However, in many cases, one can boil down to this kind of lacunary function by a simple change of variables. For instance, if the b_n grow in a geometrical ratio, $b_{n+1} \sim \gamma b_n$, $n \to \infty$, then it is not hard to verify that for exponentially localized bumps we have $d_{lac}[h] = 1$ but $d_{lac}[h \circ \log] = 1/\gamma$, that is $h(\log t)$ is lacunary.

We end this section with a negative result which allows us to restrict the set of lacunary functions.

Proposition 11.2.1 *Let $h(t)$ be a positive measurable function. If for some $m_0 \geq 0$ we have*

$$d_{m_0}^+[h] = d_{m_0}^-[h] = \alpha > 0,$$

then $d_{lac}[h] = 1$, that is h cannot be lacunary.

The proof is given in the appendix. This statement in particular excludes all the functions $h(t)$ satisfying $\langle h \rangle_T \sim T^{-D}$ with $0 < D < 1$ to be lacunary, because in that case $d_0^+[h] = d_0^-[h] = 1 - D > 0$.

11.3 Quantum chaos

We will now study the problem of intermittent time evolution in the framework of quantum mechanics. Consider a particle whose motion is now governed by the Schrödinger equation

$$\frac{\partial \psi_t}{\partial t} = -iH\psi_t,$$

where $\psi_t \in L^2(\mathbb{R}^n)$ is the wave function of the particle at time t. The Hamiltonian H is a self adjoint operator acting on the Hilbert space $\mathcal{H} = L^2(\mathbb{R}^n)$. If H is time independent, the dynamics of this system is given by the *evolution operator* e^{-iHt}

$$\psi_t = e^{-iHt}\psi_0. \tag{11.3.1}$$

The evolving state ψ_t usually spreads in configuration space and loses its initial localization. This spreading is estimated by the so-called *survival probability* $|\langle \psi_t | \psi_0 \rangle|^2$. More generally, the space time behaviour of the wave packets can be estimated by comparing ψ_t with some reference state ϕ in $\mathcal{H}$. Let us define

$$h(t) = |\langle \psi_t | \phi \rangle|^2. \tag{11.3.2}$$

This quantity is the probability for the state ψ_t to be in configuration ϕ or more simply, if ϕ is the characteristic function of some region $\Omega \subset \mathbb{R}^n$, this is the probability of finding the particle in region Ω at time t. Now let us introduce μ: the spectral measure of H associated to ψ_0 and ϕ, uniquely defined by (see e.g. [6])

$$\langle f(H)\psi_0, \phi \rangle = \int d\mu(x) f(x)$$

for all measurable functions f. From (11.3.1) and (11.3.2) it follows that

$$h(t) = |\widehat{\mu}(t)|^2,$$

where $\widehat{\mu}$ is the Fourier transform of μ

$$\widehat{\mu}(t) = \int d\mu(x) e^{-itx}.$$

Thus, the evolution of ψ_t is governed by the Fourier transform of the spectral measure. It is therefore natural to try to relate the long time behaviour of $h(t)$ to the spectral properties of the Hamiltonian. So some heuristic arguments have been given in [4] supporting the fact that the averages

$$\langle h \rangle_T = \frac{1}{T} \int_0^T dt \, h(t)$$

exhibit a scaling behaviour $\langle h \rangle_T \sim T^{-D}$ where D is a fractal dimension of the measure μ, namely the correlation dimension (e.g. [5]). In [2] some new fractal dimensions have been introduced by means of wavelet transforms, namely the *q-wavelet dimensions* $\kappa_q^{\pm}$, $q = 1, 2 \ldots$ For these dimensions it has been shown that the heuristic argument is actually true and that the long time evolution of $\langle h \rangle_T$ is governed by the upper and lower, respectively, *2-wavelet dimension* $\kappa_2^+[\mu]$ and $\kappa_2^-[\mu]$, also called upper and lower *wavelet correlation dimension*. Precisely we have $d_0^+[h] = -\kappa_2^-[\mu]$ and $d_0^-[h] = -\kappa_2^+[\mu]$. In the following, we want to show that an easy generalization of these to the *q−wavelet dimensions* makes it possible to express the exponents $d_m^{\pm}[h]$, and consequently the lacunarity dimension $d_{lac}[h]$, in terms of fractal dimensions of μ. In the next section, we introduce our main tool, the wavelet transform. Then we define a two parameter set of wavelet dimensions $\kappa_{q,m}^{\pm}$, which we relate to the exponents $d_m^{\pm}[h]$. In order not to get too far off the main flow of argument, the long or technical proofs have been relegated to the appendix.

11.4 The generalized wavelet dimensions

We now wish to introduce the wavelet dimensions. We will first make some brief recall on the wavelet analysis and list a few properties that are necessary for the following. We follow here the notations of [3]. A *wavelet* is basically a complex valued function g of zero mean ($\int g = 0$), which is well localized both in real space and Fourier space (this will soon be made more precise). The *wavelet transform* of a complex valued function s with respect to an *analysing wavelet* g is given by

$$W_g s(b, a) = \int dt \frac{1}{a} \overline{g}\left(\frac{t-b}{a}\right) s(t)$$

or in Fourier space

$$\mathcal{W}_g s(b, a) = \frac{1}{2\pi} \int d\omega \, e^{i\omega b} \overline{\widehat{g}}(a\omega) \widehat{s}(\omega), \qquad (11.4.1)$$

where $\wedge$ is the usual Fourier transform on $\mathcal{S}(\mathbb{R})$

$$\widehat{g}(\omega) = \int dx \, e^{-i\omega x} g(x).$$

This is a function over the position-scale half plane $\mathbb{H} = \mathbb{R} \times \mathbb{R}^+$. Intuitively, the wavelet transform acts as a filter selecting the details present in s at scale a

and position b. If we introduce the following notations, to be maintained in the remainder

$$g_{b,a}(t) = \frac{1}{a}g\left(\frac{t-b}{a}\right), \quad g_a(t) = \frac{1}{a}g\left(\frac{t}{a}\right), \quad \tilde{g}(t) = \bar{g}(-t),$$

then the wavelet transform may be seen as a convolution

$$W_g s(b, a) = \int dt \, \bar{g}_a(t - b)s(t) = \tilde{g}_a * s(b)$$

or a family of scalar products in $\mathbb{R}$

$$W_g s(b, a) = \int dt \, \bar{g}_{b,a}(t)s(t) = \langle g_{b,a}|s\rangle.$$

Thus, the *wavelet analysis* consists of comparing some function to a family of dilated and translated versions $g_{b,a}$ of a mother wavelet g. The *wavelet synthesis* of a function T over $\mathbb{H}$ with respect to a reconstructing wavelet h is given by

$$\mathcal{M}_h T(t) = \int \frac{da}{a} db \, T(b, a)\frac{1}{a}h\left(\frac{t-b}{a}\right).$$

This is essentially the inverse of the wavelet transform. Now let us introduce the function spaces on which the wavelet analysis is to be developed. Let $S(\mathbb{R})$ be the Schwartz space of C^∞ functions φ which, together with their derivatives, are rapidly decreasing

$$\sup_{m,n} |t^m \partial^n \varphi(t)| < \infty, \quad \text{for all } m, n > 0.$$

Denote $S_+(\mathbb{R})$ the subset of Schwartz functions having positive frequencies only ($\hat{\varphi}(\omega) = 0$ if $\omega \leq 0$). For any such function, the Fourier transform is smoothly vanishing at zero or, what amounts to the same, all the moments cancel

$$\hat{\varphi}(\omega) = O(\omega^n) \Leftrightarrow \int dt \, t^n \varphi(t) = 0, \quad n \in \mathbb{N}.$$

The reason for taking wavelets with no negative frequencies is that it considerably simplifies the computations and allows nice inversion formulae. Let us also introduce $S(\mathbb{H})$ the space of highly localized functions on the half plane, that is the functions $T(b, a)$ satisfying

$$\sup_{\mathbb{H}} |T(b, a)|(a + a^{-1})^m(1 + |b|)^m < \infty,$$

for all $m > 0$. Then the following holds true

- if g and s are in $S_+(\mathbb{R})$, then $\mathcal{W}_g s$ is in $S(\mathbb{H})$;
- if h is in $S_+(\mathbb{R})$ and $\mathcal{T}$ in $S(\mathbb{H})$, then $\mathcal{M}_h \mathcal{T}$ is in $S_+(\mathbb{R})$.

If furthermore the constant

$$c_{g,h} = \int_0^\infty \frac{d\omega}{\omega}\, \widehat{h}(\omega)\overline{\widehat{g}}(\omega) \tag{11.4.2}$$

is non-zero, then we have the *reconstruction formula*

$$c_{g,h}^{-1}\mathcal{M}_h\mathcal{W}_g = \mathbb{1}_{S_+(\mathbb{R})}, \tag{11.4.3}$$

where $\mathbb{1}_{S_+(\mathbb{R})}$ is the identity operator on $S_+(\mathbb{R})$. Now, upon reconstructing with g and analysing with h, we obtain the so-called *cross kernel equation*, which relates the wavelet transforms with respect to different wavelets g and h

$$\mathcal{W}_g s(b, a) = \int \frac{da'}{a'}\, db'\, \frac{1}{a'} P_{g\to h}\left(\frac{b - b'}{a'}, \frac{a'}{a}\right)\mathcal{W}_h s(b', a'),$$

with $P_{g\to h}(b, a) = c_{g,h}^{-1}\mathcal{W}_h g(b, a)$. If we introduce a (non-commutative) convolution on $S(\mathbb{H})$ by

$$\mathcal{T}_1 * \mathcal{T}_2(b, a) = \int_{\mathbb{H}} \frac{da'}{a'}\, db'\, \frac{1}{a'} \mathcal{T}_1\left(\frac{b - b'}{a'}, \frac{a}{a'}\right)\mathcal{T}_2(b', a'),$$

then the above equation may be more simply rewritten as

$$\mathcal{W}_g s(b, a) = P_{g\to h} * \mathcal{W}_h s(b, a), \tag{11.4.4}$$

an important equation for the following. Thus, the passage from one wavelet to another in the half plane is done by convolution with a highly localized kernel. If μ is a Borel measure on $\mathbb{R}$, its wavelet transform with respect to a wavelet $g \in S_+(\mathbb{R})$ is given by

$$\mathcal{W}_g \mu(b, a) = \int d\mu(t)\frac{1}{a}\overline{g}\left(\frac{t - b}{a}\right) = \tilde{g}_a * \mu(b),$$

and the cross kernel equation is still valid

$$\mathcal{W}_g \mu(b, a) = P_{g\to h} * \mathcal{W}_h \mu(b, a). \tag{11.4.5}$$

Since we are interested in local properties, we will now only consider finite Borel measures μ on $\mathbb{R}$. This in particular includes the case of functions in $L^1(\mathbb{R})$, which can be trivially identified with finite measures. Given some analysing wavelet $g \in S_+(\mathbb{R})$ and some real $q \geq 1$, we define

$$G_g\mu(a, q) = \|W_g\mu(\cdot, a)\|_q^q = \int db \, |W_g\mu(b, a)|^q.$$

The above quantity is finite since by Young's inequality (see the appendix)

$$G_g\mu(a, q) = \|\tilde{g}_a * \mu\|_q \leq \|\mu\|_1 \|\tilde{g}_a\|_q < \infty.$$

At small scales, a scaling behaviour of the form $G_g\mu(a, q) \sim a^{\kappa_q}$ can in general be observed giving rise to the definition of fractal dimensions κ_q. This approach has been developed in [2]. We propose to extend this definition by introducing a supplementary parameter. For $m \in \mathbb{R}$, we define the function

$$\Gamma_g\mu(t, q, m) = \int_t^1 \frac{da}{a} \, a^m \, G_g\mu(a, q), \tag{11.4.6}$$

and look at its small scale behaviour $t \to 0$. Note that $\Gamma_g\mu(t, q, m)$ is a monotone function of t. Therefore, the limit exists, but may be infinite. In the opposite case when this limit is finite, we rather look at the rate of convergence by putting

$$\Gamma_g\mu(t, q, m) = \int_0^t \frac{da}{a} a^m \, G_g\mu(a, q).$$

To summarize, we have

$$\Gamma_g\mu(t, q, m) = min\left\{ \int_0^t \frac{da}{a} a^m \, G_g\mu(a, q), \, \int_t^1 \frac{da}{a} a^m \, G_g\mu(a, q) \right\}.$$

The *generalized wavelet dimensions* $\kappa_{q,m}^{\pm}$ are now defined by

$$\kappa_{q,m}^{+}[\mu] = \limsup_{a \to 0} \frac{\log \Gamma_g\mu(a, q, m)}{\log a}, \, \kappa_{q,m}^{-}[\mu] = \liminf_{a \to 0} \frac{\log \Gamma_g\mu(a, q, m)}{\log a}.$$

These are intrinsic dimensions of the measures μ, as the following theorem shows.

Theorem 11.4.1 *The generalized wavelet dimensions $\kappa_{q,m}^{\pm}$ are well defined in the sense that they do not depend on the analysing wavelet $g \in S_+(\mathbb{R})$, provided $g \neq 0$.*

The proof of this theorem is given in the appendix.

11.5 Time evolution and wavelet dimensions

The generalized wavelet dimensions $\kappa_{q,m}^{\pm}$ can be related to the exponents $d_m^{\pm}$ introduced in section 11.1 in the following way.

Theorem 11.5.1 *Let μ be a finite Borel measure on $\mathbb{R}$ and let $h(t) = |\hat{\mu}(t)|^2$. Then we have for all integer $m \geq 0$, provided $k^{m/2}\hat{\mu}(k) \notin L^2(\mathbb{R})$,*

$$d_m^+[h] = -\kappa_{2,-m}^-[\mu], \quad \text{and} \quad d_m^-[h] = -\kappa_{2,-m}^+[\mu], \tag{11.5.1}$$

The proof is also given in the appendix. An immediate corollary is

$$d_{lac}[h] = -\lim_{m\to\infty} \frac{\kappa_{2,-m}^+[\mu]}{m}. \tag{11.5.2}$$

This shows that the lacunary long time evolution generated by the Schrödinger equation is related to the generalized wavelet dimensions of the spectral measure of the Hamiltonian H.

We wish to conclude this chapter with some remarks on the bearing of wavelet dimensions in the above time evolution problem. The reader may reasonably ask why we introduced complicated fractal dimensions $\kappa_{q,m}^\pm$ and the non-intuitive spectral measure μ to rewrite a quantity which is already physically interpretable, namely $d_{lac}[h]$. The reason is the following. To form the spectral measure, we need three ingredients: the Hamiltonian itself, the initial state ψ_0 and the reference state ϕ. Now these are *time independent* data. Thus, once the dynamics and the initial state of the system have been given, the equation (11.5.2) automatically provides the lacunarity dimension of $h(t)$. On the other hand, to compute directly the lacunarity dimension by means of the exponents $d_m^-[h]$ would require the full knowledge of $h(t)$ over a huge time span, possibly too long for measurements. Moreover, expressing the lacunarity dimension in terms of wavelet dimensions sets up a precise correspondence between the long time evolution of the dynamical system and the fractal spectral properties of its generator (the Hamiltonian). The next natural question might be why we use wavelet dimensions and not 'classical' fractal dimensions such as the correlation dimension, the box dimension, etc...The answer is simple: the usual fractal dimensions are not adapted to characterize signed or complex measures, whereas the wavelet dimensions are. For instance, the oscillating singularities appearing in 'chirps' functions such as $\sin(|x|^{-\alpha})$ are not detectable by means of the usual fractal dimensions whereas the wavelet dimensions can show them up. For positive measures, however, the wavelet dimensions can in some cases be related to better known fractal dimensions. In particular, it has been shown in [1] that for any finite positive measure μ we have

$$\kappa_2^+[\mu] = D^+[\mu] \text{ and } \kappa_2^-[\mu] = D^-[\mu],$$

where $D^+[\mu]$ and $D^-[\mu]$ are the upper respectively lower *correlation dimension* (see e.g. [5]) of the measure μ. Therefore, the lacunarity dimension in the time evolution can be related to a classical fractal dimension of the spectral measure if this latter is positive. This is, for example, the case if the reference state coincides with the initial state, that is $\phi = \psi_0$ (see section 11.3 for notations). In the general case of complex spectral measures, the correlation dimension has to be replaced by the wavelet correlation dimension.

Acknowledgements

Many thanks to Hans van den Berg for his careful and patient reading and to the Laboratoire Geosciences-Rennes, where most of this work was done, for its warm hospitality.

11.6 Appendix

Proof of theorem 11.2.1 For the purpose of the proof, we introduce the notations

$$H(T, m) = \int_1^T dt \; t^m h(t)$$

and

$$\eta(T, m) = \frac{\log H(T, m)}{\log T}.$$

With this notation we have

$$d_m^+[h] = \limsup_{T \to \infty} \eta(T, m), \text{ and } d_m^-[h] = \liminf_{T \to \infty} \eta(T, m).$$

For fixed m, $H(T, m)$ is a non-decreasing function of T such that $H(T, m) \le T^m H(T, 0)$. Therefore,

$$0 \le d_m^+[h] \le d_0^+[h] + m,$$
$$0 \le d_m^-[h] \le d_0^-[h] + m. \tag{11.6.1}$$

On the other hand $\eta(T, m)$ is, for fixed T, infinitely many times differentiable with respect to m. An elementary computation gives for $T > 1$

$$\frac{\partial \eta(T, m)}{\partial m} \ge 0, \quad \frac{\partial^2 \eta(T, m)}{\partial m^2} \ge 0,$$

that is $\eta(T, m)$ is a non-decreasing convex function of m. Thus, for any $0 \le \alpha \le 1$, we have

$$\eta(T, \alpha m) \le \alpha \, \eta(T, m) + (1 - \alpha) \, \eta(T, 0).$$

Now we use the inequalities

$$\limsup(f + g) \quad \le \limsup f + \limsup g$$
$$\liminf(f + g) \quad \le \liminf f + \limsup g$$

which yield

$$d_{\alpha m}^+[h] \le \alpha \, d_m^+[h] + (1 - \alpha) \, d_0^+[h]$$
$$d_{\alpha m}^-[h] \le \alpha \, d_m^-[h] + (1 - \alpha) \, d_0^+[h],$$

and thus

$$\frac{d_{\alpha m}^+[h] - d_0^+[h]}{\alpha m} \le \frac{d_m^+[h] - d_0^+[h]}{m},$$
$$\frac{d_{\alpha m}^-[h] - d_0^+[h]}{\alpha m} \le \frac{d_m^-[h] - d_0^+[h]}{m}.$$

Since any $m' > m$ can be expressed as m/α with $0 < \alpha < 1$, this means that $(d_m^+[h] - d_0^+[h])/m$ and $(d_m^-[h] - d_0^+[h])/m$ are non-decreasing functions of m. Now in view of (11.6.1) we have

$$0 \le \frac{d_m^-[h] - d_0^+[h]}{m} \le \frac{d_m^+[h] - d_0^+[h]}{m} \le 1.$$

It follows that the limits $\lim_{m\to\infty} d_m^\pm[h]/m$ exist and lie between zero and one. Finally, let us show that $\lim_{m\to\infty} d_m^+[h]/m$ is either zero or one. If h is of compact support, this is evident because in this case $d_m^\pm[h] = 0$. So we may suppose that h has unbounded support. Then look at

$$\limsup_{T\to\infty} \frac{\log\left(\int_T^{T+1} dt \; h(t)\right)}{\log T}.$$

If the above quantity is a finite constant, say α, then we can find a subsequence (T_n) and a constant $C > 0$ for which

$$\int_{T_n}^{T_n+1} dt \; h(t) \ge C T_n^{\alpha-1}.$$

This gives

$$\int_1^{T_n+1} t^m dt \; h(t) \ge \int_{T_n}^{T_n+1} t^m dt \; h(t) \ge C T_n^{m+\alpha-1},$$

whence $d_0^+[h] + m \geq d_m^+[h] \geq m + \alpha - 1$ and therefore $\lim_{m\to\infty} d_m^+[h]/m = 1$. In the opposite case where

$$\limsup_{T\to\infty} \frac{\log\left(\int_T^{T+1} dt\, h(t)\right)}{\log T} = -\infty,$$

it is not hard to see that $\int_1^T dt\, t^m h(t)$ is a convergent integral for all m and therefore $d_m^{\pm}[h] = 0$. This proves the theorem.

Proof of example 2 Take some γ' with $1 \leq \gamma' \leq \gamma$ and some integer N and let us estimate

$$\int_0^{b_N^{\gamma'}} dt\, t^m h(t) = \sum_{n=0}^{\infty} \int_0^{b_N^{\gamma'}} dt\, t^m \varphi(t - b_n).$$

To this end, let us look separately at each term appearing in the sum. While $n \leq N$, we have for any $\epsilon > 0$

$$\int_0^{b_N^{\gamma'}} t^m \varphi(t - b_n)\,dt = \left\{ \int_0^{b_n^{1-\epsilon}} + \int_{b_n^{1-\epsilon}}^{b_n^{1+\epsilon}} + \int_{b_n^{1+\epsilon}}^{b_N^{\gamma'}} \right\} t^m \varphi(t - b_n)\,dt = I_1 + I_2 + I_3.$$

Using the approximations $\varphi(t) \sim t^{-K}$, $t \gg 1$, we obtain the following estimates

$$I_1 \leq b_n^{(m-K+1)}$$
$$I_3 \sim b_N^{\gamma'(m-K+1)}.$$

On the other hand we have

$$c\, b_n^{m(1-\epsilon)} \leq b_n^{m(1-\epsilon)} \int_{b_n^{1-\epsilon}}^{b_n^{1+\epsilon}} \varphi(t - b_n)\,dt$$

$$\leq I_2$$

$$\leq b_n^{m(1+\epsilon)} \int_{b_n^{1-\epsilon}}^{b_n^{1+\epsilon}} \varphi(t - b_n)\,dt \leq C\, b_n^{m(1+\epsilon)},$$

for some positive constants c and C.

Thus, if we regroup the first N terms of the sum, we obtain

$$c'\, b_N^{\rho(m,\gamma')(1-\epsilon)} \leq \sum_{n=0}^{N} \int_0^{b_N^{\gamma'}} dt\, t^m \varphi(t - b_n) \leq C'\, b_N^{\rho(m,\gamma')(1+\epsilon)} \qquad (11.6.2)$$

for some other positive constants c' and C', where

$$\rho(m, \gamma') = \max\{\gamma'(m - K + 1), \ m\}.$$

The contribution of the terms with $n > N$ is negligible because

$$\int_0^{b_N^{\gamma'}} dt \ t^m \varphi(t - b_n) \leq \int_0^{b_N^{\gamma'}} dt \ t^m \varphi(b_N^{\gamma'} - b_n)$$

$$\sim b_n^{-K} b_N^{\gamma'(m+1)} \ll b_N^{\gamma'(m-K+1)}.$$

Therefore we have

$$c'' b_N^{\rho(m,\gamma')(1-\epsilon)} \leq \int_0^{b_N^{\gamma'}} dt \ t^m h(t) \leq C'' b_N^{\rho(m,\gamma')(1+\epsilon)}, \ N \to \infty, \qquad (11.6.3)$$

with c'', $C'' > 0$. Since ϵ can be choosen arbitrarily small, it follows that

$$\limsup_{N\to\infty} \frac{\log\left(\int_0^{b_N^{\gamma'}} dt \ t^m h(t)\right)}{\log b_N^{\gamma'}} = \liminf_{N\to\infty} \frac{\log\left(\int_0^{b_N^{\gamma'}} dt \ t^m h(t)\right)}{\log b_N^{\gamma'}} = \frac{\rho(m, \gamma')}{\gamma'}.$$

This yields the following estimates for $d_m^+[h]$ and $d_m^-[h]$

$$d_m^+[h] \geq \sup_{1 \leq \gamma' \leq \gamma} \frac{\rho(m, \gamma')}{\gamma'},$$

$$d_m^-[h] \leq \inf_{1 \leq \gamma' \leq \gamma} \frac{\rho(m, \gamma')}{\gamma'}. \qquad (11.6.4)$$

It turns out that the above inequalities are actually equalities. Indeed, fix some γ' and some $\epsilon > 0$. For any $T > 0$, we may find N such that $b_N^{\gamma'} \leq Tb <_N^{\gamma'(1+\epsilon)}$. Then

$$\frac{\log b_N^{\gamma'}}{\log T} \frac{\log \int_0^{b_N^{\gamma'}} dt \ t^m h(t)}{\log b_N^{\gamma'}} \leq \frac{\log \int_0^T dt \ t^m h(t)}{\log T} \leq \frac{\log b_N^{\gamma'(1+\epsilon)}}{\log T} \frac{\log \int_0^{b_N^{\gamma'(1+\epsilon)}} dt \ t^m h(t)}{\log b_N^{\gamma'(1+\epsilon)}}.$$

Taking successively the limes superior and inferior, this leads to

$$\frac{\rho(m, \gamma')}{\gamma'(1 + \epsilon)} \leq d_m^-[h] \leq d_m^+[h] \leq \frac{\rho(m, \gamma')(1 + \epsilon)}{\gamma'}.$$

Again we may choose ϵ arbitrarily small and since this holds for any γ' we have equalities in (11.6.4). Now we have to distinguish different regimes for m. If m is small enough to have $m - K + 1 \leq m/\gamma$, then $\rho(m, \gamma') = m$ for all $1 \leq \gamma' \leq \gamma$. Consequently,

$$d_m^+[h] = m, \quad \text{and} \quad d_m^-[h] = \frac{m}{\gamma}. \qquad (11.6.5)$$

If $m/\gamma \leq m - K + 1$, then $\rho(m, \gamma')/\gamma' = m/\gamma'$ if $1 \leq \gamma' \leq m/(m - K + 1)$ and $\rho(m, \gamma')/\gamma' = m - K + 1$ if $m/(m - K + 1) \leq \gamma' \leq \gamma$. This yields

$$d_m^+[h] = m, \quad \text{and} \quad d_m^-[h] = m - K + 1. \tag{11.6.6}$$

The case of exponential localization can be obtained by letting $K \to \infty$, in which case (11.6.5) is verified for all m. This concludes the proof.

Proof of proposition 11.2.1 For the proof we need the following lemma, that we give without demonstration since it is well-known.

Lemma 11.6.1

$$\liminf_{t \to 0} \frac{\log s(t)}{\log t} = \sup\{\gamma \in \mathbb{R} | s(t) \leq O(t^\gamma), \ t \to 0\},$$

$$\limsup_{t \to 0} \frac{\log s(t)}{\log t} = \sup\{\gamma \in \mathbb{R} | t^\gamma \leq O(s(t)), \ t \to 0\}.$$

We are now going to show that $d_m^+[h] = d_m^-[h] = \alpha + m$ for all $m \geq m_0$. First suppose $m_0 = 0$. Then, for all $m \geq 0$, we have $d_m^-[h] \leq d_m^+[h] \leq m + \alpha$. Now let $\epsilon > 0$. By lemma 11.6.1, we can find for all $\delta > 0$ two positive constants $0 < c < C$ such that

$$c \, T^{\alpha - \delta} \leq \int_1^T dt \, h(t) \leq C \, T^{\alpha + \delta}.$$

Rewriting this for $T^{1-\epsilon}$ in place of T and opposing the sign gives

$$-C \, T^{(\alpha + \delta)(1 - \epsilon)} \leq - \int_1^{T^{1-\epsilon}} dt \, h(t) \leq -c \, T^{(\alpha - \delta)(1 - \epsilon)}$$

and adding line by line the last two inequalities yields

$$c \, T^{\alpha - \delta} - C \, T^{(\alpha + \delta)(1 - \epsilon)} \leq \int_{T^{1-\epsilon}}^T dt \, h(t) \leq C \, T^{\alpha + \delta}.$$

Upon choosing δ small enough, we have $\alpha - \delta > (\alpha + \delta)(1 - \epsilon)$ and

$$c \, T^{\alpha - \delta} \leq \int_{T^{1-\epsilon}}^T dt \, h(t) \leq C \, T^{\alpha + \delta}.$$

Again by lemma 11.6.1, it follows that

$$\limsup_{T \to \infty} \frac{\log \int_{T^{1-\epsilon}}^T dt \, t^m h(t)}{\log T} = \liminf_{T \to \infty} \frac{\log \int_{T^{1-\epsilon}}^T dt \, t^m h(t)}{\log T} = \alpha. \tag{11.6.7}$$

Now, since $\int_1^T dt\ t^m h(t) \geq T^{m(1-\epsilon)} \int_{T^{1-\epsilon}}^T dt\ t^m h(t)$, this yields $m + \alpha \geq d_m^+[h] \geq d_m^-[h] \geq \alpha + m(1-\epsilon)$. Since ϵ is arbitrary, this shows that $d_m^+[h] = d_m^-[h] = m + \alpha$, in which case the lacunarity dimension is one. If $m_0 \neq 0$, we may apply the same reasoning to $t^{m_0} h(t)$ instead of $h(t)$ and the conclusion follows.

Proof of theorem 11.4.1 Let us begin with some comments on the definition of the function $\Gamma_g \mu$. The rate of decay of the wavelet transform $\mathcal{W}_g \mu(b, a)$ as $a \to 0$ (resp. $a \to \infty$) reflects the behaviour of the Fourier transform $\hat{\mu}$ at ∞ (resp. 0). Precisely, we have

$$\hat{\mu}(\omega) \leq O(\omega^m),\ \omega \to 0, \Rightarrow \mathcal{W}_g \mu(b, a) \leq O(a^{-m-1}),\ a \to \infty,$$
$$\hat{\mu}(\omega) \leq O(\omega^m),\ \omega \to \infty, \Rightarrow \mathcal{W}_g \mu(b, a) \leq O(a^{m+1}),\ a \to 0 \tag{11.6.8}$$

uniformly in b. (This is a consequence of (11.4.1).) Thus, if s is in $C^\infty(\mathbb{R}) \cap \mathcal{L}^1(\mathbb{R})$, then by (11.6.8), $G_g s(a, q) = ||\mathcal{W}_g s(\cdot, a)||_q^q$ is rapidly decaying at small scales. It follows that μ and $\mu + s$ have the same wavelet-dimensions $\kappa^\pm(q, m)$. Hence, if we define $\langle \mu \rangle$ as the class of equivalence of μ modulo smooth functions (that is $\langle \mu' \rangle = \langle \mu \rangle$ if $\mu' - \mu$ can be identified to a C^∞ function), then two measures belonging to the same class $\langle \mu \rangle$ have the same wavelet dimensions. Now, for a given measure μ, we always can find μ' in $\langle \mu \rangle$ whose Fourier transform is flat around 0. It suffices to takes $\mu' = \mu - \phi * \mu$ with $\phi \in \mathcal{S}(\mathbb{R})$ and $\hat{\phi}(\omega) = 1 + O(\omega^m)$, $\omega \to 0$, for all m. Therefore, we may assume that condition (11.6.8) holds when we compute the wavelet dimensions. In that case, $G_g \mu(a, q)$ is rapidly decreasing at large scales and we may thus replace $\int_t^1$ by $\int_t^\infty$ in the definition of $\Gamma_g \mu(t, q, m)$, that is we may set

$$\Gamma_g \mu(t, q, m) = \int_t^\infty \frac{da}{a} a^m\ G_g \mu(a, q).$$

With this remark in mind, we can begin the proof. Take g and h two analysing wavelets in $\mathcal{S}_+(\mathbb{R})$. Let us compare $\Gamma_g \mu(t, q, m)$ and $\Gamma_h \mu(t, q, m)$ as $t \to 0$. From equation (11.4.5) it follows that with

$$K_{a', a}(b) = \frac{1}{a'} P_{g \to h} \left(\frac{b}{a'}, \frac{a}{a'} \right)$$

the passage from $\mathcal{W}_g \mu$ to $\mathcal{W}_h \mu$ reads

$$\mathcal{W}_h \mu(\cdot, a) = \int_0^\infty \frac{da'}{a'} K_{a', a} * \mathcal{W}_g \mu(\cdot, a').$$

However we have to make sure that $K_{a',a}$ is well defined. The only possible obstruction to this is the constant $c_{g,h}$ as defined in (11.4.2) which may vanish. (Note that it is never ∞ for $g, h \in \mathcal{S}_+(\mathbb{R})$.) However it cannot vanish for all the dilated and translated versions $g_{\beta,\alpha} = \alpha^{-1} g([\cdot - \beta]/\alpha)$ of g since this would merely mean that the wavelet transform of h with respect to g vanishes, which is impossible for $h \neq 0$. Now replacing g by one of its dilated and translated versions $g_{\beta,\alpha}$ amounts to replace $\mathcal{W}_g\mu(b,a)$ by

$$\mathcal{W}_{g_{\beta,\alpha}}\mu(b, a) = \frac{1}{\alpha}\,\mathcal{W}_g\mu\left(\frac{b - \beta}{\alpha}, \frac{a}{\alpha}\right)$$

and therefore the dimensions computed with $g_{\beta,\alpha}$ instead of g are the same. We therefore may suppose that $c_{g,h} \neq 0$.

Now we have

$$\|\mathcal{W}_h\mu(\cdot, a)\|_q = \left(\int db |\mathcal{W}_h\mu(b, a)|^q\right)^{1/q}$$

$$\leq \left\{\int db \left(\int_0^\infty \frac{da'}{a'} |K_{a',a} * \mathcal{W}_g\mu(\cdot, a')(b)|\right)^q\right\}^{1/q}$$

$$\leq \int_0^\infty \frac{da'}{a'} \left\{\int db |k_{a',a} * \mathcal{W}_g\mu(\cdot, a')(b)|^q\right\}^{1/q}$$

by Minkowski's inequality

$$= \int_0^\infty \frac{da'}{a'} \|K_{a',a} * \mathcal{W}_g\mu(\cdot, a')\|_q$$

$$\leq \int_0^\infty \frac{da'}{a'} \|K_{a',a}\|_1 \|\mathcal{W}_g\mu(\cdot, a')\|_q$$

by Young's inequality.

On the other hand,

$$\|K_{a',a}\|_1 = \int_{-\infty}^{+\infty} db \frac{1}{a'} \left|P_{g \to h}\left(\frac{b}{a'}, \frac{a}{a'}\right)\right| H(a/a'),$$

with

$$H(a) = \int_{-\infty}^{+\infty} db\, |P_{g \to h}(b, a)|.$$

This is a non-negative function that is rapidly decaying as $a + 1/a$ gets large. Now set

$$\Lambda = \int_0^\infty \frac{da'}{a'} H(a/a'),$$

which is a finite constant thanks to the high localization of H, and

$$d\nu(a') = \Lambda^{-1} \frac{da'}{a'} H(a/a'),$$

which is a probability measure. Then, using Jensen's inequality, we obtain

$$\|\mathcal{W}_h\mu(\cdot, a)\|_q^q = \Lambda^q \left(\int_0^\infty d\nu(a') \|\mathcal{W}_g\mu(\cdot, a')\|_q \right)^q$$

$$\leq \Lambda^q \int_0^\infty d\nu(a') \|\mathcal{W}_g\mu(\cdot, a')\|_q^q$$

$$= \Lambda^{q-1} \int_0^\infty \frac{da'}{a'} H(a/a') \|\mathcal{W}_g\mu(\cdot, a')\|_q^q. \qquad (11.6.9)$$

Now suppose that we are in the case

$$\lim_{t \to 0} \int_t^1 \frac{da}{a} a^m G_g\mu(a, q) = \infty.$$

Then, as was explained in the last remark, we may compute the wavelet dimension with

$$\Gamma_g\mu(t, q, m) = \int_t^\infty \frac{da}{a} a^m G_g\mu(a, q).$$

With this assumption, (11.6.9) yields

$$\Gamma_h\mu(t, q, m) = \int_t^\infty \frac{da}{a} a^m \|\mathcal{W}_h\mu(\cdot, a)\|_q^q$$

$$\leq O(1) \int_t^\infty \frac{da}{a} a^m \int_0^\infty \frac{da'}{a'} H(a/a') \|\mathcal{W}_g\mu(\cdot, a')\|_q^q$$

$$= O(1) \int_0^\infty \frac{da'}{a'} H(1/a') \int_t^\infty \frac{da}{a} a^m \|\mathcal{W}_g\mu(\cdot, aa')\|_q^q$$

$$= O(1) \int_0^\infty \frac{da'}{a'} a'^{-m} H(1/a') \int_{ta'}^\infty \frac{da}{a} a^m \|\mathcal{W}_g\mu(\cdot, a)\|_q^q.$$

$$= O(1) \int_0^\infty \frac{da'}{a'} H(1/a') \Gamma_g\mu(ta', q, m)$$

$$= O(1) \int_0^\infty \frac{da'}{a'} H(t/a') \Gamma_g\mu(a', q, m),$$

that is

$$\Gamma_h\mu(t, q, m) \le O(1) \int_0^\infty \frac{da}{a} H(t/a)\Gamma_g\mu(a, q, m).$$

As can be easily checked, the same relation holds in the alternative case

$$\Gamma_g\mu(t, q, m) = \int_0^t \frac{da}{a} a^m G_g\mu(a, q).$$

Since g and h can be exchanged in the above inequality, it follows that[†]

$$\Gamma_h\mu(t, q, m) \sim \int_0^\infty \frac{da}{a} H(t/a)\Gamma_g\mu(a, q, m). \tag{11.6.10}$$

Note that the integral on the right-hand side is always finite because $\Gamma_g\mu(a, q, m)$ is of at most polynomial growth in $a + 1/a$ whereas H is rapidly decreasing in $a + 1/a$.

Now suppose that $\Gamma_g\mu(t, q, m) \le O(t^\gamma)$, $t \to 0$, for some γ. Then by (11.6.10), we have

$$\Gamma_h\mu(t, q, m) \le O(t^\gamma) \int_0^\infty \frac{da}{a} H(1/a)a^\gamma \le O(t^\gamma), \quad t \to 0.$$

Since g and h can be exchanged in (11.6.10), it follows that, for all γ

$$\Gamma_h\mu(t, q, m) \le O(t^\gamma) \Leftrightarrow \Gamma_g\mu(t, q, m) \le O(t^\gamma), \quad t \to 0. \tag{11.6.11}$$

Conversely, suppose that $\Gamma_g\mu(t, q, m) \ge Ct^\gamma, 0 < t < 1$ for some constant $C > 0$. Things are here slightly more complicated. Pick some ϵ, $0 < \epsilon < 1$, and keep it fixed. For $0 < t < 1$ we split the integral of (11.6.10) into three parts.

$$\Gamma_h\mu(t, q, m) = \left\{ \int_0^{t^{1+\epsilon}} + \int_{t^{1+\epsilon}}^{t^{1-\epsilon}} + \int_{t^{1-\epsilon}}^\infty \right\} \frac{da}{a} H(t/a)\Gamma_g\mu(a, q, m) = X_1 + X_2 + X_3$$

In the last term we may estimate $\Gamma_g\mu(t, q, m) \le O(1)$ and thus

$$X_3 \le O(1) \int_{1/t^\epsilon}^\infty \frac{da}{a} H(1/a)$$

Since $H(t)$ is arbitrarily well polynomially localized it follows that $X_3 = O(t^n)$ for all $n > 0$.

In X_1 we may estimate $\Gamma_g\mu(t, q, m) \le t^{-p}$ for some p because $\Gamma_g\mu(t, q, m)$ is rapidly decreasing in $t + 1/t$ and thus

[†] The notation $f \sim g$ means $C^{-1}f(x) \le g(x) \le Cf(x)$ for some constant $C > 0$.

$$X_1 \le O(1)t^{-p} \int_0^{t^\epsilon} \frac{da}{a} H(1/a)a^{-p}.$$

Since H is arbitrarily well polynomially localized the integral is rapidly decaying and thus again $X_1 = O(t^n)$ for all $n > 0$.

The remaining contribution is the middle term X_2. If $\Gamma_g\mu(t, q, m)$ is non-decreasing, then

$$
\begin{aligned}
X_2 &= \int_{t^\epsilon}^{t^{-\epsilon}} \frac{da}{a} \Gamma_g\mu(at, q, m)H(1/a) \\
&\ge \int_{t^\epsilon}^1 \frac{da}{a} H(1/a)\Gamma_g\mu(at, q, m) \\
&\ge \Gamma_g\mu(t^{1+\epsilon}, q, m) \int_{t^{1+\epsilon}}^1 \frac{da}{a} H(1/a) \\
&\ge \Gamma_g\mu(t^{1+\epsilon}, q, m) \int_0^1 \frac{da}{a} H(1/a) \\
&\ge C' t^{\gamma(1+\epsilon)}.
\end{aligned}
$$

If $\Gamma_g\mu(t, a, m)t$ is non-increasing, then

$$
\begin{aligned}
X_2 &\ge \int_1^{t^{-\epsilon}} \frac{da}{a} H(1/a)\Gamma_g\mu(at, q, m) \\
&\ge \Gamma_g\mu(t^{1+\epsilon}, q, m) \int_1^{t^{-\epsilon}} \frac{da}{a} H(1/a) \\
&\ge \Gamma_g\mu(t^{1+\epsilon}, q, m) \int_1^\infty \frac{da}{a} H(1/a) \\
&\ge C' t^{\gamma(1+\epsilon)}.
\end{aligned}
$$

Thus, we have for all γ and all $\epsilon > 0$

$$C t^\gamma \le \Gamma_g\mu(t, q, m) \Rightarrow t^{\gamma+\epsilon} \le C' \Gamma_g\mu(t, q, m), \tag{11.6.12}$$

and also since g and h can be interchanged

$$C t^\gamma \le \Gamma_h\mu(t, q, m) \Rightarrow t^{\gamma+\epsilon)} \le C' \Gamma_g\mu(t, q, m), \tag{11.6.13}$$

Once we have proven (11.6.11), (11.6.12) and (11.6.13), the conclusion follows from lemma (11.6.1).

Proof of theorem 11.5.1 Take some wavelet $g \in \mathcal{S}_+(\mathbb{R})$ such that $\widehat{g}$ is compactly supported. Again we may suppose in addition that $\widehat{\mu}(\omega) \le O(\omega^m)$,

$\omega \to 0$ for all m, whence $W_g \mu$ is rapidly decaying at large scale. A direct application of Parseval's equation gives

$$\int_{-\infty}^{+\infty} db\, |W_g \mu(b, a)|^2 = \int_0^\infty d\omega\, |\widehat{g}(a\omega)|^2\, |\widehat{\mu}(\omega)|^2, \qquad (11.6.14)$$

and thus, by a simple exchange of integration

$$\int_{\mathbb{H}} \frac{da}{a}\, db\, a^{-m}\, |W_g \mu(b, a)|^2 = \int_0^\infty \frac{da}{a}\, a^{-m} |\widehat{g}(a)|^2 \int_0^\infty d\omega\, \omega^m\, |\widehat{\mu}(\omega)|^2.$$

The first integral on the right-hand side is a finite constant, due to the high localization of $\widehat{g}$. The second integral is infinite by hypothesis. Then we have

$$\Gamma_g \mu(T^{-1}, 2, -m) = \int_{T^{-1}}^\infty \frac{da}{a}\, a^{-m} \int_{-\infty}^{+\infty} db\, |W_g \mu(b, a)|^2.$$

By equation (11.6.14), this can be rewritten as

$$\Gamma_g \mu(T^{-1}, 2, -m) = \int_0^\infty d\omega\, \omega^m H(\omega/T)|\, \widehat{\mu}(\omega)|^2,$$

with

$$H(t) = \int_t^\infty \frac{da}{a}\, a^{-m} |\widehat{g}(a)|^2.$$

Since H is non-negative and of compact support (since $\widehat{g}$ is), we can find numbers $\lambda > 0$ and $\Lambda > 0$ such that

$$\lambda \chi_{[0,\lambda]}(\omega) \le H(\omega) \le \Lambda \chi_{[0,\Lambda]}(\omega)$$

where χ_I is the characteristic function of I. Therefore

$$\lambda \int_0^{\lambda T} d\omega\, \omega^m |\widehat{\mu}(\omega)|^2 \le \Gamma_g \mu(T^{-1}, 2, -m) \le \Lambda \int_0^{\Lambda T} d\omega\, \omega^m\, |\widehat{\mu}(\omega)|^2,$$

and it follows that

$$d_m^+[|\,\widehat{\mu}\,|^2] = \limsup_{T \to \infty} \frac{\log(\int_0 d\omega\, \omega^m\, |\, \widehat{\mu}\,(\omega)|^2)}{\log T}$$

$$= \limsup_{T \to \infty} \frac{\log \Gamma_g \mu(T^{-1}, 2, -m)}{\log T}$$

$$= -\kappa_{2,-m}^-[\mu],$$

and

$$d_m^-[|\,\widehat{\mu}\,|^2] = \liminf_{T \to \infty} \frac{\log(\int_0^{\lambda T} d\omega\, \omega^m \,|\,\widehat{\mu}\,(\omega)|^2)}{\log T}$$

$$= \liminf_{T \to \infty} \frac{\log \Gamma_g \mu(T^{-1}, 2, -m)}{\log T}$$

$$= -\kappa_{2,-m}^+[\mu].$$

This concludes the proof.

Some useful inequalities As usual $L^p(\mathbb{R})$ is the space of measurable functions f for which

$$\|f\|_p = \left(\int dt\, |f(t)|^p \right)^{\frac{1}{p}} < \infty.$$

Hölder's inequality. If $f \in L^p(\mathbb{R})$ and $g \in L^q(\mathbb{R})$ with $1/p + 1/q = 1/r$, then we have

$$\|fg\|_r \leq \|f\|_p \|g\|_q.$$

Minkowsky's inequality. For any $p \geq 1$ we have

$$\|f + g\|_p \leq \|f\|_p + \|g\|_p.$$

Integral Minkowsky's inequality. If $f(x, y) \in L^p(\mathbb{R}) \times L^p(\mathbb{R})$ with $p \geq 1$,

$$\left\{ \int dy \left(\left| \int dx\, f(x, y) \right|^p \right) \right\}^{1/p} \leq \int dx \left(\int dy\, |f(x, y)|^p \right)^{1/p}.$$

Young's inequality. If $f \in L^p(\mathbb{R})$ and $g \in L^q(\mathbb{R})$ with $1/p + 1/q = 1 + 1/r$, then

$$\|f * g\|_r \leq \|f\|_p \|g\|_q$$

Jensen's inequality. If μ is a probability measure and φ a convex function, then we have

$$\varphi \left(\int d\mu(t) f(t) \right) \leq \int d\mu(t)\, \varphi \circ f(t).$$

References

[1] C.A. Guerin and M. Holschneider, On equivalent definitions of the correlation dimension for a probability measure, *Journ. Stat. Phys.*, **86**(3/4): 707–720, (1997)

[2] M. Holschneider, Fractal wavelet dimension and localization, *Comm. Math. Phys.*, **160**: 457–473, (1994)

[3] M. Holschneider, *Wavelets, an Analysis Tool* (Oxford University Press, Oxford, 1995)

[4] R. Ketzmerick, G. Petschel and T. Geisel, Slow decay of temporal correlations in quantum systems with Cantor spectra, *Phys. Rev. Lett.*, **69**(5): 695–698, (1992)

[5] Ya.P. Pesin, On rigorous mathematical definition of the correlation dimension and generalized spectrum for dimension, *Journ. Stat. Phys.*, **71**(3/4): 529–547, (1993)

[6] M. Reed and B. Simon, *Functional Analysis* (Academic Press, New York, 1980)

Index

Printed in the United States
By Bookmasters